S0-CED-243

RESIDENTIAL ENERGY SYSTEMS AND CLIMATE CONTROL TECHNOLOGY

RESIDENTIAL ENERGY SYSTEMS AND CLIMATE CONTROL TECHNOLOGY

OPERATION AND MAINTENANCE

Martin L. Greenwald
Montclair State College

Prentice Hall, Englewood Cliffs, New Jersey 07632

Library of Congress Cataloging-in-Publication Data

GREENWALD, MARTIN L.
Residential energy systems and climate control technology.

Includes index.
1. Dwellings—Heating and ventilation.
2. Dwellings—Air conditioning. I. Title.
TH7222.G73 1988 697 87-29102
ISBN 0-13-774811-6

Editorial/production supervision
and interior design: *Kathleen M. Lafferty*
Cover design: *20/20 Services, Inc.*
Manufacturing buyer: *Peter Havens*
Cover illustration: *Courtesy of York International Corp.*

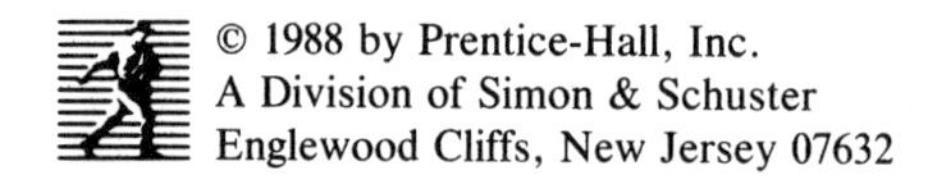

A Division of Simon & Schuster
Englewood Cliffs, New Jersey 07632

Printed in the United States of America

10 9 8 7 6 5 4 3 2 1

ISBN 0-13-774811-6

Prentice-Hall International (UK) Limited, *London*
Prentice-Hall of Australia Pty. Limited, *Sydney*
Prentice-Hall Canada Inc., *Toronto*
Prentice-Hall Hispanoamericana, S.A., *Mexico*
Prentice-Hall of India Private Limited, *New Delhi*
Prentice-Hall of Japan, Inc., *Tokyo*
Simon & Schuster Asia Pte. Ltd., *Singapore*
Editora Prentice-Hall do Brasil, Ltda., *Rio de Janeiro*

CONTENTS

PREFACE

As *Residential Energy Systems and Climate Control Technology: Operation and Maintenance* goes to press, energy production and consumption, especially in the residential sector, is in a state of flux. One fact has emerged over the past 10 years: When it comes to predicting energy availability and fuel prices, there is no such thing as an expert. Best guesses are validated and discarded based on the political and economic fortunes of those involved with the procurement and production of a variety of fossil-fuel energy sources, in conjunction with current governmental policy related to the price support of fossil-fuel and allied energy production systems.

When the manuscript for my first book was in production in 1983–1984 (*Practical Solar Energy Technology*, with Thomas K. McHugh, Prentice Hall, 1985), solar energy was riding high. Residential fuel oil prices were over $1.30 per gallon in the northeastern United States. Electric rates were approaching 15 cents per kilowatthour, and regular gasoline was priced over $1.25 per gallon. Although governmental support of solar energy was beginning to erode, the federal tax credits helped to nurture an industry that had begun to mature and offer the public a variety of reliable energy-efficient products.

Just four years later, the federal tax credits are a thing of the past. With the passing of the tax credits, the solar energy industry has dwindled to a handful of collector manufacturers. Residential fuel oil prices hover at 85 cents a gallon, and regular gas is available for less than 80 cents per gallon. This drop in fuel prices has been accompanied by a concurrent drop in interest and inflation rates as well as growth rates in the GNP.

It seems clear that fuel prices are destined to fluctuate and increase over the long term. The one definitive characteristic of all fossil fuels is their finite nature. Even wood-burning appliances are affected by the slow regeneration of the forests. It takes approximately 30 years for a tree to grow large enough to be harvested as fuel wood. The pressure on the lumber industry due to the general expansion of residential home heating has resulted in a rise in the price of lumber from approximately $40 per cord in 1975 to over $100 per cord in 1986. Residential coal prices during the same period have risen from $35 per ton to over $130 per ton in the northeastern United States.

During this time, much has been done both to decrease energy consumption and to increase energy efficiency within almost every type of heating and cooling appliance used in both residential and commercial installations. Oil-fired boilers, furnaces, and hot water heaters now operate with combustion efficiencies of over 85%, up well over 20% during the past 15 years. Gas-fired combustion technology has enabled many systems to achieve efficiencies approaching 90%, a rise of over 30% during the same period. Homeowners have been able to reduce annual heating bills by over 30% simply by upgrading the insulation characteristics of the home. At present, some studies indicate that wood- and coal-burning heaters generate more power than all the combined nuclear power plants in the United States.

Within the scope of this complex picture of energy generation and consumption, there exists a need for books that focus on the basic operating aspects of all these residential energy production technologies. A variety of textbooks are available that deal with each of these categories on a mutually exclusive basis: books on solar energy, wind energy, small-scale hydroelectric power, and residential space heating. *Residential Energy Systems and Climate Control Technology* combines for the first time a variety of energy systems and devices under one cover, focusing on both conventional and alternative forms of energy conversion.

This book is an easily understood text that is designed primarily for college-level and vocational-technical school course work; it will also be found to be useful for the person who wishes to learn more about the energy systems and climate control devices within the home with an eye toward both maximizing operational efficiency and making future modifications or additions to the home system. The book can easily be understood by a layperson with little technical background, in addition to the person who is well versed in conventional and alternative forms of energy conversion.

The subjects covered in this book are: basic principles of heat transfer; residential electrical supply systems and basic wiring configurations; classification of residential heating systems, including hot water baseboard, hot air heating, and radiant electrical heating applications; operation and maintenance of both oil- and gas-fired heating appliances; domestic hot water system design and operation, including oil, gas, and solid-fuel water heaters; solid-fuel heating appliances, including hot water and steam boilers, radiant stoves, wood gasification boilers, and hot air furnaces; residential air-conditioning and heat pump systems; small-scale hydroelectric generating facilities; and residential wind energy conversion systems.

Material presentation focuses on the practical operating aspects of the appliances, together with its piping, control, energy production, and monitoring configurations where applicable. The experience of experts within the various fields discussed is included to give the reader time-tested reliable information to maximize system operating efficiencies and to minimize system cost and maintenance.

Some of the information has been drawn in part from two previous books by me: *Practical Solar Energy Technology*, with Thomas K. McHugh, and *Residential Hot Water and Steam Heating: Gas, Oil, and Solid Fuels* (Prentice Hall, 1987). This information, in most instances, has been modified either by additions to or deletions from the original books, depending on the depth of coverage in the original texts; it is not intended as a comprehensive installation and operating guide. There is, however, sufficient coverage of all technologies to enable the homeowner or student to become well versed in each specific area. Several illustrations in Chapter 9 have been adapted from "Wind Energy Conversion Systems: A Unit of Instruction" by Martin Greenwald and Robert D. Weber (The New Jersey Division of Industrial Education and Technology, and Trenton State College, 1980).

ACKNOWLEDGMENTS

This book was prepared with the advice and help of many persons within the various disciplines of expertise covered. I am indebted to the following persons who were kind enough to review the manuscript to help ensure the validity and timeliness of information: Jack Turner, owner of Chimney Sweep Energy Corp.; Tom McHugh, coauthor of my first book and owner of Thermonitor Corp.; Robert Dorner, associate professor of Industrial Technology, Montclair State College; Phillip Le Bel, chairman and professor of Economics, Montclair State College; Leonard Litowitz, assistant professor of Industrial Technology, Millersville University; Robert Weber, chairman and professor of Industrial Education and Technology Teacher Education, Trenton State College; Joseph Rettig, director of Educational Services and Field Training, The Williamson Company; Michael Pill, president of New England Energy Development Systems; and Donald Mayer, vice-president of Northern Power Systems.

With sadness I note the passing of Joe Doak, senior field representative for The Burnham Corporation. Joe worked as a technical adviser to me not only on this text but on two previous books published by Prentice Hall. Joe's advice was sought by many individuals within the plumbing and heating trades, and he selflessly gave of his time and talents in all phases of education and consultation within the industry. His presence will continue to be felt by those who knew him and learned from him.

In addition, the support and cooperation of the many industrial firms and their representatives who supplied much of the product information in this book are greatly appreciated. The energy industry in the United States, beset by many conflicting economic pressures, continues to develop innovative, high-quality products that are, in my opinion, the best in the world.

Martin L. Greenwald

1

FUNDAMENTALS OF HEAT, ENERGY, CONVERSION, AND ELECTRICAL SUPPLY SYSTEMS

Throughout the ages, humankind has progressed from the use of energy sources that were readily available in nature to those sources that are extremely sophisticated in both theory and operational characteristics. In fact, the past 100 years have witnessed a shift from two historically traditional energy sources—wind power and coal power—to what are now regarded as traditional fossil fuels: oil, natural gas, and liquefied petroleum (LP) gas. The traditional energy sources have not been neglected, however. Modern technology has come up with revisions on some of the older energy technologies. For example, today's modern wood- and coal-fired boilers and furnaces barely resemble their earlier counterparts. Modern high-speed, low-drag wind electric generators are a far cry from the relatively low speed, high-drag water pumping windmills that are still a common sight in many parts of the country.

Prior to examining the design features of any particular energy system, we will study more closely what energy is and some of the things that various residential energy systems have in common.

LAWS OF THERMODYNAMICS

Regardless of the simplicity or the complexity of a particular source of energy, it is subject to specific laws of physics that govern its availability and conversion efficiency. The laws of physics that relate directly to heat-based energy use and efficiency are known as the **laws of thermodynamics**.

The First Law of Thermodynamics

The **first law of thermodynamics**, which describes the quality of energy production, is sometimes referred to as the principle of **conservation of energy**. This principle states that in a closed system the total amount of energy remains constant; that is, energy can be neither created nor destroyed, merely changed from one form to another. To illustrate this point, consider the familiar residential electric hot water heater. Fuel is burned at a power plant to produce electricity. Given the inefficiencies of this system, which include heat losses in the plant as well as the transmission losses that occur in bringing the electrical power from the utility company to the home, the electricity is channeled through a resistance heating element in the water heater, where it is changed to heat to raise the temperature of the water in the tank. The total amount of energy produced in this process, including all the heat losses in the utility generating and delivery system as well as heat produced within the water heater, is equivalent to the total potential energy available in the fuel that was consumed in the utility plant to produce the electricity. The hot water heater has consumed nothing; it has only changed the energy from one form to another, in this case electrical energy to both usable and rejected waste heat.

The Second Law of Thermodynamics: The Concept of Entropy

The **second law of thermodynamics** considers the **quality** of energy that is available in a closed system and introduces a term known as **entropy**. The second law states that when left to itself, heat always flows from hot to cold. To change the direction of this natural flow, work must be performed somewhere in the system, which will always produce waste heat as a by-product. This waste heat ensures that less useful energy is always available at the conclusion of the process than was available at the start; in other words, no system can be 100% efficient. The modern automobile has an overall efficiency of approximately 10% when one considers the combustion efficiency of the engine (25%), losses to the cooling system, internal frictional losses, and rolling losses of the wheels (65%). Therefore, only 10 cents of every dollar is actually used to move the automobile forward; 90 cents is used to overcome uncaptured heat and frictional forces. A modern oil-fired hot water heating system has a combustion efficiency of approximately 80 to 85%. In spite of advanced combustion technology, 15 to 20% of the heat from burning the oil, gas, or solid fuel still goes up the chimney. The concept of entropy relates to this phenomenon of waste heat. Entropy, in physical terms, defines the randomness of a system. In this case, randomness can be defined as waste heat. Once a high-quality energy source such as fuel oil is burned in a furnace, the waste heat that is produced as a by-product of this process is no longer available for use: It is unlikely not to spread out and recondense on its own. Rather, this waste heat becomes spread randomly throughout the universe. As humankind continues to consume high-quality energy sources such as nonrenewable fossil fuels (including wood and coal), entropy, or randomness, increases. Incorporating entropy into the second law of thermodynamics results in the following statement: *The entropy of a system will always increase or remain constant, but can never decrease.*

To illustrate this point, let us assume that we wish to clear a field in order to build a house. We could clear the lot using a bulldozer or backhoe. In this instance the machinery burns gasoline, a high-grade fuel source, in an internal combustion engine. The engine, in turn, produces waste heat, which is quickly dissipated. Once this heat has spread out, it is no longer available for use. Although we have decreased the randomness of the field by removing all the overgrowth, we have also increased the overall randomness of the universe due to the waste heat generated by burning gasoline, changing this high-grade fuel source to waste, random heat. When work is performed, randomness must always be increased.

Within the context of the laws of thermodynamics, let us examine the ways in which heat moves within residential heating and energy production applications. We begin our discussion with definitions of various *types* of heat, together with standard units of measurement used in a variety of applications.

MEASUREMENT OF HEAT

The standard unit for measuring heat is the British thermal unit, abbreviated Btu. The Btu combines two methods of measurement, temperature and weight. One Btu is the amount of energy required to raise the temperature of 1 pound of water 1 degree Fahrenheit (from 63°F to 64°F at sea level). Conversely, 1 Btu is released from 1 pound of water when it cools 1 degree Fahrenheit.

The Btu is the standard unit of energy used by heating contractors, architects, building contractors, and appliance manufacturers. The metric equivalent of the Btu is the calorie. One calorie is the amount of heat necessary to raise 1 kilogram of water 1 degree Celsius. One calorie is equivalent to 3.97 Btu.

Knowledge of the Btu allows the use of conversion tables that make it possible to compare a variety of fuels and energy sources (Table 1.1). Although the Btu measures heat, one must keep in mind that there are different types of heat produced in any energy system.*

Sensible Heat

Sensible heat is heat that is added to or removed from an object that changes the temperature only of a particular substance without changing the *state* of that substance. Water that is warmed from 20°F to 50°F undergoes an increase in sensible heat only.

* For further discussion of types of heat and heat measurement, consult Chapter 7.

TABLE 1.1 CONVERSION FACTORS AND BTU EQUIVALENTS OF COMMON FUELS

Fuel type	Description
Natural gas	1 therm = 1,000,000 Btu = 100 ft^3
Fuel oil	138,000–144,000 Btu/gal
LP gas (propane)	93,000 Btu/gal
Electricity	3412 Btu/kilowatthour
Mixed hardwoods	24 million Btu/cord
Mixed softwoods	15 million Btu/cord
Coal (anthracite)	12,500 Btu/lb

Note: The energy sources that are the oldest geologically are also the densest in terms of energy per unit of measurement.

Latent Heat

Latent heat changes the state of a substance without causing any temperature change. Heat that changes the state of water from a solid (ice) to a liquid (water), or from a liquid to a vapor (steam), without changing the temperature of the water is latent heat. To change 1 pound of ice at 32°F to 1 pound of water at 32°F requires the addition of 144 Btu of latent heat per pound. To change water at 212°F to steam at 212°F requires the addition of 970 Btu of latent heat per pound.

Superheat

Superheat is sensible heat that is added to a vapor above its boiling point at existing pressure conditions. For example, water boils at 212°F at atmospheric pressure. If 970 Btu is added to 1 pound of the water at 212°F, the water will change to steam at 212°F. If we add any heat to the vapor at this point, the additional heat will raise the temperature of the vapor, superheating it (in addition to raising the temperature of the water, the pressure of the water vessel is raised as well). Superheat is always sensible heat, in that the vapor does not change state but merely experiences a rise in temperature.

Specific Heat

The **specific heat** of a substance is the number of Btu required to raise or lower 1 pound of that substance 1 degree Fahrenheit. The specific heat of water is 1 Btu, so 1 Btu will raise 1 pound of water 1 degree Fahrenheit.

RADIATION, CONDUCTION, AND CONVECTION

Heat is transferred from one object to another in three fundamentally distinct ways, known as radiation, conduction, and convection. Although each of these methods of heat transfer is different from the others, most heating, cooling, and energy production/consumption systems use at least one, and in most cases all, of these heat transfer methods.

Radiation

Energy moves through space as waves made up of electrical and magnetic forces called **electromagnetic radiation**. Radio waves, light waves, x-rays, and the heat one feels in front of a burning fireplace are all electromagnetic waves; they differ from one another in the rate at which the waves rise and fall (frequency), as well as the distance between repetitions of wave shape (wavelength). Radiated energy does not require a substance such as air to travel through. For example, the earth is heated by the sun through the vacuum of space. However, certain substances can block radiation, such as clouds in the sky, clothing, and so on. Light, shiny surfaces reflect radiant heat energy, whereas dark, dull surfaces absorb energy. Figure 1.1 illustrates the wide range of the electromagnetic spectrum, together with the associated frequencies and wavelengths of the common classifications of electromagnetic energy. Figures 1.2 and 1.3 illustrate the characteristics of frequency and wavelength on which the spectrum is based.

Conduction

Conductive heat transfer takes place between two objects that are in contact with one another. Heat will flow from the hotter to the cooler object in relationship to the amount of contact surface area, the temperature difference between the two objects, and the thermal conductivity of each substance. Materials that are good electrical **conductors**, such as copper and aluminum, are also good conductors of heat, since their atomic structure supports the movement of excess free electrons. In relation to conductivity, materials are classified according to their thermal conductivity, *K*, and are rated as either conductors as insulators. **Insulators** are materials that are poor conductors of heat, such as fiberglass, concrete, and oil. Thermal conductivity is generally expressed in Btu per inch per square foot per hour per degree Fahrenheit (Btu/in./ft^2/hr/°F). *K* is different for each substance.

Rather than refer to the thermal conductivity of a substance, it has become more common to refer to a material's ability to resist the flow of heat, a value referred to as the ***R* value** of a substance. *R* values are relative numbers. For example, a material with an *R* value of 10 will resist the flow of heat twice as effectively as will a material with an *R* value of 5. Table 1.2 lists the *R* values of common building materials.

Convection

Convection is the transfer of heat via the action of a moving fluid, which may be either a gas or a liquid. Convection may take place by either forced or natural means. Forced convection relies on the use of a fan or circulating pump to phys-

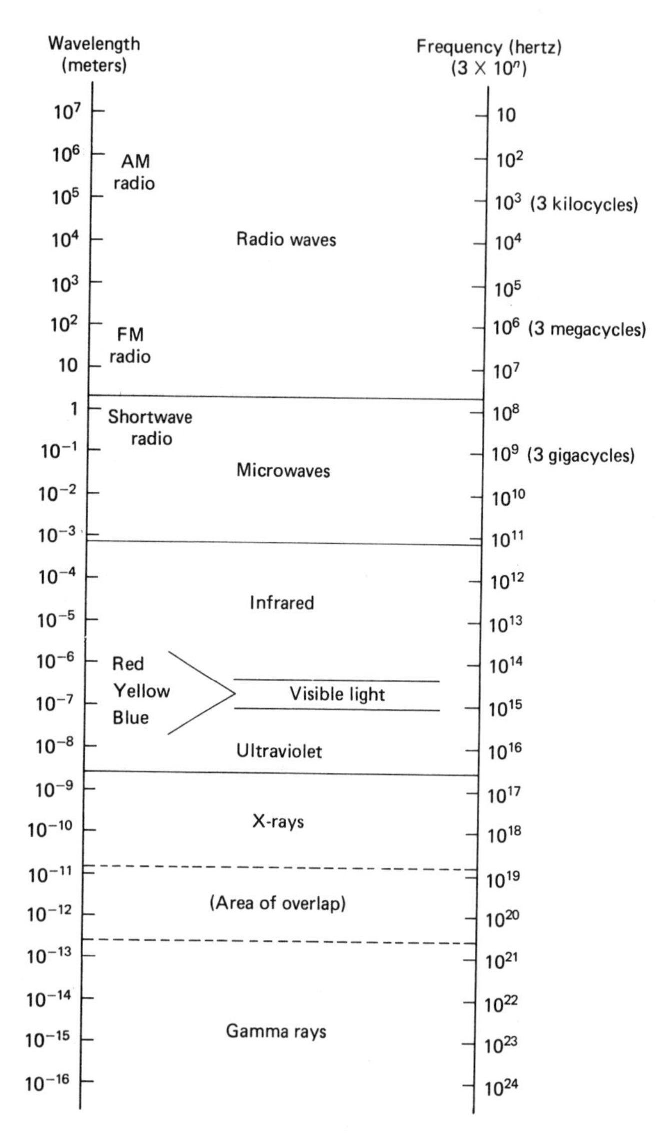

Figure 1.1 Electromagnetic spectrum.

ically move the air or liquid throughout the system to aid in the transfer of heat. Natural convection relies on the use of currents of moving air to transfer heat without the use of electrically powered fans or water circulators. For example, the use of a circulating pump in a hot water baseboard system illustrates the principle of forced convection to move heat from the hot water boiler throughout the piping circuit and the terminal heating units. Natural convection can be illustrated by the flow of room air around room radiators for space heating and relies on natural air currents to transfer heat to the room air (Figure 1.4).

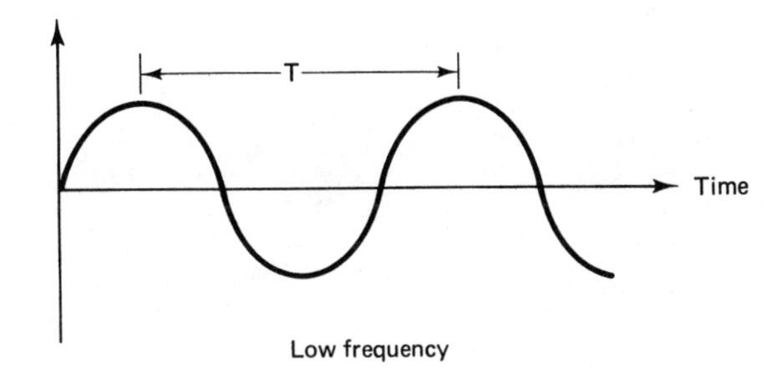

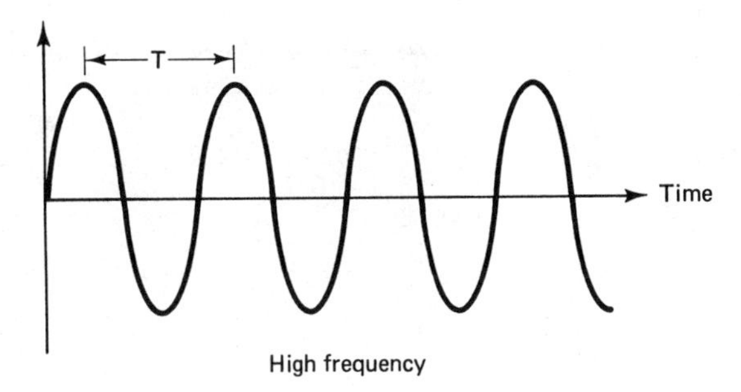

Figure 1.2 Wave frequency is the number of repetitions that occur each second. Wave frequency is different for each material. Note that low frequency is, of consequence, associated with long wavelengths, and vice versa. (From M. Greenwald and T. McHugh, *Practical Solar Energy Technology,* Prentice-Hall, Englewood Cliffs, N.J., 1985.)

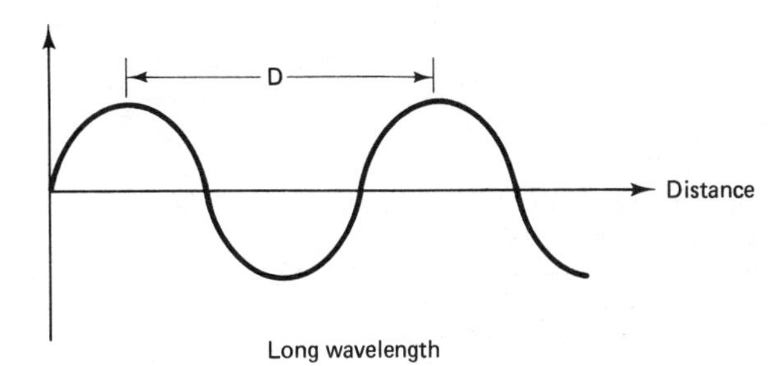

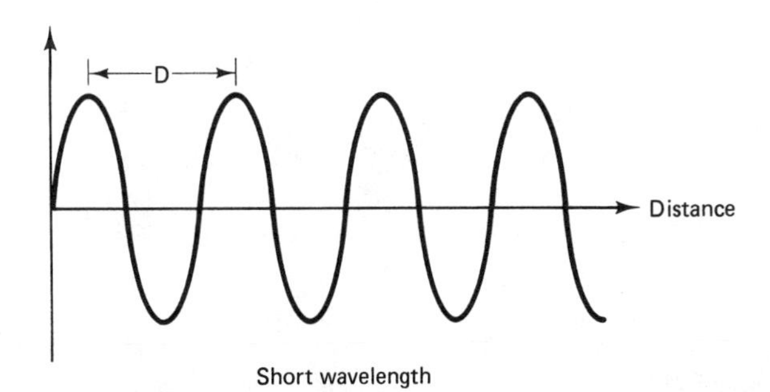

Figure 1.3 Wavelength, the distance between repetitions of wave shape. Wavelengths are measured in meters. (From M. Greenwald and T. McHugh, *Practical Solar Energy Technology,* Prentice-Hall, Englewood Cliffs, N.J., 1985.)

TABLE 1.2 *R* VALUES OF COMMON BUILDING MATERIALS AND INSULATION

Material description	*R* value, per inch
Fiberglass insulation	3.7
Cellulose insulation	3.7
Vermiculite	3.5
Polystyrene	4.0
Urethane foam	7.5
Building brick	0.20
Wood siding	0.85
Concrete block	0.20
Asphalt shingles	0.40 (common thickness)
Wood shingles	0.94 (common thickness)
Insulating sheathing board	2.64
Gypsum board (Sheetrock)	0.45 (standard $\frac{1}{2}$ in.)
Plywood	0.63 (standard $\frac{1}{2}$ in.)
Plaster	0.20
Cement	0.20
Ceiling tile	1.20 (standard $\frac{1}{2}$ in.)
Solid wood door	2.00 (standard $1\frac{3}{4}$ in.)
Combination storm windows	1.80

The molecules of air are heated as they come into contact with the baseboard heating unit. The heated air molecules become less dense and rise. Cooler air from below moves in to take its place, thus establishing a natural convective airflow.

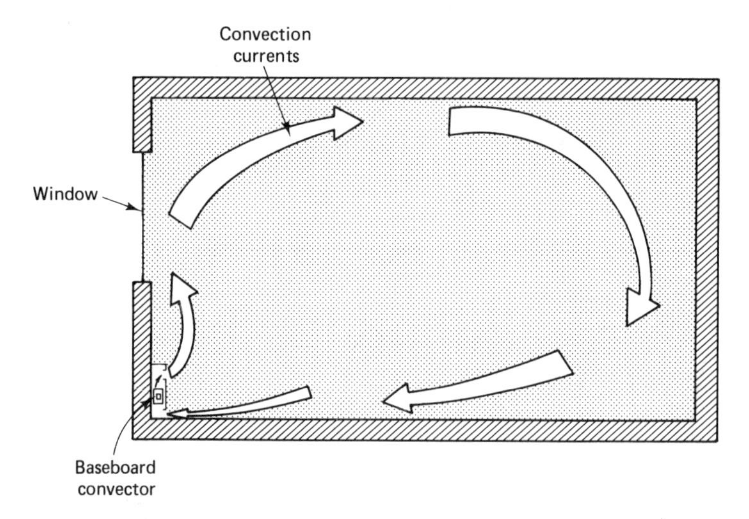

Figure 1.4 Transfer of heat via natural-convection air currents. (From M. Greenwald and T. McHugh, *Practical Solar Energy Technology*, Prentice-Hall, Englewood Cliffs, N.J., 1985.)

Understanding the principles of radiation, convection, and conductive heat transfer can help to optimize the efficiency of all heating and energy systems.

FUNDAMENTALS OF RESIDENTIAL ELECTRICAL ENERGY

An understanding of basic electrical principles and their application in residential installations is prerequisite to understanding and analyzing any energy production or consumption technology. Electrical energy comes from the flow of electrons from one particle to another within a closed system. Electrical power is measured according to factors that express the pressure of the electrical flow, its quantity, and the resistance offered to that flow. These terms are comparable to the flow of water in a closed water system. Table 1.3 illustrates the terminology involved in measuring electrical energy together with the terms often used to compare electrical flow to the flow of water.

Voltage

The unit of electrical pressure is the **volt**. Just as the pressure in a water system is the force, and determines the amount of the water that will flow through any given-diameter pipe, electrical pressure, or voltage, determines the amount of electrical energy that will be able to flow through the wiring in a given installation.

Amperage

The rate of electrical flow is measured in units called **amperes**. Referring to the analogy of a water system to explain electrical relationships, water flow through any pipe or valve depends on the pressure of the system, the diameter of the pipe, the length of time that the water is flowing, and the total length of the pipe in the system. Piping systems that are excessively long, or those that contain small-diameter pipe, offer a greater amount of resistance to water flow than do those that have large-diameter pipe with relatively short runs. Similarly, in electrical supply systems, large-diameter wires allow electricity to flow more freely than do small-diameter wiring circuits. Also, the shorter the length of the wiring, the

TABLE 1.3 COMPARISONS OF ELECTRICAL FLOW AND WATER FLOW

Function	Electrical	Water
Pressure	volt	lb/in.^{2}
Rate of flow	ampere	gal/min
Power (rate of doing work)	watt	ft-lb/min or horsepower
Work or energy	watthour	ft-lb

Source: The Williamson Company.

less electrical resistance there will be in the system. The term "ampere" is often used to describe the overall supply characteristics of a residential electrical feed. A 200-ampere system refers to the total amperage that can be delivered to the home service panel, and to appliances, at any one time. A 200-ampere service system will deliver 200 amperes at 220 volts ac (alternating current). Most electrical appliances list their ampere draw on the manufacturer's identification plate to help the user avoid overloading circuits.

Resistance

The length of a wiring system, the diameter of the wire, and the type of wire affect the resistance offered to the flow of electrical current. The unit used to measure electrical resistance is the **ohm**.

The flow of electricity through a circuit with resistance produces heat. Sometimes this heat is desirable. For example, a hot water heater uses high-resistance elements that heat up when an electrical current is passed through. The heat from the element is transferred to the surrounding water, resulting in a temperature increase of the water. Similarly, electrical baseboard heating units can raise room temperatures by exposing the heating element to room air. At times, however, excessive heat produced inadvertently in an electrical system can lead to overheated wiring, resulting in a branch circuit failure. In properly protected branch circuits, fuses and circuit breakers prevent electrical wires from overheating. However, improperly fused or wired electrical systems can cause electrical fires, a leading cause of fire in residential structures.

Voltage, amperage, and resistance are used to calculate power, or wattage, which determines the amount of work that can be accomplished by a particular appliance in the electrical system.

Power and Wattage

Power is the rate at which work is accomplished over time. The unit of power measurement with which we are all familiar is the **horsepower**. One horsepower is equivalent to 33,000 foot-pounds (ft-lb) of work per minute (550 ft-lb/sec), equivalent to lifting 1 ton $6\frac{1}{2}$ ft/min. Power, in electrical terms, is the product of the electrical appliance voltage multiplied by its amperage, expressed by the term **wattage** and measured in watts (W).

To illustrate this concept, assume that a light bulb is rated at 120 volts (V) with a current draw of 0.5 ampere (A). The wattage of the bulb is calculated as follows:

$$\begin{aligned} \text{watts} &= \text{volts} \times \text{amperes} \\ &= 120 \times 0.5 \\ &= 60 \text{ W} \end{aligned}$$

If the wattage of the appliance is multiplied by the amount of time that the unit operates, the energy consumption value, or amount of work accomplished,

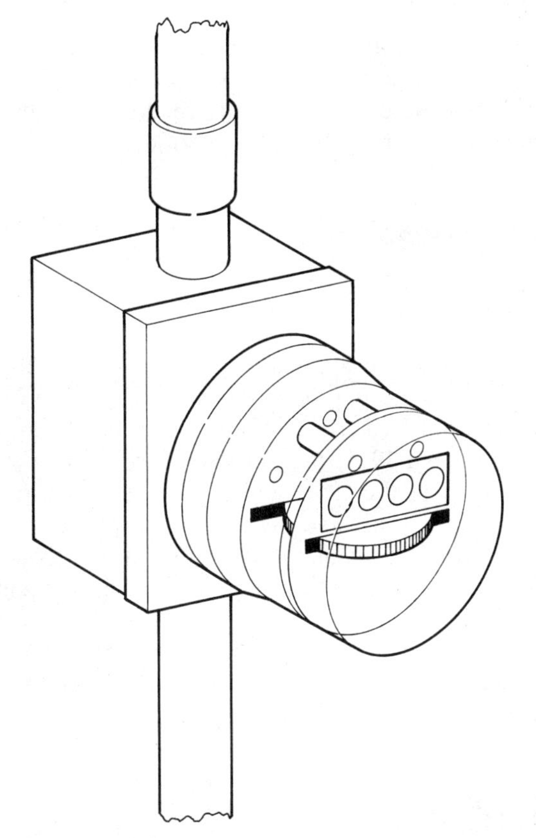

Figure 1.5 Kilowatthour meter.

is determined. The measurement of electrical energy consumption, the rate at which power is being consumed over a specified period of time, either in minutes or hours, results in **watthours** of energy. The standard billing for energy consumption is the kilowatthour, which is equivalent to 1000 watthours.

Energy consumption is recorded on a kilowatthour meter (Figure 1.5) that tabulates total energy consumption in a specific service location. The standard kilowatthour meter consists of separate dials, illustrated in Figure 1.6. To use the meter, the lowest number on each dial is used, and the total energy consumption

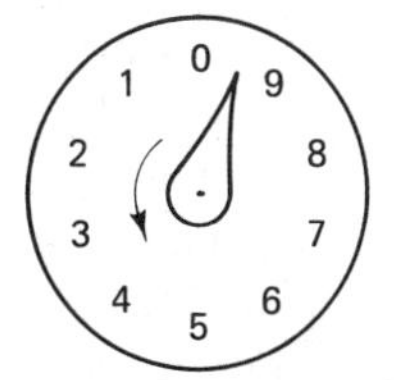

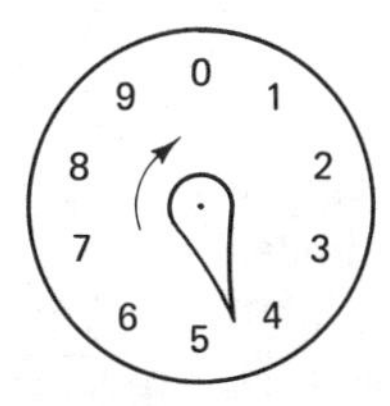

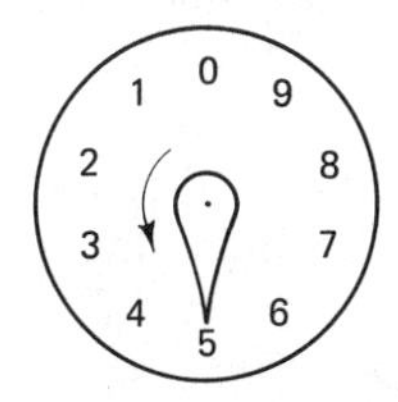

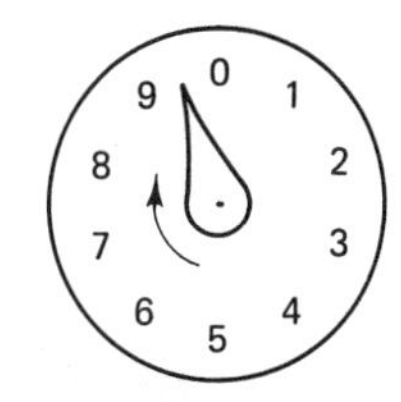

Figure 1.6 Separate dials on a kilowatthour meter that keep track of electrical consumption. The dials are commonly read from right to left, using the smallest number on each dial to determine the final reading. The dials here show a reading of 9449 kilowatthours.

is read from left to right. For example, the meter shown reads 9449. Monthly billing procedures used by local utilities consist of subtracting the previous monthly meter reading from the current monthly reading to determine the total number of kilowatthours used during the billing period.

RESIDENTIAL WIRING APPLICATIONS

National Electrical Code®

To ensure that safety standards are applied to both the manufacture and installation of electrical appliances, the *National Electrical Code®* (NEC) was established in 1897 through the combined efforts of insurance, electrical, architectural, and other interested groups. In 1911, the National Fire Protection Association (NFPA) became the sponsor of the Code and continues to act in this capacity. The Code is revised every three years to ensure that only the latest safety methods and procedures are utilized in all installations. Although the Code itself has no legal authority, most states and localities require that all residential electrical wiring conform to the Code as a *minimum* requirement. In some localities, the Code requirements are superseded by local guidelines. Therefore, prior to undertaking any electrical installation or modification, it is necessary to consult with the local electrical inspector to determine all applicable requirements relating to the project at hand.

Service Entrance Requirements

The **service entrance** of the residential electrical system is the wiring system that connects the home electrical breaker panel to the utility-supplied wiring from the nearby service transformer. These wiring systems are subjected to rigid inspection by the local electrical inspector and must meet all the latest applicable NEC requirements. Residential electrical service entrance systems can be wired using either entrance cable (Figure 1.7) or conduit (Figure 1.8).

When installing either conduit or service cable, the entrance head must be installed at least 10 ft above the ground and must be at least 3 ft from doors, windows, or emergency fire escapes. Note the use of drip loops on each of the entrance wires shown in Figure 1.7. These loops are formed to prevent water from entering the electrical service system. The Code requires that to ensure a properly configured drip loop, the entrance head must be installed above the top insulator of the incoming service lines. When using service cable, it must be secured at least every 4 ft with metal straps. Watertight connectors are installed where the cable enters and exits the meter box. For safety and proper appliance operation, the service entrance system must be properly grounded and bonded. Figure 1.9 illustrates three residential bonding arrangements. It should be noted that all such procedures must strictly conform to all applicable code regulations.

Sizing the electric service. The electric service that enters a building is sized according to the maximum number of amperes that must be delivered by the system to the branch circuits within the home. Over the years the size of the residential electric service has grown along with the increase in per capita electrical consumption. The most common ampere service capabilities are discussed below.

30-Ampere Service. Thirty-ampere service consists of two No. 8 conductors supplying 110-V service to a small 30-A service panel. This type of service is restricted to small outbuildings and barns where only lighting and small portable appliances are used. This service size is not suitable for residential applications and is usually provided to the outbuilding as a subpanel branch circuit from the household breaker panel.

60-Ampere Service. Sixty-ampere service, for many years the standard-size residential service, consists of three-wire No. 6 conductors supplying a combination of 110- and 220-V power to the home. This type of service is adequate for residential lighting and small appliances, including an electric range and dryer, but due to the limited amperage that is available, no additional major appliances can be added to the system. Sixty-ampere service is now considered inadequate for single-family residential applications and has not been installed in new residential construction for many years.

100-Ampere Service. One hundred–ampere service is the minimum size permitted by the Code in single-family residences of up to 3000 ft^2. One hundred–ampere service is adequate to power a wide assortment of appliances, including an electric range, hot water heater, washer and dryer, fossil-fuel heating system, and water supply. This service capacity, although adequate for most applications, is insufficient for use in homes that have electric heating or central air conditioning.

150-Ampere Service. One hundred fifty–ampere service offers all the advantages of 100-A service and in addition permits the use of central air conditioning. This type of service is flexible, in that it allows for expansion in the number of electrical appliances that can be used in the future, and is adequate for most residential applications, with the exception of electric space heating.

200-Ampere Service. Two hundred–ampere service, the highest-capacity residential service commonly available, is required for homes using electric space heating or heat pump systems. Two hundred–ampere service is also common in rural and farm applications, where the service panel from the barn can be used as a central supply for distribution to smaller outbuildings using distribution poles.

When designing the electrical service for any purpose, both the electrical inspector and the local utility must be consulted prior to defining any system configuration. Also, the consumer should have the latest Code book available as a reference tool.

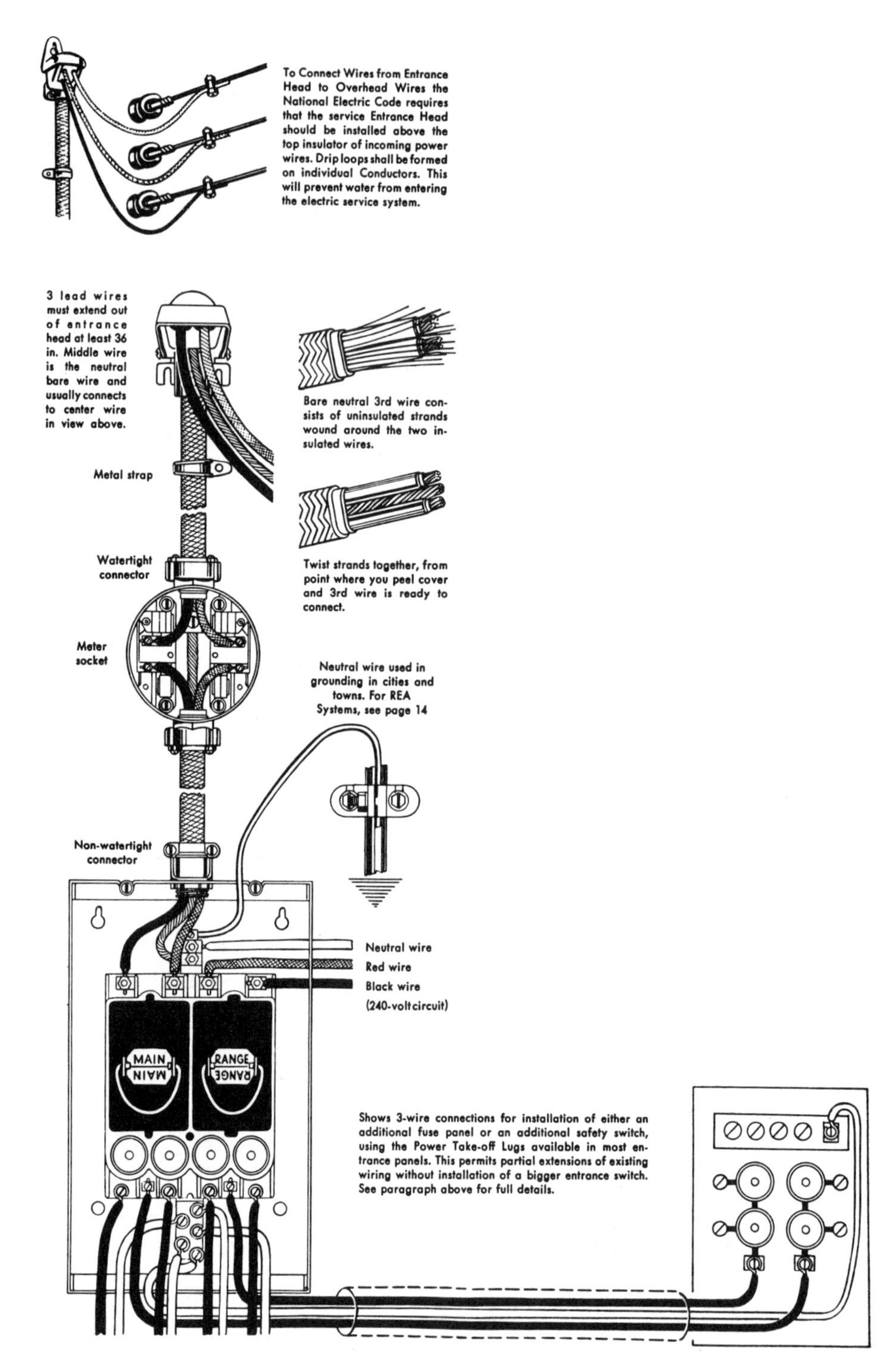

Figure 1.7 Residential electric service using entrance cable. (Courtesy of Sears, Roebuck, and Company.)

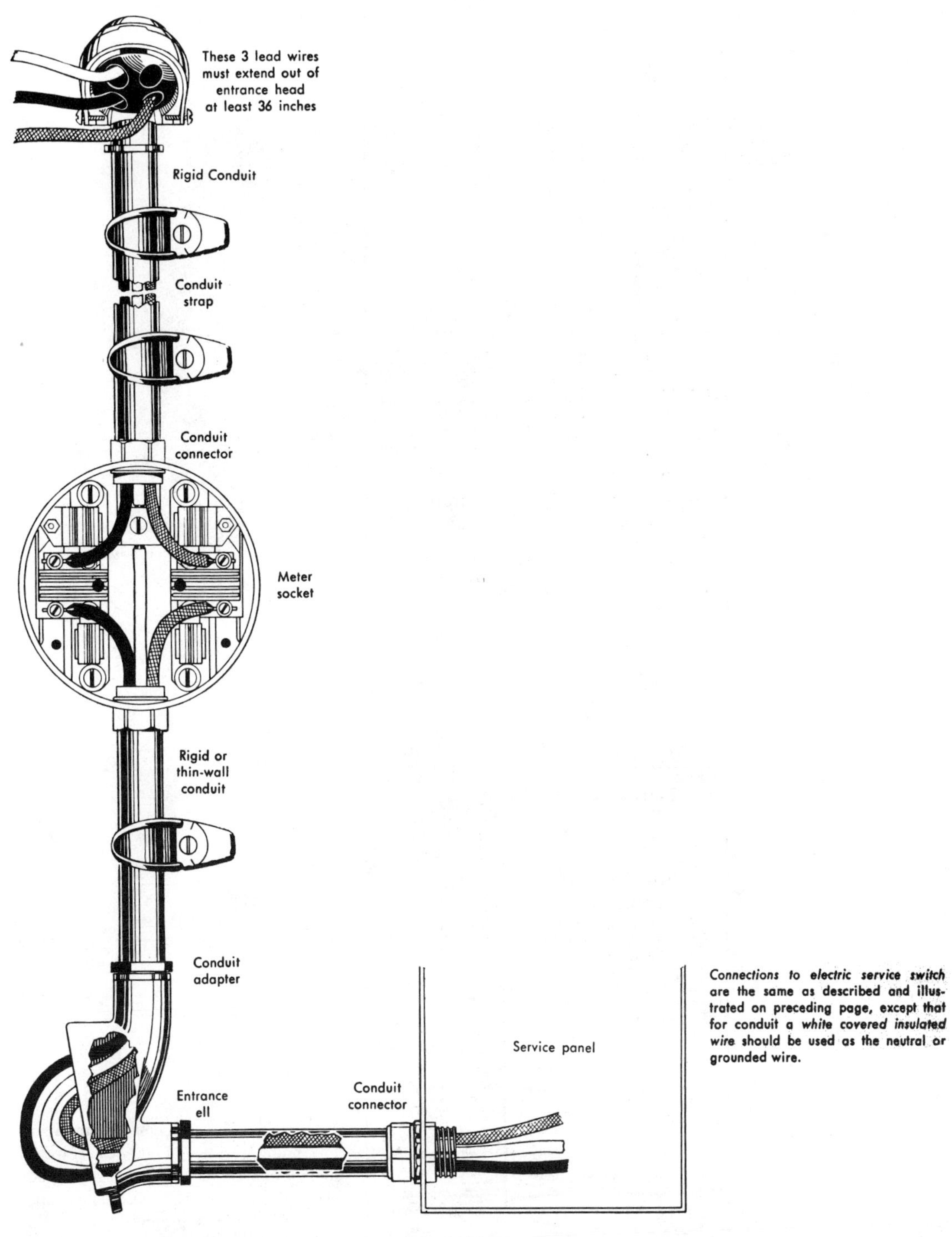

Figure 1.8 Residential electric service using conduit. (Courtesy of Sears, Roebuck, and Company.)

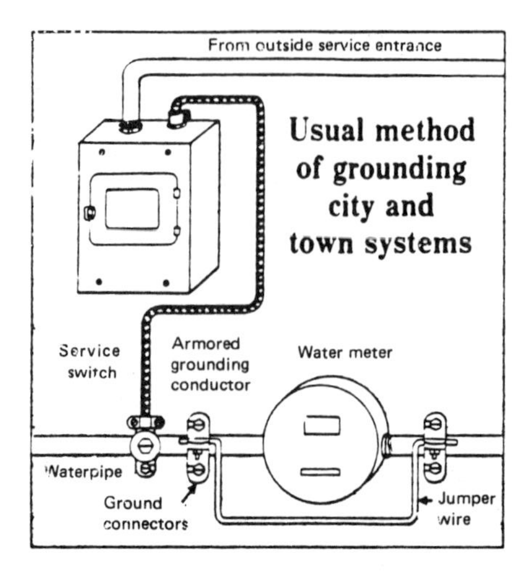

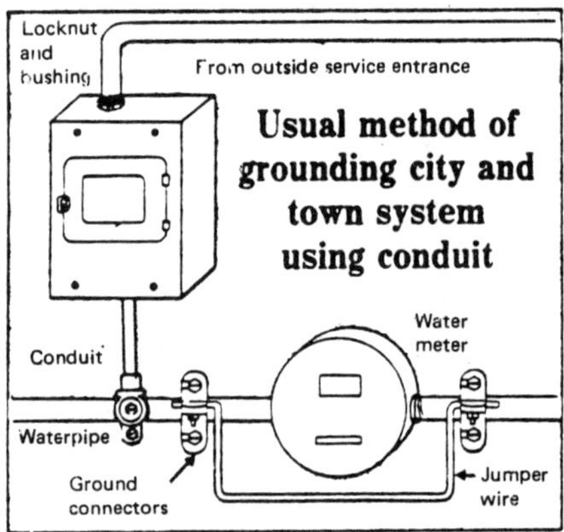

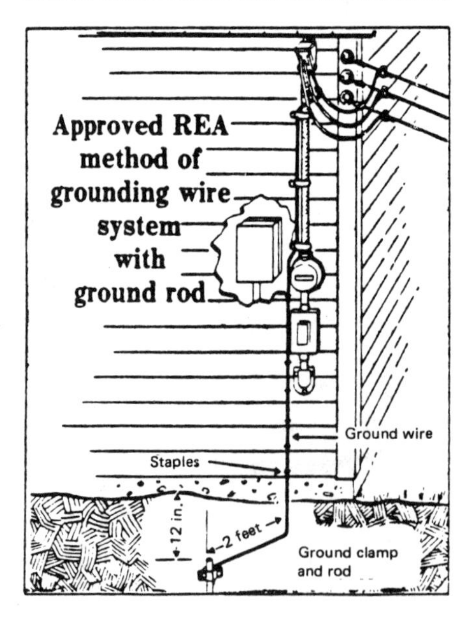

Figure 1.9 Three common methods for grounding residential service entrance installation. *Note:* All applicable Code regulations must be investigated prior to undertaking a grounding or wiring configuration. (Courtesy of Sears, Roebuck, and Company.)

WIRE CLASSIFICATIONS AND SIZING

A great variety of wire types and sizes are available for use in residential applications. Each is designed for a specific set of operating conditions and environments and needs to be selected based on the proposed end use. We will now examine the methods used to both size and differentiate electrical wire.

Wire Sizing

All electrical wire used in residential and commercial applications is sized according to the American Wire Gauge (AWG) standard. Wire sizes are based on the cross-sectional area of the wire measured in units called circular mils (0.001 in.). The size of the wire determines its current-carrying capability or ampacity. The greater the cross-sectional area of the wire, the smaller its AWG number. The smaller the AWG wire gauge number, the greater its ampacity. The wire that is used for a particular purpose must be of sufficient size to carry the maximum load required for a particular circuit, with adequate fuse or circuit breaker protection. The dimension of rubber- and plastic-covered conductors used in residential wiring applications is illustrated in Table 1.4.

TABLE 1.4 DIMENSIONS OF RUBBER- AND PLASTIC-COVERED CONDUCTORS, FROM NO. 0000 TO NO. 40 AMERICAN WIRE GAUGE SIZES

		Cross section		
Gauge number	Diameter (mils)	Circular mils	Square inches	Pounds per 1000 ft
0000	460.0	212,000.0	0.166	641.0
000	410.0	168,000.0	0.132	508.0
00	365.0	133,000.0	0.105	403.0
0	325.0	106,000.0	0.0829	319.0
1	289.0	83,700.0	0.0657	253.0
2	258.0	66,400.0	0.0521	201.0
3	229.0	52,600.0	0.0413	159.0
4	204.0	41,700.0	0.0328	126.0
5	182.0	33,100.0	0.0260	100.0
6	162.0	26,300.0	0.0206	79.5
7	144.0	20,800.0	0.0164	63.0
8	128.0	16,500.0	0.0130	50.0
9	114.0	13,100.0	0.0103	39.6
10	102.0	10,400.0	0.00815	31.4
11	91.0	8,230.0	0.00647	24.9
12	81.0	6,530.0	0.00513	19.8
13	72.0	5,180.0	0.00407	15.7
14	64.0	4,110.0	0.00323	12.4
15	57.0	3,260.0	0.00256	9.86
16	51.0	2,580.0	0.00203	7.82

(*table continues*)

TABLE 1.4 DIMENSIONS OF RUBBER- AND PLASTIC-COVERED CONDUCTORS, FROM NO. 0000 TO NO. 40 AMERICAN WIRE GAUGE SIZES (*continued*)

Gauge number	Diameter (mils)	Cross section Circular mils	Cross section Square inches	Pounds per 1000 ft
17	45.0	2,050.0	0.00161	6.20
18	40.0	1,620.0	0.00128	4.92
19	36.0	1,290.0	0.00101	3.90
20	32.0	1,020.0	0.000802	3.09
21	28.5	810.0	0.000636	2.45
22	25.3	642.0	0.000505	1.94
23	22.6	509.0	0.000400	1.54
24	20.1	404.0	0.000317	1.22
25	17.9	320.0	0.000252	0.970
26	15.9	254.0	0.000200	0.769
27	14.2	202.0	0.000158	0.610
28	12.6	160.0	0.000126	0.484
29	11.3	127.0	0.0000995	0.384
30	10.0	101.0	0.0000789	0.304
31	8.9	79.7	0.0000626	0.241
32	8.0	63.2	0.0000496	0.191
33	7.1	50.1	0.0000394	0.152
34	6.3	39.8	0.0000312	0.120
35	5.6	31.5	0.0000248	0.0954
36	5.0	25.0	0.0000196	0.0757
37	4.5	19.8	0.0000156	0.0600
38	4.0	15.7	0.0000123	0.0476
39	3.5	12.5	0.0000098	0.0377
40	3.1	9.9	0.0000078	0.299

Source: J. Markell, *Residential Wiring*, Reston, Reston, Va., 1984, Fig. 4.2, p. 68.

TABLE 1.5 CONDUCTOR APPLICATIONS CHART FOR THE MOST COMMONLY USED SIZES OF RESIDENTIAL ELECTRICAL WIRE

Wire gauge size	Common usage
No. 20–No. 22	Electronic circuits, phone extensions
No. 16–No. 18	Low-draw extension cords, lamp cords, bell wiring, small appliance cords
No. 12–No. 14	Normal branch circuits in the home wiring system, including appliances and lighting loads
No. 6–No. 10	Large electrical appliances, such as hot water heaters, clothes dryers, air-conditioning equipment, water pumps, etc.
Nos. 3, 2, 1, 1/0, 2/0, 3/0, and larger	Service entrance cable and cable used for feeding electrical subpanels

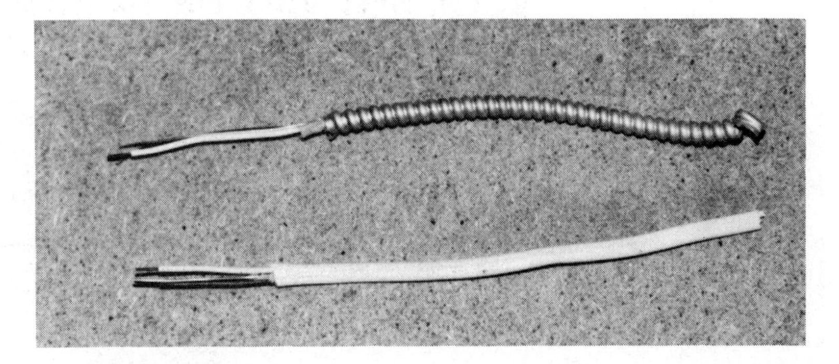

Figure 1.10 Armored cable, commonly referred to as BX, consists of the electrical conductors encased within a spiral metal outer enclosure. The metal armor is accepted as the grounding conductor, although an internal bonding conductor is included to increase the grounding effectiveness of the outer metal sheathing. Romex, plastic-sheathed cable, is somewhat easier to work with, although its use is not permitted in as many locations as is armor-sheathed wire.

Note: The use of aluminum conductors is limited to terminals that show "AL" markings. If a terminal is not listed as being suitable for aluminum conductors, under no circumstances should aluminum be used.

Although wire is manufactured in a great variety of gauge sizes, wire for residential uses is generally restricted to the smaller gauge sizes. Table 1.5 shows the normal range of applications for the most commonly used wire gauge sizes.

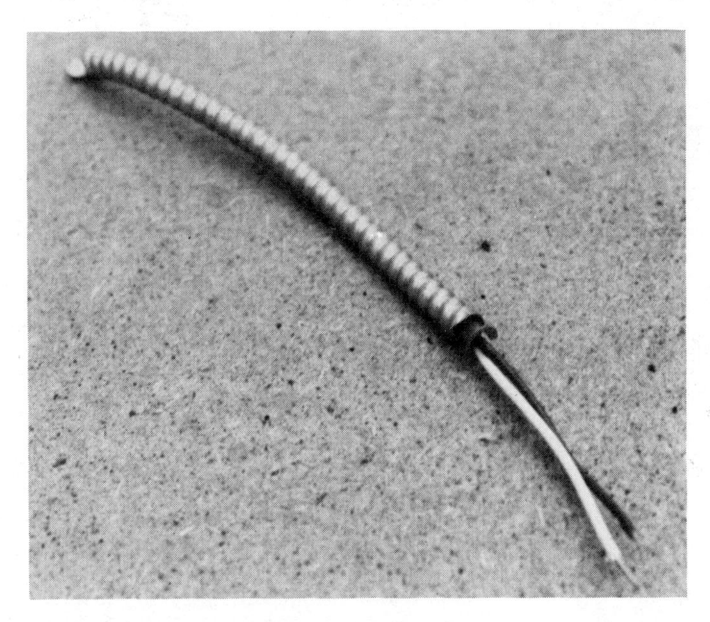

Figure 1.11 Antishort bushing inserted into the end of the BX cable to reduce the possibility of sharp metal edges piercing the insulation of the electrical conductors where they exit the metal sheathing.

1 You must install electric service with capacity to meet present and future needs

Never forget the importance of providing plently of EXTRA capacity when you plan your wiring. The size of the service entrance switch, the size of the connecting wires, determine the total amount of electricity you can use at any one time. With extra capacity, you can keep on adding new electrical servants as you need them, without fear of overloading wires or blowing fuses. A properly planned "Service Entrance" (as shown at left) including switch and connecting wires will relieve you of the annoyance and loss of power and current caused by overheated wires. Proper capacity assures you of getting top performance from your electrical tools at all times. For further details and specific recommendations, see pages 8-9.

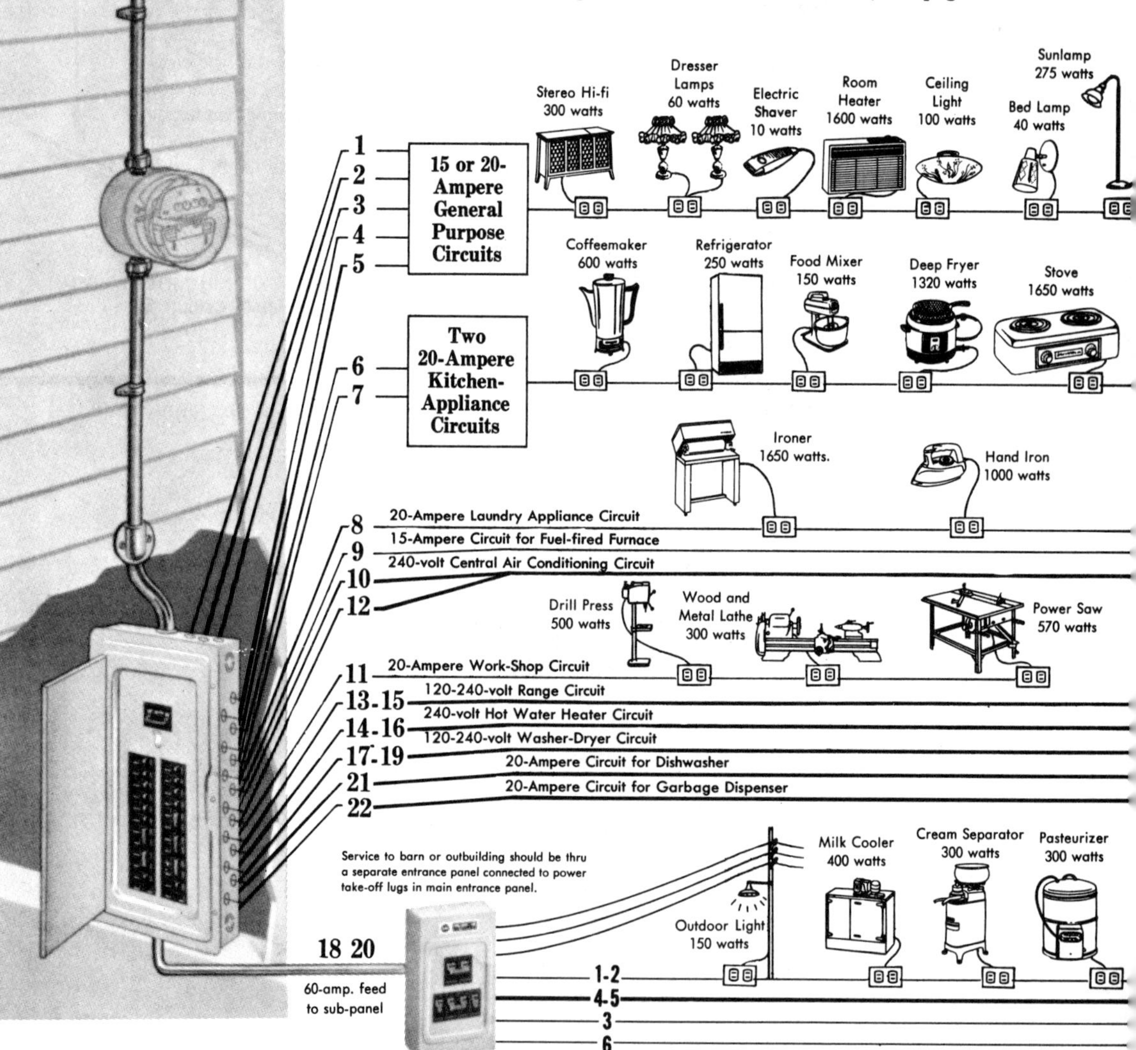

Figure 1.12 Branch circuit arrangement typical of most single-family and small multiple-family residences. (Courtesy of Sears, Roebuck, and Company.)

2 You must plan enough circuits to deliver full power always

Divide lights and outlets into various branch circuits as shown.

CODE REQUIREMENTS: all receptacles must be grounded type for new work or replacement on grounded systems. At least TWO 20-amp appliance grounded type circuits for kitchen, dining room and one for the laundry, independent of lighting fixtures.

A separate 20-amp general purpose circuit is recommended for every 500 square feet or a 15-amp circuit every 375 feet of floor space.

3 You must provide enough outlets on each circuit for convenience

A convenience outlet located every 12 feet of running wall space is required to provide complete flexibility in furniture placement, prevents unsightly long extension cords, assures better lighting.

IN KITCHENS, an outlet every 4 feet of counter space provides quick plug-in of appliances without moving them around, lets you make the most of your work space.

Be sure to plan a few outdoor outlets for holiday lighting, appliances or summer fun.

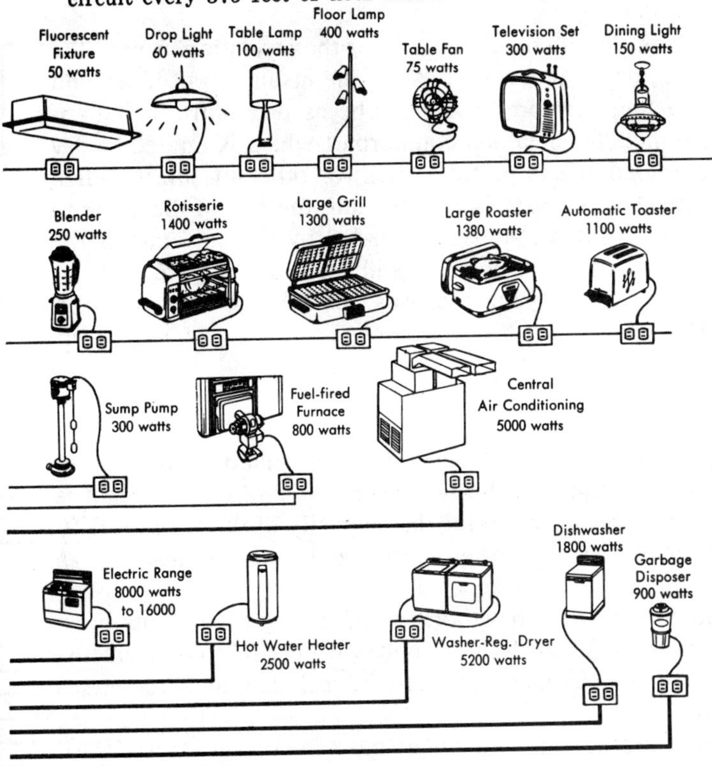

Figure total wattage for each circuit

NOTICE: Good planning insures that all outlets on a floor are not on one circuit. This prevents one blown fuse from throwing a whole floor into darkness. Note 4 circuits on 2nd floor. Note also that dining room, kitchen and laundry are on separate circuits. Put most circuits where load is heaviest.

CAUTION!

Maximum carrying capacity of a 20-ampere general-purpose or appliance circuit with Number-12 wire is 2400 watts. Check wattages shown here. No combination of appliances exceeding that wattage should be used on any one circuit at any one time.

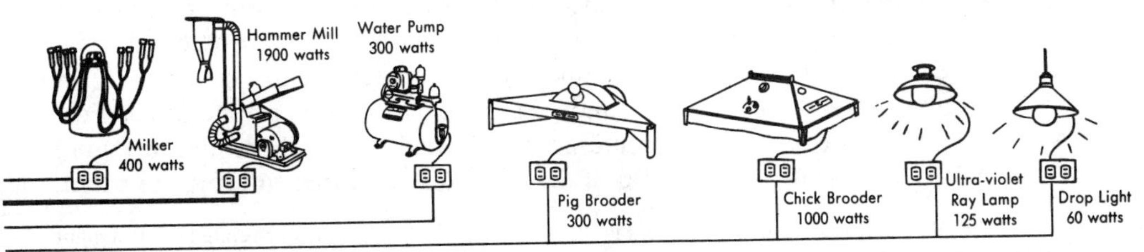

Figure 1.12 *(continued)*

Cable Types

Electric wire and cable can be supplied encased in a variety of outer sheathings, depending on the wiring application. For most residential applications, electric wire and cable can be classified as either nonmetallic or armor-sheathed cable. **Nonmetallic sheathed cable** is classified as type NM or NMC, commonly called **Romex**. This type of cable consists of two or more conductors enclosed within a moisture-resistant, flame-retardant outer sheathing.

Armored cable, classified by the Code as either AC or ACT, is usually referred to by electricians as **BX cable**. BX cable consists of two or more conductors enclosed within a flexible metallic enclosure. Figure 1.10 illustrates both Romex and BX cable.

Armored or BX cable is classified as type ACT if the insulation around the conductors is thermoplastic, or simply as type AC if the insulation used around the conductors is thermosetting. Armored cable can be used in more wiring applications than can its non-metallic-sheathed counterpart which is limited to dry locations only, including embedding in plaster, brick, or other masonry. When using armored cable, an **antishort bushing** is inserted into the end of the cable to prevent the conductors from accidentally being abraded by the cut ends of the armor shielding. The use of an antishort bushing is illustrated in Figure 1.11.

For a complete listing of cable types and uses, consult the latest issue of the National Electric Code.

FUSED BRANCH CIRCUITS

The wiring system in most homes consists of several circuits that branch out from the main service panel—hence the term **branch circuit**. Each branch circuit is protected by a fuse or circuit breaker rated for the ampacity of the wire used. A full branch-circuit arrangement in a typical residential application is illustrated in Figure 1.12.

When designing the circuitry in a new home or analyzing the circuits in an older one, adequate-size wiring for all fixtures is based on Code requirements with regard to the number of fixtures used, type of wire installed, and techniques used to install the system.

Adequate overcurrent protection is the single most important safety factor in the design and installation of any electrical system. Overcurrent protection is based on the ampacity of the wire in each branch circuit. This protection is installed in each circuit by the use of either circuit breakers or fuses.

Circuit breaker panels are the standard type of overcurrent protection device. A typical circuit breaker panel, including wired breakers, is illustrated in Figure 1.13.* Note from the illustration that of the three service wires entering the panel,

* The use of ground-fault interrupters is required in any water- and heater-related environment within the home. Consult the NEC for all applicable ground-fault locations.

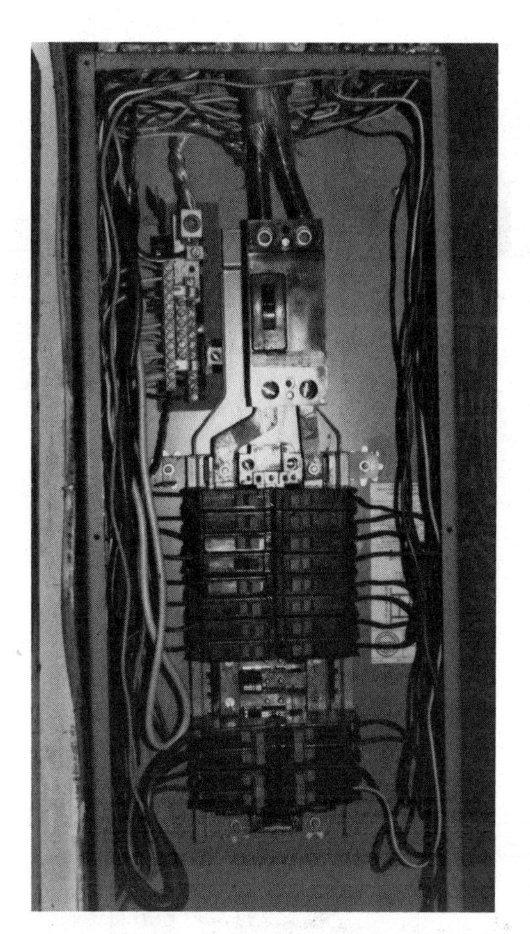

Figure 1.13 Circuit breaker panel with typical branch circuit breakers installed.

two of the conductors are rated for 220 V; the third wire is the ground wire and is connected to the grounding lug in the panel. Two hundred forty volts of voltage potential lies between the two "hot" wires; 120 V of potential exists between either hot line and the neutral. Individual circuits branch off to the various circuit breakers in the panel, based on the specific loads to be served.

A **fuse box**, typical of many older electrical installations, is illustrated in Figure 1.14. Fuses are excellent safety devices. Their main advantage over circuit breakers is that they fail only once, at which point they must be replaced. Circuit breakers are more convenient than fuses in that they are easily reset rather than being replaced. However, circuit breakers tend to lose some degree of sensitivity each time they trip, as well as with age. Therefore, they should be replaced periodically to ensure maximum overcurrent protection.

Now that some of the basic concepts dealing with energy production and

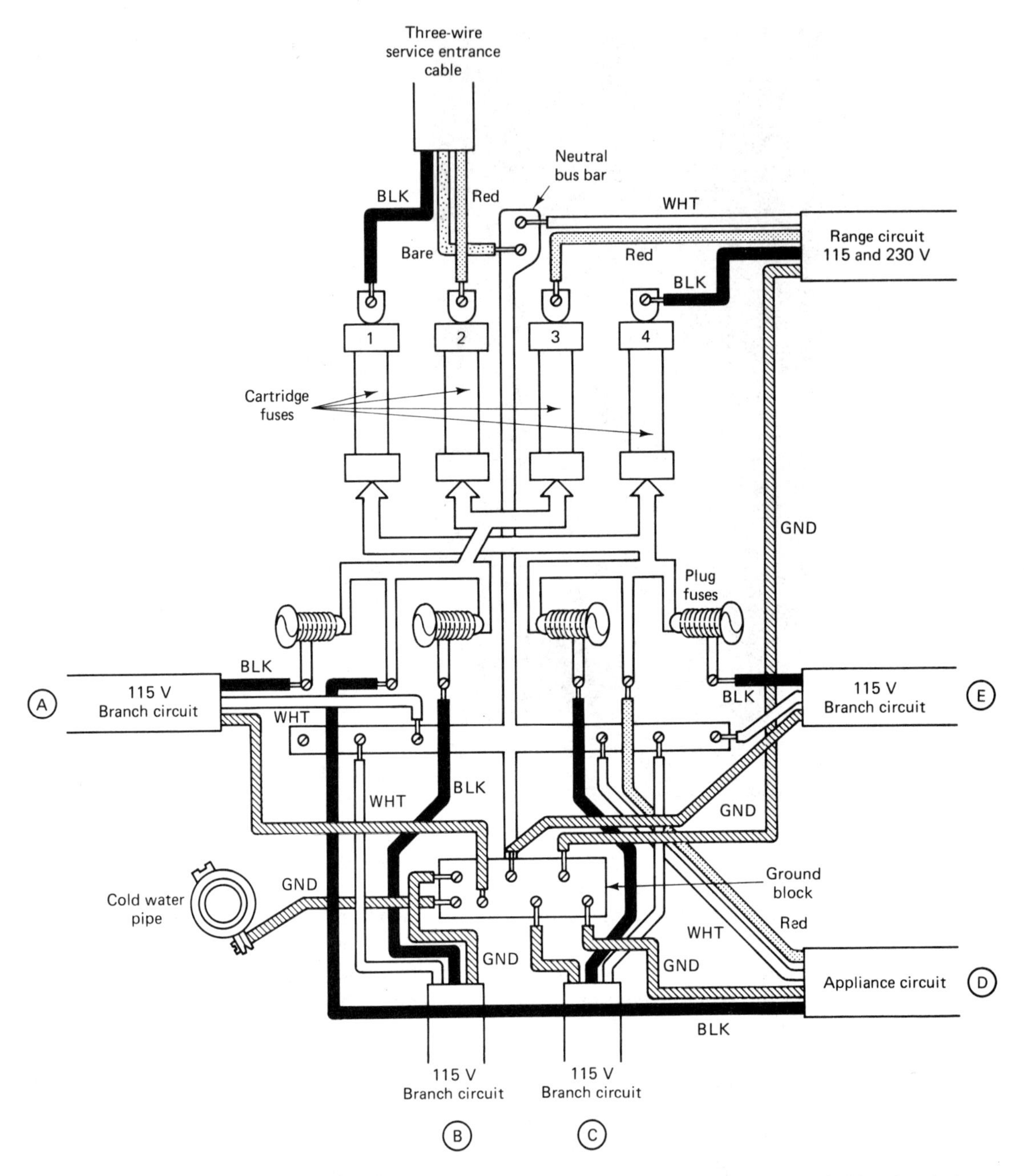

Figure 1.14 Pictorial and schematic of fuse-type service panel. Typical of older electrical installations, fuse panels are often upgraded to modern circuit breaker panels when the old fuse boxes can no longer handle the number of branch circuits within a home. (From *Home Wiring* by R. Graf and G. Whalen. Published by Craftsman Book Company, Box 6500, Carlsbad, CA 92008.)

consumption have been discussed, we can move on to an investigation of the various energy systems within the home: their design and method of operation. This will lead to the development of an overall energy strategy designed to ensure an efficient system of energy delivery and consumption designed to maximize benefits and minimize both cost and waste.

2

HEATING SYSTEM CLASSIFICATION

Residential space heating is provided by one of four basic types of systems: hot air forced convection, hot water baseboard, steam heating, and electric radiant heating systems, typified by wall-mounted electric baseboard units or radiant electric heating panels. Although both hot air and hot water baseboard heating systems can operate in the gravity-fed mode, this discussion is limited to systems that employ forced convection, since this is the most widespread and efficient type of heat transfer available. Also, natural-convection or gravity-fed systems are usually installed in older, smaller homes or weekend residences that do not require a conventional centralized space heating system. In addition to focusing on the design concepts of the four basic types of heating systems, a section on active solar heating as it applies to hot water and hot air applications is provided that illustrates the interfacing of the solar heating system with a backup or conventional heating system. Specific information relating to the operation and maintenance of oil and gas heating appliances is presented in Chapters 3 and 4.

Before installing any heating system, the homeowner must have a basic understanding of residential heat-loss analysis and how this affects the size and complexity of any space heating system.

RESIDENTIAL HEAT LOSS AND SYSTEM SIZING

The size or heating capacity of any heating system should be based on the maximum amount of heat that it will be required to furnish during the coldest days of the season. Thus the size of the heating systems for two identical houses located in different parts of the country will be different, based on local climate conditions. Generally, a design temperature for a particular area that specifies the lowest temperature that will be encountered during the coldest winter months is selected. A properly sized heating system should be able to maintain an interior temperature of 68°F based on the outside design temperature. For example, it might be found that in an area with a design temperature of −10°F, a particular home has a heat loss of 100,000 Btu/hr. In this situation, the heating system must be capable of delivering a *net* heating capacity of 100,000 Btu/hr. To determine exact residential heating loads, a detailed heat-loss analysis should be calculated. This can be done by using a manual step-by-step procedure or, more easily, by the use of computer programs that have built into them all the factors necessary to compute an exact heat-loss figure for almost any residential situation.

Many heating contractors can determine heat-loss figures for their prospective customers. Also, most large wholesale supply houses selling plumbing and heating equipment will provide this service for their customers for a nominal charge.

There are methods for quickly determining the approximate heat loss for different homes that, although they lack the high degree of accuracy of a computerized or detailed manual calculation, will provide a reasonably accurate estimate of the heat loss of a home. In this method, the home is placed into one of three basic categories: modern, tightly constructed, and superinsulated homes: for example, passive solar designs; contemporary homes equipped with conventional insulation characteristics, such as $3\frac{1}{2}$ in. of insulation in the side walls and 9 in. in the ceiling, with storm windows; and older, uninsulated homes, such as farms and weekend summer cottages. To each category we assign an approximate heat-loss figure per square foot of living area, based on the category assigned to the home. This figure is then multiplied by the square footage of the home to obtain the approximate heat loss.

Table 2.1 lists the heat-loss figures for each of the home categories discussed.

TABLE 2.1 CATEGORIES FOR SIMPLIFIED RESIDENTIAL HEAT-LOSS ANALYSIS

Category	Description	Heat loss (Btu/ft^2)
1	Older, uninsulated homes	60–75
2	Modern construction, standard insulation	40–50
3	Modern, tight construction	25–35

Note: Figures are approximate and will vary based on individual buildings and design temperature differentials.

This type of procedure should not take the place of a detailed heat-loss analysis; it should merely supplement or help initially to determine the basic sizing of the heating system.*

THERMOSTATS

Thermostats are a frequently overlooked component in most heating systems and are often added as an afterthought after completion of the basic system design and installation. Despite their small size, the entire system function and dependability focus on this device. If the thermostat ceases to function properly, the house will either overheat or freeze. Several factors must be taken into account when selecting and installing a thermostat (Figure 2.1).

Location of Thermostats

Thermostats should be located on interior building walls about 5 ft off the floor. Locations to be avoided are those directly over baseboard, steam radiation, or hot air supply ducts, which will cause readings that are not indicative of the major portions of the room or home.

Types of Thermostats

Single-stage thermostats. Single-stage thermostats are so named because they open and close a set of electrical contacts at one set temperature point for either a heating or a cooling system. Conventional thermostats contain a coiled bimetallic spring connected to a mercury switch (Figure 2.2). In thermostats equipped with contact points, a drop in room temperature causes the bimetallic coil to contract and rotate, which closes the contact points, allowing electricity to flow in the low-voltage thermostat circuit, which closes a relay to ignite the burner and auxiliary circulating pumps or fans when the heating system reaches operating temperature. In thermostats equipped with a mercury switch, rotation of the coiled bimetallic spring tilts a capsule filled with a small amount of mercury. As the glass-enclosed switch tilts, mercury falls from one side of the capsule to the other, engulfing two electrical contacts, closing the low-voltage thermostat circuit.

Single-stage thermostats are also capable of activating both a heating system and a cooling system from the same unit. In this type of configuration, called a **single-pole, double-throw circuit**, a rise in room temperature sensed by the thermostat activates one electrical circuit that controls the air-conditioning system, whereas a drop in temperature activates a second set of electrical contacts to turn on the heating system.

* For additional techniques used in determining heating and cooling loads, see Chapter 7.

Figure 2.1 Standard residential room thermostat. (Courtesy of Honeywell, Inc.)

In single-zone heating systems, only one thermostat is used to control the system. In multiple-zone heating systems, one thermostat is required for each heating zone. Figure 2.3 illustrates a simplified single-zone hydronic system, where the thermostat energizes the boiler aquastat to turn on the burner. Figure 2.4 illustrates a multizone heating system that uses three separate thermostats to control each heating zone through an auxiliary zone valve (5).

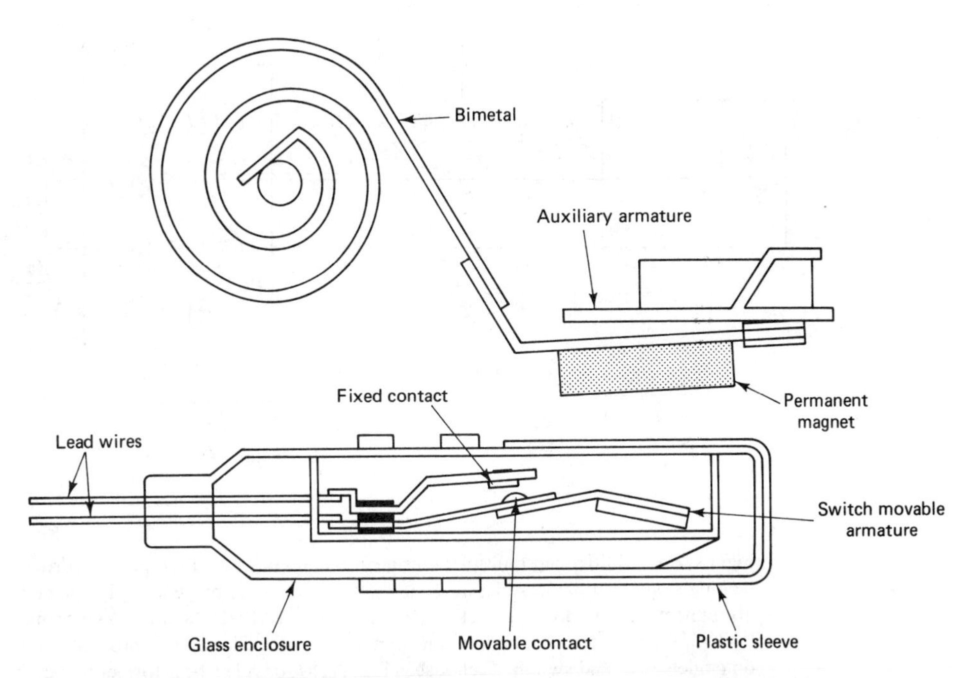

Figure 2.2 Bimetallic spring coil and switch contact assembly in typical house thermostat. (Courtesy of Robertshaw Controls Company.)

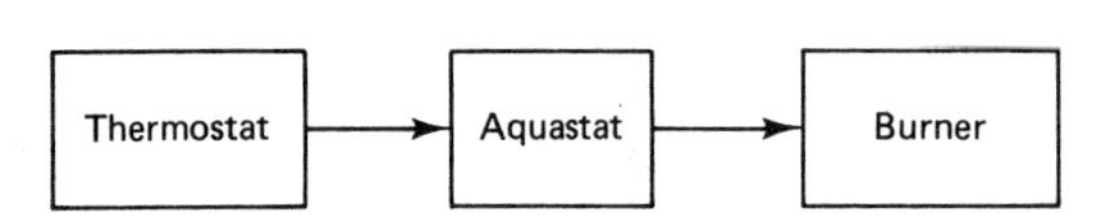

Figure 2.3 Single-zone heating sequence. Most single-zone residential heating systems utilize the thermostat to energize a low-voltage circuit that controls the burner mechanism. This concept is identical in gas- and oil-fired systems.

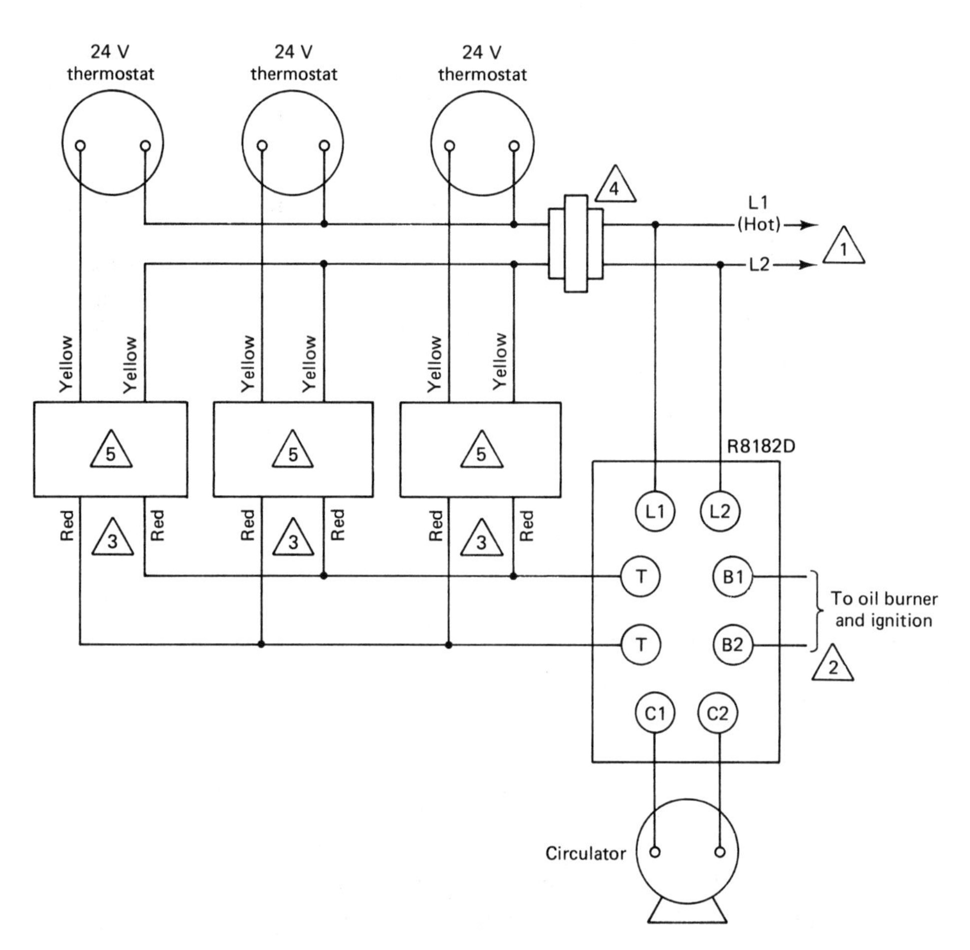

Figure 2.4 Multizoned/multiple thermostat configuration in typical hydronic heating system utilizing separate zone valve controls: 1, power supply—supply disconnect and overload protection, as required; 2, R8182H has white and orange lead wires for oil burner and ignition connections; 3, red lead wires provided only on models with end switch; 4, choose AT72, AT87, or AT88 transformer to match maximum system electrical load; 5, use V8043, V8044, or V8343 zone valves with auxiliary end switch only. (Courtesy of Honeywell, Inc.)

Heat anticipators. Thermostat heat anticipators are small electrical devices built into the conventional thermostat that anticipate residual heat from the heating system that has not been sensed by the thermostat. In effect, this allows the thermostat to turn off the heating system in anticipation of heat yet to be delivered, allowing careful control of room temperatures.

Heat anticipators are adjustable and are set based on the current draw in the circuit. Most boiler and furnace operating controls list the current draw requirements for the heat anticipator circuit on the relay cover box or in the instruction sheet for the unit. These values should be used to set the anticipator. If a heating system consistently raises the room temperature above the thermostat set point or shuts off before the room has been heated sufficiently, the anticipator circuit should be checked to make sure that it has been set properly. A typical heat anticipator circuit is illustrated in Figure 2.5.

Automatic set-back thermostats. Since less heat is required in the home when it is empty as well as in the late evening hours, thermostats that automatically set back the heating levels at required intervals can save significant amounts of money during a typical heating season. Most set-back thermostats either have pins that are placed on a rotating clock wheel to raise and lower the set temperature, or use solid-state microprocessor logic to program the set points. Figure 2.6 illustrates a solid-state set-back thermostat.

Multistage thermostats. Multistage thermostats control two or more individual set points within a heating or cooling cycle. For example, in a typical hot water heating system using a multistage thermostat, the first stage might turn on the circulating pump to move water throughout the heating system. If the room temperature continues to drop to the second-stage set point, for example, 3°F below the first-stage set point, the burner will be energized to raise the temperature of the water within the boiler. The same logic would apply to a hot air heating system: Stage 1 activates the circulating fan, and stage 2 activates the burner to increase the temperature of the heated air.

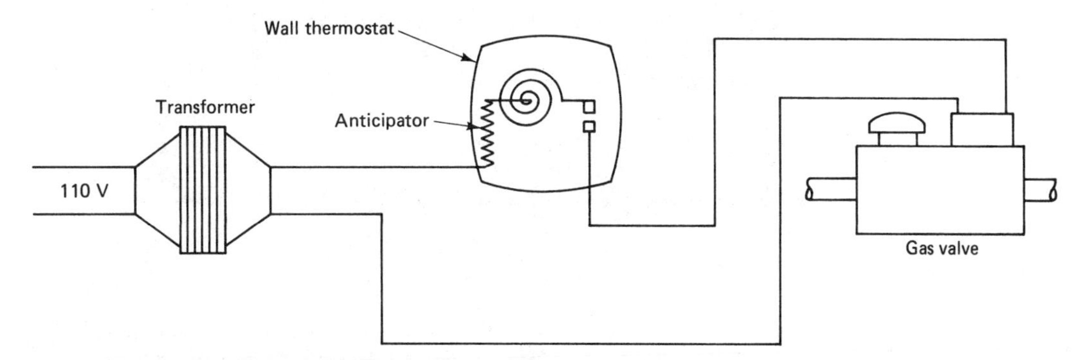

Figure 2.5 Heat anticipator circuit (typical). (Courtesy of Robertshaw Controls Company.)

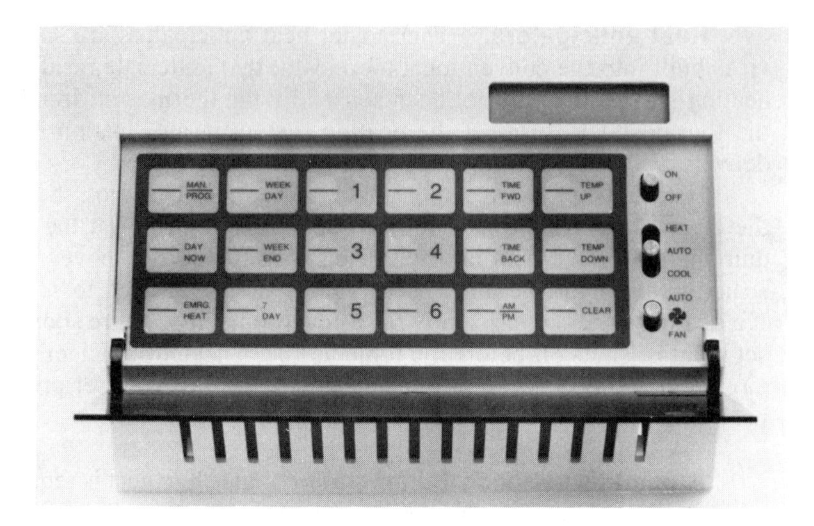

Figure 2.6 Automatic set-back thermostat. (Courtesy of American Stabilis.)

Multistage thermostats are also used in solar heating applications. In these systems, the first stage activates the solar system to supply heat to the residence. If the solar system is not capable of supplying the necessary amount of heat to the home, the room temperature begins to drop and the second stage of the thermostat activates the conventional backup heating system. Another use for mul-

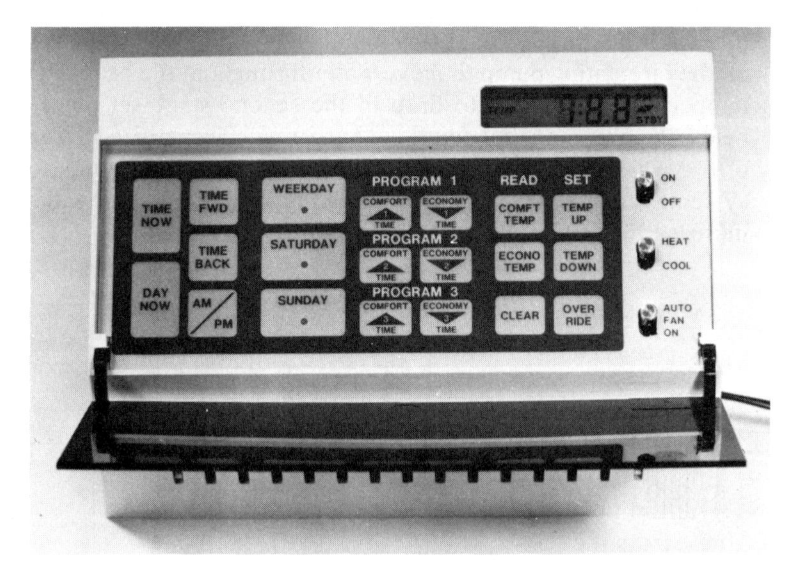

Figure 2.7 Programmable multistage thermostat. These units feature microprocessor-controlled functions to monitor and direct complex heating and cooling functions. (Courtesy of American Stabilis.)

tistage thermostats are in those heating systems equipped with a multifuel boiler or furnace. In multifuel systems, the first stage of the thermostat controls the solid-fuel combustion. As long as the heat produced by the solid-fuel system is sufficient to supply the required heat, the room temperature will be stable and the heating system will operate from the solid-fuel combustion. If the solid-fuel fire dies down, the room temperature drops, causing the second stage of the thermostat to activate the fossil-fuel ignition system in the heating unit. Figure 2.7 illustrates a solid-state programmable multistage thermostat suitable for these complex heating/cooling functions. Note from Figure 2.7 that this type of thermostat allows for individual daily heating and cooling levels. A digital display is incorporated to show day/date/time and temperature levels.

FORCED HOT AIR HEATING SYSTEMS

Forced hot air heating systems rely on a central furnace to provide heat through the combustion of a fuel, which heats via a heat exchanger, delivering it to various parts of the home using an electric blower and associated ductwork. This arrangement in a typical residential installation is illustrated in Figure 2.8. Note from the figure that the system is composed of three basic parts: the hot air furnace, the hot air supply system, and the return-air ductwork.

A typical hot air furnace in cross section is illustrated in Figure 2.9. There are various configurations of warm air furnaces, based on airflow patterns in the furnace. In a typical warm air furnace the fuel combustion chamber is located in the bottom of the unit. This combustion chamber can be designed to burn either fuel oil, gas (natural or LP), or solid fuels such as coal and wood. Some units can burn a combination of fuels, such as wood/coal and either oil or gas, and are referred to as **multifuel furnaces**. Figures 2.10 and 2.11 illustrate a conventional oil-fired hot air furnace and a multifuel hot air furnace, respectively. Note from the illustrations that the multifuel unit is larger than its fossil-fuel counterpart, due to the larger combustion chamber necessary to accommodate solid-fuel burning.

The combustion of the fuel takes place within a sealed heat exchanger. A flue pipe is connected to the heat exchanger to remove the effluents of combustion. The gauge of the flue pipe used to vent the furnace into the chimney will vary based on the type of furnace used in the installation. An oil-fired furnace will normally use 26-gauge pipe; a multifuel furnace requires 22- or 24-gauge black pipe or stainless steel flue pipe. The cold return air from the home is channeled over the heat exchanger, moved by the blower in the furnace, where it picks up heat from the combustion of the fuel. From the heat exchanger, the blower forces the heated air through the ductwork to the rooms in the house and then back through the return air ducts to the furnace, where the cycle is repeated until the temperature required in the home has been reached.

The sequence of operation of a typical hot air furnace is as follows. Upon

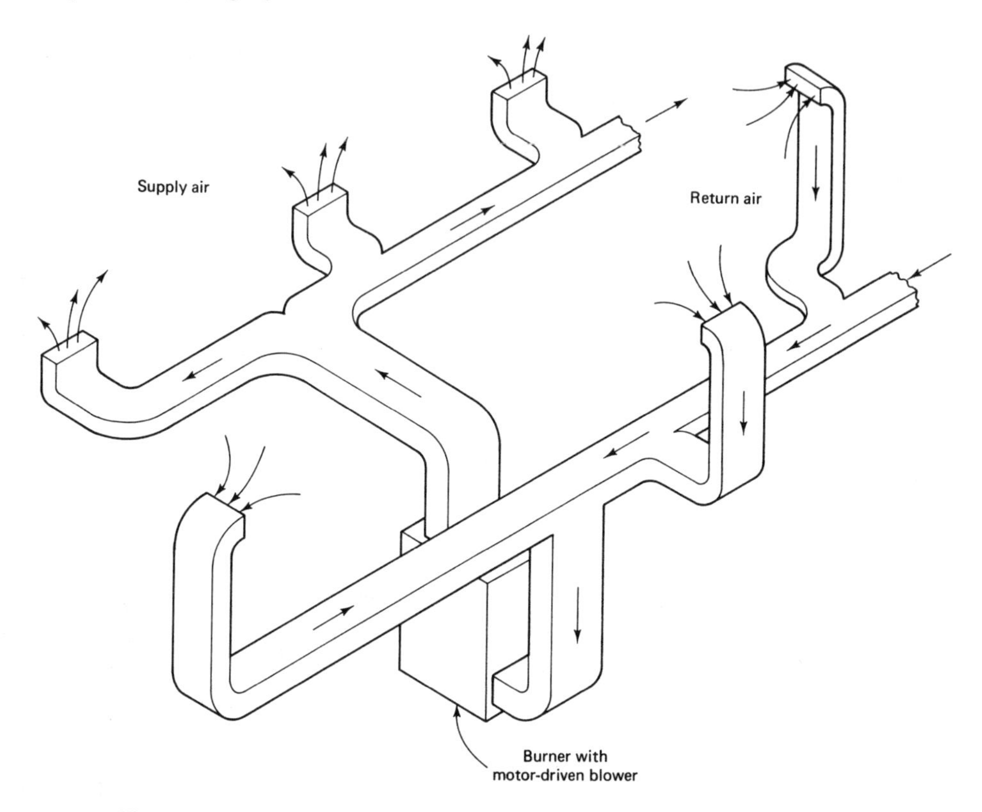

Figure 2.8 Forced-air central heating system. (From M. Greenwald and T. McHugh, *Practical Solar Energy Technology,* Prentice-Hall, Englewood Cliffs, N.J., 1985.)

a call for heat from the room thermostat, the burner mechanism is activated (in the case of a solid-fuel unit, additional combustion air is introduced into the combustion chamber to build up the intensity of the solid-fuel fire). When the air temperature within the plenum has risen to a preset point (usually 150 to 160°F), the blower is turned on. The movement of air throughout the heating system both heats the house and cools the air in the plenum as heat is withdrawn from the heat exchanger. After several minutes of operation, the temperature within the plenum drops to a low-limit set point, which turns off the blower. The burner will continue to operate, building up heat within the heat exchanger to allow the cycle to continue. Should the temperature rise above what is considered a safe high-limit setting, the fan-limit switch will automatically turn off power to the burner while allowing the fan to run in order to cool down the furnace. When the temperature falls below the high-limit setting, normal burner operation will resume.

Zoned Hot Air Systems

Configuring a heating system into separate heating zones is desirable since it allows certain portions of the home to be heated while leaving unoccupied or

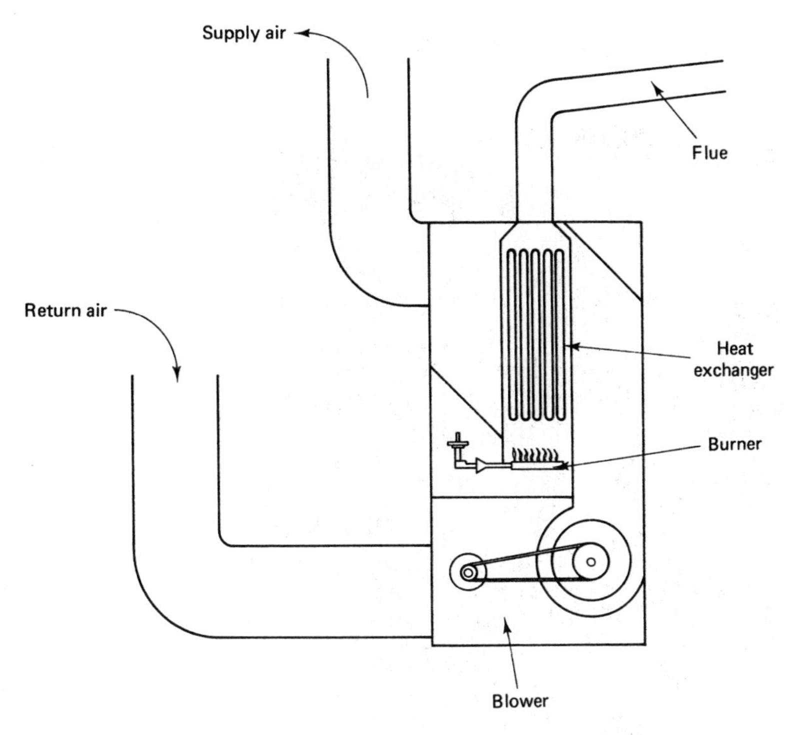

Figure 2.9 Sectional arrangement of a residential hot air furnace. (From M. Greenwald and T. McHugh, *Practical Solar Energy Technology*, Prentice-Hall, Englewood Cliffs, N.J., 1985.)

Figure 2.10 Oil-fired hot air furnace. (Courtesy of The Williamson Company.)

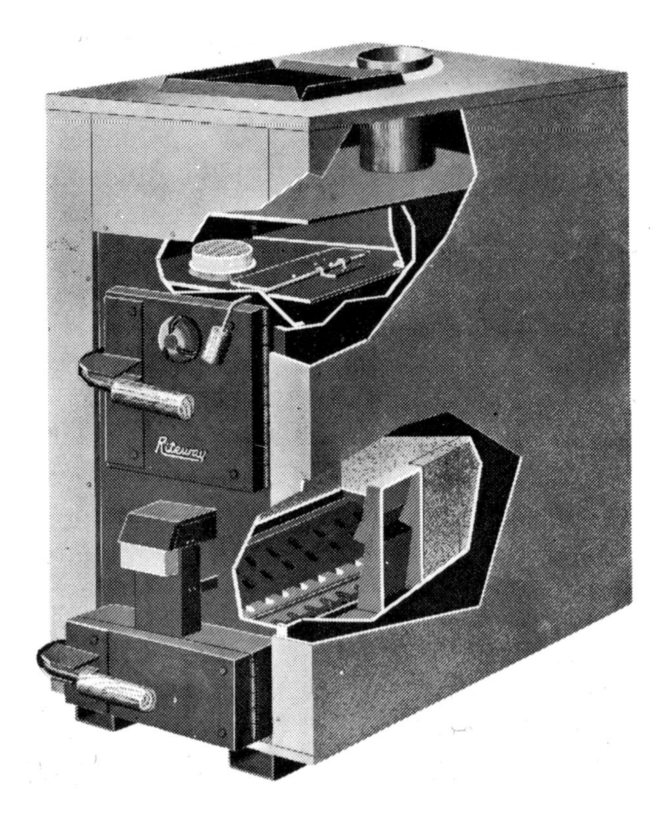

Figure 2.11 Multifuel hot air furnace. (Courtesy of Riteway-Dominion Manufacturing, Inc.)

unused areas at cooler temperatures. In zoned hot air systems, certain trunk lines within the duct system contain an electrically operated air duct (Figure 2.12).

The sequence of operation in a multizone hot air system is similar to that of a single-zone system. The major difference is that when the thermostat in a particular zone calls for heat, the specific zone damper is opened. At the same time, the burner and blower are energized to provide hot air for the particular zone. Each heating zone works independently of the others. Warm air circulates in a particular zone based on the position of the zoned air damper, which is either opened or closed depending on the zone thermostat setting.

Air Cleaners and Humidification

Warm air heating systems lend themselves to simple, effective installation of electrostatic air cleaners and power humidifiers. Since the system contains a trapped moving stream of air that can be channeled through humidifiers and air cleaners, virtually all the air within the home will pass through the system several times per hour, depending on the capacity of the furnace and associated blower unit (Figure 2.13).

Electrostatic **air cleaners** work by imparting an electrical charge on dust and

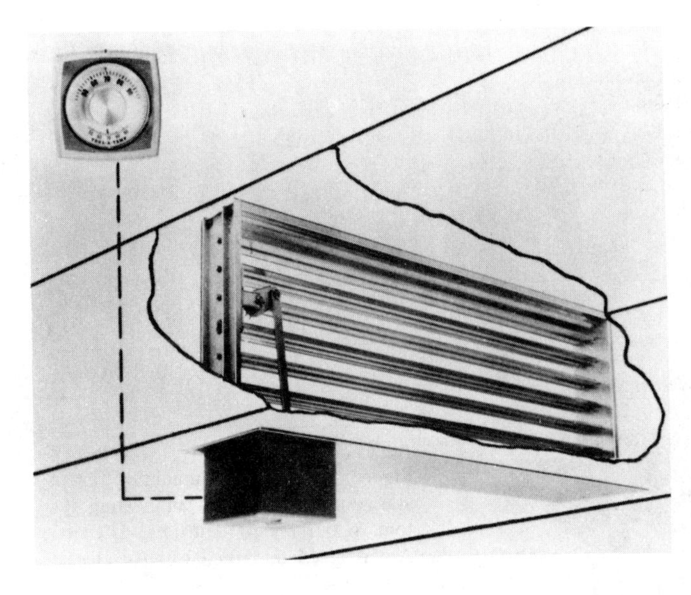

Figure 2.12 Electrically operated hot air zone dampers. (Courtesy of Trol-A-Temp.)

dirt particles in the airstream and an opposite electrical charge on plates within the cleaner. As the airstream moves through the unit, the charged particles of dust and dirt are attracted to the oppositely charged particles, removing them from the room air. Air cleaners such as these are highly effective in providing relief for people who suffer from a variety of allergies.

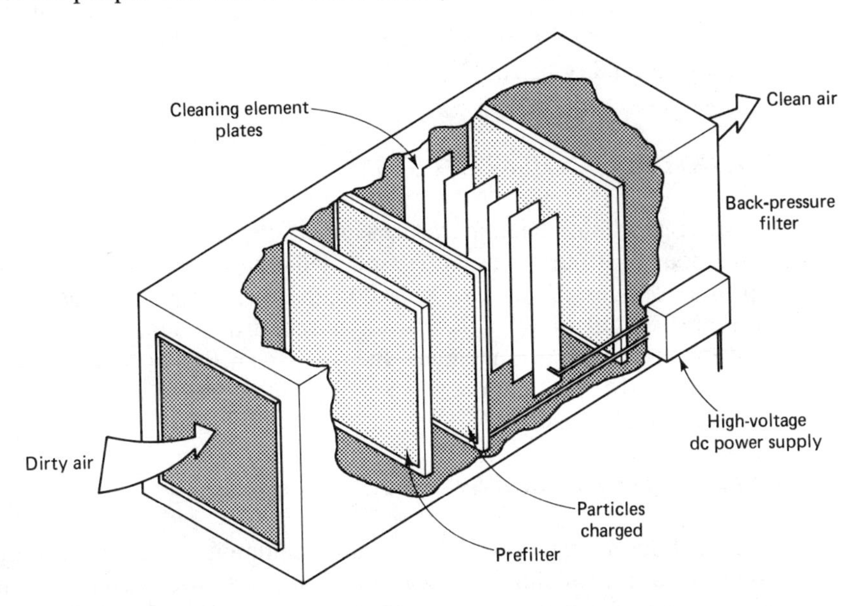

Figure 2.13 Electrostatic air cleaner. (From M. Greenwald and T. McHugh, *Practical Solar Energy Technology*, Prentice-Hall, Englewood Cliffs, N.J., 1985.)

Figure 2.14 Power humidifier for installation in forced hot air heating system. (Courtesy of Auto Flo Division, American Metals, Inc., Los Angeles.)

Power **humidifiers** raise indoor humidity levels by exposing the passing airstream in the heating system to a water-soaked wheel. The wheel revolves in a trough of water, continually picking up a fresh supply of water, which is given up to the air as it passes over the wheel. A typical power humidifier is illustrated in Figure 2.14. Since relative humidity levels during the winter can often drop to between 10 and 20%, humidifiers can provide a high degree of comfort by raising the indoor humidity to comfortable levels.

HOT WATER (HYDRONIC) HEATING SYSTEMS

Hot water baseboard heating systems are efficient, low in cost, and reliable. They consist of a hot water boiler with associated controls connected to a closed piping loop that contains the heating units located throughout the home (Figure 2.15). There are several types of hot water baseboard systems. These variations are based on the piping arrangement that is used in the system. What follows is a discussion of the most common types of piping configurations used in residential hydronic installations.

Series-Loop Baseboard Systems

The **series-loop** piping system is the most common type of baseboard hot water system used in residential applications. This system consists of a single loop of pipe run throughout the house, with the baseboard heating units placed in series with the piping (Figure 2.16). The single-zone series-loop system is perhaps the

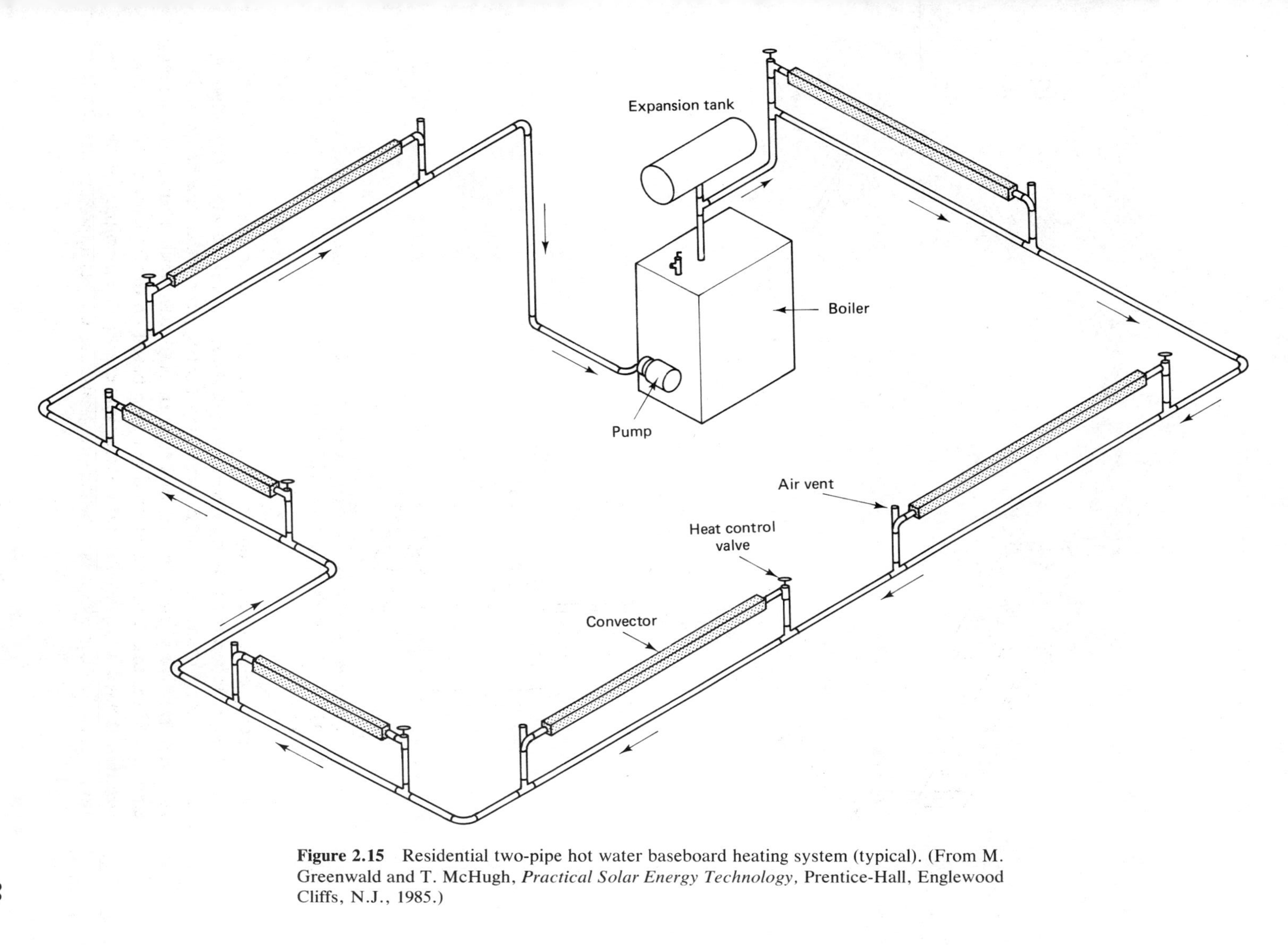

Figure 2.15 Residential two-pipe hot water baseboard heating system (typical). (From M. Greenwald and T. McHugh, *Practical Solar Energy Technology*, Prentice-Hall, Englewood Cliffs, N.J., 1985.)

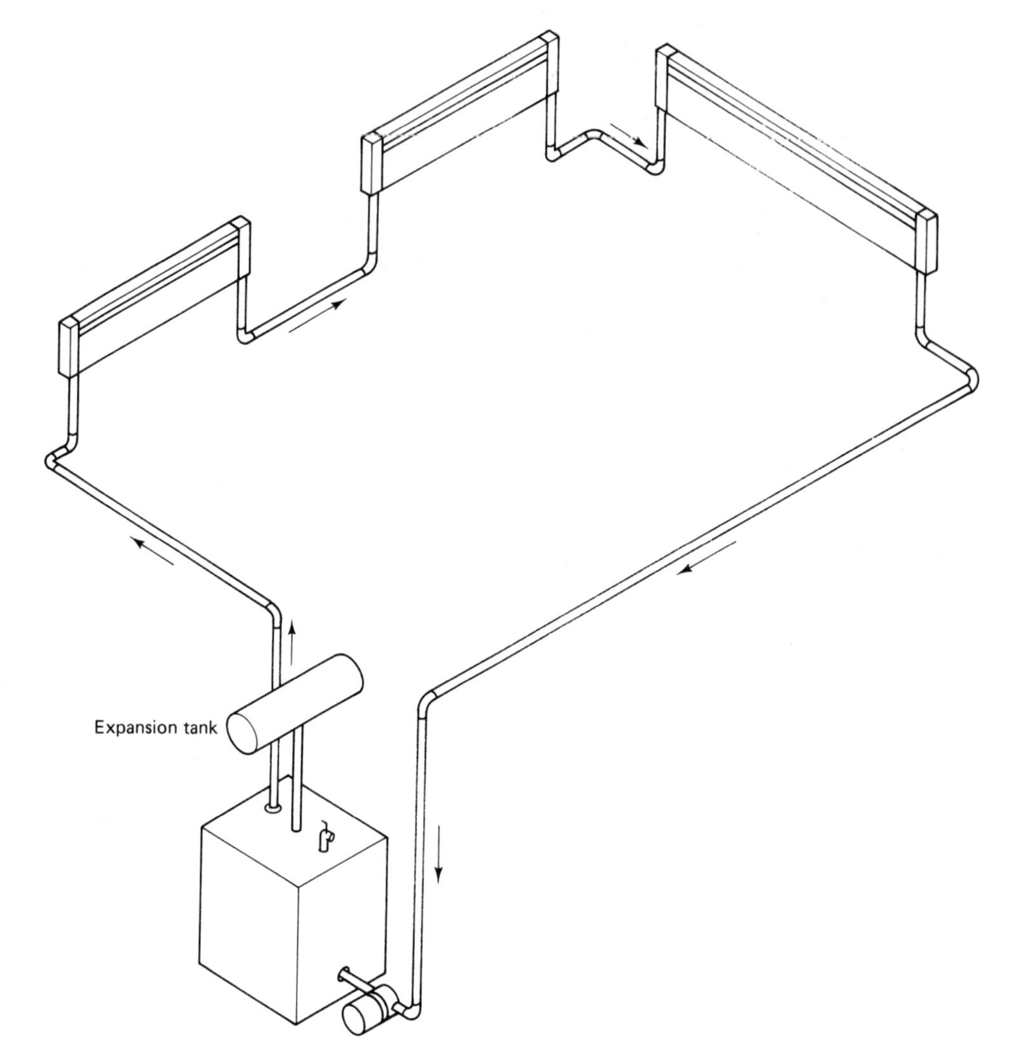

Figure 2.16 Series-loop hot water baseboard configuration. (From M. Greenwald and T. McHugh, *Practical Solar Energy Technology*, Prentice-Hall, Englewood Cliffs, N.J., 1985.)

least expensive baseboard option. The heating zone begins and ends at the hot water boiler. When the heating system is operating, the circulating pump delivers hot water throughout the home, providing heating to all the baseboard units. Although relatively inexpensive, this type of system is limited to the maximum amount of heat that can be delivered per loop based on the limitations of the pipe size that is used in laying out the system. For example, $\frac{3}{4}$-in. pipe, the most commonly used size in residential heating situations, can deliver approximately 40,000

Btu/hr based on a circulating water flow rate of 4 gallons per minute. Therefore, in a large home, one series loop zone cannot be used if the total heat loss in the home is greater than 40,000 Btu/hr. To overcome this limitation, if the installer wishes to keep the single-zone concept, the heating circuit can be split into two equal segments, which effectively doubles the delivered heat capacity of the system (Figure 2.17). Table 2.2 illustrates the maximum Btu capacities for various size of pipe used in residential heating installations.

Supply-Loop Systems

A modification of the single-zone series-loop design is known as the **supply-loop** system (Figure 2.18). In a supply-loop piping configuration, a single pipe runs around the circumference of the heating system. Where needed, baseboard convectors tap off of the supply line with two pipes: One pipe feeds hot water to the baseboard; the second pipe returns the water from the baseboard heater back to the supply loop through a flow-diversion fitting. In this way, each baseboard heater draws its own supply of heating water from a ready reserve that is constantly circulating. This type of system is flexible in that each baseboard heater can be equipped with its own shutoff valve. If no heat is required in a particular room or area of the home, the appropriate valves can be closed, which gives the system room-by-room zoning capability. The zoning can be made fully automatic by equipping each terminal heating unit with its own zone valve, which can be located within the heating unit (Figure 2.19).

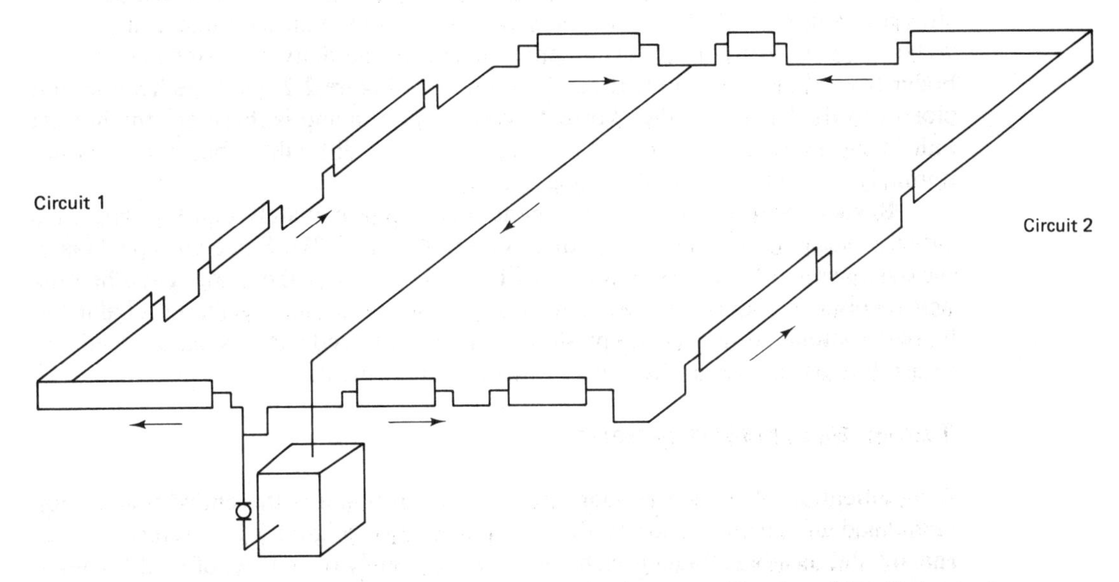

Figure 2.17 Double-circuit single-zone hydronic system. (Courtesy of The Hydronics Institute.)

TABLE 2.2 MAXIMUM BTU CAPACITIES BASED ON PIPE DIAMETER

Baseboard size (in.)	Total load (Btu/hr)	Velocity based on 20°F temperature drop
Cast-iron baseboard		
$\frac{3}{4}$	40,000	2.4
Steel pipe baseboard		
$\frac{1}{2}$	18,000	1.9
$\frac{3}{4}$	40,000	2.4
1	72,000	2.7
$1\frac{1}{4}$	160,000	3.5
$1\frac{1}{2}$	230,000	3.7
2	450,000	4.6
Copper tube baseboard		
$\frac{1}{2}$	15,000	2.1
$\frac{3}{4}$	35,000	2.3
1	75,000	3.0
$1\frac{1}{4}$	130,000	3.5

Source: Taco, Inc.

Direct/reverse-return piping. Piping systems are also classified as either direct or reverse return, depending on the path of water flow between the feed and return lines in the system. A supply-loop system with **direct-return piping** is shown in Figure 2.20. In a direct-return system, the main feed line delivers hot water to each terminal heater; a separate return line delivers water back to the boiler from the heating unit. It can be seen from Figure 2.20 that the heating unit closest to the boiler has the shortest overall pipe run and is therefore the hottest unit in the system. Since direct-return piping is so difficult to balance, it is not normally used in residential heating systems.

Reverse-return piping arrangements offer approximately equal piping runs for all baseboard units in the heating system (Figure 2.21). Since all pipe runs in the system to and from the baseboard heaters are about the same, each heating unit receives an equivalent amount of heated water, which results in a balanced heating system. Almost all supply-loop systems plumbed with a separate feed and return line are set up in the reverse-return arrangement.

Radiant Slab Heating Systems

A modification of the series-loop heating system features the entire piping loop embedded within the concrete floor of a structure (Figure 2.22), referred to as **radiant slab heating**. Radiant slabs provide a uniquely even level of heat throughout the home. The boiler operates in a conventional manner, circulating hot water throughout the embedded loop and heating the concrete slab in the process. Once the slab has reached a working temperature of approximately 90°F, it will radiate

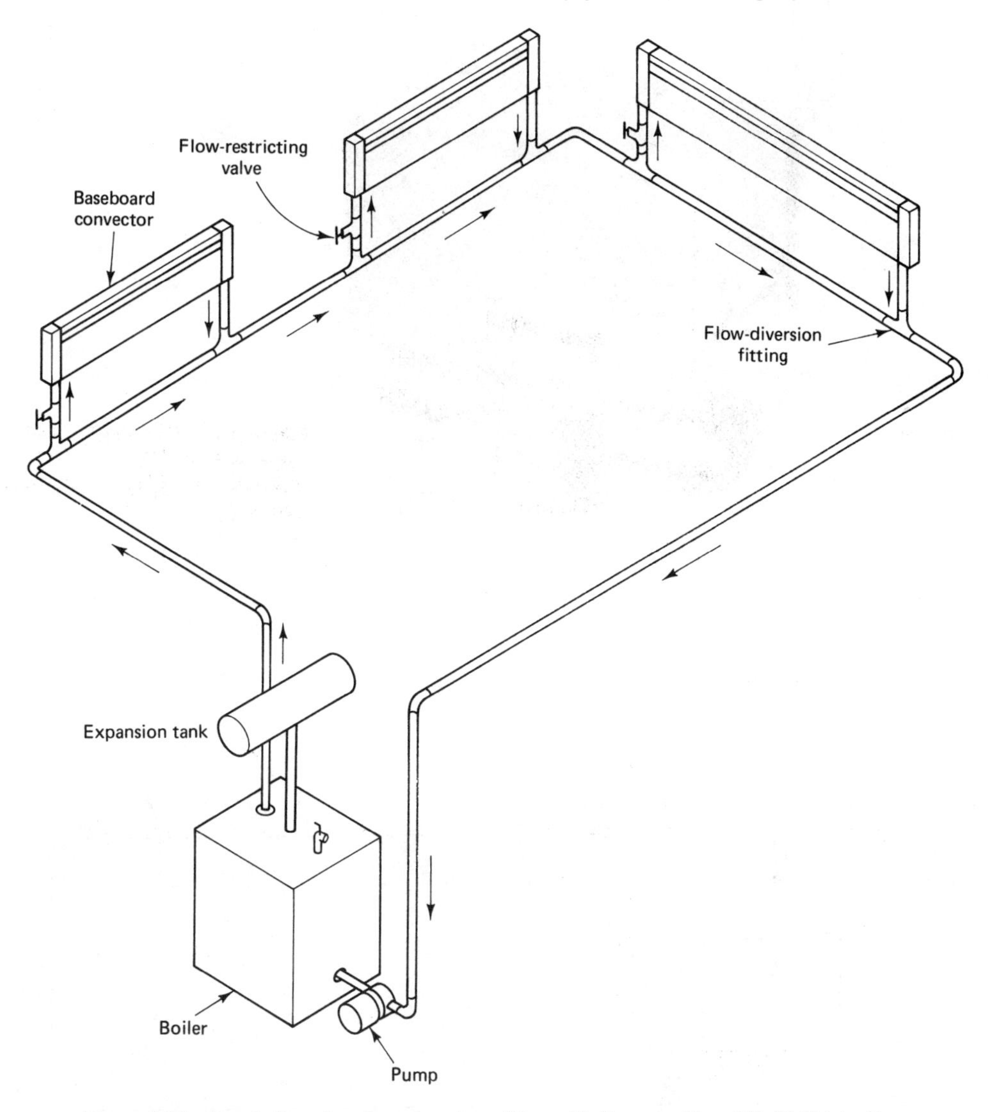

Figure 2.18 Supply-loop baseboard system. (From M. Greenwald and T. McHugh, *Practical Solar Energy Technology,* Prentice-Hall, Englewood Cliffs, N.J., 1985.)

heat to the home for several hours from the thermal mass of the concrete. One prerequisite for this type of system is that the slab must be well insulated. This prevents heat loss from the slab to the earth surrounding the slab. Since the mass of the slab is very large, this type of system does not cycle as quickly as does a hot water baseboard system with conventional terminal heaters, but it will provide even, comfortable heat for the building occupants long after the boiler has shut off. Also, since the piping is buried in the floor, no baseboard units are visible.

Figure 2.19 Electrically operated zone valve for hydronic heating systems. (Courtesy of Flair International Corporation.)

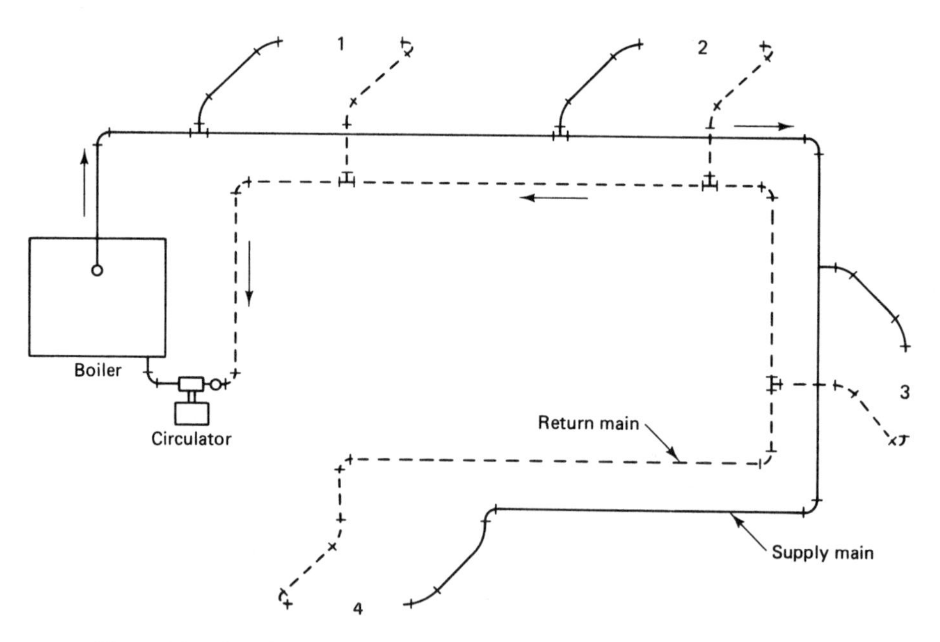

Figure 2.20 Supply-loop direct-return hydronic system. (Courtesy of The Hydronics Institute.)

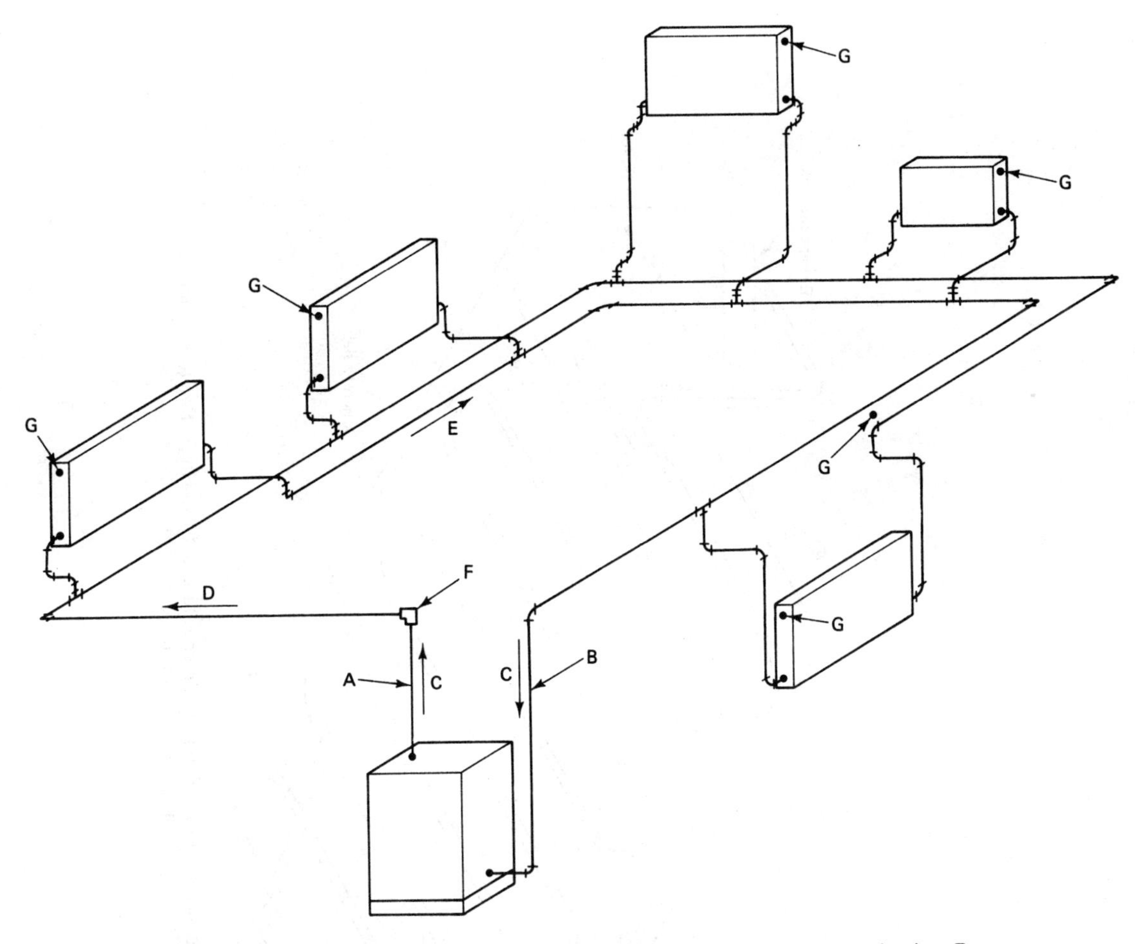

Figure 2.21 Supply-loop two-pipe reverse-return hydronic system. A, supply pipe; B, return pipe; C, direction of flow of water; D, pitch supply pipe upward; E, pitch return pipe downward; F, flow control valve required if boiler has a tankless coil and optional if no coil is present; G, air vent on each terminal heating unit and at end of supply main. (Courtesy of The Hydronics Institute.)

Multiple-Zone Series-Loop Systems

Provision can easily be made within the series-loop circuit to divide the heating system into separate zones, which increases system efficiency and allows for sections of the home to be heated only when and where necessary. Figure 2.23 illustrates a multizone hot water baseboard system that is controlled by a separate circulator in each zone. The operation of the two-zone system is simple. When either of the zone thermostats call for heat, both the circulator for that zone and the burner mechanism in the boiler are turned on. Heat is delivered only to the

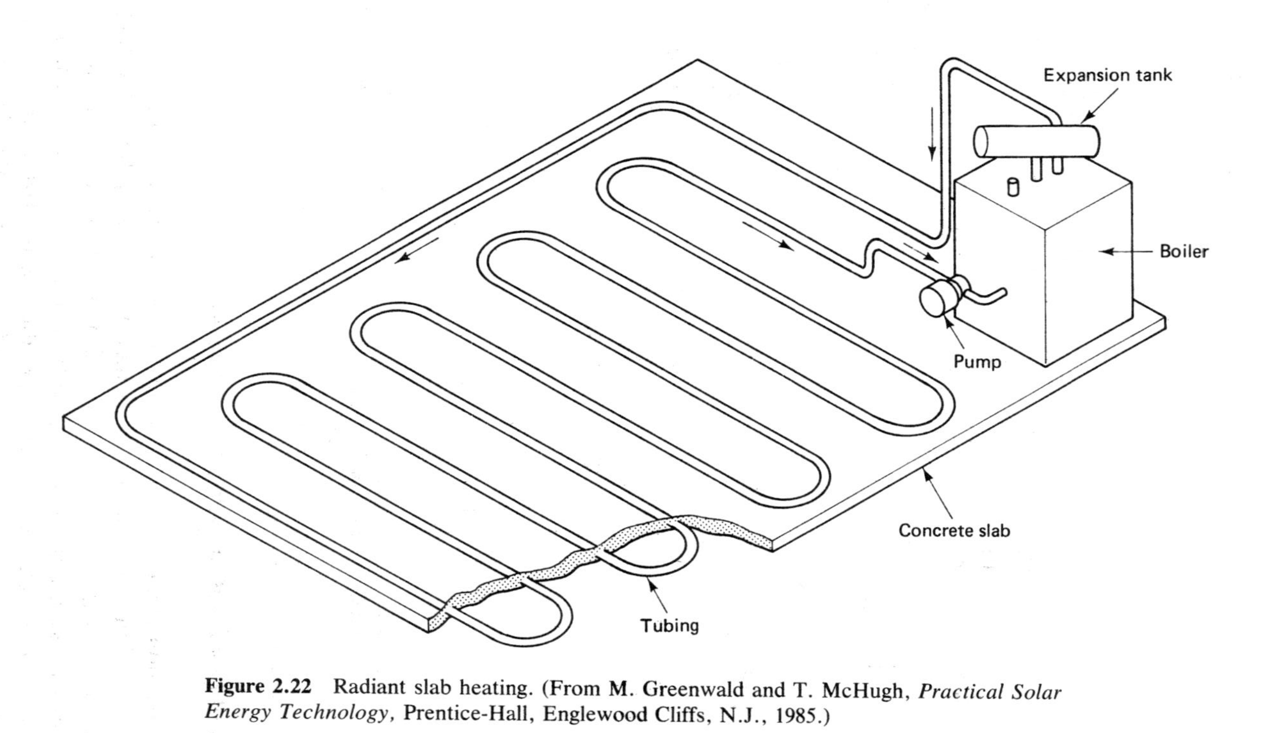

Figure 2.22 Radiant slab heating. (From M. Greenwald and T. McHugh, *Practical Solar Energy Technology*, Prentice-Hall, Englewood Cliffs, N.J., 1985.)

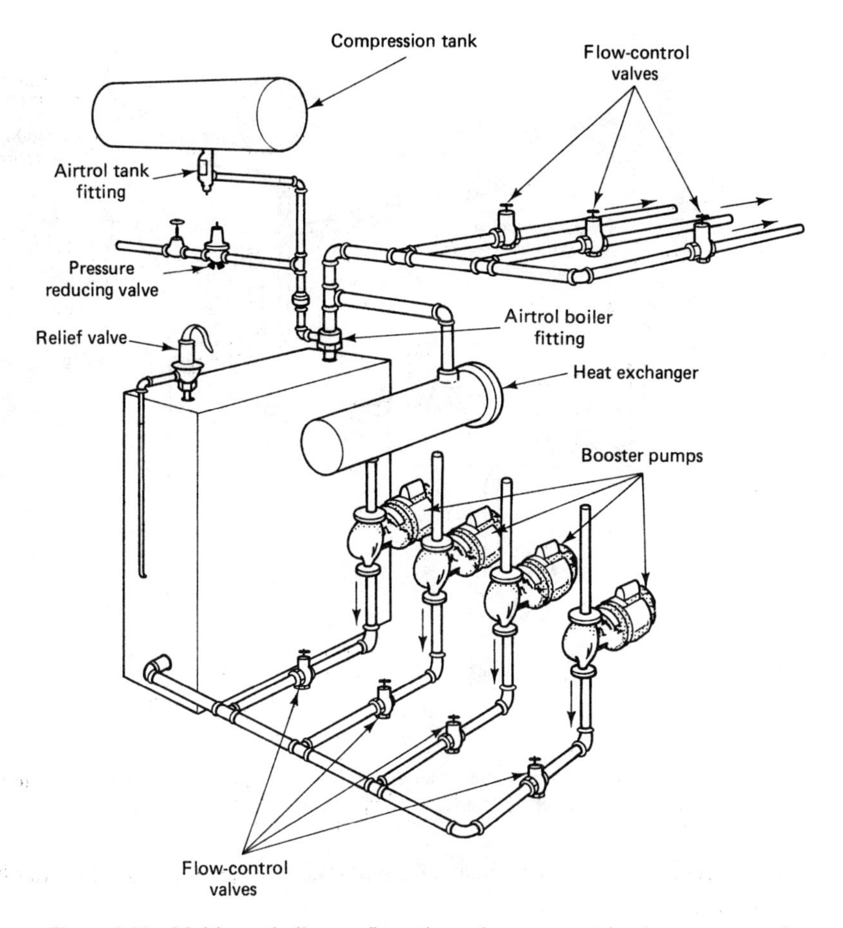

Figure 2.23 Multizone boiler configuration using separate circulators to control each heating zone. (Courtesy of ITT Fluid Handling Division.)

zone that requires it, although all the zones may operate simultaneously. The water temperature in the boiler is controlled by a conventional aquastat that turns the burner on and off, depending on the boiler water temperature. Note the use of flow-check valves, which prevent the gravity flow of water in the zone when the circulator either is not operating or when the boiler water is heated for domestic hot water without a call for space heating.

Another modification of the zoned hydronic system uses zone valves rather than separate circulators for each heating zone (Figure 2.24). When using zone valves, one circulator controls all water movement within the heating system. The zone valve is an electrically operated water valve which, when it opens, provides power to the burner mechanism and water circulator by closing an end-switch contact in the valve. Since the zone valve completely closes the water line

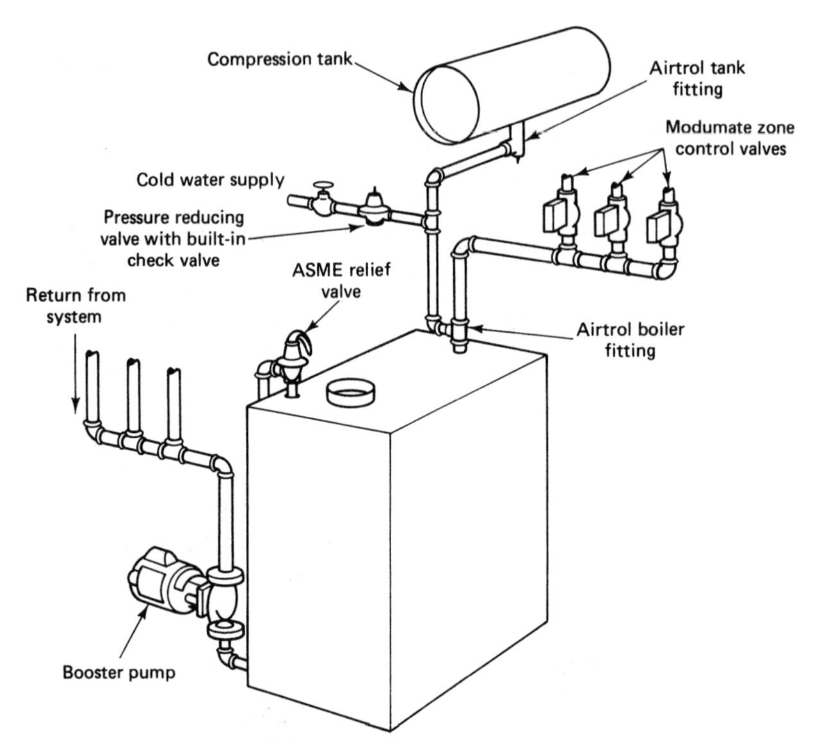

Figure 2.24 Multizone boiler configuration using separate zone valves to control each heating zone. (Courtesy of ITT Fluid Handling Division.)

when there is no call for heat in a particular zone, the use of a check valve is not required.

Baseboard Heaters

Every hot water baseboard heating system (with the exception of the radiant slab system) depends on a terminal heating unit, commonly referred to as either a **baseboard** or a **terminal heater**. The typical heater is composed of a copper tube to which a series of radiant aluminum fins is attached. This tube is located within an enclosure which is fastened to the wall. A typical residential baseboard heater is illustrated in Figure 2.25.

In addition to standard fin-type baseboard, two other types of heat distribution units are used in residential applications: case-iron radiators and convectors. **Convectors** are heating units that contain finned piping enclosed within a metal cabinet. The metal cabinets contain openings at the top and bottom for cold air intake at the bottom and heated air exhaust at the top. Convectors are usually installed either fully or partially recessed into a wall. A unit of this type is illustrated in Figure 2.26.

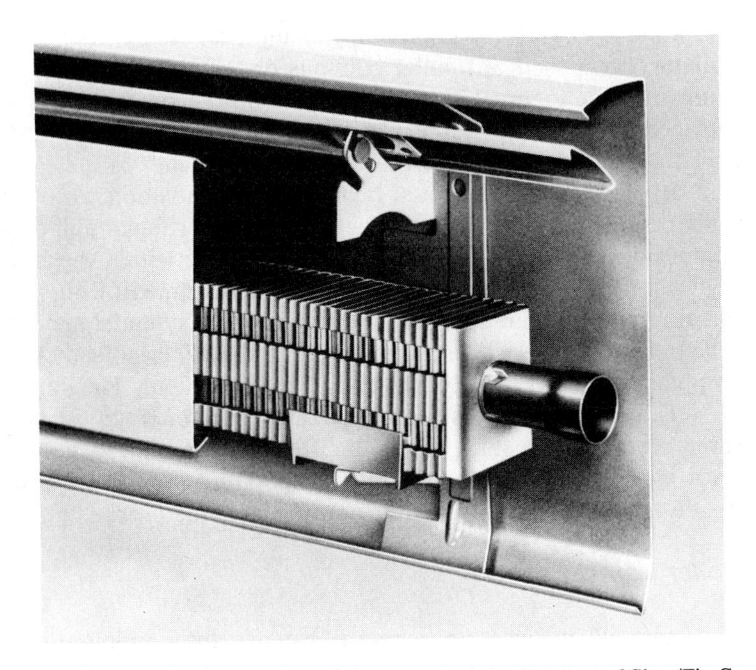

Figure 2.25 Residential hydronic baseboard heater. (Courtesy of Slant/Fin Corporation.)

Figure 2.26 Hydronic convector. (Courtesy of The Burnham Corporation.)

Cast iron was the choice in radiation for many years. A typical **cast-iron radiator**, composed of tubular columns or sections which are fastened together, is illustrated in Figure 2.27. The advent of fin-type baseboard radiation has largely replaced cast-iron radiation in most residential construction, duc to casc of installation and lower cost per unit of delivered heat.

Baseboard heating units, be they finned radiation, convectors, or cast-iron radiators, are rated by the manufacturer of the particular unit to deliver a specified amount of heat based on the water temperature within the heating system. The Hydronics Institute, formerly known as the Institute of Boiler and Radiator Manufacturers (IBR), publishes a list of ratings of residential and commercial finned-tube radiation. These ratings, which are updated periodically by the Institute, can be obtained from: The Hydronics Institute, Berkeley Heights, NJ 07922.

Table 2.3 illustrates the IBR sizing chart for a typical baseboard manufacturer. Note that the chart shows the Btu output per linear foot of baseboard based on a variety of system operating temperatures for two conventional types of finned-tube radiation available from this manufacturer.

Boiler Configuration

In almost all hot water baseboard systems, the configuration of the hot water boiler is similar; the major differences are based on the type of fuel that the boiler is designed to burn. However, all the basic control and piping configurations, referred to as **trim**, are fairly uniform. Figure 2.28 illustrates a typical boiler and trim arrangement (the burner has been omitted from the illustration for clarity).

Figure 2.27 Cast-iron radiator. These units are suitable for either hydronic or steam heating applications. When sizing for either type of heating system, the manufacturer's literature should be consulted to ensure installation of the proper amount of radiation. (Courtesy of The Burnham Corporation.)

TABLE 2.3 IBR SIZING CHART FOR TYPICAL BASEBOARD MANUFACTURER

Element[a]	Water flow (gal/min)	Pressure drop	170°F 76.7°C	180°F 82.2°C	190°F 87.8°C	200°F 93.3°C	210°F 98.9°C	215°F 101.6°C	220°F 104°C	230°F 110°C	240°F 116°C
No. 15-75E baseboard with E-75 element; $\frac{3}{4}$-in. nominal copper tubing, $2\frac{5}{8}$-in. × $2\frac{1}{8}$-in. × 0.009-in. aluminum fins bent to $2\frac{19}{64}$ in. × $2\frac{1}{8}$ in., 55 fins/ft	1	47	480	550	620	680	750	780	820	880	950
	4	525	510	580	660	720	790	820	870	930	1000
No. 15-50 baseboard with E-50 element; $\frac{1}{2}$-in. nominal copper tubing, $2\frac{5}{8}$-in. × $2\frac{1}{8}$-in. × 0.009-in. aluminum fins bent to $2\frac{5}{16}$ in. × $2\frac{1}{8}$ in., 55 fins/ft	1	260	490	550	610	680	740	770	800	860	920
	4	2880	520	580	640	720	780	810	850	910	970

Source: Slant/Fin Corporation.

Note: IBR-approved hot water ratings for Fine/Line 15 series [Btu/hr per linear foot with 65°F (18.3°C) entering air]. Ratings are for element installed with damper open, with expansion cradles. Ratings are based on active finned length (5 to 6 in. less than overall length) and include a 15% heating effect factor. Use 4-gal/min ratings only when flow is known to be equal to or greater than 4 gal/min; otherwise, 1-gal/min ratings must be used.

[a] End fins are of tinned steel for extra ruggedness.

Various types of combustion systems used in hydronic systems are discussed in subsequent chapters.

The boiler in Figure 2.28 illustrates the normal assortment of control devices and piping, although the specific installation configuration and position of some components vary from one manufacturer to another. Note that the circulator is

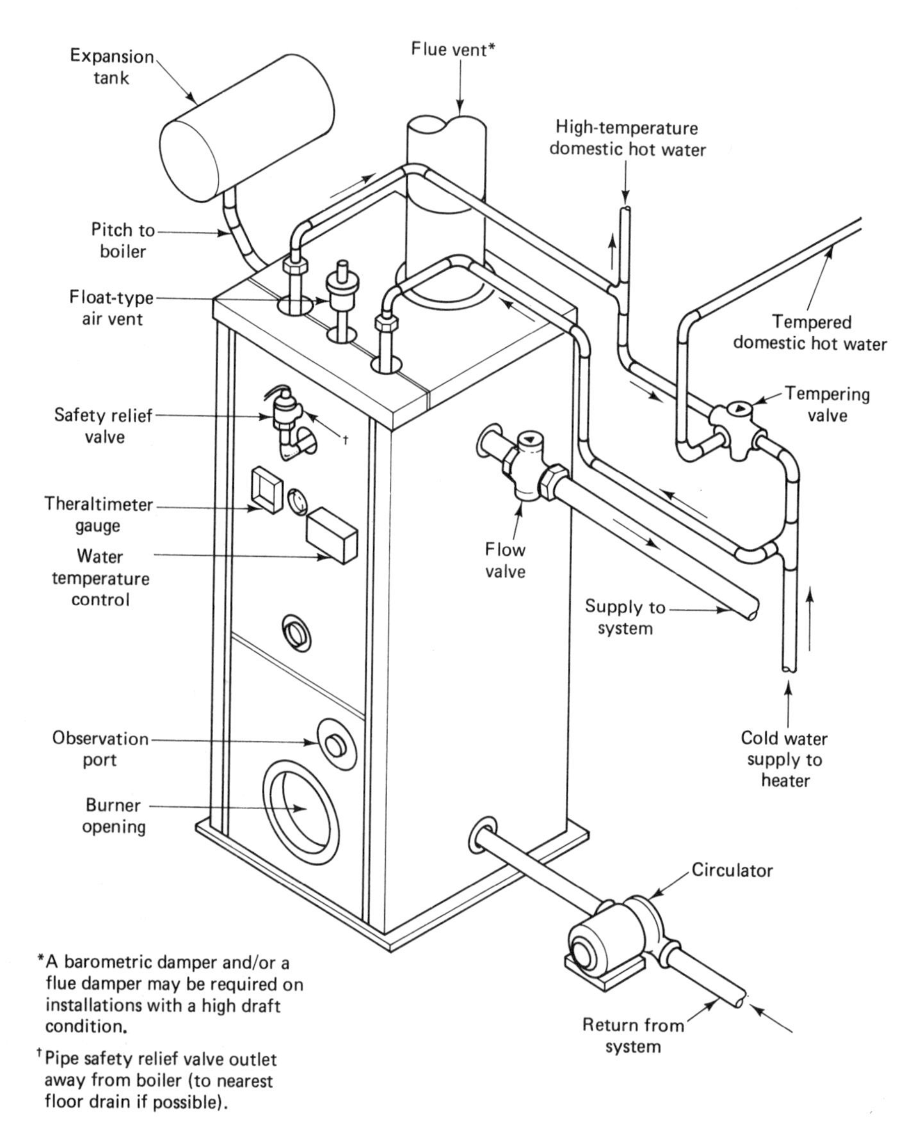

Figure 2.28 Typical piping and trim configuration for hot water boiler. (Courtesy of The Burnham Corporation.)

usually installed on the return line from the baseboard circuit in close proximity to the boiler. The supply line to the baseboard contains a flow control valve that prevents hot water from circulating through the heating system when there is no call for space heat but when the boiler is maintaining standby temperature for domestic hot water heating via the tankless coil (Figure 2.29).

The tankless coil piping circuit includes a hot water tempering valve to prevent scalding water from entering the hot water supply lines. The valve allows a set temperature to be selected by turning an adjustment knob. A bimetallic spring will adjust to changing water temperatures to maintain the set temperature selected as long as the cold water and hot water temperatures are within the operating range of the valve, illustrated in Figure 2.30.

The **theraltimeter gauge**, sometimes referred to as a **tridicator**, indicates water temperature, pressure, and altitude pressure within the boiler. The water temperature control is usually referred to as an **aquastat** and controls the burner mechanism based on boiler water temperature and the heating demand as sensed by the room thermostat(s).

The safety relief valve required on all boilers prevents excessive pressures from building up within the unit. In residential hydronic boilers, these valves are set to discharge when boiler pressure exceeds 30 pounds per square inch (psi); in steam boilers the discharge pressure is 15 psi. This valve should always be

Figure 2.29 A flow control valve, sometimes called a flow-check valve, is used in zoned hydronic systems that use separate circulators for zone control. The valve contains a weighted seat that raises and allows water to flow only when the circulator is activated. (Courtesy of Taco, Inc.)

Figure 2.30 Hot water tempering valves protect against scalding hot water entering the hot water supply system. The valves can be set to the desired water temperature by adjusting a control knob which adjusts a bimetallic spring-loaded valve that continually adjusts to provide the correct water temperature. (Courtesy of Taco, Inc.)

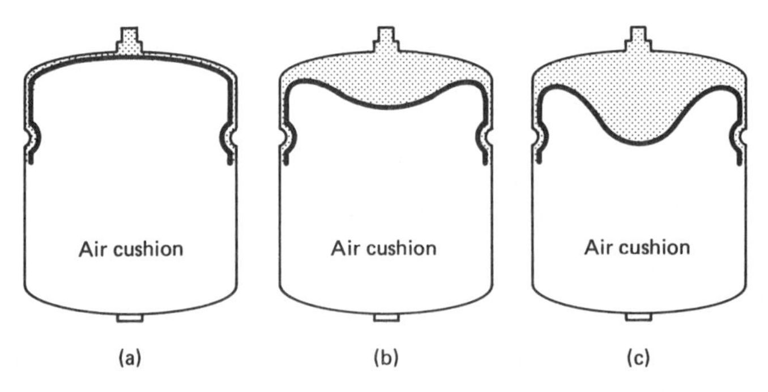

Figure 2.31 Operation of a diaphragm-type expansion tank. The tank is precharged with an air cushion below a flexible diaphragm (a). As system pressure increases, the diaphragm will continue to expand to accommodate the increased volume of water (b and c). This action prevents continual discharge of the pressure relief valve due to rising system pressures when the boiler comes up to operating temperature. (Courtesy of Amtrol, Inc.)

piped to within 6 in. of the floor to prevent personal injury should the valve discharge.

An expansion tank is provided in all hydronic systems to absorb the continual expansion and contraction of the water within the heating system during normal cycling. Since most heating systems are "closed" in the sense that pressure built up as a result of expansion of the water as it is heated cannot back out of the system (due to check valves in the cold water feed lines) the expansion tank will accommodate these fluctuating pressures. Figure 2.31 illustrates the operation of an expansion tank as the system pressure increases, causing an internal diaphragm to flex and accommodate the increased pressure. As the pressure in the system decreases, the air cushion within the expansion tank pushes back against the diaphragm and the tank returns to normal position (a). These tanks are available in different sizes based on the water capacity of the heating system. The manufacturer's instruction sheet should be consulted to properly size all expansion tanks.

Most expansion tanks are installed in conjunction with an air scoop, which is designed to isolate trapped air bubbles within the heating system water and purge them through a special air purge valve (Figure 2.32). Many expansion tank assemblies also incorporate an automatic cold water feed valve. This conveniently combines expansion tank, air removal, and cold water feed functions into one unit. Figure 2.33 illustrates two variations of the automatic feedwater function in combination with the expansion tank.

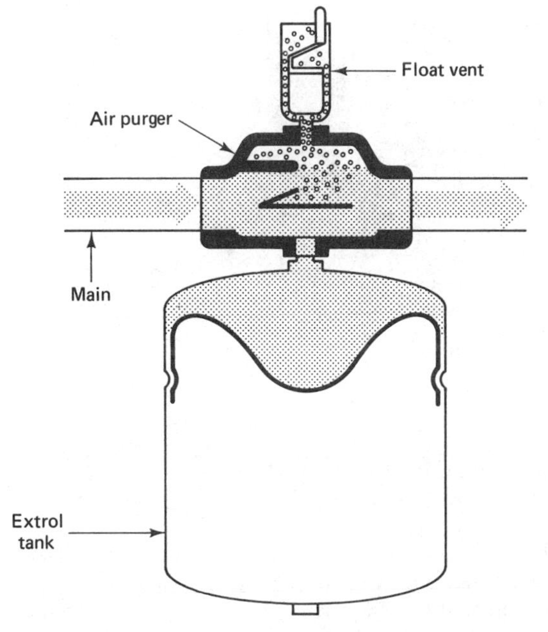

Figure 2.32 Combination expansion tank/air scoop assembly with float vent for continual discharge of trapped air within the heating system. (Courtesy of Amtrol, Inc.)

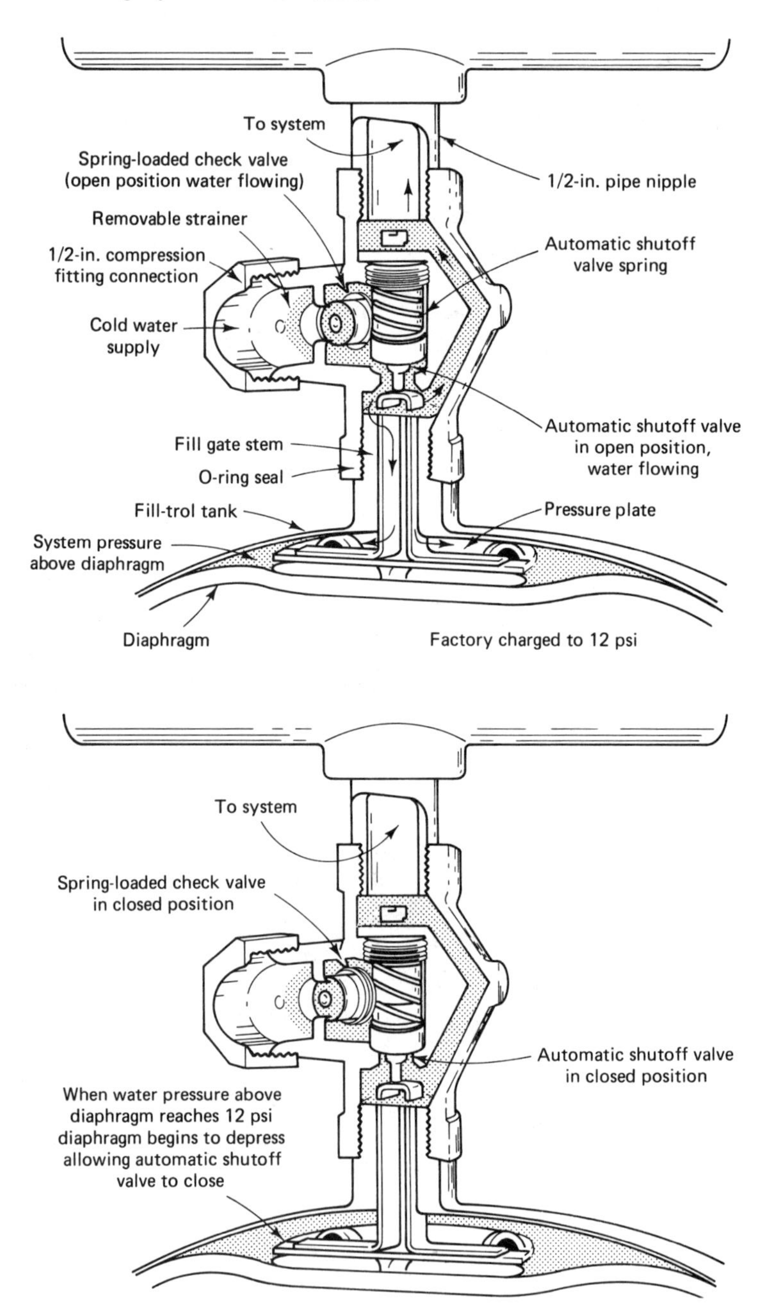

Figure 2.33 Automatic feedwater valve built into expansion tank assembly. This valve will automatically bring the cold water system pressure up to 12 psi, and it is adjustable. Also, the feed valve acts as a check valve to prevent system water from entering the residential water system. (Courtesy of Amtrol, Inc.)

STEAM HEATING SYSTEMS

Steam heating was the system of choice for most residences prior to the emergence of hot water baseboard systems. Thus steam heating is usually found in older homes, and it is not unusual to find them operating satisfactorily after 50 or more years of continuous service.

The sequence of operation of these systems is as follows. Water is heated, turning it into low-pressure steam in a boiler that is located at the lowest part of the system. The steam is channeled through the piping system into radiators located in each room. The latent heat of vaporization of the steam is released as the steam condenses back into water and returns via the piping system back into the bottom of the boiler, where the cycle continues until the desired room temperature is reached.

Steam systems are classified according to the piping configuration between the boiler and radiation, which is examined below in greater detail.

Steam Piping Configurations

One-pipe steam systems. In a one-pipe steam heating system, a single pipe serves a dual function by supplying steam to the radiation as well as serving as the return pipe for the condensate back to the boiler. A one-pipe steam system is illustrated in Figure 2.34. Note from the illustration that the service pipe is sloped upward from the boiler. This facilitates both the travel of the steam in the system and positive condensate drainback. Each radiator is equipped with its own shutoff valve. In effect, this allows for manual zoning of the system. Radiators are usually placed along outside walls and under windows. This placement allows for convective air currents to move sufficient quantities of air around the radiators, which increases conductive and convective heat transfer in the heating system.

Two-pipe steam systems. A two-pipe steam system utilizes a separate feed and return line for each radiator in the heating system. This type of system maximizes the flow of both steam and condensate by separating the condensate and steam lines. A two-pipe system is illustrated in Figure 2.35. Aside from the piping configuration, the sequence of operation of both one- and two-pipe systems is identical.

Wet return condensate piping. Condensate classification is yet another method for identifying steam heating systems. This method involves classification of the heating system based on whether the return lines are "wet" or "dry."

A wet return system is one in which a portion of the return line is located below the waterline of the boiler. This type of arrangement is illustrated in Figure 2.36. The wet return lines are completely filled with water and manifolded together

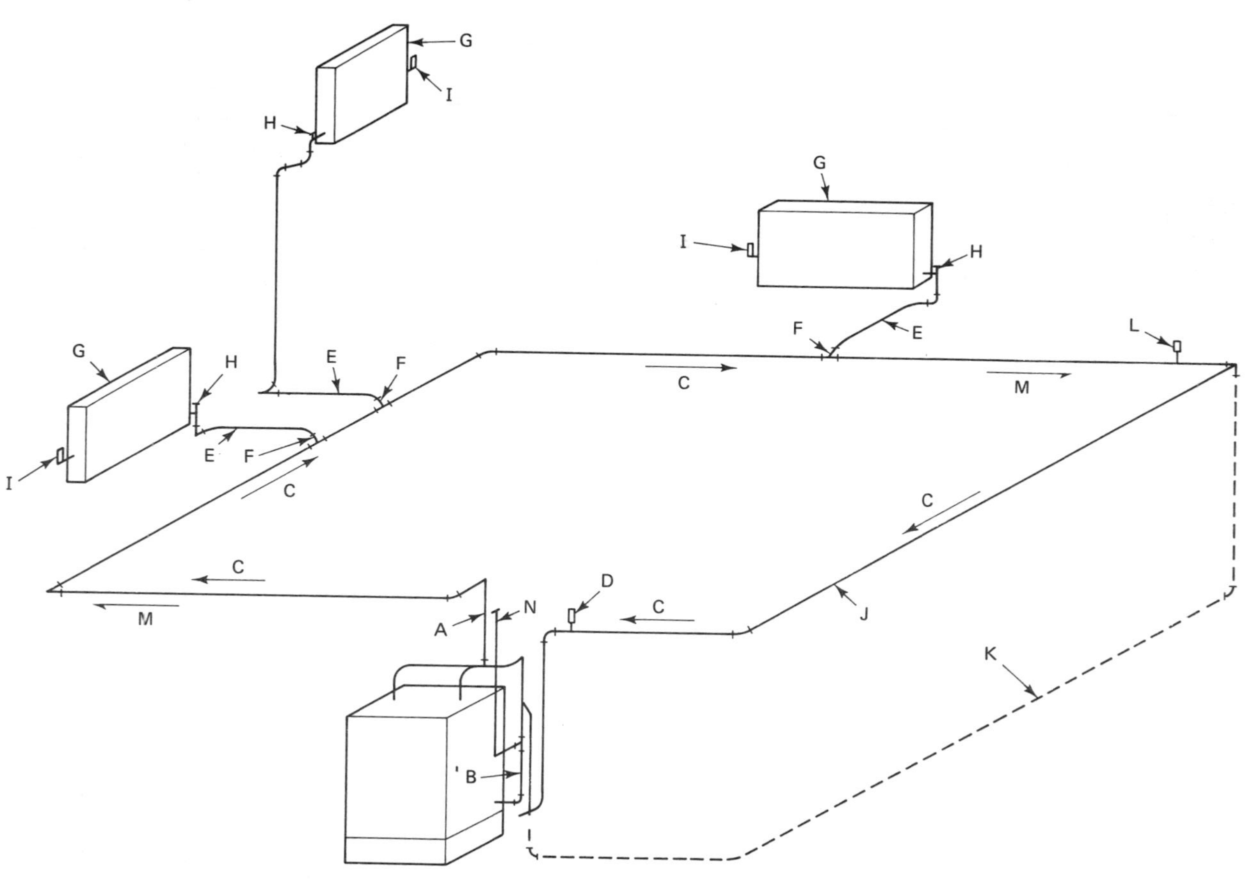

Figure 2.34 Layout of a one-pipe steam heating system: A, supply pipe; B, return pipe; C, pipe pitches downward in direction of arrow not less than $\frac{1}{4}$ in. in 10 ft; D, main vent; E, branches; F, 45-degree elbow; G, heat distribution units; H, radiator valve; I, air vent; J, dry return; K, wet return; L, main vent located at this position when wet return is used; M, direction of steam flow; N, makeup water line. (Courtesy of The Hydronics Institute.)

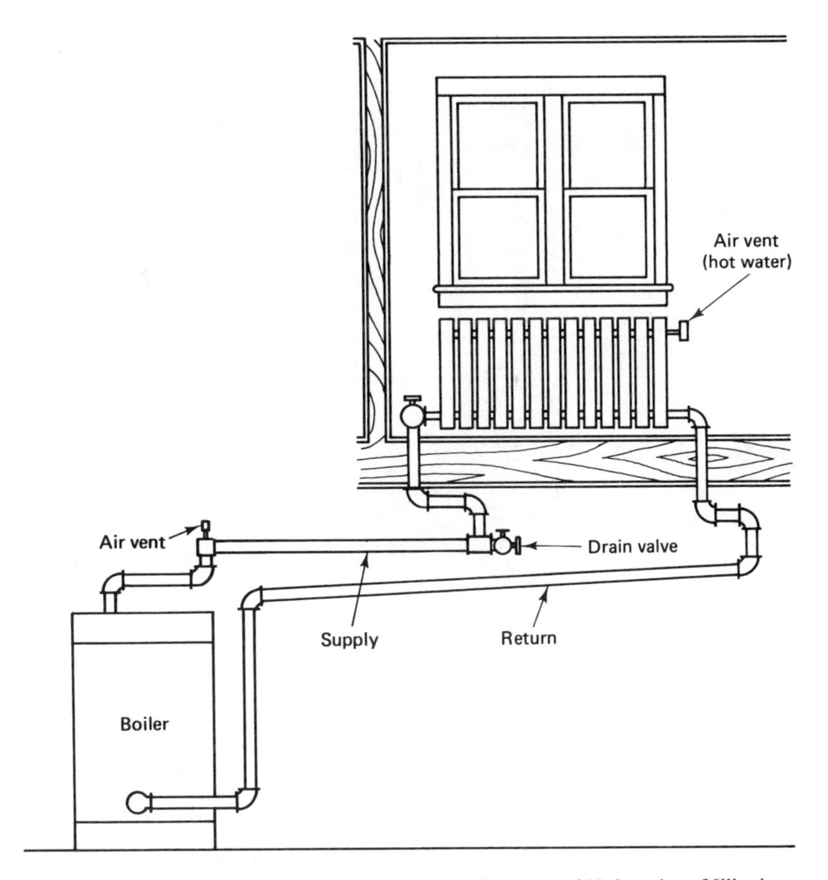

Figure 2.35 Two-pipe steam heating system. (Courtesy of University of Illinois, Small Homes Council.)

below the waterline of the boiler. Dry returns differ from the wet return system in that the return lines are located above the waterline in the boiler. Dry return systems are most often of the one-pipe design. Note that the wet return design features a configuration known as a **Hartford loop**. This loop is designed to equalize pressure between the boiler and the return mains in the heating system, preventing pressure in the boiler from pushing water up into the steam piping. Figures 2.37 and 2.38 illustrate the effect of pressure on the steam system with and without a Hartford loop.

Note that the crossover in Figure 2.37 equalizes the pressure between the boiler and return lines. Since the operating pressure in the boiler is higher than the pressure in the return lines, the tendency without the equalization loop would be for water to be forced back into the heating system due to the higher boiler pressure, as illustrated in Figure 2.38. A Hartford loop piping configuration is illustrated in detail in Figure 2.39.

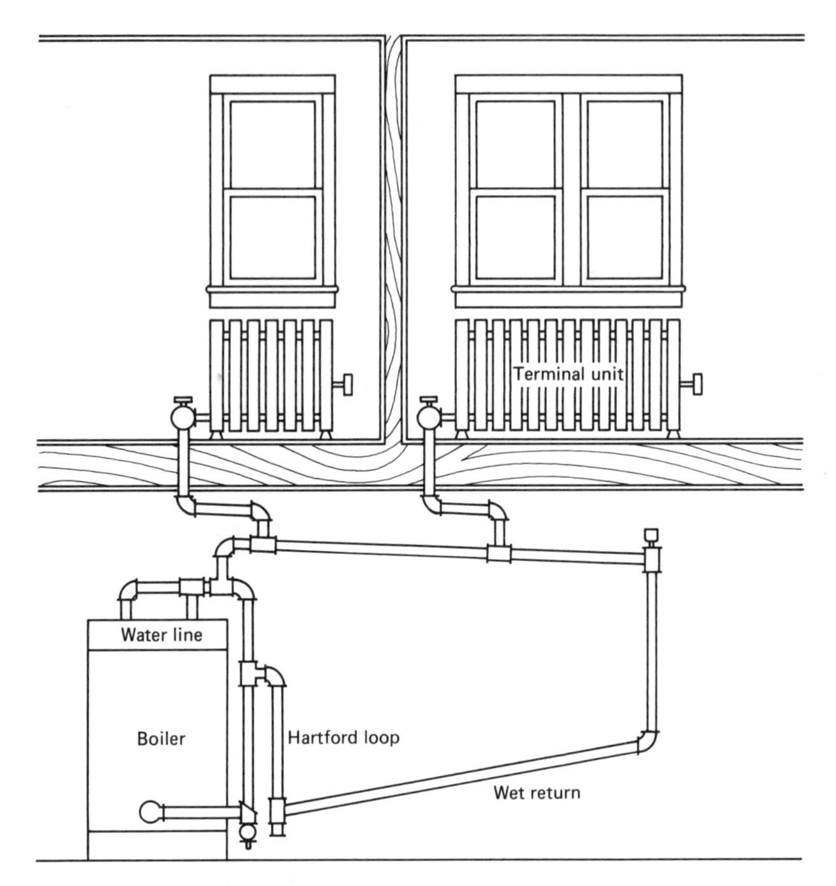

Figure 2.36 Wet return configuration on steam heating system. (Courtesy of University of Illinois, Small Homes Council.)

Operating Controls

The operating controls on steam and hot water boilers are quite different, due to the operating characteristics and pressures of steam versus hot water. Steam systems operate at lower pressures than do hot water boilers: 15 psi maximum for steam versus 30 psi maximum for hydronic systems.

Steam systems contain low-water cutoff devices that stop all fuel combustion should the water level in the boiler drop to a predetermined low limit. Loss of water in a steam boiler, or operating the boiler with water below recommended levels, can cause severe damage. Also, since scale and sediment tend to build up in steam systems due to the constant boiling and condensation of the water, steam boilers are equipped with devices that handle both low-water and scale problems.

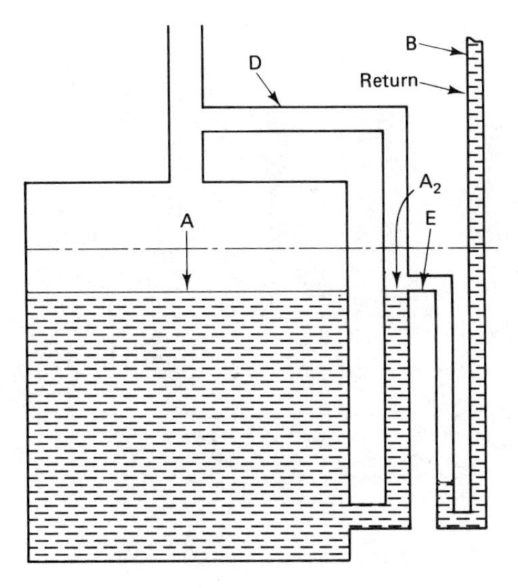

Figure 2.37 Steam boiler equipped with Hartford loop piping configuration. (Courtesy of The Hydronics Institute.)

A low-water cutoff (Figure 2.40) is designed to shut down the boiler automatically should the water level fall below the recommended level.

The cutoff operates on a float principle. When the float level drops below a certain point, a set of electrical contacts that are connected to the power line serving the burner unit are opened, preventing the boiler from firing. When the

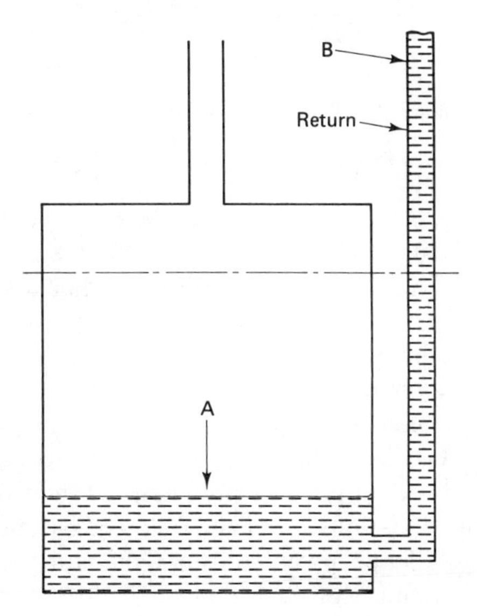

Figure 2.38 Steam boiler without Hartford loop configuration. (Courtesy of The Hydronics Institute.)

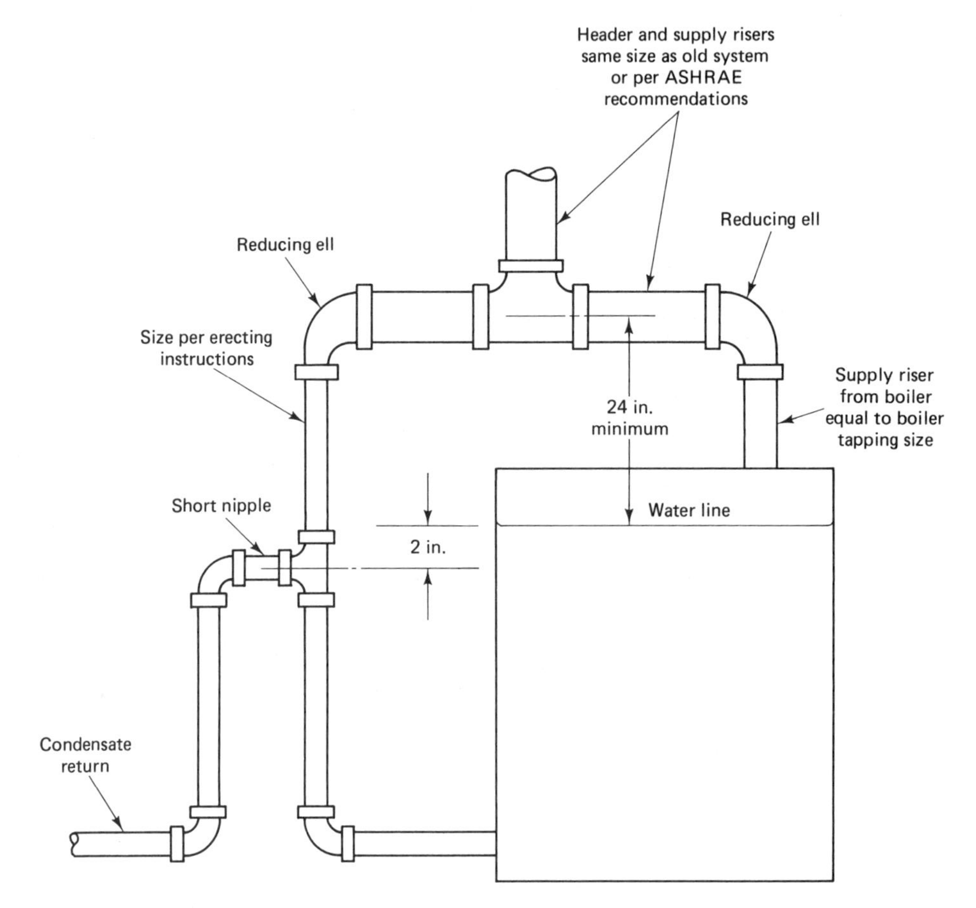

Figure 2.39 Detailed piping of Hartford loop. (Courtesy of Weil-McLain, a Marley Company.)

water level is again raised, the float closes the electrical contacts, allowing the burner to resume normal firing. Many low-water cutoffs are equipped with a water column and gauge assembly, which gives the homeowner a visual indication of the water level in the boiler (Figure 2.41). The water column alerts the homeowner to the need to add water to the boiler, which can be accomplished either manually or through the use of an automatic feedwater mechanism.

The boiler should be flushed periodically to remove any scale or sediment, which builds up over a period of time. Steam boilers are equipped with flush fittings (Figure 2.42) for this maintenance procedure. When flushing the boiler, the manufacturer's recommendations should be followed carefully. The low-water cutoff should also be flushed once a week during the heating season. Consult the manufacturer's recommendations for each unit.

Figure 2.40 Low-water cutoff for steam boiler. (Courtesy of ITT Fluid Handling Division.)

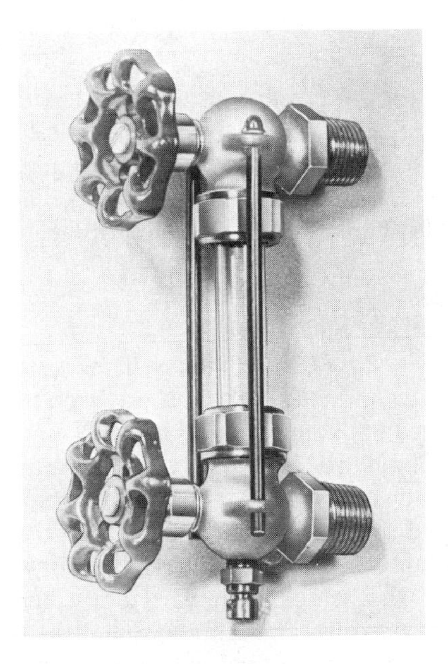

Figure 2.41 Water column and glass gauge assembly. The level of water in the gauge indicates the water level in the steam boiler. The gauge should be flushed periodically to keep the valves free of sediment. (Courtesy of Conbraco Industries, Inc.)

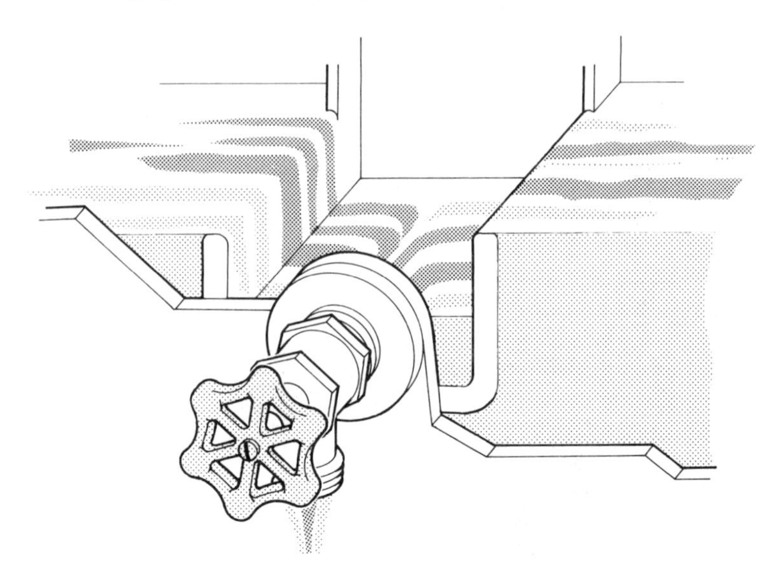

Figure 2.42 Flush fitting on steam boiler. When flushing the boiler, if the boiler contains a plug rather than a special flushing fitting, the plug is removed and a suitable bushing with a garden hose is attached. When reinserting the plug, a combination of cotton wicking, Teflon tape, and Teflon pipe dope should be used to limit the possibility of leaks around the plug fitting. (Courtesy of Slant/Fin Corporation.)

RESISTANCE HEATING SYSTEMS

The term **resistance heating** usually refers to terminal heating units and devices that are electrically powered. Electric resistance heating falls into three common types of heating systems: the electric baseboard heater, including oil-filled units; radiant panels, which make up portions of the walls or ceiling of a residence; and central heating boilers or furnaces, which rely on strip-heating elements.

Electric Baseboard Heaters

Electric baseboard heaters are modular units, sold in a variety of sizes. They fasten to the walls in each room, as do their hot water counterparts. Each baseboard heater or series of heaters is supplied by a 220-V line controlled by a line voltage thermostat for regulating room heat. A thermostat is usually provided for each room in the home, which allows for room-by-room zoning. A typical electric baseboard unit is illustrated in Figure 2.43. The baseboard is made up of a strip of high-resistance metal, surrounded by fins to aid in convective heat transfer. Electricity passing through the heater is converted into heat and transferred to the room by convective air currents circulating around the heating elements.

One modification of the conventional baseboard heater is the closed-loop

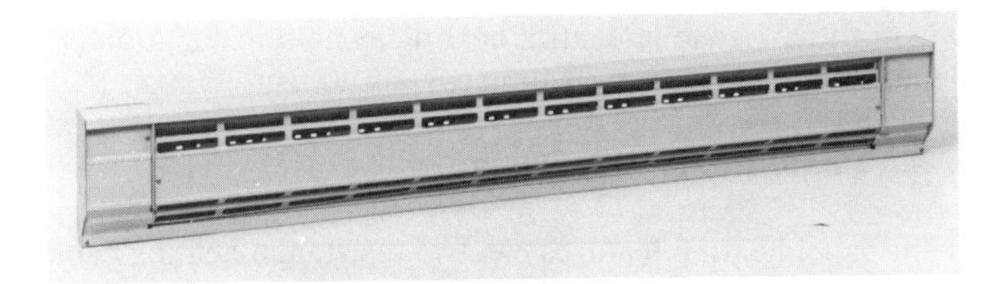

Figure 2.43 Electric baseboard heater. (Courtesy of Electromode, Inc.)

oil- or water-filled heater (Figure 2.44). Electricity is used to heat the sealed liquid within the heater, which then circulates through the heater columns by natural convection. Heat is also transferred from the fluid to the room via natural convection. These units provide some residual heat after the electricity is turned off.

Electric baseboard units are easy to install and are much cheaper than central fossil-fuel or solid-fuel hydronic or hot air heating systems. Also, electric heaters are virtually maintenance free and require only occasional dusting to keep the air louvers on the baseboard free of dirt.

Rising utility rates have made electric baseboard heating far more expensive to operate on an annual basis than are their fossil- and solid-fuel counterparts. In areas where utility costs are high, these types of systems are ideal for weekend and vacation homes where the operation of a central heating system is not required

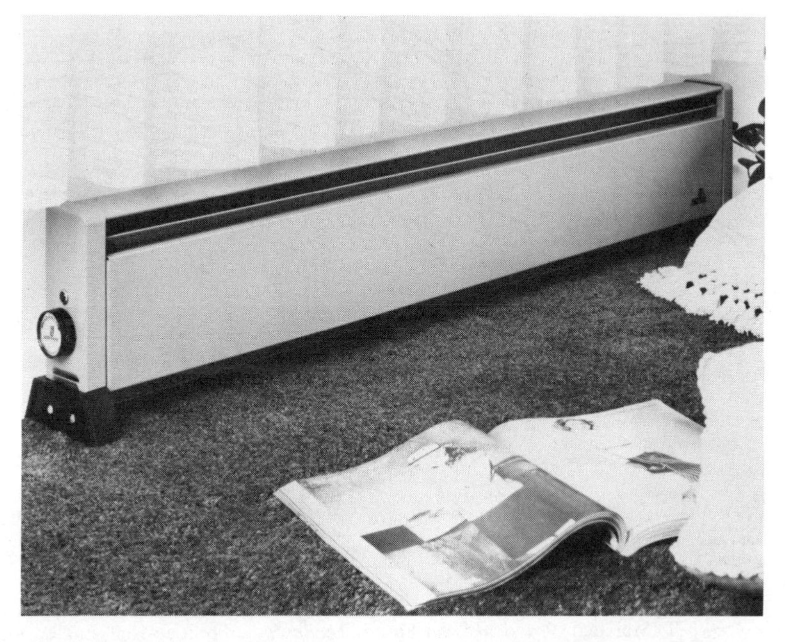

Figure 2.44 Liquid-filled water heater. (Courtesy of Intertherm Corporation.)

or where the expense in central heating system installation is a significant factor in determining the type of heating system to use.

Radiant Panel Heating Systems

Radiant panel heating features electric resistance heating cable embedded within wall or ceiling panels of the room. The panels are assembled and wired together to form one continuous heating panel. A system of this type is illustrated in Figure 2.45. A line voltage thermostat controls power to the panels in each room. The wall or ceiling panels are conventional in appearance. On a call for heat the entire panel radiates heat into the room. These types of systems are more expensive than their electric baseboard counterparts but less expensive than conventional central heating systems.

Electric Resistance Central Heating Systems

There are a variety of electrically powered central heating boilers and furnaces available for use in conventional hydronic and hot air heating applications. Figure 2.46 illustrates an electric hydronic boiler. The boiler in the figure contains a series of resistance heating elements that are energized in sequence when there is a call for heat. The elements turn on in sequence rather than all at once, to

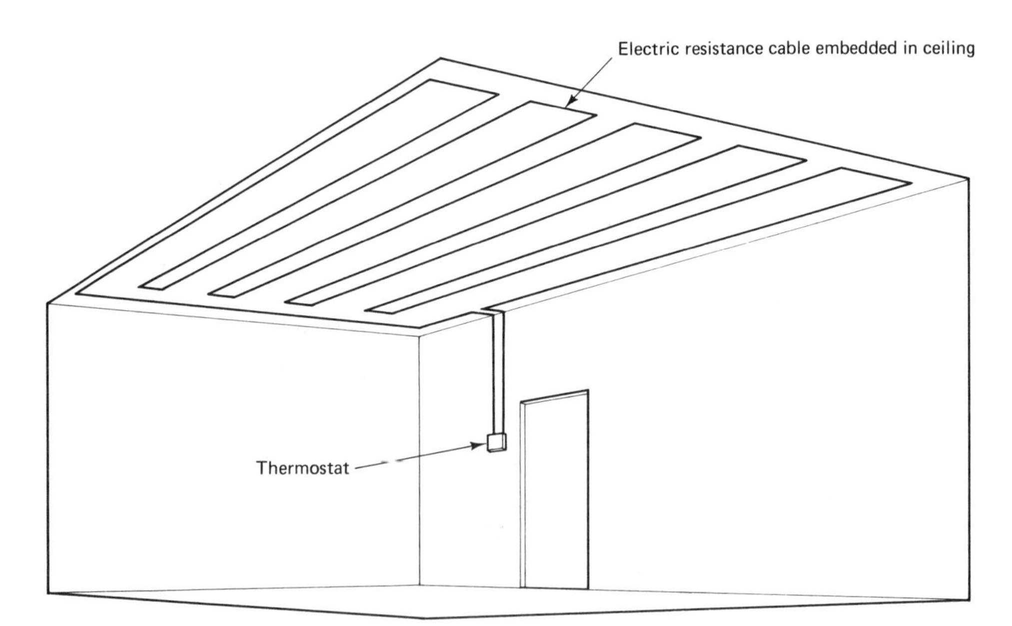

Figure 2.45 Radiant ceiling electric resistance heating. (From M. Greenwald and T. McHugh, *Practical Solar Energy Technology*, Prentice-Hall, Englewood Cliffs, N.J., 1985.)

Figure 2.46 Electric boiler. Most electric boilers are small and compact in size. Their small water capacity, coupled with the immersed resistance heating coils, enables these units to reach operating temperature quickly. In areas where utility costs are high, these units can be very expensive to operate on a yearly basis. (Courtesy of The Burnham Corporation.)

minimize current draw on the wiring system, bringing the boiler on line gradually. Electric furnaces employ resistance heating strips that are exposed to the airstream for heat transfer.

The economics of electric furnaces and boilers is similar to that for electric baseboard and radiant panel heating. These systems are expensive to operate on a yearly basis when the electric unit is the only source of heat in the home. However, in geographic areas where the heating season is minimal and the homeowner wishes to avoid annual maintenance and the installation costs of other types of systems, these units are ideal. Also, they can be more cost-effective and flexible than fossil-fuel systems for use in vacation and weekend homes since electric boilers and furnaces can cycle from a cold start to a fully heated condition in seconds. They can remain inoperative for extended periods with no adverse operating effects.

SOLAR SPACE HEATING SYSTEMS

Solar space heating systems are classified as active or passive. **Active solar space heating** systems use a fluid to transfer heat from the solar collectors to a storage medium and from this medium to the interior of the house using a fan or circulating pump. **Passive solar heating** is a design feature integrated into the architecture of the home and is designed and constructed to capture incident solar radiation during the day, store whatever excess energy may be available in addition to providing space heating during the day, and then radiate this heat into the residence during sunless periods or in the evening.

We first examine active solar space heating systems that employ an array of solar collectors, storage tanks, and heating systems for distribution of the stored energy.

Active Solar Space Heating

Most active solar space heating systems are installed in conjunction with conventional fossil- or solid-fuel backup central heating appliances. Operating in this manner, the solar system provides first-stage residential heat. After the solar system has exhausted its heat or is unable to keep up with the heat loss in the home, the backup heating system will be called on to provide whatever additional heat is required to satisfy the thermostat setting.

Solar heating systems are not normally sized to provide 100% of the annual heat required by the residence. To size the solar system at 100% of the annual heating requirements would require large solar collector arrays and storage tanks, which would be far too expensive to be cost-effective. Therefore, most solar space heating systems are sized to provide between 40 and 60% of the annual space heating requirements of the home. Sizing in this way ensures that all the heat collected by the solar system will be transferred to the residence and not end up as stored surplus heat that eventually dissipates by conductive or convective losses somewhere in the system.

Determining the size of the solar system is dependent on the heat loss of the home and is influenced by the type and size of backup heating system. In new construction a detailed heat-loss analysis of the residence should be calculated.* The solar system and conventional backup heating system must be designed to operate efficiently with one another. In older homes the solar system must be sized according to both the heat loss of the home and the size of the existing central heating system.

Although conventional heat-loss analyses are usually calculated by architects and heating contractors, solar system sizing is best done by a solar professional. The procedure used to size the solar system and determine the most cost-effective size of a proposed solar installation is referred to as an ***f*-chart** calculation. *f*-charts are most easily prepared on a microcomputer using the appropriate software program, available from a variety of solar equipment distributors and mail-order software firms. A sample *f*-chart calculation illustrating most of the variables that must be entered in the program to determine the operational characteristics of a particular solar system is illustrated in Figure 2.47.

Lacking the sophistication of an *f*-chart computer program, system size can also be made by using basic assumptions about the residence: the size of the

* See Chapter 7 for additional information on both heat-loss and heat-gain calculations.

```
COLLECTOR PARAMETERS
 C1. COLLECTOR AREA ................................  256.00 FT2
 C2. FR-UL PRODUCT .................................    0.76 BTU/HR-FT2-DEG F
 C3. FR-TAU-ALPHA (NORMAL INCIDENCE) ...............    0.77
 C6. NUMBER OF COVERS ..............................    1.00
 C7. INDEX OF REFRACTION ...........................    1.53
 C8. EXTINCTION COEFFICIENT X LENGTH (KL)...........    0.04
 C9. INCIDENCE ANGLE MODIFIER CONSTANT .............    0.14
 C10. COLLECTOR FLOW RATE * SPECIFIC HEAT/AREA ......  12.60 BTU/HR-FT2-DEG F
 C12. COLLECTOR SLOPE ..............................  56.00 DEGREES
 C13. COLLECTOR AZIMUTH ............................   0.00 DEGREES
 C14. GROUND REFLECTANCE ...........................   0.20

COLLECTOR-STORE TRANSFER PARAMETERS
 T1. EPS-CMIN OF COLLECTOR-STORE HX/COLLECTOR AREA ..  11.50 BTU/HR-FT2-DEG F

STORAGE UNIT PARAMETERS
 S1. TANK CAPACITY/COLLECTOR AREA ..................  11.00 BTU/DEG F-FT2
 S2. STORAGE UNIT HEIGHT/DIAMETER RATIO ............   2.00
 S3. HEAT LOSS COEFFICIENT .........................   0.09 BTU/HR-FT2-DEG F
 S4. ENVIRONMENT TEMPERATURE (-1000 FOR TENV=TAMB) ..  68.00 DEG F
 S5. HOT WATER AUXILIARY TANK UA ...................   0.00 BTU/HR-DEG F
 S6. HOT WATER AUX TANK ENVIRONMENT TEMPERATURE .....  68.00 DEG F

DELIVERY DEVICE PARAMETERS
 D1. EPS-CMIN OF LOAD HEAT EXCHANGER ............... 1750.00 BTU/HR-DEG F
 D2. MINIMUM TEMPERATURE FOR HX OPERATION ..........  90.00 DEG F

LOAD PARAMETERS
 L1. BUILDING UA ...................................  588.00 BTU/HR-DEG F
 L2. ROOM TEMPERATURE ..............................  68.00 DEG F
 L3. HOT WATER USE .................................  40.00 GALLONS/DAY
 L4. HOT WATER SET TEMPERATURE .....................  130.00 DEG F
 L5. WATER MAINS TEMPERATURE .......................  50.00 DEG F

AUXILIARY PARAMETERS
 A1. AUXILIARY FUEL TYPE (1=GAS.2=ELEC.3=OIL) .......   3.
 A2. AUXILIARY DEVICE EFFICIENCY ...................   0.60
 A3. HOT WATER AUXILIARY FUEL (1=GAS.2=ELEC.3=OIL) ..   3.
 A4. AUXILIARY WATER HEATER EFFICIENCY .............   0.45
 ?
flist fldata

 INPUT NOT RECOGNIZED -FLIST
 ?
list fldata


        FUEL COST

ELEC (BLOCK RATE STRUCTURE)
     ANNUAL INFLATION RATE  10.0 %
     BLOCK                  1
     COST ($/MMBTU)         6.3
     MAX. USE (MMBTU) 9477.7

OIL  (BLOCK RATE STRUCTURE)
     ANNUAL INFLATION RATE  10.0 %
     BLOCK                  1
     COST ($/MMBTU)         8.0
     MAX. USE (MMBTU)10000.0
```

Figure 2.47 f-chart for sizing solar heating system. (For a more detailed analysis of the f-chart method, the reader is referred to the classic text on the subject: W. A. Beckman, S. A. Klein, and J. A. Duffie, *Solar Heating Design by the F-Chart Method,* Wiley-Interscience, New York, 1977.)

```
****************   FCHART  ANALYSIS   (VERSION 4.0)   *****************

 NEW YORK(CEN PK)NY      LATITUDE 40.5
```

THERMAL PERFORMANCE

	HT (MMBTU)	TA (DEG-F)	SHLOAD (MMBTU)	HWLOAD (MMBTU)	QU (MMBTU)	QLOSS (MMBTU)	DELTE (MMBTU)	FNP	FDHW
JAN	6.83	32.0	14.35	0.83	2.67	0.11	0.00	0.17	0.50
FEB	7.22	33.8	12.50	0.75	2.78	0.10	0.00	0.20	0.50
MAR	9.30	41.0	10.47	0.83	3.71	0.11	0.00	0.32	0.50
APR	9.72	51.8	5.46	0.80	3.91	0.12	0.00	0.60	0.50
MAY	10.58	62.6	1.93	0.83	2.94	0.26	0.12	0.93	0.77
JUN	10.13	71.6	0.00	0.80	1.57	0.49	0.22	1.00	1.00
JUL	10.57	77.0	0.00	0.83	1.32	0.60	-0.11	1.00	1.00
AUG	10.33	75.2	0.00	0.83	1.50	0.58	0.09	1.00	1.00
SEP	9.77	68.0	0.41	0.80	1.56	0.53	-0.17	1.00	1.00
OCT	9.44	59.0	2.95	0.83	3.08	0.28	-0.15	0.78	0.83
NOV	6.58	48.2	7.44	0.80	2.93	0.11	0.00	0.34	0.50
DEC	5.74	35.6	12.90	0.83	2.28	0.11	0.00	0.16	0.50
YR	106.21	54.7	68.41	9.74	30.24	3.40	0.00	0.34	0.72

PURCHASED ENERGY SUMMARY

	GAS	ELECTRIC	OIL	TOTAL
USE (MMBTU)	0.00	0.00	87.14	87.14
COST ($)	0.00	0.00	697.09	697.09

Figure 2.47 *(continued)*

proposed solar system and basic parameters concerning the residential heat loss. The following steps illustrate the manual calculations that can be used to size various types of solar installations with a reasonable degree of accuracy in predicting system performance and economic benefits.

The solar collector array can be sized only after the residential space heating load has been determined. After the solar system has been sized, the storage requirements of the system are decided upon. The step-by-step procedure for this analysis is as follows:

1. Determine the heat loss of the residence at a specified design temperature.
2. Determine the north latitude of the residence.
3. Determine average Btu/ft^2/day of solar insolation based on a specified period during the heating season.
4. Convert the Btu heat loss of the residence into Btu/degree-day and determine the total heat loss for the period selected in step 3.
5. Determine the delivered Btu of solar energy, and from this figure calculate the percentage of energy delivered by the solar system that comprises the total energy required to heat the residence.

For purposes of this illustration, let us assume that our proposed house has a Btu heat loss of 65,000 Btu/hr at a design temperature of 0°F as calculated by the builder. From a map we determine that the home is located at 32° north latitude.

To calculate the Btu/ft²/day at our proposed house, consult the ASHRAE chart in Appendix G. Note from this table that the solar insolation values are based on the degree of tilt of the collector surface. For solar space heating systems, it is common practice to increase the tilt angle of the collector array approximately 10° above the north latitude of the location. Therefore, the amount of solar insolation available at the home is approximately 2118 Btu/ft²/day during the month of January (this value is an approximate extrapolation of the data given in the chart for January 21).

To convert the heat loss of the residence from Btu/hr to Btu/degree-day at the specified design temperature, we use the following formula:

$$\frac{\text{design heat loss of house}}{\text{design temperature}} \times 24 \text{ hr} = \text{heat loss per degree-day}$$

Substituting the appropriate values in the equation above, we get

$$\frac{65{,}000}{65°\text{F}} \times 24 = 1000 \times 24 = 24{,}000 \text{ Btu/degree-day}$$

To make our calculations for a period of one month (January) during the winter, we need to know the number of degree-days for this location. Appendix H lists the number of degree-days for most locations in the United States. The amount of heat required for the residence during the month of January is found by multiplying the number of degree-days in the month by the calculated heat loss in Btu/degree-days:

936 degree-days in month × 24,000 Btu/degree-days

= 22,464,000 Btu of heat required in January to heat the residence

With this information, we can now determine the percentages of space heating that we can expect from our proposed solar system. For purposes of this analysis, let us assume that we wish to install an array of ten 4 ft × 8 ft solar collectors. We first calculate the amount of solar energy that the system can collect on a daily basis:

daily collectable energy = ft² collector area × available solar energy

Thus

320 ft² collector area × 2118 Btu/ft²/day = 677,760 Btu

This figure multiplied by 31 days equals the amount of solar energy collected by the system for the month of January, on which our calculations are based. Therefore,

$$31 \text{ days} \times 677{,}760 \text{ Btu/day} = 21{,}010{,}560 \text{ Btu collected per month}$$

To determine the percentage of the total heat required by the residence that this figure represents, we first determine the net output of the solar system and divide this figure by the total heat required for the residence:

$$\text{net solar system output} = \text{gross output} \times \text{system efficiency}$$

We can assume that an overall solar system efficiency is approximately 65%. This figure includes all heat losses from the collectors, associated piping, storage, and delivery system components. Therefore, the net heat output from the solar system is

$$21{,}010{,}560 \times 0.65 \text{ efficiency} = 13{,}656{,}864 \text{ net Btu for January}$$

This figure is then used to calculate the percentage of solar heating to the residence during the month as follows:

$$\frac{\text{net solar system output}}{\text{heat required for month}} = \text{percentage of solar contribution}$$

Substituting our calculated figures, we proceed:

$$\frac{13{,}656{,}864}{22{,}464{,}000} = 60.7\% \text{ solar contribution}$$

Once these figures have been determined, a number of different solar scenarios can quickly be investigated to decide on the most cost-effective solar system to install. Based on the overall installed cost of the system including storage, it is most cost-effective to try to keep the solar contribution to the total heating load between 50 and 60%.

The approximate cost-effectiveness of the solar system can be determined quickly by performing a **return on investment** (ROI) calculation. In its simplest terms, the ROI is calculated by dividing the first year's energy savings based on solar system performance by the net cost of the solar system:

$$\text{return on investment} = \frac{\text{first year's energy savings}}{\text{net system cost}}$$

It should be kept in mind that a simple ROI calculation leaves out many variable factors that make extrapolation of the ROI inaccurate: For example, inflation rates, cost of fuel, and variations in solar system maintenance costs from one year to the next are not accounted for. For a complete economic analysis of the system over a 20-year period, an *f*-chart performance will need to be calculated. ROI calculations are useful, however, in helping to pinpoint the various costs and benefits to be derived from solar system installations.

Solar Heating System Interfacing

Almost all solar energy systems are designed to operate in conjunction with backup fossil central heating systems. Thus the configuration of the system will be based on the type of heating system in the residence. The most common types of solar space heating installations are those installed in conjunction with hot air heating, hot water baseboard, and individual fan-type convectors, each of which will now be examined.

Solar system with hot water central heat. Figure 2.48 illustrates a solar heating system installed in conjunction with a backup conventional hot water boiler. The collector array operates in a closed loop with the heat transfer medium (usually, nontoxic antifreeze) circulating between the collectors and internal heat exchangers within the solar storage tanks. This system is designed to operate in a two-stage heating configuration. When the thermostat calls for heat, solar-heated water in the storage tanks is circulated throughout the baseboard heating circuit, provided that the temperature of the water is above a predefined low limit (usually, 85 to 100°F). Note the installation of isolation solenoid valves (number 17) that prevent the solar-heated water from circulating through the boiler or from operating due to a low solar water temperature. Even though the solar-heated water is lower in temperature than the boiler would ordinarily furnish, it is still capable of providing significant amounts of heat to the home (approximately 60% based on our sample calculations). This mode of heating will continue until either the temperature of the solar-heated water drops below the low-limit setting of the control aquastat or the heat loss in the home is too great for the solar system to keep up with. The system incorporates a two-stage thermostat that can activate two separate electrical circuits at two temperature settings. These settings are separated between 3 and 8°F. When the room temperature drops to the lower of the two settings, the backup boiler is activated. When the backup system is on, the solar system circulation stops* and the isolation solenoid valves prevent any water in the solar storage tanks from entering the heating system. This type of design prevents the boiler from heating the water in the solar storage tanks (in this instance, approximately 360 gal, which would place a significant drain on the boiler and be very wasteful of fuel). Note, too, from Figure 2.48 that a domestic hot water heater is plumbed into the collector loop. During operation, the collector fluid will circulate through the heat exchanger in the domestic hot water tank as long as the temperature in the loop is higher than the temperature of the water in the tank. An isolation valve (number 17) is placed in the domestic hot water piping to prevent collector fluid from circulating in the loop should the tank temperature rise above that of the collector fluid. A single-zone system would operate similarly. The number of zones is not a factor.

* Only circulation of solar water in the heating system is halted. Circulation within the collector loop continues if there is a positive temperature differential between the collectors and the storage tanks.

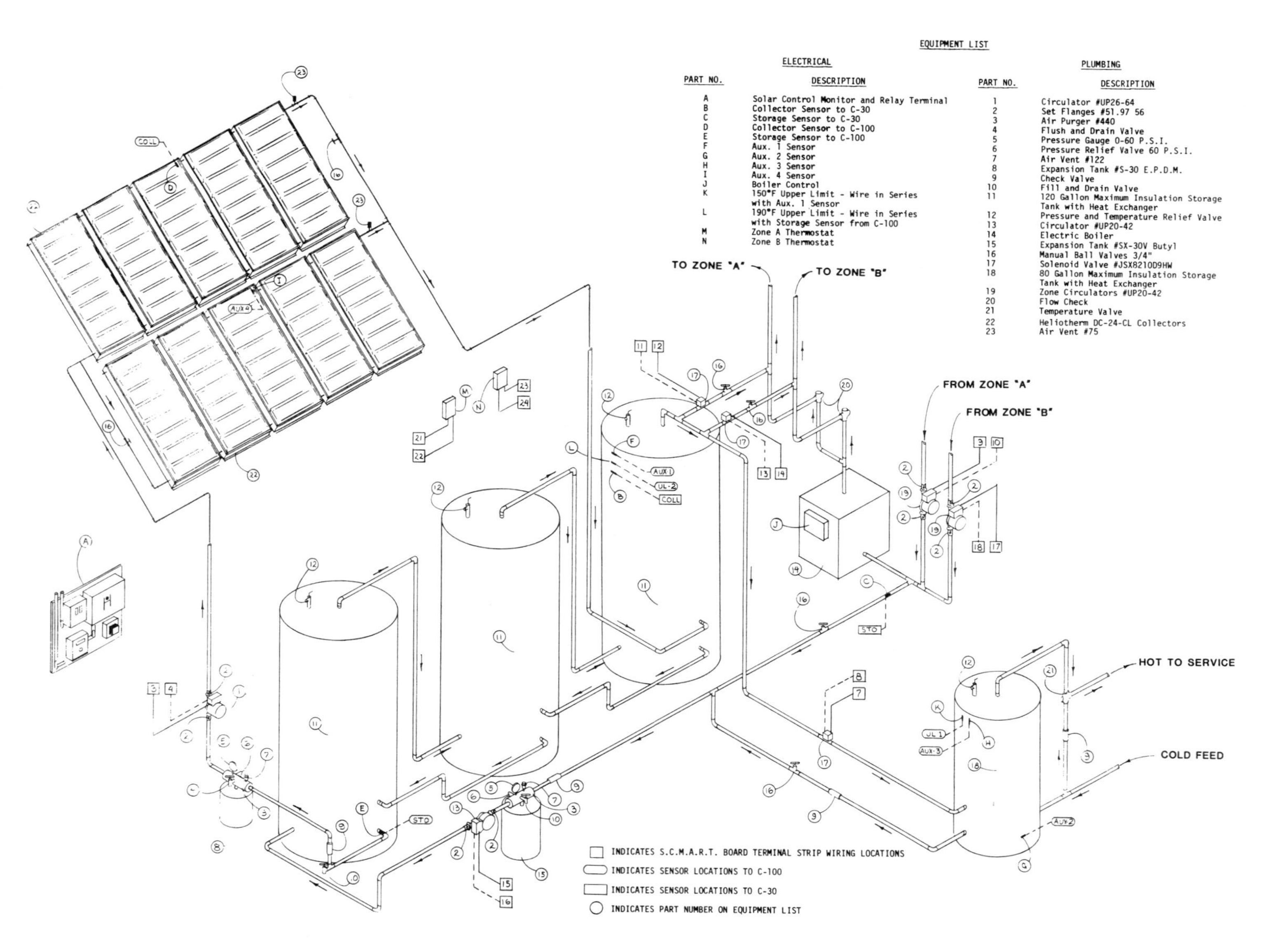

EQUIPMENT LIST

ELECTRICAL

PART NO.	DESCRIPTION
A	Solar Control Monitor and Relay Terminal
B	Collector Sensor to C-30
C	Storage Sensor to C-30
D	Collector Sensor to C-100
E	Storage Sensor to C-100
F	Aux. 1 Sensor
G	Aux. 2 Sensor
H	Aux. 3 Sensor
I	Aux. 4 Sensor
J	Boiler Control
K	150°F Upper Limit - Wire in Series with Aux. 1 Sensor
L	190°F Upper Limit - Wire in Series with Storage Sensor from C-100
M	Zone A Thermostat
N	Zone B Thermostat

PLUMBING

PART NO.	DESCRIPTION
1	Circulator #UP26-64
2	Set Flanges #51.97 56
3	Air Purger #440
4	Flush and Drain Valve
5	Pressure Gauge 0-60 P.S.I.
6	Pressure Relief Valve 60 P.S.I.
7	Air Vent #122
8	Expansion Tank #S-30 E.P.D.M.
9	Check Valve
10	Fill and Drain Valve
11	120 Gallon Maximum Insulation Storage Tank with Heat Exchanger
12	Pressure and Temperature Relief Valve
13	Circulator #UP20-42
14	Electric Boiler
15	Expansion Tank #SX-30V Butyl
16	Manual Ball Valves 3/4"
17	Solenoid Valve #JSX8210D9HW
18	80 Gallon Maximum Insulation Storage Tank with Heat Exchanger
19	Zone Circulators #UP20-42
20	Flow Check
21	Temperature Valve
22	Heliotherm DC-24-CL Collectors
23	Air Vent #75

Solar system with hot air central heating. Figure 2.49 illustrates a solar heating system installed in conjunction with a central hot air furnace. Even though the heating system is hot air, water is still used as the storage medium for the solar collectors, due to its excellent economic and space-saving characteristics for storing heat. The solar system interfaces with the furnace via a heat exchange coil inserted into the return-air plenum (number 20). The sequence of operation of the system is as follows. During the day, the solar collectors heat the storage tanks via internal heat exchangers operating under the influence of the solar differential controller, which controls the solar circulating pump. When there is a call for heat by the room thermostat, water from the solar storage tanks is circulated through the return-air plenum heat exchanger, assuming that the solar storage water is above the low-limit setting for the system (usually between 100 and 125°F). At the same time, the furnace fan is turned on to move heated air throughout the home within the duct system. The return air is heated as it passes through the solar heat exchanger. This system also makes provision for heating domestic hot water from the solar collector loop.

The heat exchanger coil in the return-air plenum must be sized for the ductwork in the home as well as the maximum number of Btu that the solar system can be expected to produce at any given time. This system also uses a two-stage thermostat similar to that used in the hot water baseboard system. A low-limit aquastat serves to prevent circulation of water through the heat exchanger loop in the plenum if the solar water temperature is too low for effective space heating.

Solar heating with fan convection. Individual fan convectors are used with solar heating systems when it is neither desirable nor possible to interconnect the solar system with a conventional central heating system (such as central steam heating or electric baseboard or radiant ceiling heating). In the case of steam heating, the temperature of the steam in the system is far too high for the solar system to provide first-stage space heating. In the case of an electric baseboard installation, no central heating system exists for the solar system to be piped into. In these instances, separate fan convectors are installed to deliver heat to the home. A typical fan convector is illustrated in Figure 2.50.

A fan convector operates on a principle similar to an automobile heater: Hot water circulates through a coil; a fan located behind the coil blows room air across the coil transferring heat from the coil to the passing air via forced convection. Each convector can be controlled by its own room thermostat. A solar system using fan convectors is illustrated in Figure 2.51. The system illustrated in the figure is somewhat different from other central heating solar applications in that the solar storage tanks are filled with potable water rather than closed-loop heating system water. Thus this system can supply both central space heating and do-

Figure 2.48 (*Opposite page*) Solar heating system interfaced with an existing fossil-fuel hot water heating system. (From M. Greenwald and T. McHugh, *Practical Solar Energy Technology*, Prentice-Hall, Englewood Cliffs, N.J., 1985.)

EQUIPMENT LIST

PLUMBING

PART NO.	DESCRIPTION
1	Circulator #UP26-64
2	Set Flanges #51.97 56
3	Air Purger #440
4	Flush and Drain Valve
5	Pressure Gauge 0-60 P.S.I.
6	Pressure Relief Valve 50 P.S.I.
7	Air Vent #122
8	Expansion Tank #S-30
9	Check Valve 3/4"
10	Fill and Drain Valve
11	120 Gallon Maximum Insulation Storage Tank with Heat Exchanger
12	Pressure and Temperature Relief Valve
13	Circulator #UP20-42
14	Air Purger #440
15	Expansion Tank #SX-30V
16	Manual Ball Valves 3/4"
17	Solenoid Valve #JSX8210D9HW
18	80 Gallon Maximum Insulation Storage Tank with Heat Exchanger
19	Tempering Valve 1/2"
20	Forced Air Heat Exchanger
21	Existing Forced Air Furnace

ELECTRICAL

PART NO.	DESCRIPTION
A	Solar Control Monitor and Relay Terminal
B	Thermostat #T675A1540
C	Remote Bulb Controller #T675A1417
D	To Collector Sensor
E	STO. Sensor
F	Aux. 1 Sensor
G	Aux. 2 Sensor
H	Aux. 3 Sensor
I	Aux. 4 Sensor
J	Existing Furnace Control
K	150°F Upper Limit-Wire in Series with Aux. 1 Sensor
L	190°F Upper Limit-Wire in Series with Storage Sensor

Figure 2.50 Fan convector. These units are ideal for adding localized separate zoned heat from a central hydronic system. The convector contains its own water pump and operating thermostat, allowing it to function independently of the central heating system. (Courtesy of Turbonics, Inc.)

mestic hot water. Since the system operates with potable water, the operating pressures within the fan convectors and solar storage tank are at house or well pressure. Only the solar collector loop operates on low pressure. A low-limit switch is incorporated into the system to prevent convector operation if the temperature of the storage tanks falls to a low limit between 85 and 100°F.

When sizing water storage for the system, each fan convector should be supplied by one 120-gal solar storage tank, which in turn is heated by three or four 3 ft × 8 ft or 4 ft × 8 ft solar collectors. This arrangement prevents the storage tank temperature from being drawn down too quickly when the convectors are in use.

Passive Solar Space Heating

Heating a home by using solar radiation that penetrates into the interior of the home requires special architectural design features that are capable of trapping, storing, and distributing this energy. Most solar homes make a definitive design statement and have many features in common. Passive solar design is something that must be incorporated into the architecture of the home before the first board is put in place, since these architectural fundamentals are pervasive throughout the structure of the home.

Passive solar energy is one of the most cost-effective options for home heating, since good passive design saves money in two ways: It lowers the overall heating and cooling costs of the home and provides for varying portions of the space heating and cooling requirements. Although these modifications add 10 to 15% to the overall cost of the home, the additional cost will be more than compensated for in first-year energy savings.

Figure 2.49 (*Opposite page*) Solar heating system interfaced with an existing fossil-fuel central hot air heating system. (From M. Greenwald and T. McHugh, *Practical Solar Energy Technology,* Prentice-Hall, Englewood Cliffs, N.J., 1985.)

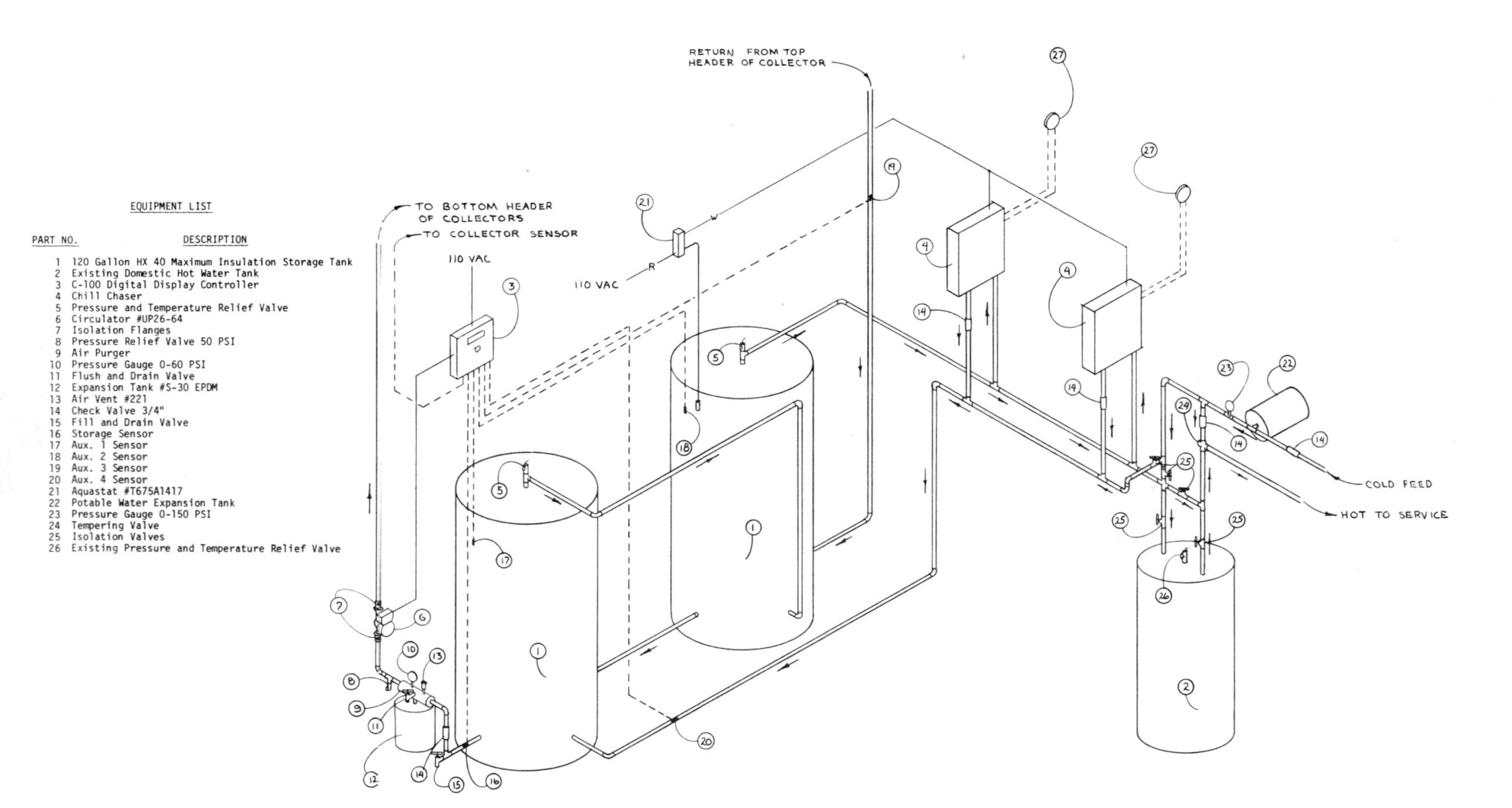

Figure 2.51 Solar heating system using separate fan convectors. This type of solar application is generally installed when the house has either a central steam heating system or is equipped with electric resistance heating. (From M. Greenwald and T. McHugh, *Practical Solar Energy Technology*, Prentice-Hall, Englewood Cliffs, N.J., 1985.)

Passive solar, by definition, refers to the capture, distribution, and storage of solar energy with a minimum of moving parts and electrically operated equipment. In this chapter we examine some of the more prominent design aspects of passive solar houses together with implications for their implementation into existing housing where applicable.

Design statement. All passive solar houses are unique in design compared to conventionally designed houses. One such passive solar residence is illustrated in Figure 2.52. The major surface area of the roof of the residence should face true south. Since true south deviates from magnetic south in most areas of the United States, an isogonic chart (Figure 2.53) should be consulted to determine the correction angle that should either be added to or subtracted from the magnetic compass reading to obtain the true south reading in a particular location.

In many localities, it is customary for the roofline of the house to face the street. For passive solar to work, the north/south line of the roof must face east/west (in this way, the major roof area will be facing south). In certain instances, therefore, positioning the house in this way will require a variance. In new construction, the building inspector should be consulted to determine if this will cause a problem in building on a specific lot.

Figure 2.52 Passive solar home (typical).

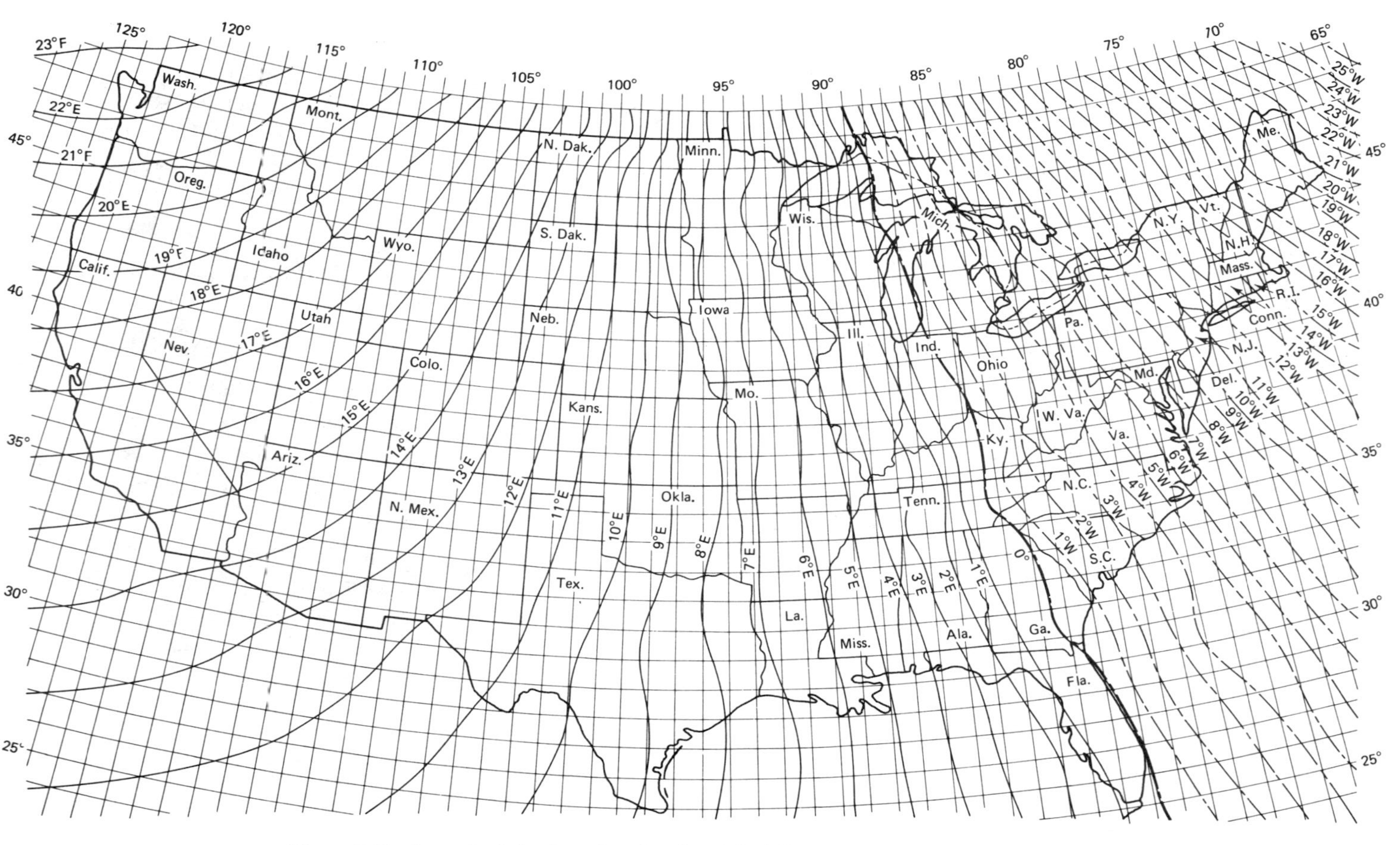

Figure 2.53 Isogenic deflection chart for the United States. (From U.S. Department of Commerce, Coast and Geodetic Survey, 1965.)

Use of glass. The southern exposure of the major portion of the solar home allows collection glass to be placed along the southern face of the building. Conversely, a minimum amount of glass is generally used along the northern width of the building, since the majority of winter winds strike the building from this direction. Viewed from the inside of the home, this glass gives an open, unobstructed feeling to the interior of the house (Figure 2.54). The north-facing side of most passive solar homes contains few windows. In the illustration in Figure 2.55 the window provided is for emergency egress only and is not installed to provide any significant amount of natural lighting.

Insulation and bermed walls. Features of all passive solar and most energy-efficient homes include a lot of insulation in the walls and ceilings, the use of insulated windows and doors, and the use of earth-bermed walls. Insulation upgrading is one area not restricted to new passive solar construction, but an energy-efficient upgrading that can be incorporated into practically any residential structure, regardless of age. When older homes are upgraded with new insulation, the insulation is usually blown in through holes in the walls and covered with fill plugs (Figure 2.56).

Most passive solar homes feature wall, ceiling, and floor insulation values that restrict normal heat losses to a minimum. Wall studs in many of these homes are made from 6-in. lumber as opposed to the $3\frac{1}{2}$-in. framing lumber used in conventional residential construction. The use of the thicker studs in both the walls

Figure 2.54 Solar homes are light and airy in appearance due to the large area of south-facing glass.

Figure 2.55 Emergency egress windows on north wall of house. The window units are specially designed for emergency purposes only and are not used for significant lighting or ventilation purposes.

Figure 2.56 Insulation is normally added to older homes through holes drilled in the exterior siding and plugged after the insulation has been blown into the exterior wall cavities.

and ceilings enables the use of high *R*-value insulation. For example, a conventionally built well-insulated home of 1500 ft^2 will have a heat loss of between 60,000 and 75,000 Btu at 0°F. The same-size passive solar home with upgraded insulation will have a heat loss of approximately 25,000 to 30,000 Btu/hr. Not only are the annual heating and cooling costs for the passive solar structure much less than those of the conventional home, but the low-Btu heat-loss figure allows for the installation of smaller, less expensive heating equipment, which would be too small for use in other, more traditionally constructed homes.

Figure 2.57 Insulated window quilts. Good-quality window quilts not only reduce heat loss through the window glazing, but greatly reduce drafts and air infiltration, a major cause of window heat loss. (Courtesy of Appropriate Technology Corporation.)

Older, drafty windows can also be upgraded with newer insulated units that offer superior resistance to both air infiltration and heat loss. Although advances in window and glazing technology have increased the energy efficiency of double- and triple-glazed windows, the heat loss through them is still considerably more than that through a well-insulated wall or ceiling. To help overcome these deficiencies, the use of insulated decorator window quilting material is recommended. These quilts come in a variety of colors and sizes and are adaptable to a variety of windows and doors in both new and older homes (Figure 2.57). Although most insulated window quilts and shades are operated by hand, the use of motorized drives is recommended for the larger window sizes. Motorized installations can be equipped with the ability to raise or lower automatically, depending on available sunlight, a feature that maximizes solar gain and minimizes heat loss through the glazing.

The incorporation of earth-bermed walls helps to insulate much of the structure against infiltration and heat loss. Also, the use of massive quantities of earth helps to promote even temperature levels in the home, since ground temperature below the frostline stabilizes at approximately 50°F. Note in Figure 2.58 that the berming covers virtually the entire lower portion of the home, ending just below window level.

Figure 2.58 Earth-bermed wall. Earth is an ideal insulating material and helps maintain even temperatures against the exterior wall of the home.

Roof angles and overhangs. The roof structure in all passive solar homes is one of the most important design structures of the house. The roof should be designed to accommodate active solar collectors whether or not they are part of the original structure. This necessitates a roof angle that is approximately equal to, or about 10° greater than, the north latitude of the location. This angle ensures maximum solar radiation at a nearly perpendicular level during the winter months, when the sun is low in the sky. Also, an overhang should be designed onto the roof, to enable the winter sun to penetrate the interior of the house while providing adequate shading during the summer. An overhang of this type is shown in Figure 2.59.

In addition to active solar space heating systems, large, unobstructed roof angles will allow for possible use of photovoltaic electric panels if their use becomes economically feasible in the future.

All roof structures should be well insulated. If the home is designed with an attic, the attic floor joists should be insulated; the space between the roof rafters should not. This will prevent condensation from forming within the roof, resulting in water damage to the structure. If, however, the ceiling of the room

Figure 2.59 A properly designed roof overhang will effectively shade the home during summer, while admitting virtually all the available solar radiation during the winter months.

is the finished roof, all insulation is placed between the finished ceiling and the interior of the roof sheathing and should contain a vapor barrier.

Open interior design. Solar houses are relatively free and unobstructed within the structure. This allows for free convective air currents to establish even heating throughout the house.

Many solar homes incorporate a ceramic tile floor that is located to take maximum advantage of the incoming winter solar radiation and to provide a reasonable amount of thermal mass for heat radiation during the evening hours. The placement of the tile can be determined either by calculating the maximum penetration of the sun during the winter through drawings based on roof and sun angles, or simply by observing how far the sun penetrates during the winter (sometimes this may not be possible, although it is the most accurate method of determining how far the sun penetrates into the dwelling). Given the unobstructed space in many of these homes, heat radiation from the floor tile will circulate freely due to an absence of conventional wall partitions. If the home contains wall partitions, the use of small transfer fans placed on doorways or in walls can help move the air from one room to another (Figure 2.60).

Air transfer systems. In many instances, solar homes are constructed so tightly that circulation of fresh air within the home is restricted to the point

Figure 2.60 Small transfer fans are effective in moving the air between rooms and interior partitions to increase even interior temperatures and heat flow. (Courtesy of Oak Ridge Products.)

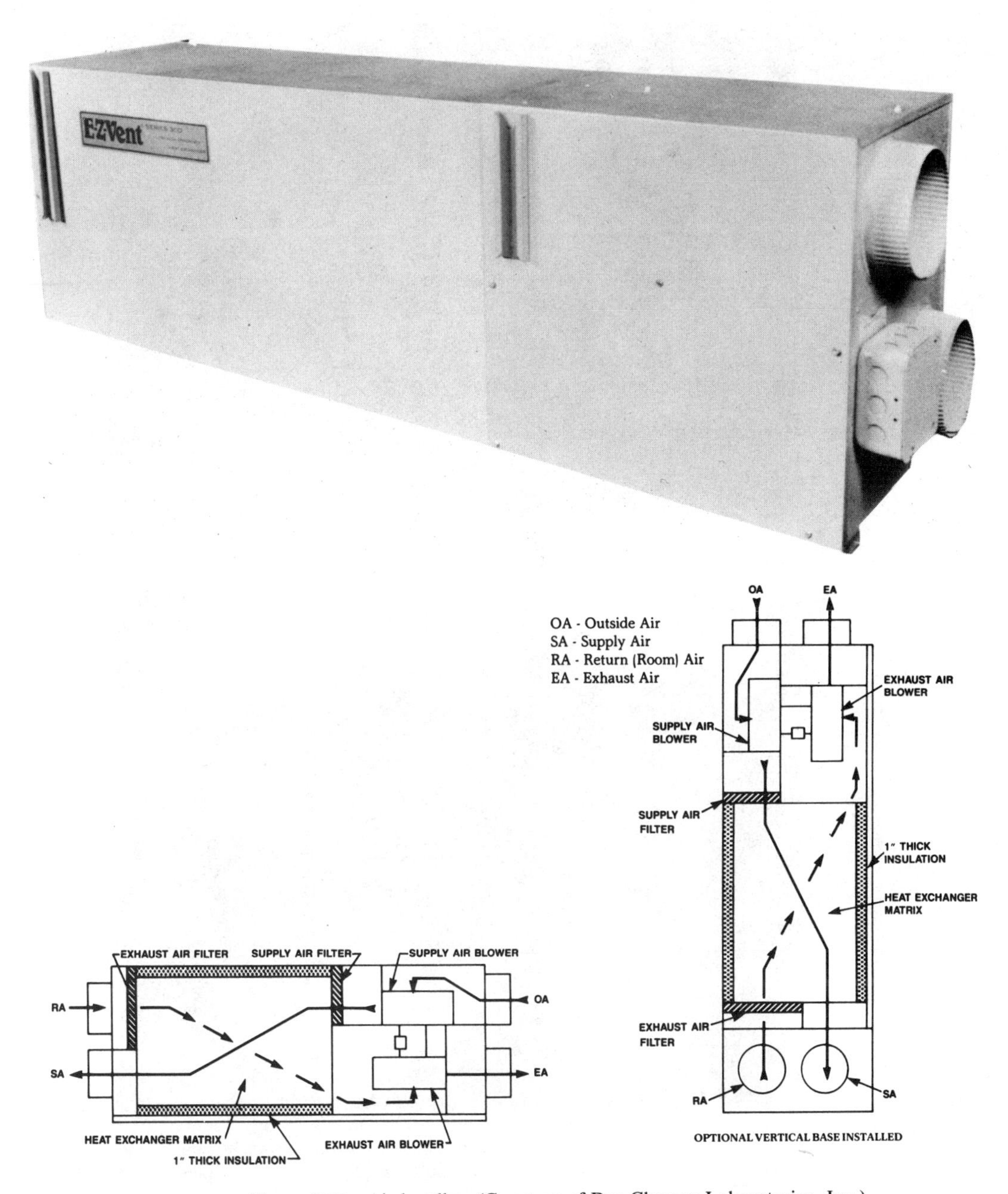

Figure 2.61 Air handler. (Courtesy of Des Champs Laboratories, Inc.)

Figure 2.62 Add-on sunroom. Such units have become increasingly popular and offer a year-round activity center. It should be kept in mind that depending on the size of the room, additional heat might be required for the add-on, due to insulation characteristics, amount of glazing, and so on. (Courtesy of Sunplace, Inc.)

that provision should be made to bring outside air into the residence for safety purposes. Whereas many conventional homes have two to four air changes per hour, it is not uncommon for solar homes to have less than one air change per hour. This can lead to a dangerous buildup of airborne chemicals and other toxic substances within the interior of the home that would otherwise be vented to the outdoors. To overcome this problem, the use of an air handler is recommended (Figure 2.61).

An air handler is unique in that it simultaneously removes interior air while drawing in exterior air, transferring up to 85% of the interior heat to the fresh incoming air in the process. Not only is fresh air supplied to the home, but the money and energy spent in heating the interior air is recaptured to a large degree. A good deal of condensate is produced by the air handler in this process; therefore, provision must be made not only to drain the condensate from the air handler, but to provide an overflow removal system in case the primary drain lines should become inoperative.

Passive solar additions and retrofits. The use of add-on solar rooms has become a big business. When it is neither possible nor desirable to consider a comprehensive solar retrofit, the installation of a small sun space within the existing home can provide some of the benefits of passive solar heating within the confines of the existing house design.

Most solar rooms are supplied in kit form and are available in a variety of sizes and shapes to match available space and funds for their installation. An add-on sunroom of this type is illustrated in Figure 2.62.

The installation of most add-on sun spaces involves the construction of suitable footings and floors. Most kits can be bolted together in just a few days. The homeowner should not expect significant amounts of residential space heating from the sunroom: Most of these units act as buffers, with the total amount of heat gain about equal to the net heat loss of the sunroom during the average year. If the add-on sunroom is large, provision must be made to heat it, either by extending the existing home heating system or by incorporating a small space heater into the design of the room. Some sunrooms feature movable window insulation panels that are used to cover the windows during evenings and extended sunless periods, to limit heat loss.

3

OPERATION AND MAINTENANCE OF OIL-FIRED APPLIANCES

In this chapter we cover the general operation and maintenance of residential boilers, furnaces, and water heaters powered by fuel oil. Our discussion of the basic operating principles of these systems is designed to give the reader a broad understanding of how oil-fired appliances operate and how to keep them in proper working order.

Most service and installation manuals that are supplied with these appliances describe simple service and maintenance procedures to be performed by the homeowner. With proper understanding of operating principles and some experience, most homeowners can keep their oil-fired appliances operating at peak efficiency.

This chapter focuses on the combustion characteristics of fuel oil, oil burner operating theory, appliance configuration, and maintenance of a variety of oil combustion appliances.

FUEL OIL

Combustion Characteristics

Residential-grade fuel oil is a clear, refined product that constitutes one of several grades of crude oil. It is classified as No. 2 fuel oil (this is the same classification that is given to diesel fuel, which is also No. 2 crude oil, with different additives for use in internal combustion engines).

Fuel oil contains between 135,000 and 144,000 Btu/gal when burned completely. The oil is composed of approximately 84% carbon and 14% hydrogen, together with trace elements. It yields carbon dioxide (CO_2) and water when burned in a conventional heating appliance: the combustion of carbon accounts for approximately 75,000 Btu; the hydrogen in the fuel oil accounts for the remaining 65,000 Btu.

To ensure complete combustion, between 50 and 100% excess combustion air must be supplied to the burner mechanism. Although excess air lowers the combustion temperature of the fuel oil in the combustion chamber, lowering combustion efficiency somewhat, this is desirable since relatively complete combustion of the fuel oil can only take place given sufficient quantities of excess combustion air. Fuel oil combustion in the presence of 100% excess air at the burner head is illustrated in Figure 3.1.

Fuel Storage and Supply Systems

Fuel oil is stored in a tank located either within the home or buried in the ground adjacent to the residence. These tanks are made either from heavy-gauge steel supplied with a corrosion-resistant coating or from fiberglass.

Underground storage tanks. Underground oil tanks are used when the homeowner either wishes to store more than 275 gal of oil in one tank without

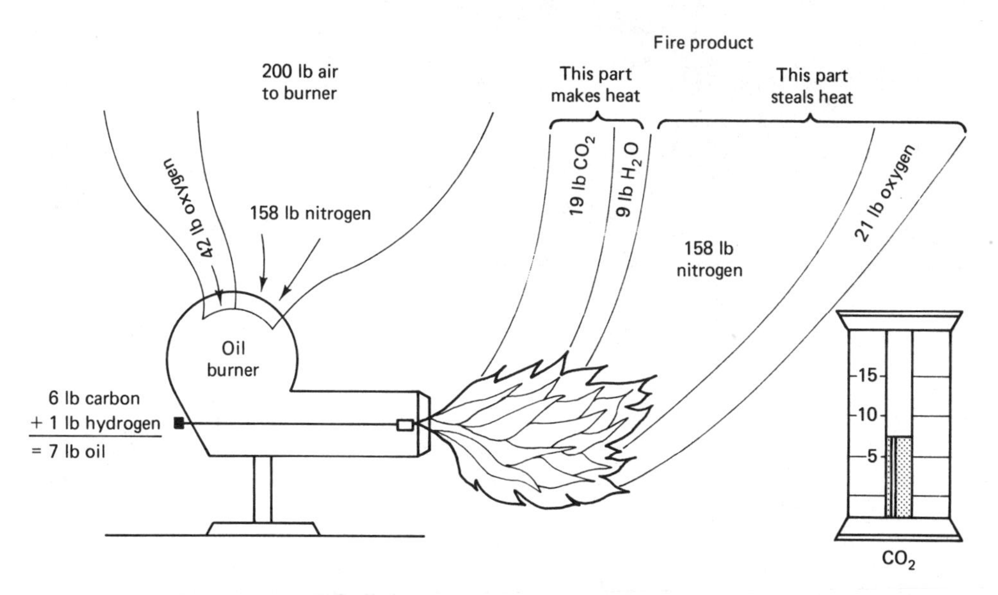

Figure 3.1 Products of oil-fired combustion with 100% excess air supplied to the burner. (Courtesy of Dwyer Instruments, Inc.)

taking up additional space in the home or because local building codes specify the use of underground tanks. Underground storage tanks are usually supplied in capacities of 550 or 1000 gal. The distance between the storage tank and the foundation wall of the residence is also specified in many local building codes. Underground tanks require a two-line oil feed system; one of the lines is a feed line, the other is a return line. The two-line configuration, illustrated in Figure 3.2, is also referred to as a self-purging system and is required on any installation in which the oil in the storage tank must be lifted from below ground level to the combustion head of the burner.

Note from Figure 3.2 that one of the feed lines supplies the oil burners; the other is the return line, which delivers excess fuel from the fuel pumps back to the storage tank. The oil tank should slope away from the feed lines to prevent

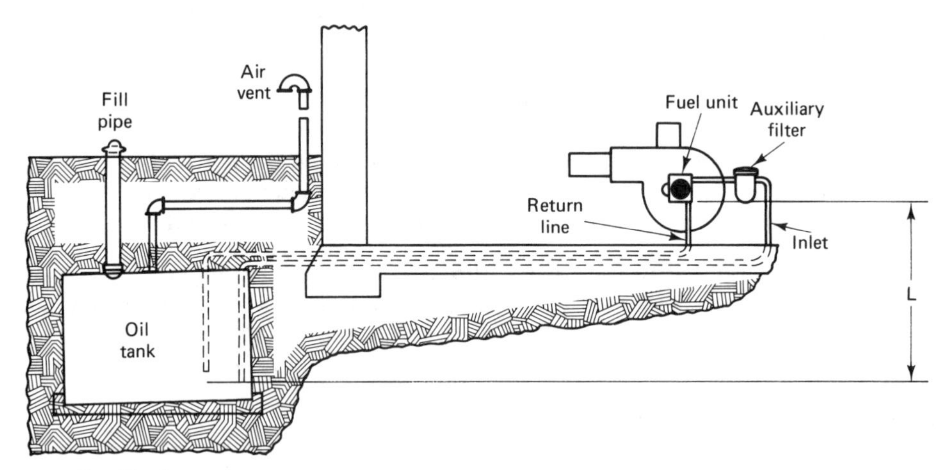

Two-Pipe Systems

Lift L (ft)	Length of Tubing[a] (ft)			
	Single-Stage Units		Two-Stage Units	
	3/8 in. OD	1/2 in. OD	3/8 in. OD	1/2 in. OD
0	53	100	68	100
2	45	100	63	100
4	37	100	58	100
6	29	100	53	100
8	21	83	48	100
10	13	52	42	100
12	Not recommended; use two-stage fuel unit		37	100
14			32	100
16			27	100
18			22	88

[a] Line lengths include *both* vertical and horizontal lengths

Figure 3.2 Two-line self-purging oil supply system. Note that the fuel oil tank is sloped away from the oil feed and return lines to prevent sediment accumulation from entering the oil lines. (Courtesy of Suntec Industries.)

Figure 3.3 Oil filter housing with filter elements. Oil filters should be replaced every time the burner is serviced, or a minimum of once each heating season. (Courtesy of General Filters.)

any debris or sediment buildup within the tank from entering the oil supply line. All lines within the home should be installed to minimize any possibility of damage. The lines should contain a minimum number of unions to minimize the possibility of oil leaks or to prevent air from entering the fuel lines. A fusible shutoff valve should be installed at the point where the fuel lines enter the house, along with an appropriate filter assembly (Figure 3.3).

Aboveground tank installation. When oil tanks are located above ground, the bottom of the tank must be less than 8 ft below the fuel pump to achieve proper lift in the oil line. (If the oil must be lifted more than 8 ft to the fuel pump, a two-stage pump may be required instead of the single-stage pump normally supplied with almost all residential heating equipment.) Aboveground tanks can be installed either inside or outside the house. Indoor installations are preferable since the possibility of the tank or oil line freezing is minimized during the winter (fuel oil waxes at approximately 6°F). When installed indoors, local building codes should be consulted, which may specify the location of fireproof partitions between the tank and adjacent areas, if required, and the minimum distance that the tank must be located away from the oil burner. A typical aboveground indoor tank installation is illustrated in Figure 3.4.

Note from the figure that a shutoff valve is installed where the fuel line exits from the oil tank. These valves contain a fusible link which is designed to melt and close the oil line automatically if the ambient temperature surrounding the tank rises above 165°F. Figure 3.5 illustrates the proper location of fusible-link shutoff valves in a typical residential installation, together with recommended fire and heat-sensing alarms.

Note from the figure that a typical aboveground tank installation uses only one feed line, as opposed to the two-line system usually installed in underground tank installations. In these instances, one-line systems can be used since the lift required in an aboveground installation is minimal. Although a one-line system is not self-purging (air must be bled manually from the oil lines through a special

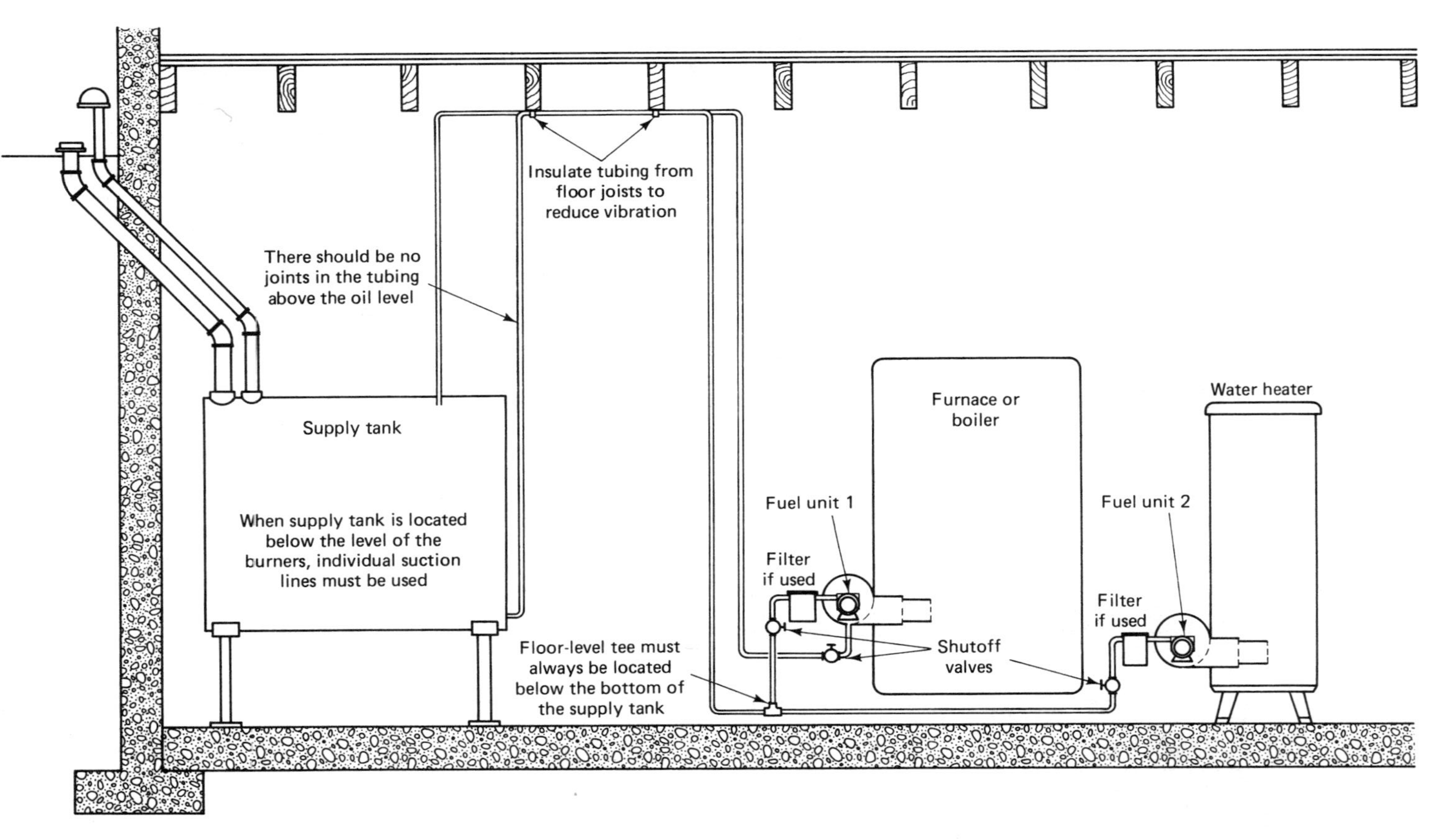

Figure 3.4 Aboveground oil tank installation. Vent line should be installed at a point higher than the winter snow line. (Courtesy of Suntec Industries.)

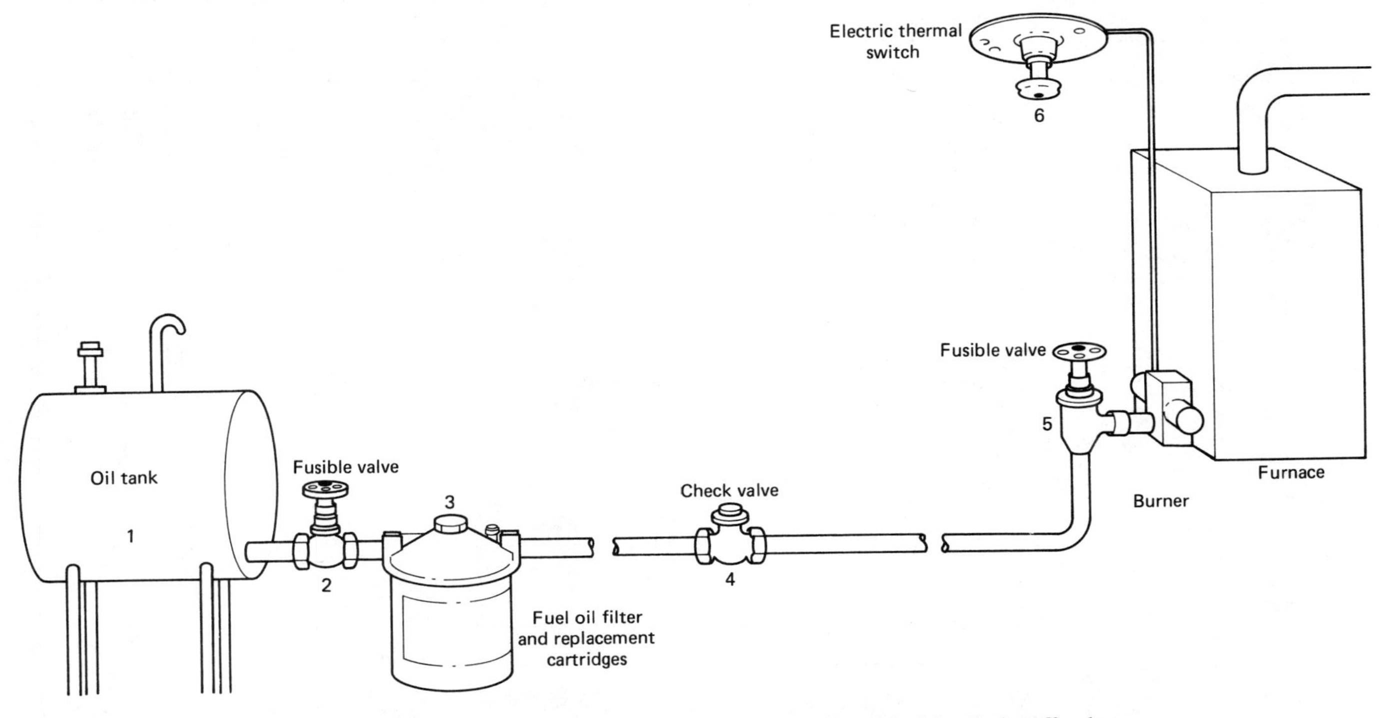

Figure 3.5 Recommended configuration for the installation of fusible oil shutoff valves (Courtesy of Highfield Manufacturing Company, Firomatic Products.)

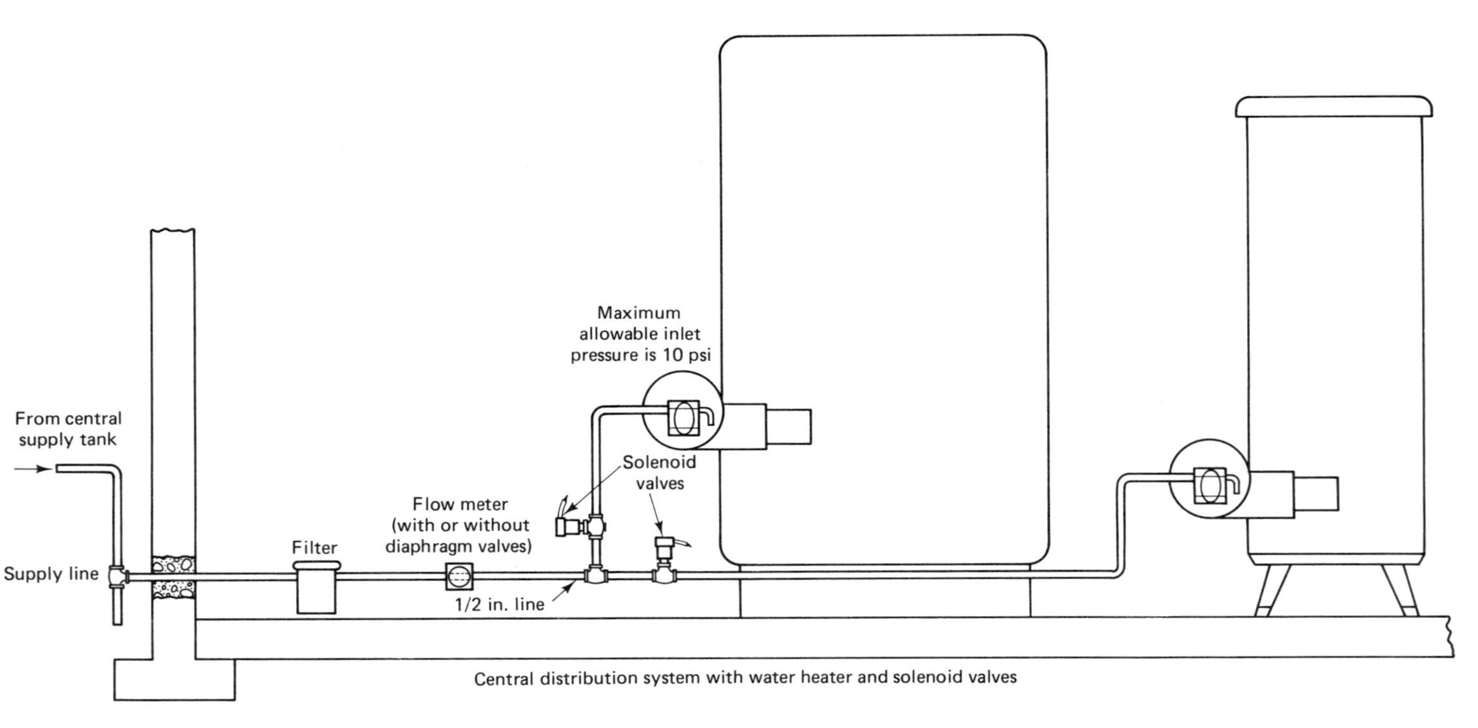

Figure 3.6 Single oil supply line feeding two appliances. Note the use of solenoid valves to help prevent loss of prime in the fuel pumps when they are not operating. (Courtesy of Suntec Industries.)

fitting on the fuel pump prior to starting the burner), it is adequate in most instances. Sometimes, a two-line self-purging system is installed in aboveground tanks to allow the system to prime itself, in addition to keeping the oil moving in the system during cold weather.

If two appliances are supplied by one oil tank, the supply piping circuit illustrated in Figure 3.6 is used, which allows both burners to operate simultaneously. Problems associated with two appliances fed from one storage tank are usually connected to either a loss of prime that can occur in the fuel pumps when they are not operating, or a condition known as pulsed burning, caused by oil starvation in one or both fuel pumps due to an insufficient delivery of fuel through the supply system. This makes burner startup difficult if a single-line piping design is used. If a two-line system cannot be installed, a solenoid valve should be installed on the feed line of each oil burner and wired to energize simultaneously with each burner motor. When the burner is shut off, the solenoid valve closes, which will prevent a loss of prime in the fuel pump (Figure 3.6).

OIL BURNER OPERATION

Oil burners have undergone significant design changes during the past 15 years, which has resulted in modern units that are capable of delivering between 80 and 88% combustion efficiency. This is in marked contrast to earlier oil burners, which averaged combustion efficiencies of 50 to 75%.

Modern oil burners are referred to as high-speed flame retention burners (to distinguish them from earlier low-speed units). Flame-retention burners operate at rotational speeds of 3450 rpm, delivering combustion air to the burner head at high velocities. The combustion head is designed to hold (or retain) the flame against the end of the burner with a predefined shape, resulting in a very hot controlled flame pattern (Figure 3.7). A high-speed flame-retention burner is illustrated in Figure 3.8.

The key to the operation of any oil burner is the burner nozzle and electrode assembly, including the specially designed combustion head (Figure 3.9). Note from the illustration that the electrodes are positioned directly above, and in line with, the nozzle. During operation, a spark jumps the gap between the two electrodes. Combustion air supplied by the blower fan (Figure 3.10) combines to ignite with the atomized oil spray from the nozzle in a highly efficient combustion pattern. The high voltage that is required to produce the spark to jump the gap between the electrodes is supplied by a special ignition transformer. Two spring terminals extend from the transformer and contact the electrodes to deliver the necessary voltage for spark generation.

The combustion head of the burner is specially designed to promote both turbulence and recirculation of the air within the flame pattern inside the combustion chamber, resulting in high combustion efficiency.

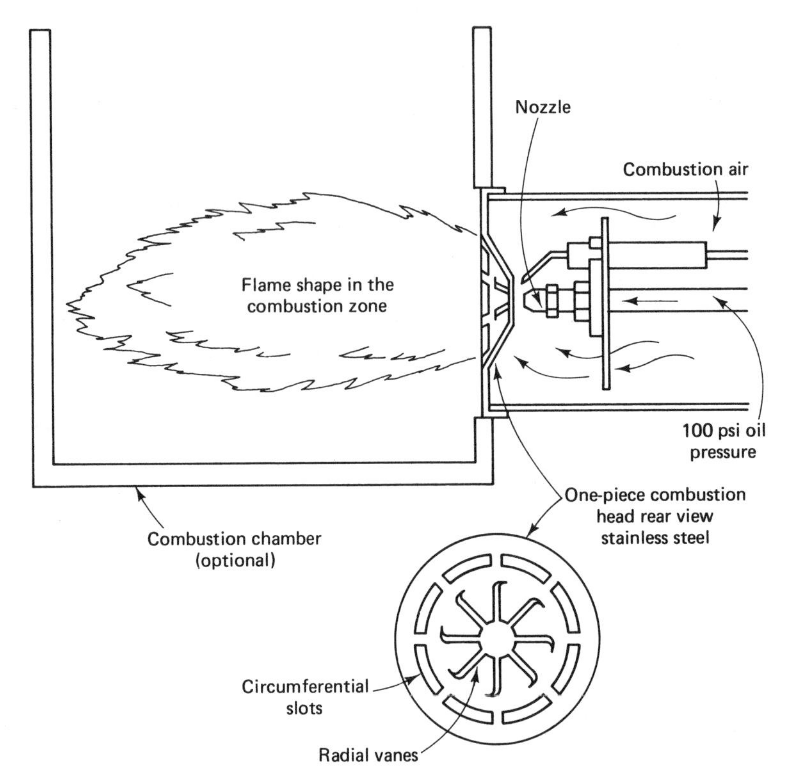

Figure 3.7 Combustion pattern of high-speed flame-retention burner. (Courtesy of R. W. Beckett, Inc.)

The most important component for proper operation of the burner is the nozzle itself. An oil burner nozzle is illustrated in Figure 3.11. Nozzles are sized to deliver the proper amount of fuel to the combustion chamber, based on the manufacturer's Btu input rating. If, due to improper sizing, the nozzle delivers too much fuel, overheating and failure of the integrity of the appliance can result. If too little fuel is delivered, not only will the appliance not heat properly, but poor combustion efficiency can lead to sooting and plugging of the interior flue and heat transfer passages.

Fuel oil nozzles are rated according to three major criteria: the amount of fuel delivered, stated in gallons per hour; the spray pattern delivered by the nozzle—solid, hollow, or combination pattern; and the angle of the spray pattern. The delivery rate of the nozzle is the criterion used to match the correct nozzle with the stated Btu input for the appliance. To determine the proper nozzle size for any appliance, the following formula should be used:

$$\text{nozzle size (gal/hr)} = \frac{\text{rated Btu input of appliance}}{140{,}000}$$

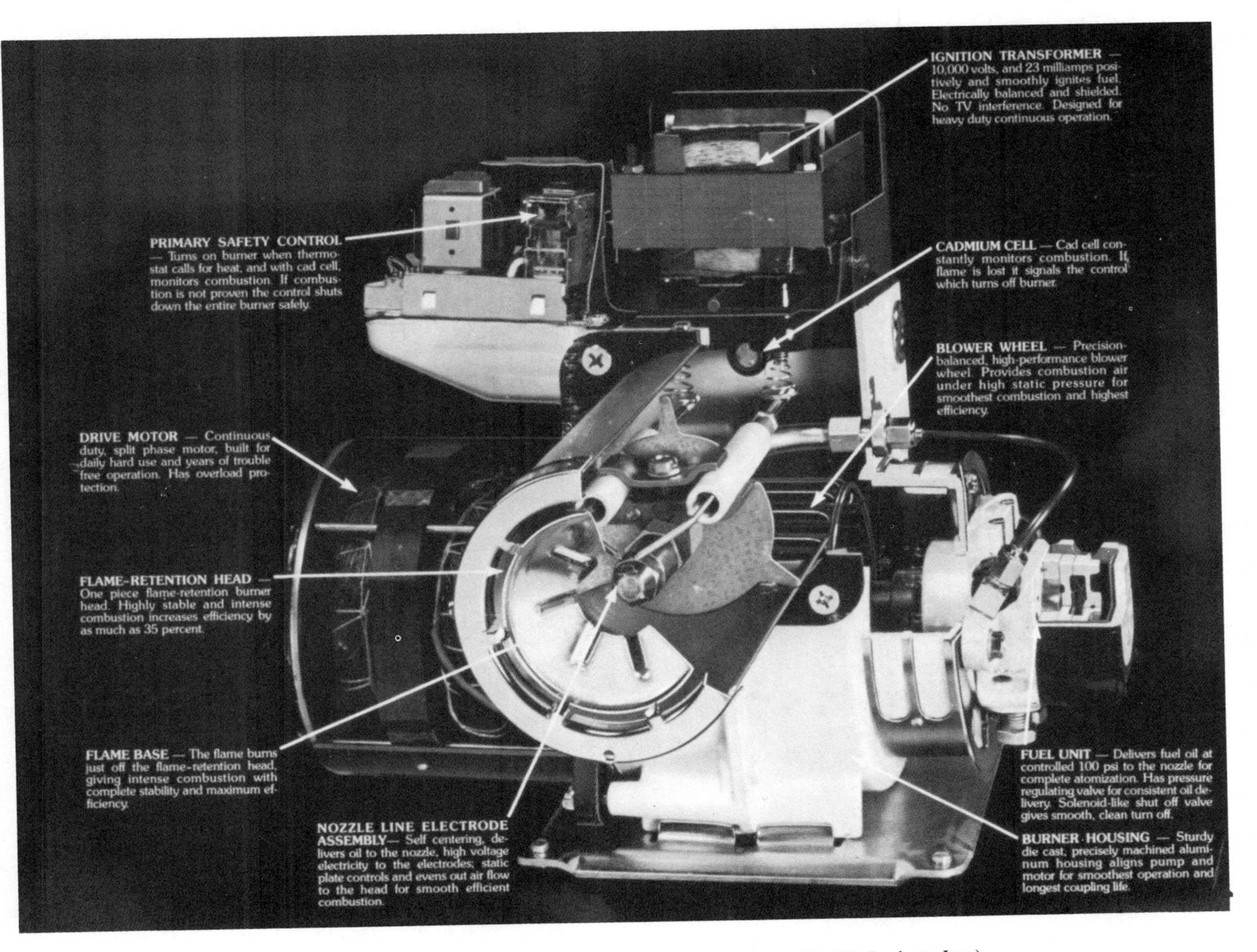

Figure 3.8 High-speed flame-retention oil burner. (Courtesy of R. W. Beckett, Inc.)

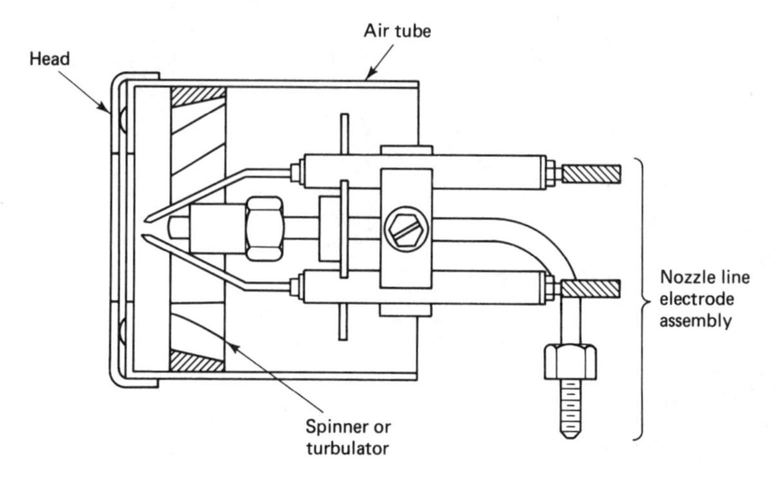

Figure 3.9 Combustion head assembly of typical high-speed flame-retention oil burner.

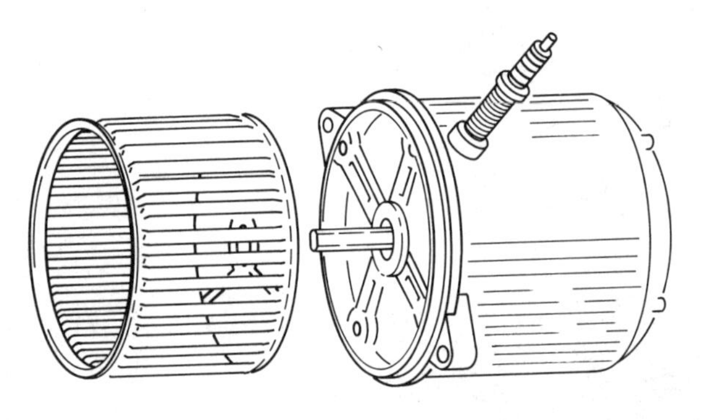

Figure 3.10 Motor and air blower used on high-speed oil burners (typical). These assemblies will usually be interchangeable on several different oil burner assemblies.

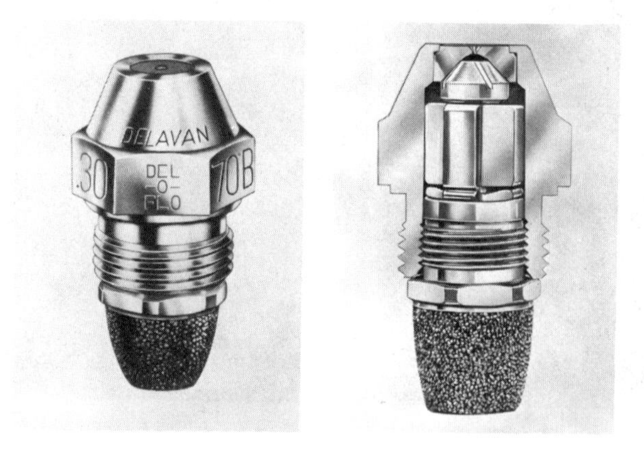

Figure 3.11 Full and sectional views of oil burner nozzle. Note the marking on the nozzle indicating firing rate, spray angle, and spray type. (Courtesy of Delavan Inc.)

The spray pattern required for a particular unit depends on its Btu rating as well as the shape and size of the combustion chamber. The three spray patterns commonly used in residential heating equipment are illustrated in Figure 3.12.

The spray pattern and atomization of the fuel oil is produced by the design configuration of the nozzle (Figure 3.13). During operation, oil at 100 psi is delivered by the fuel pump to the swirl chamber of the nozzle, where it is thrown against the walls of the chamber, leaving an air gap in the center. As the oil exits the nozzle assembly, it breaks into thousands of tiny droplets, a process referred to as **atomization**, which is necessary to expose all the fuel oil to sufficient air,

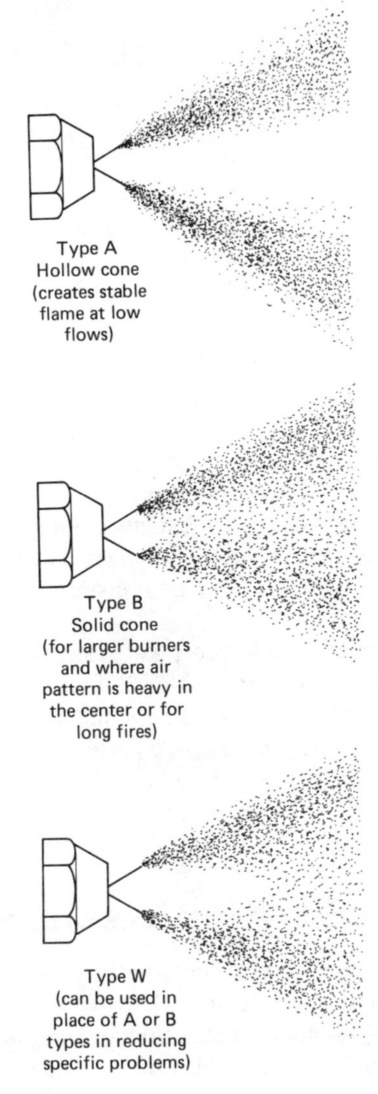

Figure 3.12 Oil burner nozzle spray patterns. (Courtesy of Delavan Inc.)

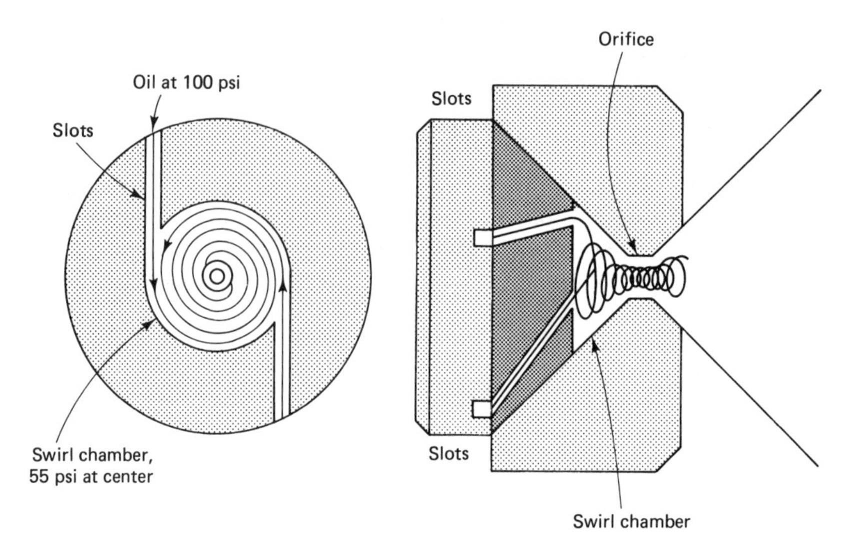

Figure 3.13 Operation of the swirl chamber in the oil burner nozzle. This swirling action helps to promote atomization of the fuel oil as it exits the nozzle directly into the combustion flame path. (Courtesy of Delavan Inc.)

allowing complete combustion to take place. The exact angle of the spray shape is determined by the nozzle design configuration.

In general, hollow-cone nozzles are used in residential heating equipment with firing rates under 1 gal/hr. A highly stable flame pattern is produced where the droplet sizes of the oil after atomization by the nozzle sometimes tend to be irregular.

Solid-cone nozzles are used in appliances calling for firing rates between 2 and 3 gal/hr. These nozzles tend to produce longer flame patterns than the hollow nozzles that are required by the larger combustion chambers in these heating units.

Combination nozzles are used over a wide range of firing conditions. At the lower firing ranges (e.g., 0.5), the spray pattern tends to be hollow. At firing ranges above 5 gal/hr, the spray pattern tends to be solid.

When sizing and matching nozzle spray patterns for a particular application, the appliance manufacturer should be consulted to make sure that the proper nozzle configuration will be installed. The procedures for nozzle servicing are described later in the chapter.

OIL-FIRED APPLIANCE CONFIGURATION AND OPERATION

Oil is used as a primary heat source for a variety of residential hot water and steam boilers, furnaces, and water heaters. The nature of the combustion process results in appliances that yield high combustion and operating efficiencies, several of which are now examined, together with their basic operating characteristics.

Oil-Fired Hot Water and Steam Boilers

The operating characteristics of oil combustion in both hot water and steam boilers are similar. Each of these boilers is illustrated in Figures 3.14 and 3.15.

Most residential boilers are sold as packaged units; the oil burner, aquastat, circulator, electrical wiring, and other major components are factory installed. These boilers are constructed from either cast iron or steel and can be of either wet or dry base construction. Most cast-iron boilers are of the **wet base** design, in which water circulates above and below the combustion chamber. Most steel

Figure 3.14 Oil-fired hot water boiler. Note the raised pinned surface of the metal castings to increase the surface area exposed to the combustion gases, promoting efficient heat transfer from the combustion chamber to the surrounding water jacket. (Courtesy of Weil-McLain, a Marley Company.)

Figure 3.15 Oil-fired steam boiler. Note the location of the glass gauge assembly on the side of the boiler to determine water level within. (Courtesy of The Burnham Corporation.)

boilers are of the **dry base** design, in which water does not circulate below the combustion chamber. Figure 3.16 illustrates dry base construction of both cast-iron and steel units. Note the location of the tankless coil, which is immersed in the hottest water at the top of the boiler.

Steam boiler configuration is somewhat different from that of hot water design. Steam boilers have a low-water cutoff and glass gauge assembly, which allows visual inspection of the water level within the boiler. The cutoff also provides for emergency shutdown of the burner should the water level within the boiler fall to a predetermined dangerous level.

Sequence of operation for hot water boilers. When the room thermostat calls for heat, the aquastat relay on the boiler closes the electrical circuit to the burner to begin combustion. When the burner comes on, a safety circuit within the aquastat is activated, which is controlled by a cad cell, located within the line of sight of the combustion flame. The cad cell is wired into the safety circuit of the aquastat and will shut down the oil burner if a flame is not sensed by the cell after approximately 20 seconds of energizing the burner. Figure 3.17 illustrates a typical main boiler aquastat (a) and cad cell (b), along with the location within the burner housing.

After a stable flame pattern has been established, the cad cell disengages the safety circuit, which allows continued burner operation until the high-limit setting of the boiler is reached. Most aquastats also contain a low-limit setting that will prevent the circulator from operating below a preset low-limit temper-

(a)

(b)

Figure 3.16 (a) Dry base steel boiler (courtesy of The Burnham Corporation); (b) dry base cast-iron boiler (courtesy of Slant/Fin Corporation).

ature. This feature prevents cold water from circulating within the heating system until the water temperature reaches approximately 140 to 160°F. The circulator continues to run until the thermostat setting has been satisfied, and the oil burner will normally cycle on and off in response to the drop in boiler water temperature. As long as the thermostat calls for heat, the burner will turn on when the water temperature drops approximately 10°F below the high-limit setting and will shut off when this high-limit setting has been reached.

If the boiler contains a tankless coil for producing domestic hot water, the low-limit temperature setting on the aquastat will be maintained within the boiler, regardless of whether or not the thermostat is calling for heat. This ensures that the boiler water will always be hot enough to furnish sufficient hot water for the home.

Steam boiler operation. Oil-fired steam boilers operate somewhat differently from their hot water counterparts. Rather than responding to boiler-water temperature, the steam system cycles on and off based on pressure within the boiler. When the room thermostat calls for heat, the burner begins to operate.

(a)

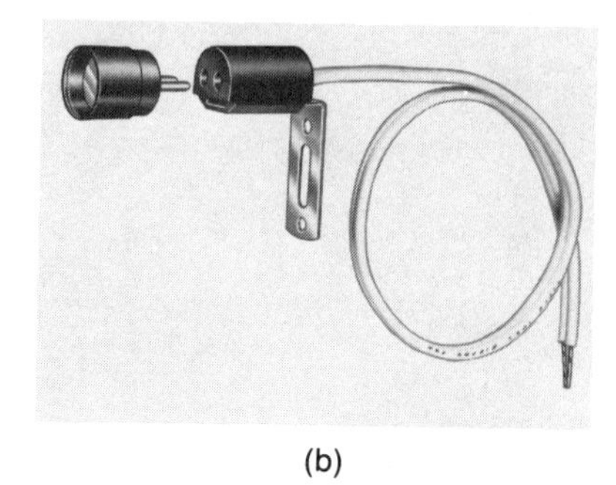

(b)

Cad cell

Ignition transformer

Air tube

Ignition electrode

Nozzle

(c)

Figure 3.17 (a) Oil burner aquastat. The cad cell is connected to the terminals marked F-F; the thermostat is connected to the terminals marked T-T (courtesy of Honeywell, Inc.); (b) cad cell (courtesy of Honeywell, Inc.); (c) cad cell location (typical).

The cad cell safety circuit is identical in both steam and hot water oil burners. Assuming that a flame has been established, the burner continues to operate, increasing the temperature within the boiler. As the water is turned into steam, pressure within the system rises from 0 psi at the beginning of a cold-start cycle to between 3 and 7 psi as steam is produced at full operating pressure. A pressure switch (Figure 3.18) shuts the boiler off when the high-pressure setting (usually 5 psi) has been reached. The burner turns back on when the system pressure drops to the cut-in setting (usually 3 psi below the cutout pressure, which is adjustable). This switch is set by selecting the cut-in pressure plus a specified differential (usually 3-psi cut-in with a 2- or 3-psi cutout differential). In this way, the burner will turn on when the boiler pressure reaches 3 psi (assuming that the thermostat is calling for heat) and will shut the burner off when the boiler water pressure reaches 5 psi. Operating pressures within steam boilers are limited to a maximum of 15 psi, compared to maximum pressures of 30 psi for hot water boilers.

Operation of the oil burner is also dependent on a minimum water level in the boiler. An automatic low-water cutoff (Figure 3.19) opens the electrical circuit to the burner should the boiler water fall below a safe level.

(a)

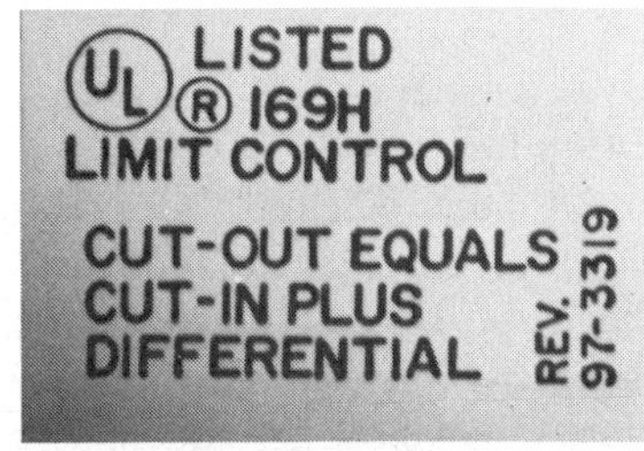

(b)

Figure 3.18 (a) Steam boiler pressure switch cut-in setting; (b) determination of steam pressure cutout setting. (Courtesy of Honeywell, Inc.)

Figure 3.19 Low-water steam cutoff switch. (Courtesy of ITT Fluid Handling Division.)

Depending on whether the steam piping is of the one- or two-pipe design, steam rises into the radiators, where it condenses, releasing its latent heat of vaporization, and returns to the boiler via the piping circuit.

If the steam boiler has a tankless coil for domestic hot water production, the boiler will maintain a low-limit temperature of approximately 140 to 160°F, unless the thermostat calls for space heating.

Oil-Fired Hot Water Heaters

Perhaps the most efficient way to heat domestic hot water, oil-fired water heaters use high-speed flame retention burners firing at low-gal/hr rates that yield very high operating efficiencies. Although oil-fired water heaters are available in tank sizes ranging from 30 to 100 gal, most residential heaters use 30-gal storage tanks and are capable of recovering from 100 to 120 gal of hot water per hour. A typical residential oil-fired water heater is illustrated in Figure 3.20.

When water temperature falls below the aquastat setting on the heater, con-

Figure 3.20 Residential oil-fired water heater. (Courtesy of Bock Water Heaters, Inc.)

tacts within the aquastat close the electrical circuit to the oil burner. Wired through a cad cell safety circuit, burner operation will continue assuming that a flame has been established, until the water within the heater reaches the aquastat set temperature. This aquastat also contains a high-limit setting that will turn off the burner should the temperature rise to this point.

Heat distribution from the combustion chamber is aided by the use of baffles within the flue passage of the appliance (Figure 3.21). The heater should be vented into a suitable chimney for safe removal of effluents.

Figure 3.21 Baffles used in water heater flue pipe to promote turbulence of the exhaust gases and to aid in transferring heat to the water within the heater (typical).

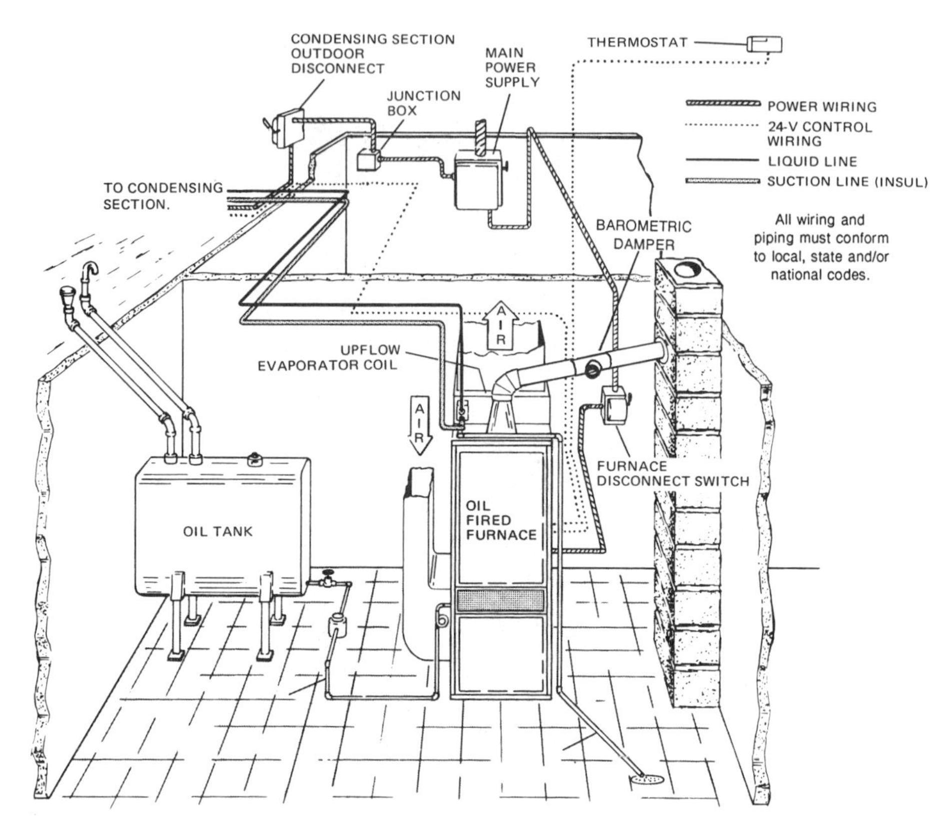

Figure 3.22 Installation of a typical residential hot air oil-fired furnace equipped for central air conditioning as well. (Courtesy of York International Corp.)

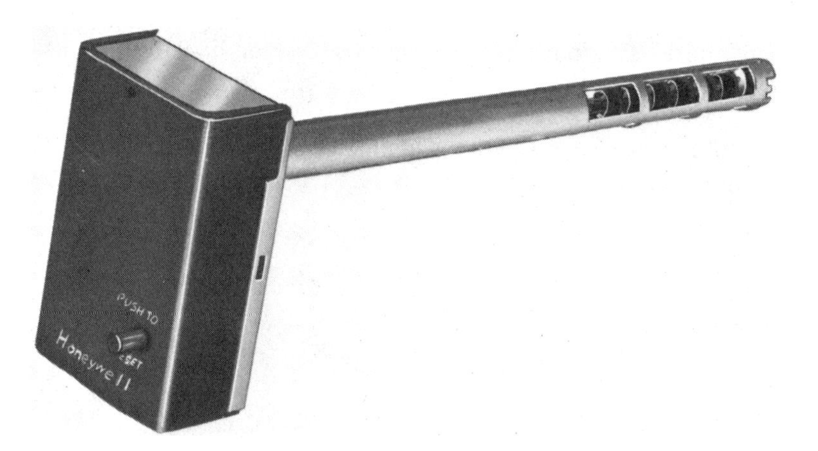

Figure 3.23 Fan-limit switch used on hot air furnaces. These switches serve to turn the blower on and off, based on the temperature of the air within the hot air plenum. Most of these switches also incorporate a high-limit contact that will deenergize the burner should the temperature reach the high-limit set point. (Courtesy of Honeywell, Inc.)

Oil-Fired Hot Air Furnaces

Figure 3.22 illustrates the installation configuration of a typical residential oil-fired hot air furnace. Due to the relatively large heat exchanger surfaces and room required by the blower and associated ductwork, hot air furnaces require a great deal more space than their hydronic or steam counterparts.

During operation, the thermostat initiates burner ignition to raise the temperature of the heat exchanger. A fan limit switch (Figure 3.23) senses an increase in heat exchanger temperature and activates the blower when temperature in the hot air plenum reaches approximately 150 to 160°F.

The ignition safety circuit in the burner relies on either a cad cell or a stack switch. The cad cell operates in the same way in either hot water, steam, or hot air oil burners. The stack switch is somewhat different in operating design. The switch is located on the exhaust, or stack pipe, of the oil burner. A bimetallic helical coil rotates as it is heated by the exhaust temperature in the stack pipe. As the coil rotates due to the heat of combustion from the oil burner, it disengages a safety circuit within the switch, which allows burner ignition to continue. Should a loss of flame occur, the coil will rotate in the opposite direction as it cools, energizing the safety circuit to open the electrical supply line that supplies the oil burner. These switches have a manual reset button for use when the safety circuit locks out the electrical supply. Figure 3.24 illustrates the helical coil arrangement that is common to many stack switch safety circuits.

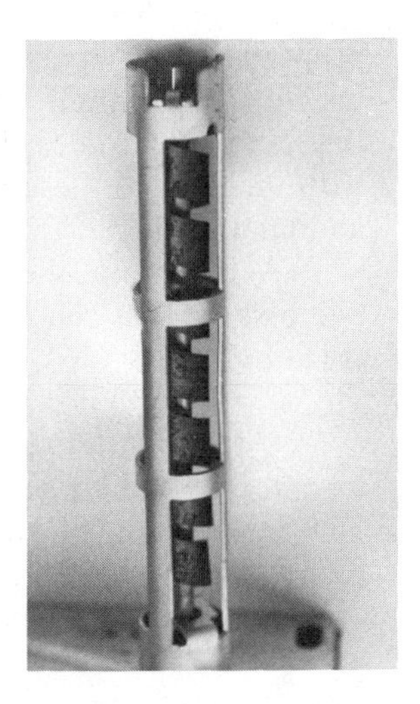

Figure 3.24 Stack switch used for proving ignition. These switches incorporate a safety circuit that will automatically lock out the burner electrical circuit if heat is not sensed by the switch within approximately 30 seconds of burner initiation (note the reset button on the switch box cover). (Courtesy of Honeywell, Inc.)

MAINTENANCE PROCEDURES FOR OIL-FIRED APPLIANCES

Of the various types of residential fossil-fuel heating technologies, oil-fired appliances require the most maintenance. First and foremost of these procedures is the tuning and servicing of the burner, including a combustion efficiency test to ensure maximum operating efficiency. Other maintenance procedures involve examination and replacement of components associated with the specific appliance. Our discussion begins with the maintenance of the oil burner and its component parts.

Oil Burner Maintenance

Oil burner maintenance, usually performed by experienced service personnel, can also be done by the homeowner once the required techniques are understood.* The discussion will focus on the various parts of the oil burner that must be inspected and serviced on an annual basis.

Ignition system. The oil burner nozzle is the most critical component in the system for achieving efficient oil combustion. It is good practice to replace the nozzle on a yearly basis, regardless of the amount of fuel used. The nozzle is removed by first turning off the oil supply and electricity to the burner, raising the transformer cover, and disconnecting the nozzle and electrode assembly from the burner housing (Figure 3.25).

The nozzle can be removed with an ordinary open-end wrench, although a nozzle wrench simplifies this procedure. The replacement nozzle must be the proper size and spray pattern configuration for the appliance. If in doubt as to nozzle specifications, the manufacturer of the unit should be consulted. The setting of the electrodes should be done at this time, by consulting the instructions with the unit. If these are unavailable, those in Figure 3.26 can be followed for an initial setting.

At this time, the air intake grates located on the mounting base of the oil pump should be checked to make sure that they are free of dirt or obstructions (Figure 3.27).

Oil filter assemblies. All oil burners should be equipped with a filter assembly, usually located either at the output of the oil tank (assuming an indoor tank installation) or plumbed into the oil feed pipe circuit upstream of the fuel pump (see Figure 3.3).

To gain access to the filter, the housing assembly must be opened. Filters

* Some localities may have local ordinances that govern the servicing of home heating equipment. The building inspector should be consulted to determine if any of these ordinances are applicable in the specific locality of residence.

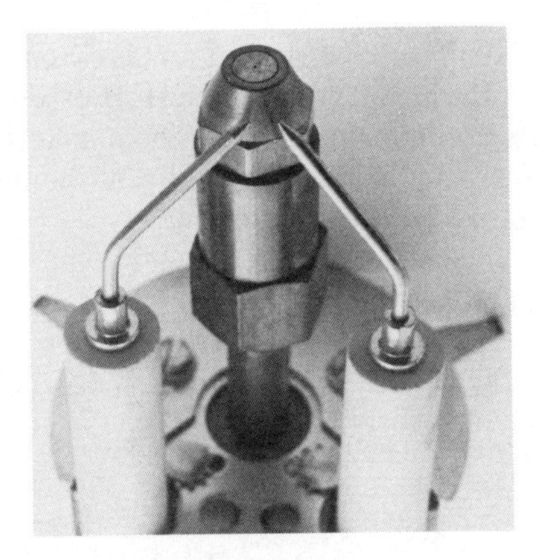

Figure 3.25 Nozzle and electrode assembly removed from burner unit. When removing this assembly, care must be taken not to apply excess pressure to electrodes, which can cause misalignment of electrode tips and cracked porcelain insulators, rendering the unit inoperable.

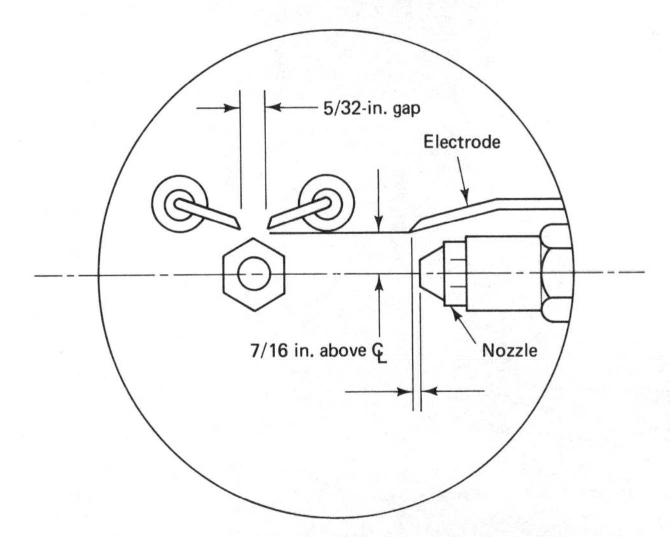

Figure 3.26 Gapping and set dimensions of typical oil burner nozzle/electrode assembly. Exact settings may vary from one manufacturer to another. Figures are composites.

Figure 3.27 Air intake damper on oil burner motor must be cleaned periodically to promote unrestricted combustion air being delivered to the combustion chamber.

should be changed annually. The use of clean filters is essential for proper system operation, since removal of the smallest dirt particles is essential to prevent blockage of the narrow passages within the nozzle. The smallest amount of dirt can cause a malfunction. Depending on the complexity of the heating system, some units may have more than one filter in the circuit.

Combustion Efficiency Testing

Testing the combustion efficiency of the oil burner is done after the components have been serviced. Efficiency tests should be taken at least once during each

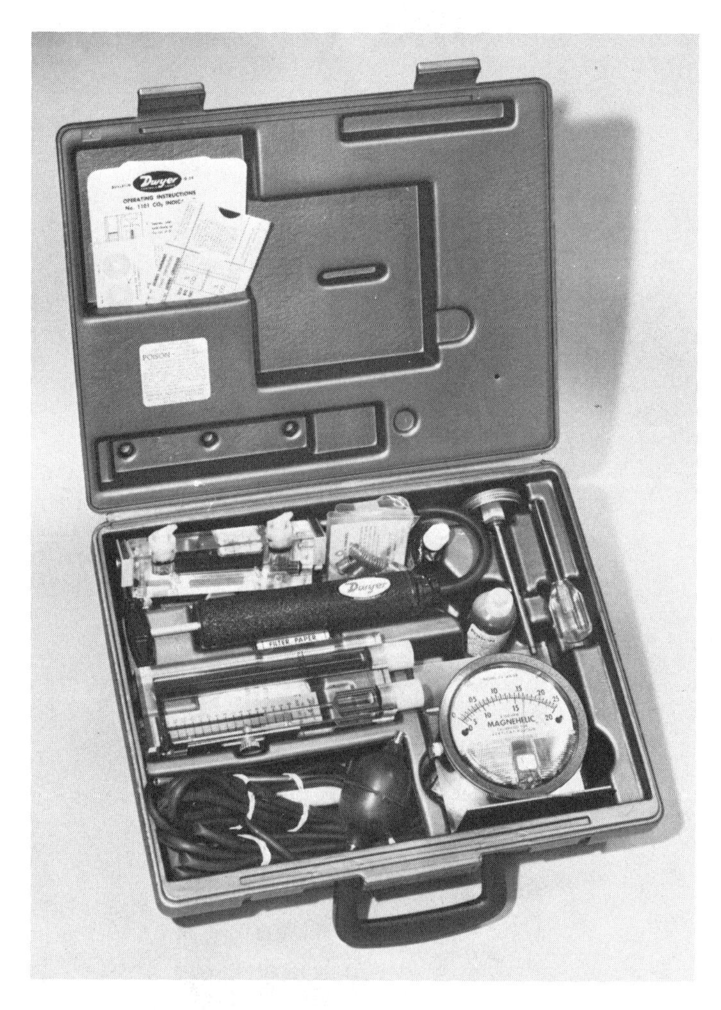

Figure 3.28 Combustion efficiency test kit. (Courtesy of Dwyer Instruments, Inc.)

heating season. The use of a special combustion efficiency test kit is required for these procedures (Figure 3.28). To aid in performing these efficiency tests, a small sampling hole should be made in the stack or vent pipe located between the draft damper and the flue collar on the unit (Figure 3.29). Maximum efficiency is based on the results of the carbon dioxide test, with the burner operating at minimum smoke levels at proper stack temperature and draft conditions. A draft measurement is first taken after the burner has been serviced.

Draft measurements. Draft within the chimney is dependent on the evacuation or negative pressure within the chimney, which is responsible for removing the effluents and by-products of combustion. If chimney draft is excessive, excess heat will be sucked out of the appliance, reducing overall heating efficiency. Conversely, if chimney draft is not strong enough, backpuffing of the unit can occur. Also, built-up heat and combustion by-products will not be removed at the proper rate, causing poor combustion and eventual overheating of the unit. Draft is meas-

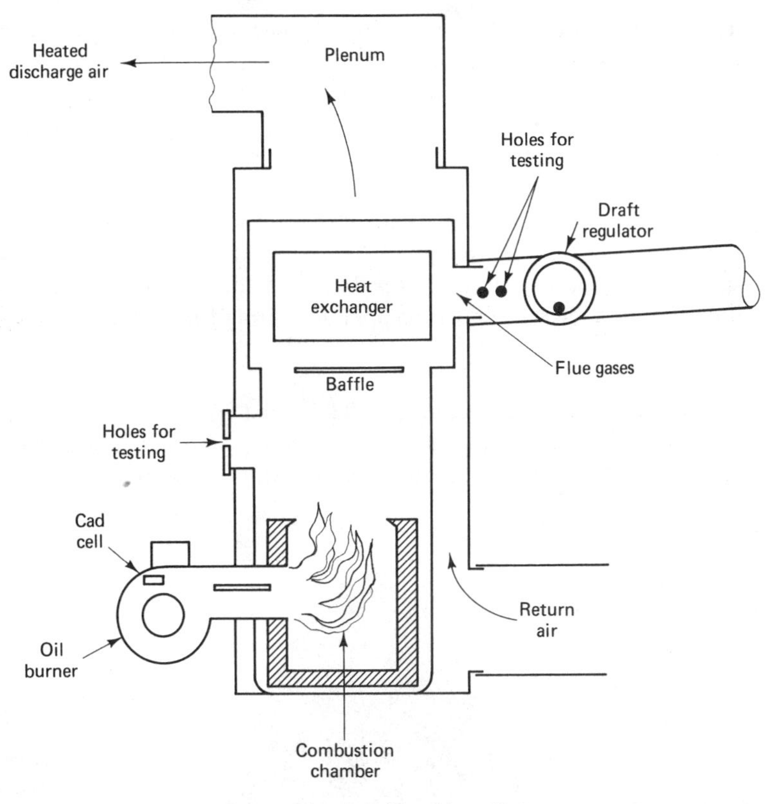

Figure 3.29 Location of sampling hole for taking efficiency test exhaust gas samples. (From W. B. Cooper et al., *Warm Air Heating for Climate Control,* Prentice-Hall, Englewood Cliffs, N.J., © 1980, Fig. 10-7, p. 199. Reprinted by permission of Prentice-Hall, Inc.)

Figure 3.30 Manometer used for taking appliance draft measurements. (Courtesy of Dwyer Instruments, Inc.)

ured in inches of water column (w.c.), with a device called a **manometer**. A manometer for draft measurement testing is illustrated in Figure 3.30.

To perform this test, the burner should be turned on and allowed to run for several minutes. The sensing probe of the manometer is inserted into the sampling hole and the reading is taken. The proper draft for most oil-fired appliances falls between 0.02 and 0.03 in. of water column. If the draft reading falls outside these limits, the barometric draft damper should be reset until the proper readings can be obtained. A barometric draft damper of this type is illustrated in Figure 3.31. The location of the draft damper is critical for proper operation. Figure 3.32 illustrates the proper and improper locations for conventional barometric dampers.

Carbon dioxide testing. The most important test of all the efficiency procedures is the carbon dioxide test, which is an indication of how much excess air is present in the combustion chamber during burner operation. When burned

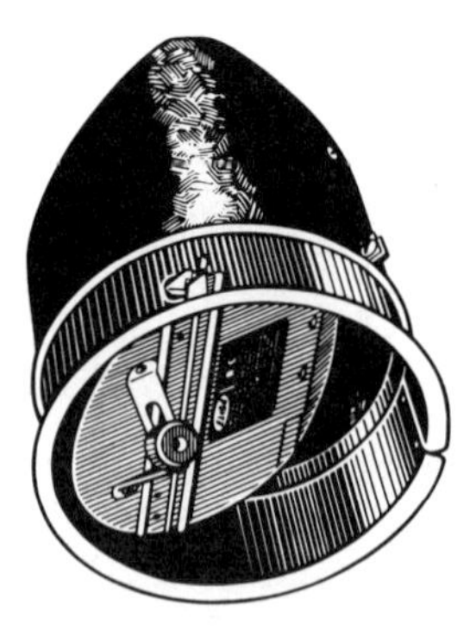

Figure 3.31 Barometric draft damper. These dampers are adjustable by changing the location of a small counterweight on the door. The damper should be adjusted to maintain an overfire draft of 0.02. (Courtesy of Field Controls Corp.)

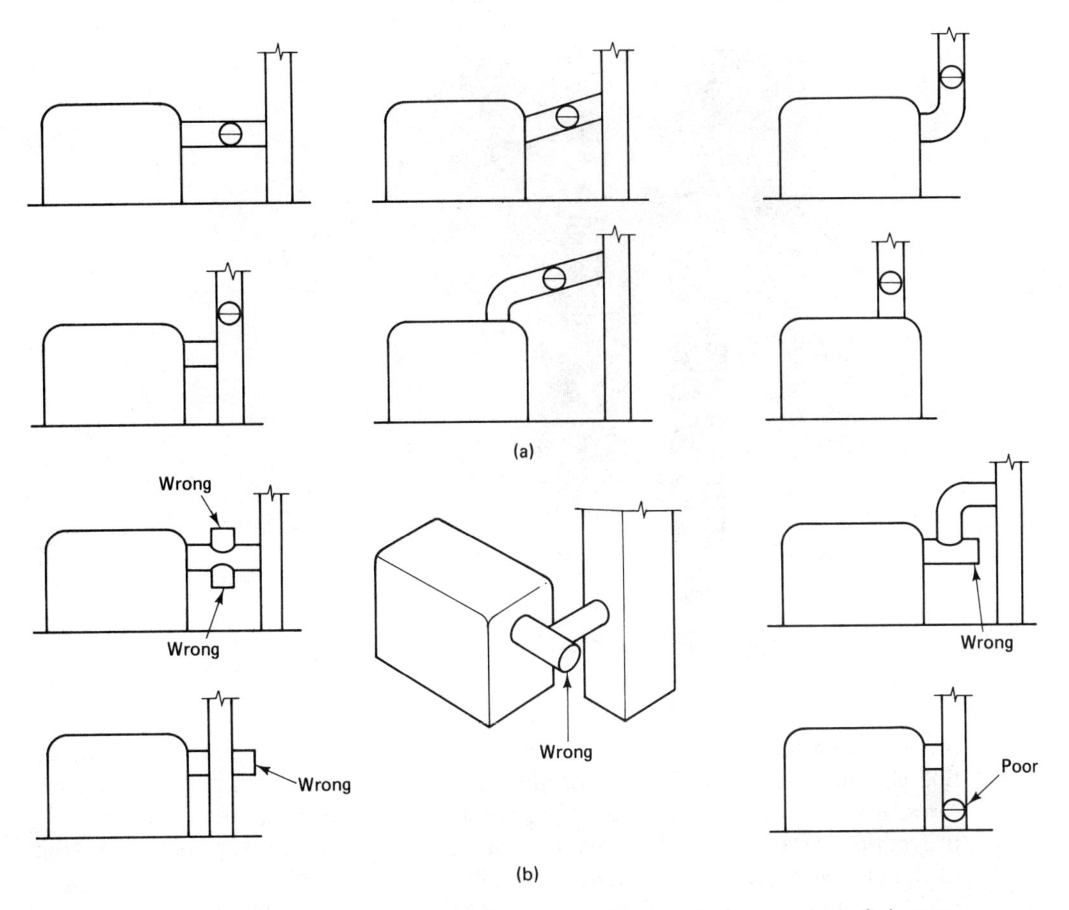

Figure 3.32 (a) Proper and (b) improper locations for barometric draft dampers.

in air composed of 21% carbon and 79% nitrogen, oil yields 21% carbon dioxide by volume in the stack gases, assuming a combustion efficiency of 100%. The efficiency of the oil-burning process, however, depends on excess air being available at the nozzle. If 100% excess air is now supplied to the burner, the resulting percentage of carbon dioxide will be 10.5% (once again assuming 100% combustion efficiency).

The carbon dioxide test is performed by passing a known quantity of flue gas through a solution that can absorb large quantities of carbon dioxide (the fluid used in most carbon dioxide testers is potassium hydroxide). A change in test fluid volume occurs in proportion to the amount of carbon dioxide absorbed by the fluid. The percentage of carbon dioxide in the flue gas is then indicated on a calibrated scale. A carbon dioxide absorption tester of this type is illustrated in Figure 3.33.

When performing this test, the sampling tube of the tester is inserted into

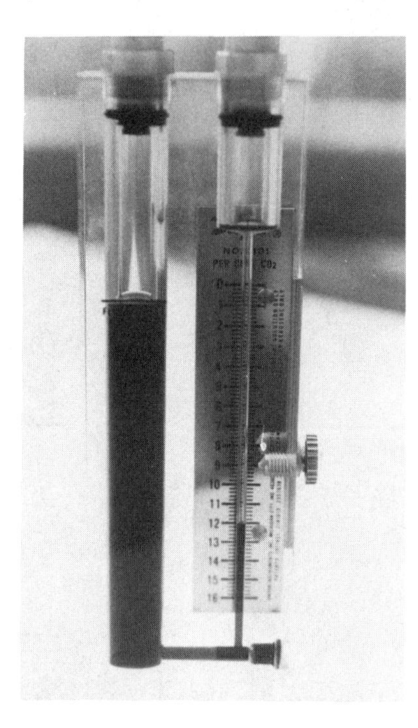

Figure 3.33 Carbon dioxide tester. The tester in this illustration reads 12%. Prior to each sample, the tester must be recalibrated to 0%. (Courtesy of Dwyer Instruments, Inc.)

the sampling hole and a small hand pump is used to draw a specific volume of flue gas into the tester. It is important to follow the manufacturer's instructions to ensure that the precise amount of gas is pumped into the tester during this procedure. After drawing the gas, the test fluid is swirled to ensure proper mixing of the gas with the test fluid prior to reading the test results.

Interpretation of the carbon dioxide test can help to identify specific combustion problems with the oil burner relating to excess air adjustment. If too much excess air is present, the carbon dioxide levels in the flue gases will be relatively low (between 5 and 6%); flame temperature will be low; and fuel consumption will be high. The proper amount of excess air present for combustion is a balancing act: Too little excess air results in high smoke levels, causing sooting of internal heat transfer passages, reducing the efficiency of the unit; too much excess air results in low flame temperature and excess smoke as well. To determine the smoke levels present in the combustion process, a smoke test should be taken with the appropriate sampling device (Figure 3.34).

Smoke testing. The smoke test is similar in principle to the test for carbon dioxide: A sampling device is used to pump a specified amount of flue gas through a piece of sampling paper, leaving a residue on the paper that is proportional to the amount of smoke in the sampling gas. The shade of the residue is then compared to a standardized smoke-level chart to determine the exact amount of smoke contained in the gas, measured on a scale from 1 to 10. A sampling chart of this

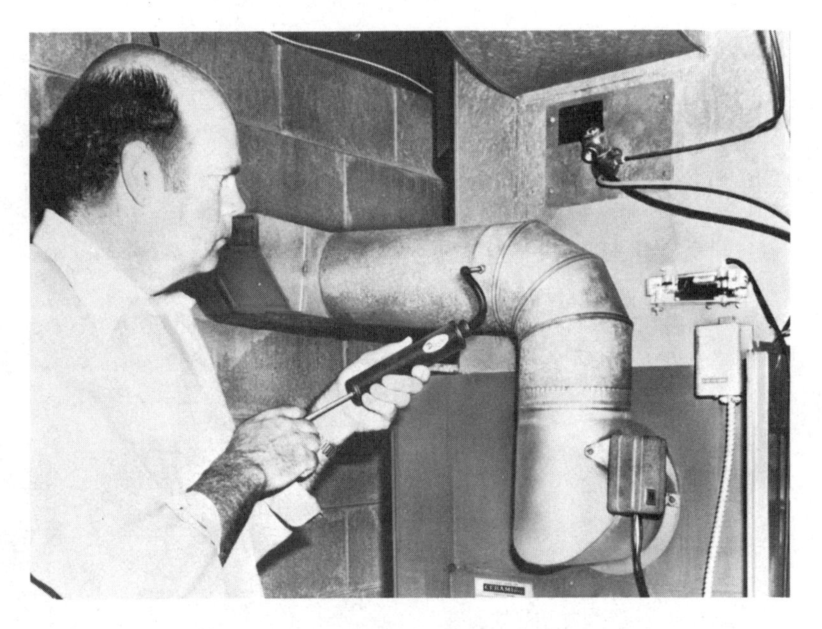

Figure 3.34 Using a smoke sampler to determine smoke levels in exhaust gases. (Courtesy of Dwyer Instruments, Inc.)

type is illustrated in Figure 3.35. No. 1 smoke is the maximum allowable smoke for pressurized oil-fired boilers and hot water heaters; No. 2 is the maximum for hot air furnaces. Older oil-fired units may read even higher.

Stack temperatures. After the smoke test has been taken, the stack temperature of the unit should be measured using a specially designed stack thermometer (Figure 3.36). Stack temperatures should be taken only after the appliance has come up to full operating temperature. When testing oil-fired water heaters, the water within the tank should be hot; boilers and furnaces should run through several heating cycles prior to measuring stack temperatures.

Determining combustion efficiency. After all measurements have been taken, the combustion efficiency is calculated by positioning a slide rule calculator supplied with all test kits at the proper settings based on the results of the carbon dioxide and stack temperature readings. The combustion efficiency is then displayed on the calculator (Figure 3.37).

Appliance Maintenance

Maintenance procedures on oil-fired appliances exclusive of the burner are simple to perform. To keep the appliance in top operating condition, these procedures should be performed periodically.

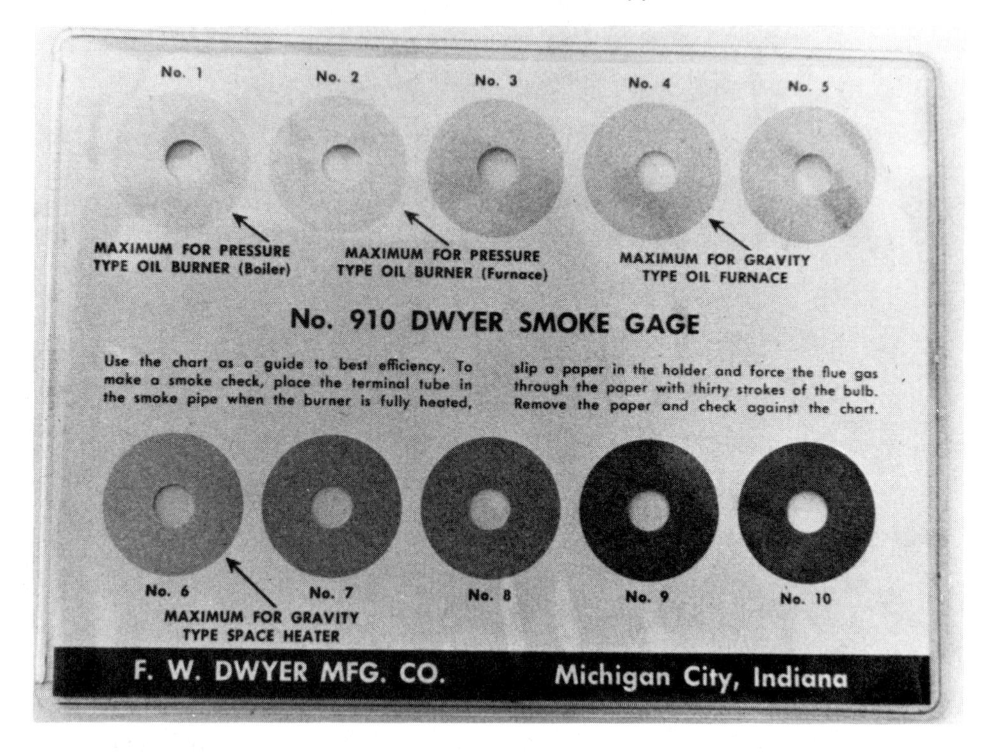

Figure 3.35 Smoke chart. To determine smoke levels, the sample paper from the sampling device is placed underneath the smoke chart until the smoke level shade of the sample and the gauge number is the same. Most residential heating appliances should operate with no more than No. 1 smoke. (Courtesy of Dwyer Instruments, Inc.)

Figure 3.36 Stack thermometer.

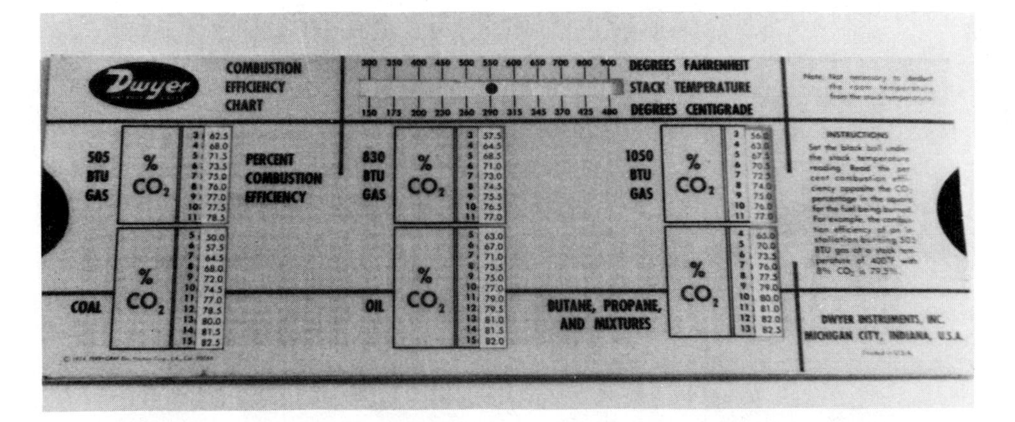

Figure 3.37 Combustion efficiency calculator. The stack temperature reading, based on a maximum of No. 1 smoke, is set on the top of the calculator. The appliance efficiency is then read from the oil/CO_2 scale based on the carbon dioxide level measured.

Boilers. All boilers require periodic cleaning of the flue passages and fire tubes to remove soot buildup. A removable panel that allows access to these passages is identified in the literature that accompanies most boilers when they are installed (Figure 3.38). To clean these passages properly, a flue brush sized for the boiler is used to loosen soot and ash from the internal baffles, and the residue is then removed with a vacuum cleaner.

If the system uses an oil-lubricated circulating pump, it should be oiled annually, according to the manufacturer's instructions. Circulators should not be overoiled. To do so can lead to premature failure of the circulator motor.

The proper operation of all safety valves, aquastats, and fill valves in the system should be checked periodically. The cold water system pressure should be between 12 and 15 psi. If the pressure is not within these limits, the fill valve should be adjusted according to the manufacturer's instructions to bring the system pressure to within acceptable limits. If, after adjusting the fill valve, the pressure still falls outside the settings noted above, the valve should be replaced.

The safety relief valve should be inspected to make sure that it is not leaking. It can be released manually occasionally to make sure that it will automatically close after discharge. If the valve shows any signs of deterioration or if it leaks, it should be replaced. The boiler aquastat should be checked to make sure that the burner shuts off at the specified high-limit setting. The unit should be observed through two or three complete heating cycles. A comparison of the readings between the aquastat and temperature gauge should be made to make sure that there is no great disparity in temperature readings (minor variations in temperature readings between the gauge and aquastat turn on/turn off points are normal, how-

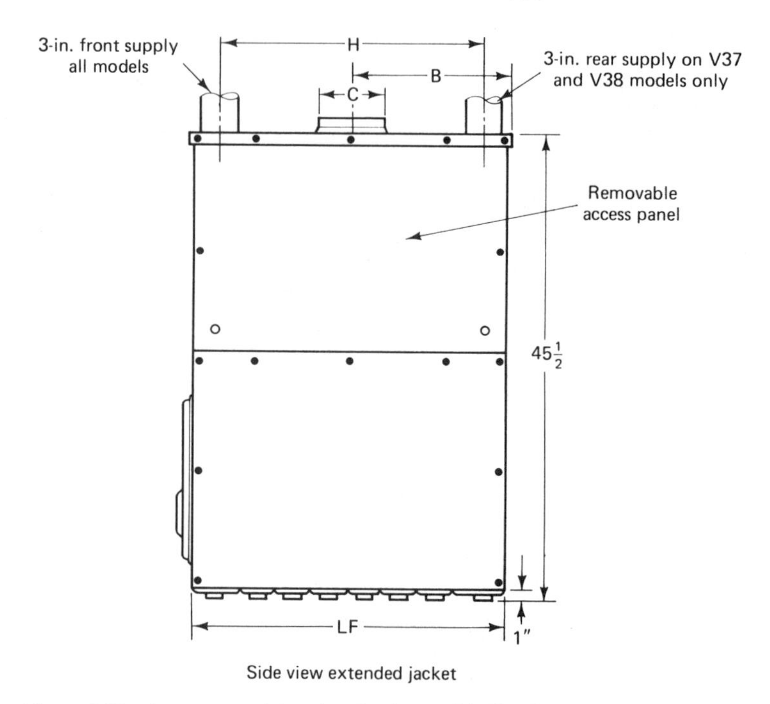

Figure 3.38 Access panel for cleaning internal boiler flue passages (typical).

ever). Differences between 5 and 10°F are not uncommon. Differences greater than this, however, should be investigated and the defective component replaced.

Steam boilers require a weekly flushing of the low-water cutoff to rid the unit of dirt and sediment that will accumulate in the operating environment of the system. The flushing procedure (Figure 3.39) ensures that sediment is removed and that the cutoff is functioning properly. To ensure proper operation of the low-water cutoff, the burner is turned on and while it is running the low-water cutoff is flushed. If the cutoff is working properly, lowering the water level within the cutoff will open the electrical contacts, turning off the burner. After the flushing is complete, if the cutoff valve contains an automatic-fill feature, the boiler will refill and the ignition sequence will resume when the water level reaches the cut-in point of the low-water safety valve. If the boiler requires manual refilling, the oil burner should resume operation when the water level reaches the required limit within the cutoff valve.

The steam boiler must be flushed periodically to prevent a buildup of scale and sediment. Most steam boilers have a flushing plug or fixture located at the top of the unit for this purpose. Flushing is done by using special additives recommended by the boiler manufacturer that are added to the boiler as it is filled and allowed to drain through the flushing fixture. Specific instructions should be followed carefully before putting the boiler back into service. When replacing the

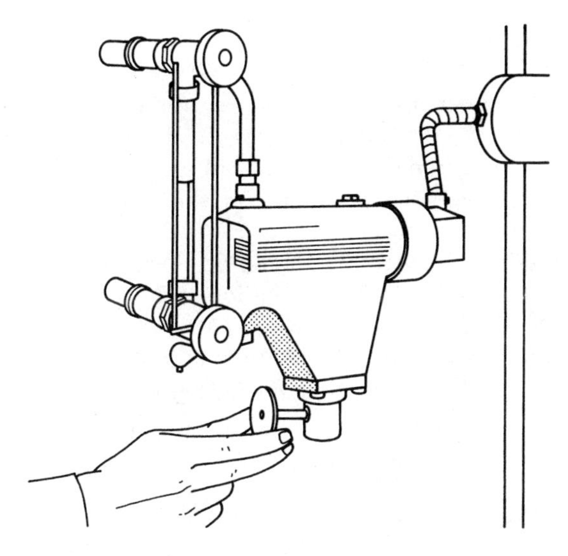

Figure 3.39 Flushing low-water steam cutoff.

threaded flush plug, use a combination of cotton wicking, Teflon tape, and Teflon pipe dope on all threaded connections greater than $\frac{3}{4}$ in. in diameter to make sure that the plug will not leak after it is replaced.

Furnaces. Maintenance of oil-fired furnaces is much less complicated than that of hot water or steam boilers. On a regular basis, air filters should be replaced to keep the static air pressure on the blower motor and fan assembly at a minimum, in addition to keeping the circulating air clean (Figure 3.40).

The heat exchangers on hot air furnaces, similar to the flue baffles on boilers, must be kept free of soot and ash and should be vacuumed annually. Access to

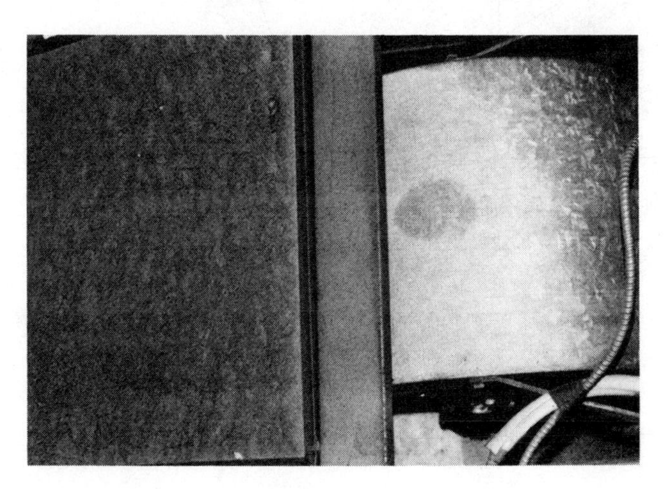

Figure 3.40 Replacing furnace air filters.

the heat exchanger is usually through a coverplate on the furnace or through the flue pipe box (Figure 3.41).

If the furnace contains either a humidifer or electronic air cleaner, these units should be serviced whenever the furnace is cleaned. Humidifier water pads should be cleaned to remove mineral deposits, which will clog the pad, preventing full water absorption. Air cleaner plates and filters should be inspected and cleaned on a weekly basis.

The blower and blower motor assembly should be checked monthly during the heating season. Areas of concern are the lubrication of the motor, if required, along with lubrication of the fan bearings and proper tensioning of the drive belt. Extra drive belts should be kept on hand. Drive belts tend to loosen up over a period of time, and the tension will need to be readjusted periodically. As a general

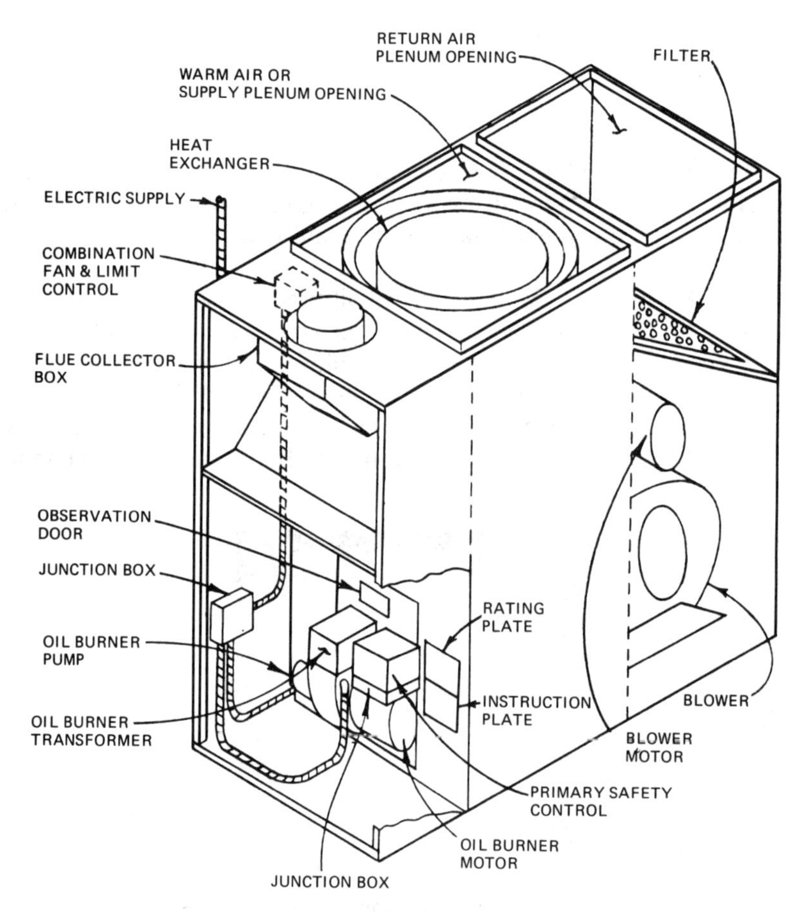

Figure 3.41 Locations of heat exchanger on (a) low-boy furnace and (b) counterflow furnace. (Courtesy of The Williamson Company.)

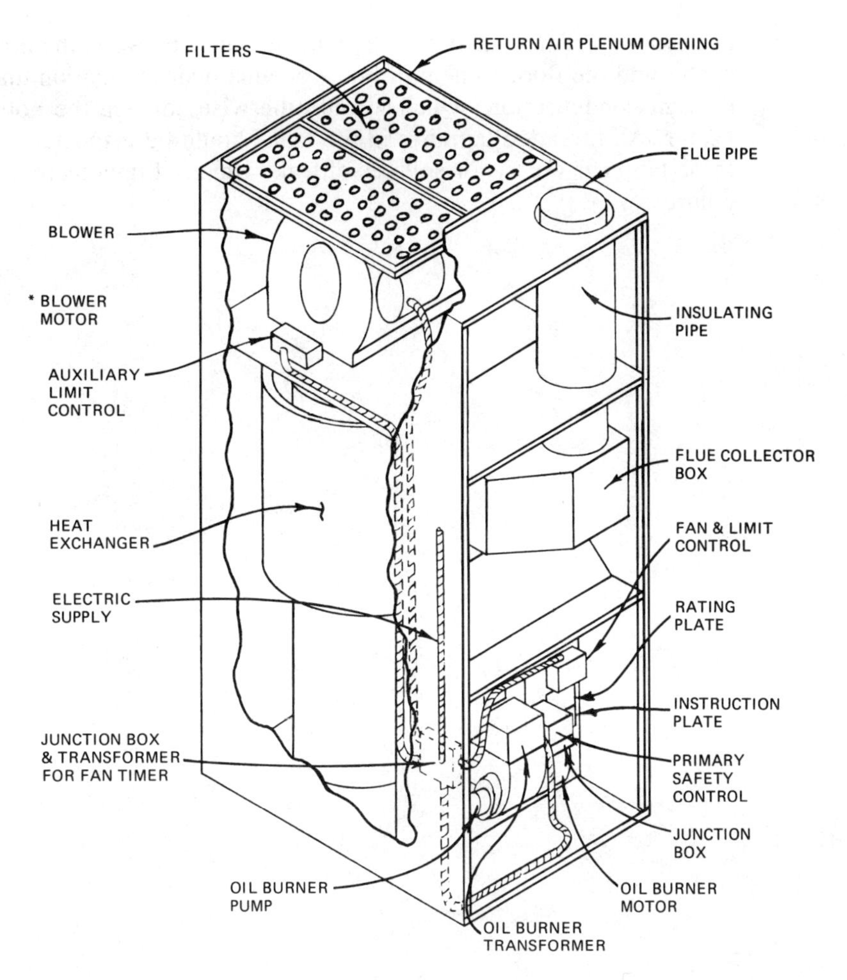

Figure 3.41 *(continued)*

rule the belt should have a flex in the middle of between $\frac{1}{2}$ and 1 in. During operation it should run smoothly without wobble or deflection. Overtightening of the belt, however, can place undue stress on the bearings in both the fan assembly and drive motor, causing premature bearing failure. A belt should be replaced when it can no longer be properly tensioned or when small cracks develop.

Hot water heaters. Oil-fired water heater maintenance requires annual cleaning of the flue baffle and combustion chamber. The tank should be flushed monthly to prevent a buildup of sediment, which would otherwise impede proper heat transfer. Also, the anode rod in the top of the heater should be changed occasionally. There must be sufficient clearance between the bottom of the water heater and the floor to help promote adequate air circulation under the heater to prevent condensation, which would otherwise corrode the bottom of the water heater. All threaded tappings in the tank should be inspected occasionally since these tappings are major sources of leaks which, if undetected, could cause tank failure.

4

OPERATION AND MAINTENANCE OF GAS-FIRED APPLIANCES

Gas-fired appliances differ markedly from oil-fired units with regard to combustion technology. Although the basic sequence of operation of gas-fired boilers, furnaces, and water heaters is essentially the same as with oil units, the nature of gas along with the variety of combustion systems available presents many different types of device configurations.

FUEL GASES

There are two types of fuel gas used in residentual heating appliances: natural gas and liquefied petroleum (LP) gas. Natural gas is made up largely of methane, developed over a period of millions of years from the decay of organic material. Although some natural gas deposits are discovered separately, most accompany crude oil deposits. LP gas is composed largely of propane and is a by-product of the refinement of crude oil. The characteristics of both natural and LP gas are now examined in more detail.

Characteristics of Natural Gas

Natural gas has a heating capacity of approximately 1000 Btu/ft^3. As it comes from the well, the gas has no odor, which is added later to enable leak detection. During the combustion of natural gas two major by-products are formed: carbon dioxide (CO_2) and water vapor. If the gas is not completely oxidized during the combustion products, carbon monoxide, an odorless lethal gas, is produced, along with other objectionable aldehydes and soot. Complete combustion is essential if the gas is to be safely used for residential heating applications.

Natural gas will burn only under the proper set of conditions relating to the amount of air and gas available for combustion. These upper and lower limits of the combustion mixture range from 4 to 14% gas to air. At gas concentrations of less than 4%, too few gas molecules are interspersed within the oxygen molecules for continued flame propagation; in mixtures of greater than 14% gas, too many gas and too few oxygen molecules are available to support combustion.

Although there are many ways to burn gas, most of it is burned under atmospheric conditions in one of two ways, known either as a yellow flame atmospheric burner or a blue flame atmospheric burner.

A typical **yellow flame atmospheric burner** is illustrated in Figure 4.1. All combustion air in a yellow flame burner is drawn from the air surrounding the flame. No air is premixed with the gas prior to combustion. Yellow flame burners are most commonly used for low-Btu-input pilot burners and small illuminating lamps. It is important that the flame not come into contact with any surface cooler than 1200°F, to prevent sooting and incomplete combustion.

Blue flame burners (Figure 4.2) premix primary air with the gas prior to its combustion in the burner. The flame draws surrounding secondary air as needed. Blue flame burners usually have an air shutter arrangement that adjusts the amount of primary air that can be mixed with the incoming fuel gas and are commonly found on residential hot water heaters, gas stoves, and heating boilers and furnaces. Whereas the outer flame of the burner can come into contact with surfaces cooler than 1200°F, the inner blue flame should not.

Characteristics of Liquefied Petroleum Gas

LP gas varies significantly from its natural gas cousin. Essentially odor-free in its natural state, LP gases are chemically odored for leak detection. The heating content of LP gas ranges between 2000 and 2500 Btu/ft^3. Since the heating value

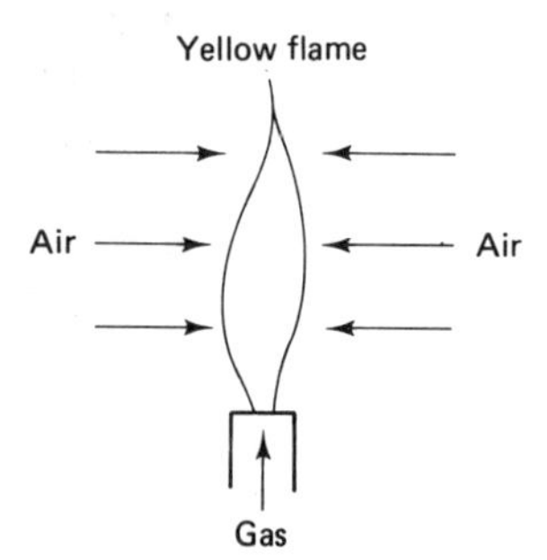

Figure 4.1 Yellow flame atmospheric burner.

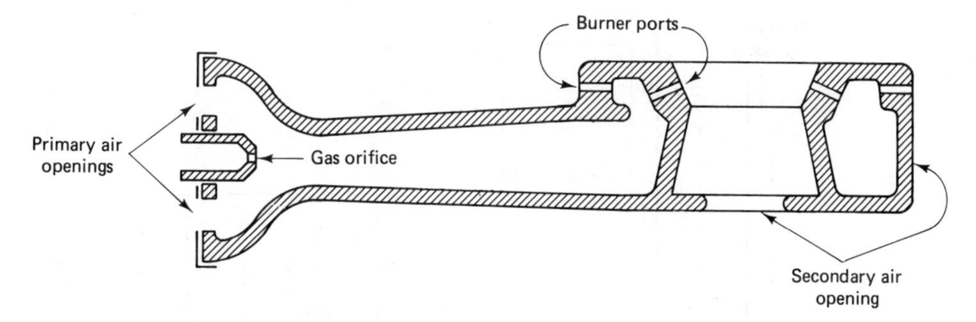

Figure 4.2 Blue flame atmospheric burner.

of the gas is approximately twice that of natural gas, smaller appliance orifices are used when LP is the fuel gas to obtain equivalent Btu appliance inputs.

The limits of the combustion mixture for LP gas range from 2.4 to 10%, with a concentration of approximately 5% gas yielding maximum flame propagation. Both yellow and blue flame burners are used in LP appliances; however, the burner orifices along with the pressure within the gas manifold are modified to take into account the higher Btu rating per gallon of LP gas.

One other major difference in the two gases lies in their relative densities. Natural gas is lighter than air and therefore disperses quickly. LP gas is heavier than air and will gather in the lower confined areas, making LP gas less safe than natural gas to handle and use.

Natural gas is the fuel of choice in urban and high-density suburban areas that are equipped with centralized pipeline distribution facilities. LP gas is generally classified as a rural fuel and must be stored in tanks adjacent to the residence. The tanks are filled by large delivery trucks on an automatic basis at periodic intervals calculated to keep the customer supplied with adequate fuel. In many respects, the use of LP gas and fuel oil are similar, given their need for on-site storage and automatic delivery.

Gas Pressure Requirements

All gas-fired burners depend on delivering the gas to the burner orifices at the correct operating pressure. The pressure requirements for natural gas and LP gas appliances are different, a function of their respective heating values and different densities. Gas pressure is measured and calculated in inches of water column (w.c.) in a device called a manometer. Figure 4.3 illustrates a manometer used to take the pressure measurement from the gas pressure regulator.

Natural gas can feed into the house at approximately 25 psi; the residential pressure regulator will reduce this to 7 in. (w.c.). Individual appliance regulators will reduce this pressure to between 3 and 5 in. w.c.

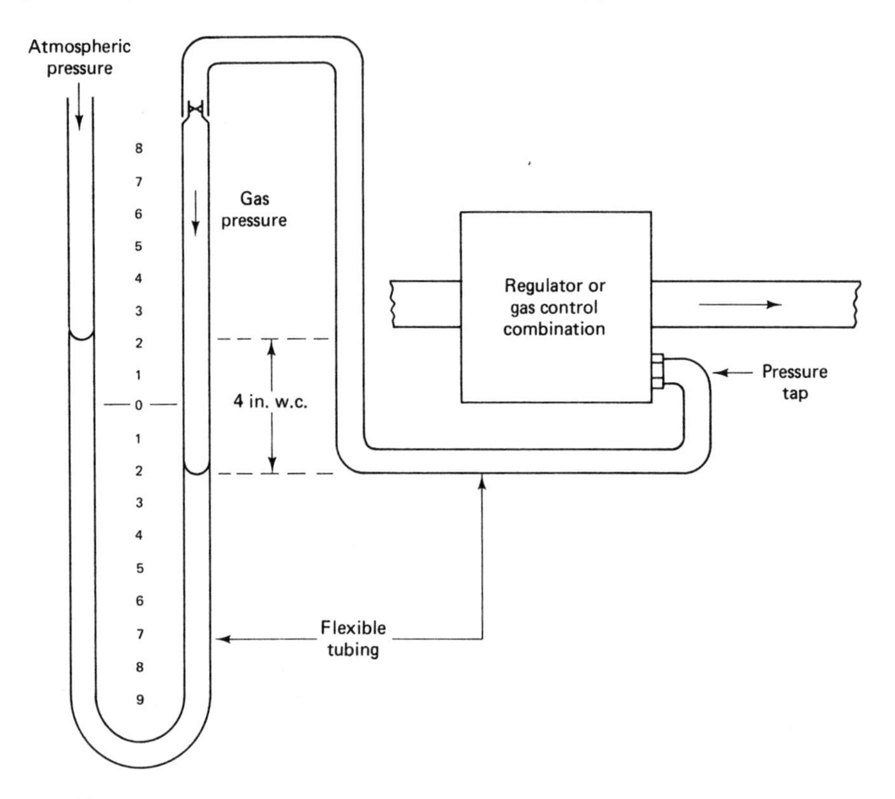

Figure 4.3 Principle of operation of a manometer used to take gas pressure measurements. Here the manometer reads 4 in. of water column (w.c.).

LP gas is stored at approximately 125 psi in the storage tank. This pressure will vary based on the outside temperature. The regulator on the storage tank reduces this pressure to approximately 11 w.c. for individual appliances. Most LP burners have individual regulators to maintain gas pressure at 11 w.c. The higher pressure in LP gas burners versus natural gas systems is due to the greater density of the LP gas, which requires an increase in gas velocity as it enters the burner orifice. This in turn increases the amount of primary air injected into the air/gas mixture, promoting turbulence and combustion efficiency.

CLASSIFICATION OF GAS COMBUSTION APPLIANCES

Gas can be combusted in a number of ways. The variations of combustion technology apply to most conventional boilers and furnaces. Gas water heaters are almost always of the conventional atmospheric type, which is examined next.

Conventional Atmospheric Combustion Technology

The most prevalent types of combustion technology used in gas appliances is of the atmospheric design configuration. A hot air furnace utilizing this type of burner is illustrated in Figure 4.4. Atmospheric burners are available in several designs, shown in Figure 4.5.

Most appliances using atmospheric combustion incorporate several burners that draw their gas from a supply manifold. Almost all burners of this type are blue flame burners, and each is equipped with its own air shutter for primary air adjustment. The flame pattern of these burners is illustrated in Figure 4.6. Note from the figure that the flame pattern is stable and is characterized by a clearly defined inner blue cone, an outer blue cone, and a mantel with some yellow tipping. The flame itself should not be yellow, but distinctly blue in color. Yellow tipping of the flame is normal, as are occasional yellow streaks caused by dust particles in the gas stream.

Atmospheric combustion uses one of two types of ignition systems which initiate main burner operation and monitor the burner, shutting off the gas flow in an emergency loss of flame situation. The two types of ignition systems in use are classified as either standing pilot or spark ignition.

Standing pilot ignition systems. A **standing pilot** ignition system is illustrated in Figure 4.7. The standing pilot light serves a dual function: It lights the main burners on a call for heat by the appliance and serves to shut off gas flow should a loss of pilot light occur.

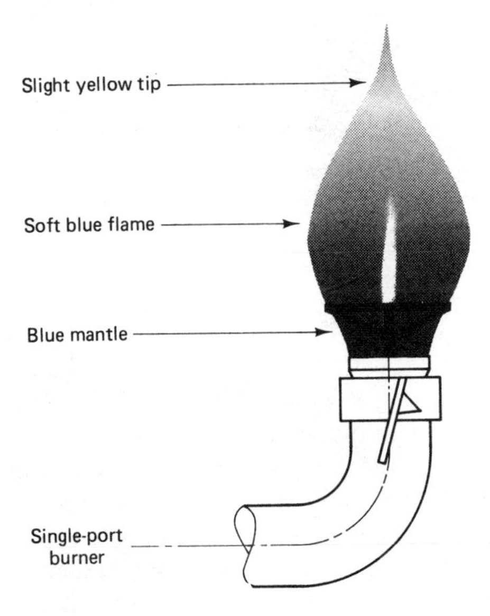

Figure 4.4 Atmospheric combustion burner of the type used on gas-fired warm air furnaces. (Courtesy of The Williamson Company.)

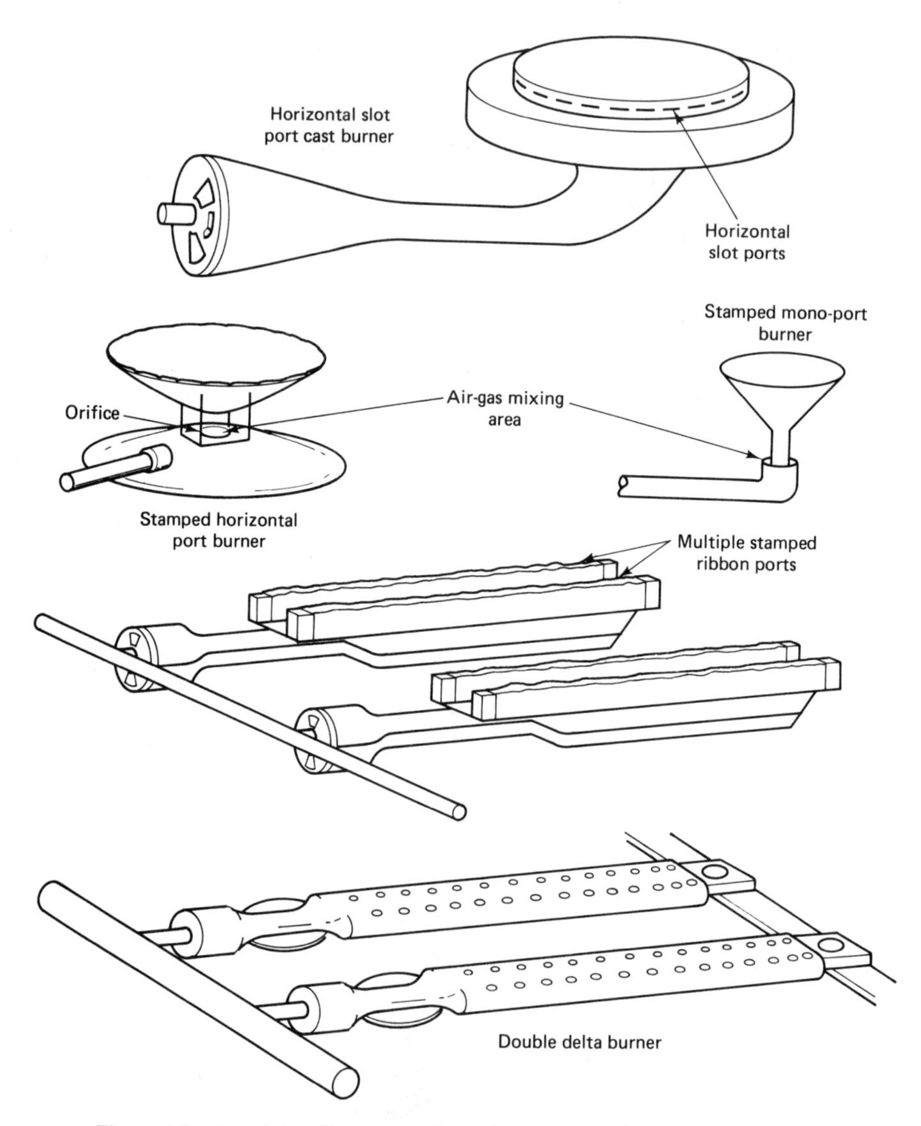

Figure 4.5 A variety of burner configurations are used in gas-fired appliances. (Courtesy of Robertshaw Controls Company.)

During normal operation, gas passes through the automatic pilot valve on its way to the main burner manifold. Should the pilot light be extinguished for any reason, the valve will shut off the gas flow to both the main burners and the pilot light. This is known as 100% shutdown and is required on all LP gas appliances. Since the LP gas is heavier than air, uncombusted gas from the pilot assembly would collect at the bottom of the appliance rather than dispersing, as natural gas would do in the same situation. Non-100% shutdown is sometimes used in natural gas appliances, where only the main burner gas is shut off, leaving the pilot gas flowing. In most instances, however, 100%-shutdown arrangements are the norm.

The conventional standing pilot ignition system is of the autopilot type, in which the pilot flame is directed against a thermocouple. The thermocouple generates a small amount of electricity of sufficient power to energize an electromagnet in the gas valve assembly. As long as the pilot light burns against the thermocouple, electricity is generated to keep the gas valve open. If the pilot light goes out, electrical power from the thermocouple ceases, deenergizing the gas valve, shutting off the flow of gas to the main and pilot burners. Figures 4.8 and 4.9 illustrate the thermocouple/valve assembly during both standing-flame and loss-of-pilot-flame conditions.

The action of the standing pilot ignition system is triggered by operation of the thermocouple. The thermocouple is a device composed of two dissimilar metals joined at each end. If one of the ends is maintained at a higher temperature than the other, an electrical current is established in the thermocouple. This electricity is used to power a small electromagnet that holds the gas valve open during normal operation. A simple thermocouple is illustrated in Figure 4.10.

A variation of the standing thermocouple pilot is known as a thermopile pilot system. A **thermopile** is a design configuration consisting of several thermocouples wired in series to increase the electricity generated by the system so that it has sufficient power to operate the entire gas valve assembly without external elec-

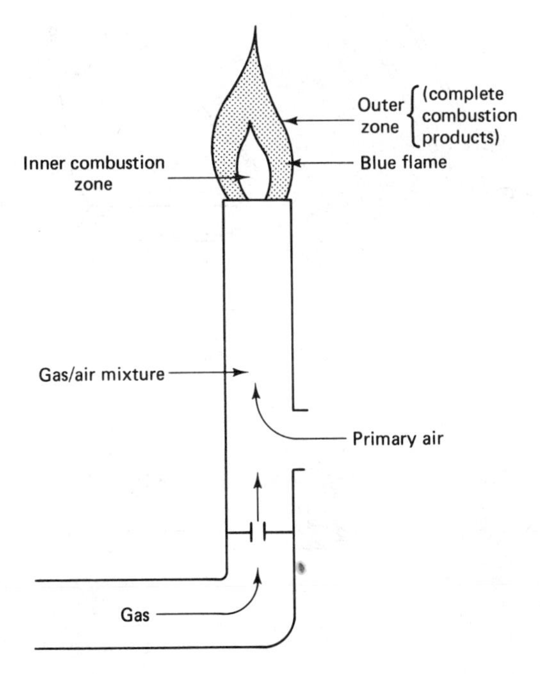

Figure 4.6 Blue flame atmospheric combustion flame pattern. (Courtesy of Robertshaw Controls Company.)

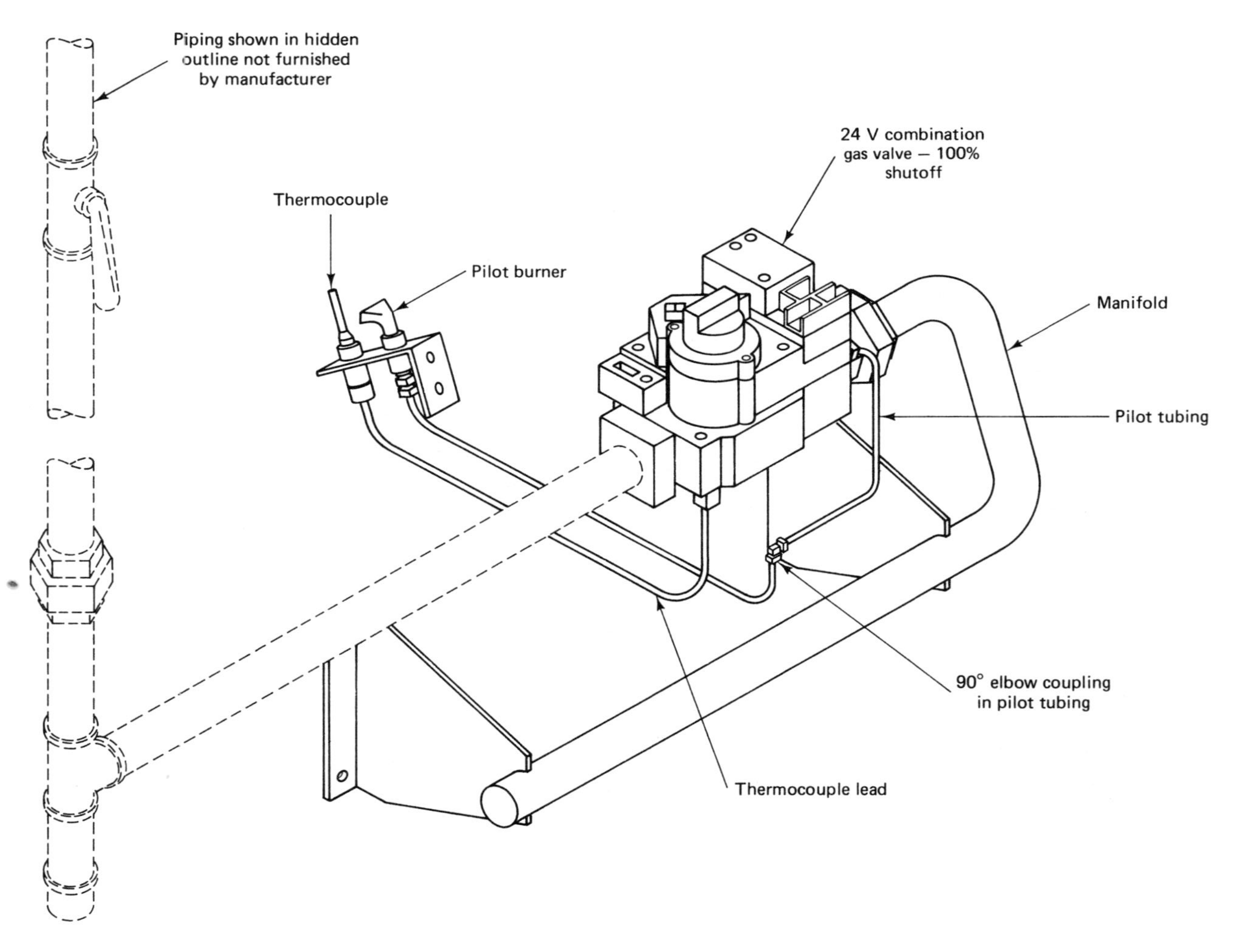

Figure 4.7 Standing pilot ignition system. (Courtesy of The Burnham Corporation.)

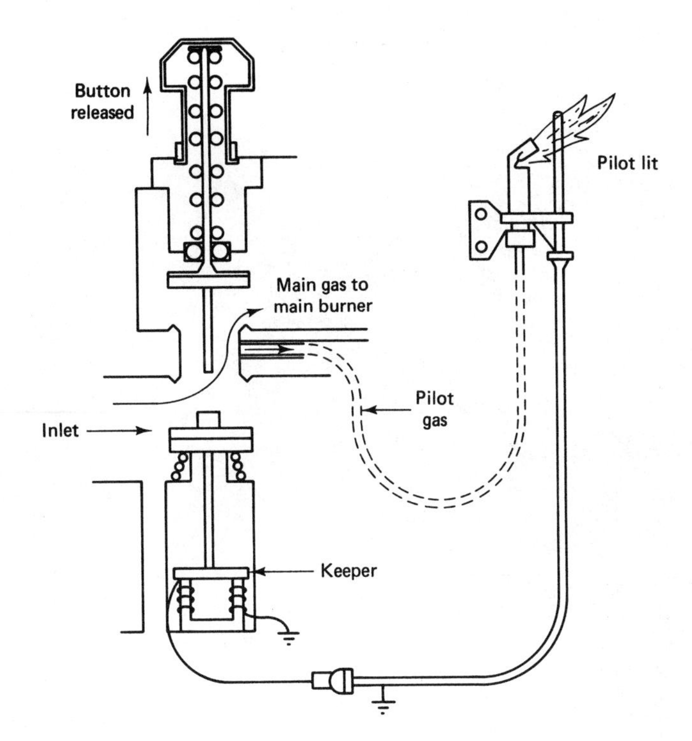

Figure 4.8 Normal operation of pilot valve assembly. Gas will travel to the main burner as long as the pilot remains lighted. (Courtesy of Robertshaw Controls Company.)

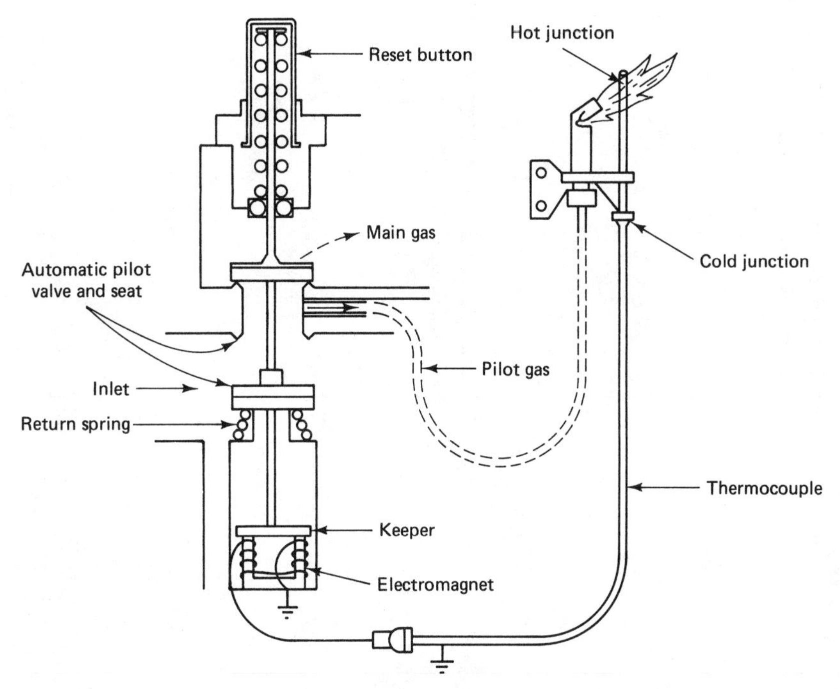

Figure 4.9 Operation of pilot valve during loss of pilot flame condition. (Courtesy of Robertshaw Controls Company.)

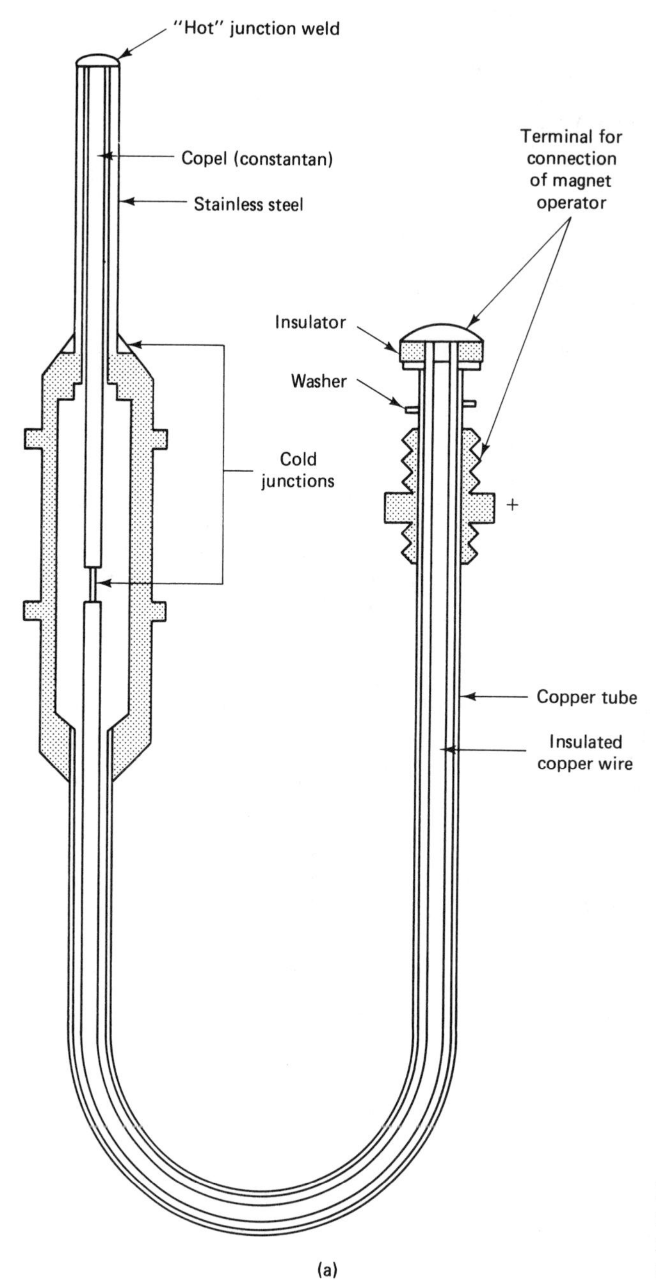

Figure 4.10 (a) Thermocouple cross section (courtesy of Robertshaw Controls Company); (b) thermocouple (courtesy of Honeywell, Inc.).

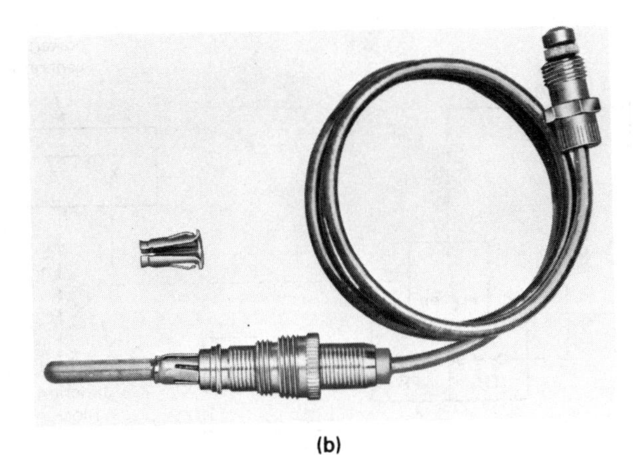

(b)

Figure 4.10 *(continued)*

trical power. An operating schematic of a thermopile system is illustrated in Figure 4.11.

A proper flame pattern produced by the pilot assembly directed against the thermocouple is essential if the unit is to operate correctly. Figures 4.12 and 4.13 illustrate the different types of pilot flames produced by both natural and LP gas. The pilot flame must be in direct contact with the thermocouple, as illustrated.

Intermittent spark ignition systems. In intermittent or spark ignition systems the pilot light burns only when there is a call for heat, achieving lower gas consumption than that of standing pilot systems. In the spark ignition system, voltage from the control mechanism is applied to an ignition electrode, causing a spark to jump the gap between the ignition electrode and pilot burner. This establishes a pilot flame, as gas is simultaneously released to the pilot burner during the operation. A typical spark ignition system is illustrated in Figure 4.14.

A safety circuit consisting of a flame sensor located adjacent to the pilot burner will turn off all gas to the system in less than 1 second should a loss of pilot flame occur. In this instance, automatic circuitry within the gas control module will attempt to reestablish the pilot light. If this attempt fails, the safety circuit will lock out, requiring the valve to be reset manually before the ignition sequence can be repeated. The pilot light, once established, serves to ignite the main gas burners.

Two-stage burner systems. To enhance energy efficiency further, some newer gas appliances feature a two-stage burner. During milder weather, the appliance operates on a low burner flame; the high flame is switched on when the outside temperature drops to a lower specified level. This two-stage design is illustrated in Figure 4.15.

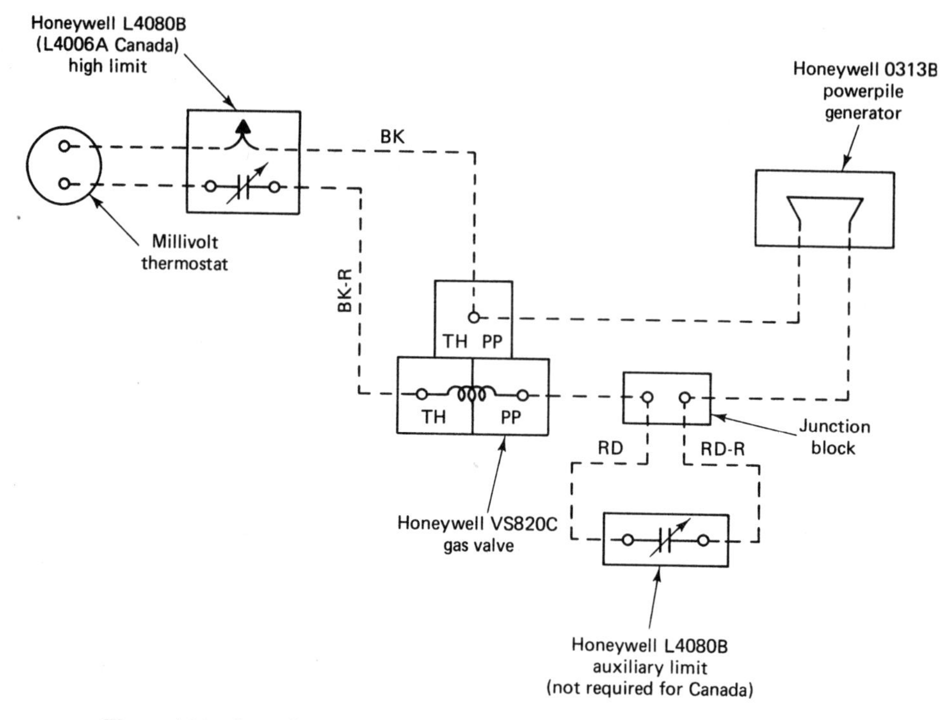

Figure 4.11 Operating schematic of thermopile generator. (Courtesy of Honeywell, Inc.)

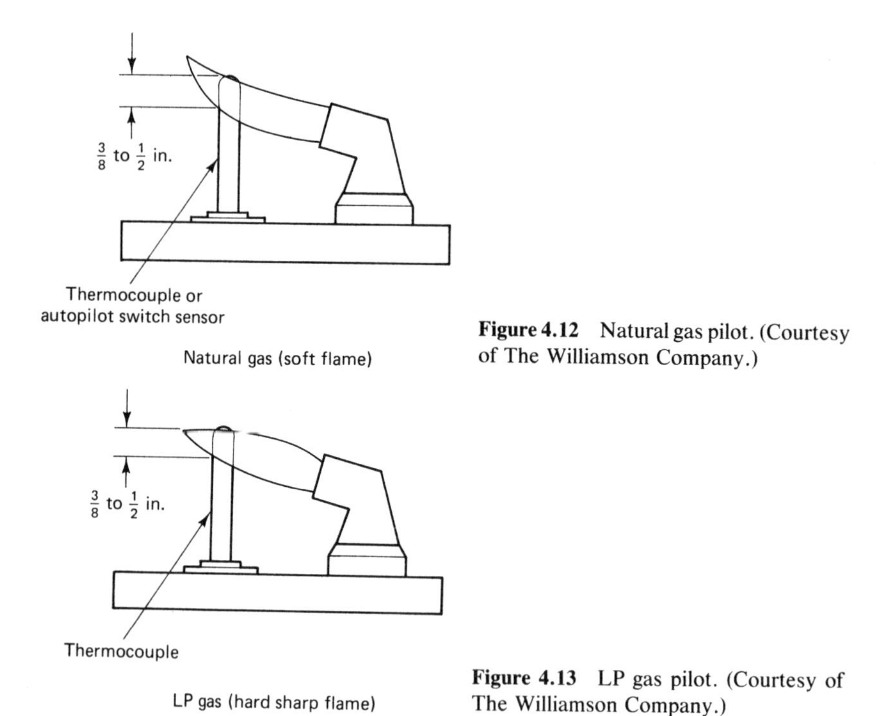

Figure 4.12 Natural gas pilot. (Courtesy of The Williamson Company.)

Figure 4.13 LP gas pilot. (Courtesy of The Williamson Company.)

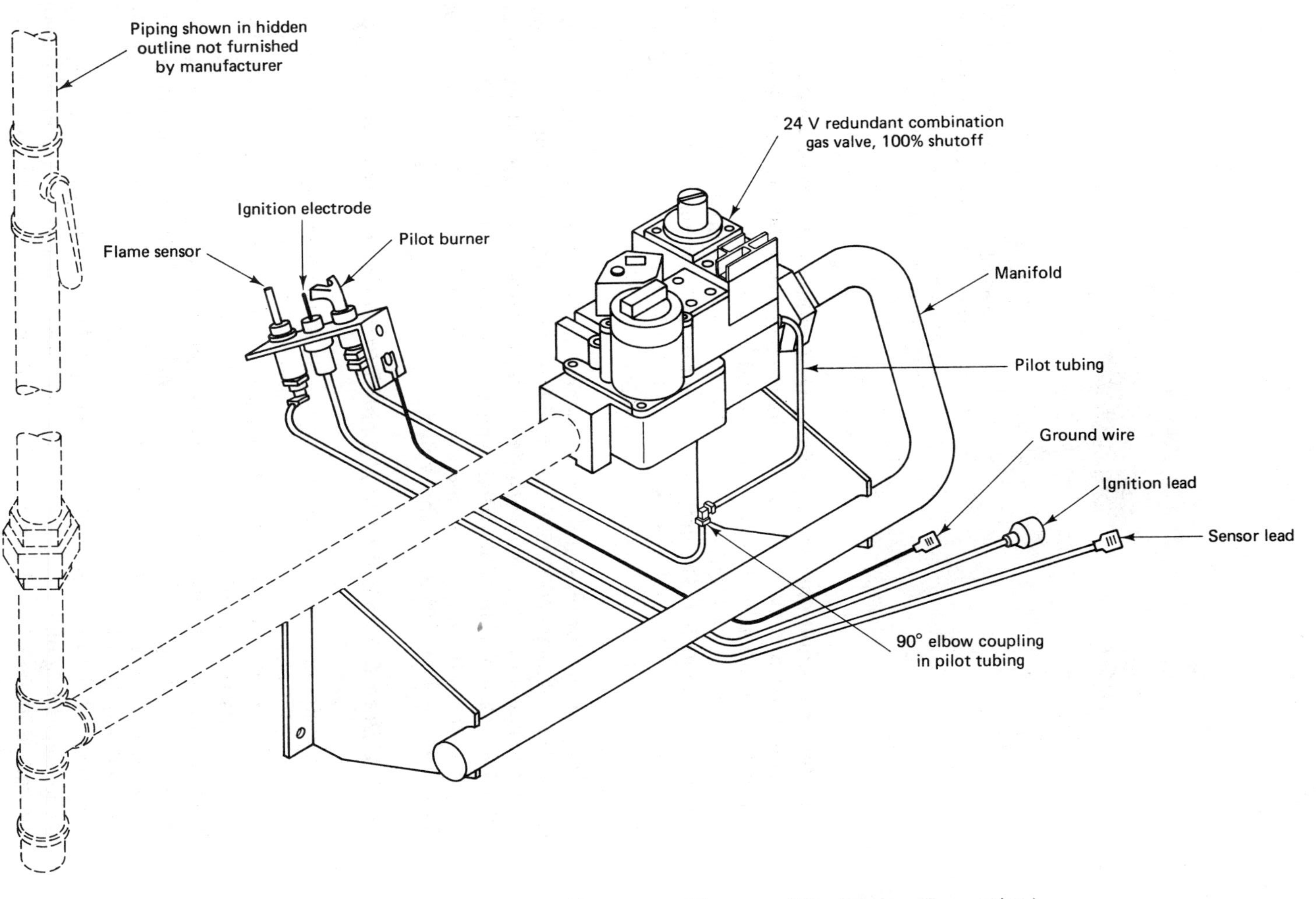

Figure 4.14 Intermittent spark ignition system. (Courtesy of The Burnham Corporation.)

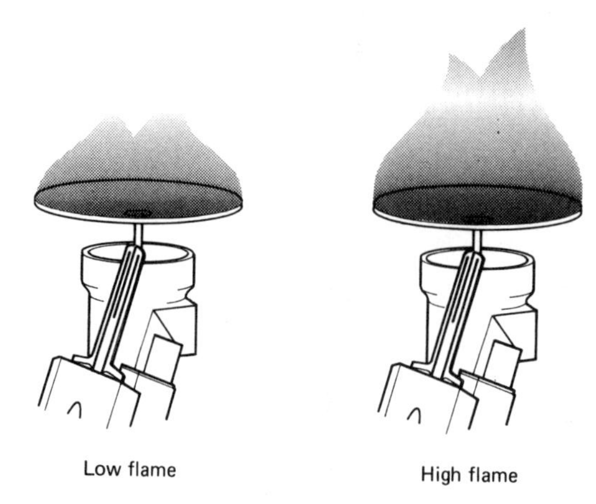

Figure 4.15 Two-stage burner design commonly used in gas-fired hot air furnaces. This burner configuration allows for varying heat output, depending on outdoor air temperatures. (Courtesy of The Williamson Company.)

Venting requirements for atmospheric combustion appliances. The vent system on a conventional atmospheric combustion heating appliance is illustrated in Figure 4.16. The appliance should be vented into a separate chimney or separate flue in a multiflue chimney. Note the use of a draft hood to which the vent pipe is connected. The draft hood draws dilution air into the vent system to lower the temperature of the stack gases and also acts to divert momentary downdrafts away from the gas combustion chamber. This action is illustrated in Figure 4.17.

Horizontal pipe runs in the vent system should be kept as short as possible and sloped a minimum of 1 in. for every 4 ft of run. Each joint should be secured with a minimum of three noncorrosive fasteners. The seams should always be placed at the top of the pipe. High-temperature cement should be used to seal the joint where the pipe enters the thimble of the chimney. The pipe should be flush with the inside of the chimney flue.

Induced-Draft Condensing and Noncondensing Combustion Technology

Induced-draft combustion technology relies on a high-speed blower that pushes air and gas through the unit and out the venting system. High-speed air injection results in very high combustion efficiencies: 85 to 90% in some instances. This is due to relatively complete mixing of the fuel gas and combustion air and the length of travel of the flue gases within the appliance. Figure 4.18 illustrates the relatively long length of travel of flue gas in induced-draft appliances, which maximizes heat transfer, as opposed to the relatively short heat transfer passages in conventional designs.

Induced-draft appliances operate in either the condensing or noncondensing mode, depending on whether or not the flue gases condense during operation of the unit. Both boilers and furnaces can operate in either mode, depending on the design of the appliance.

Induced-draft condensing appliances. In an induced-draft condensing system, the flue gases are cooled to the dew point of the vapors within the stack gases. As the gases condense, the latent heat of vaporization is recovered and transferred to either the airstream of a furnace or the water within a boiler. A typical boiler of this type is illustrated in Figure 4.19.

Most condensing units feature either a separate condensate removal system or a condensate trap built into a heat extractor or secondary heat exchanger. The condensate produced in this manner is relatively caustic, with a low pH (3.5 to 6.5 on a scale of 0 to 14). For this reason, only specially constructed stainless steel or PVC plastic pipe should be used for condensate removal. Where possible, a neutralizer cartridge should be installed in the condensate line for detoxification

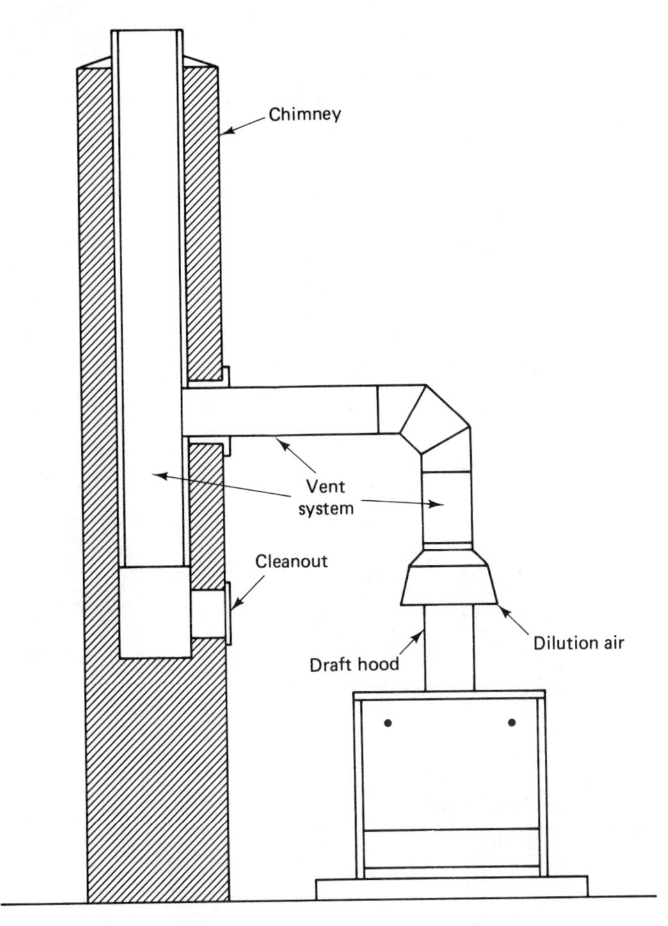

Figure 4.16 Venting configuration commonly used in atmospheric combustion appliances. Note the use of a draft hood above the heater for dilution air and diversion of momentary downdrafts in the chimney. (Courtesy of The Burnham Corporation.)

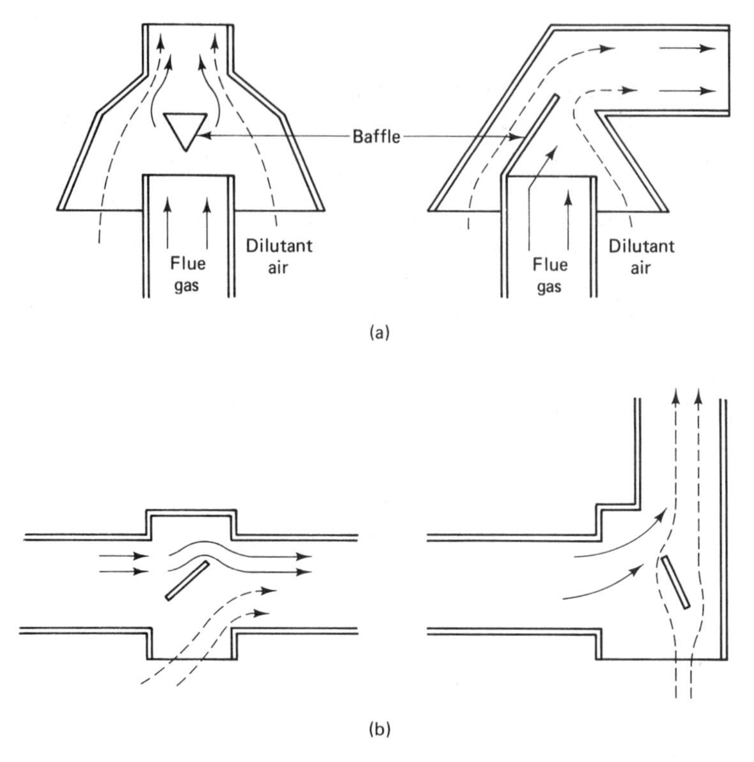

Figure 4.17 Action of draft hood in diverting momentary downdrafts away from the exhaust hood and flue passages of the appliance.

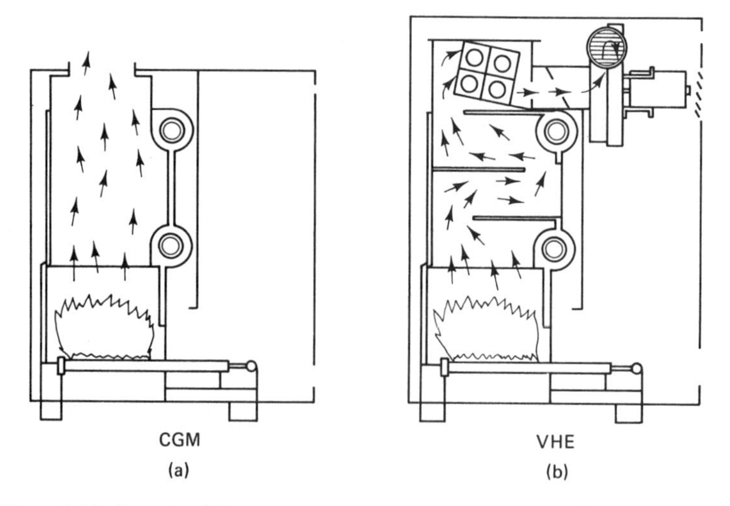

Figure 4.18 Length of flue gas travel in (a) conventional boiler and (b) induced-draft unit. Increasing the length of flue gas travel, coupled with internal baffles, greatly increases the efficiency of heat transfer between the flue gas and surrounding boiler water. (Courtesy of Weil-McLain, a Marley Company.)

(a)

(b)

Figure 4.19 (a) Induced-draft condensing boiler; (b) component arrangement illustrating induced-draft fan, burners, and gas controls. (Courtesy of Weil-McLain, a Marley Company.)

of the liquid prior to releasing it into a drainage or sewerage system. This arrangement is illustrated in Figure 4.20.

Induced-draft noncondensing appliances. Induced-draft appliances may also operate in the noncondensing mode, which uses induced draft for increased combustion efficiency, eliminating the need to deal with any condensate. Figure 4.21 illustrates the cross section of a noncondensing induced-draft hot water boiler. Although similar in most respects to their condensing counterparts, noncondensing units operate at somewhat higher stack temperatures since the temperature of the flue gases remains below the condensation point.

Venting requirements for induced-draft appliances. Induced-draft appliances that operate in the condensing mode require special venting systems.

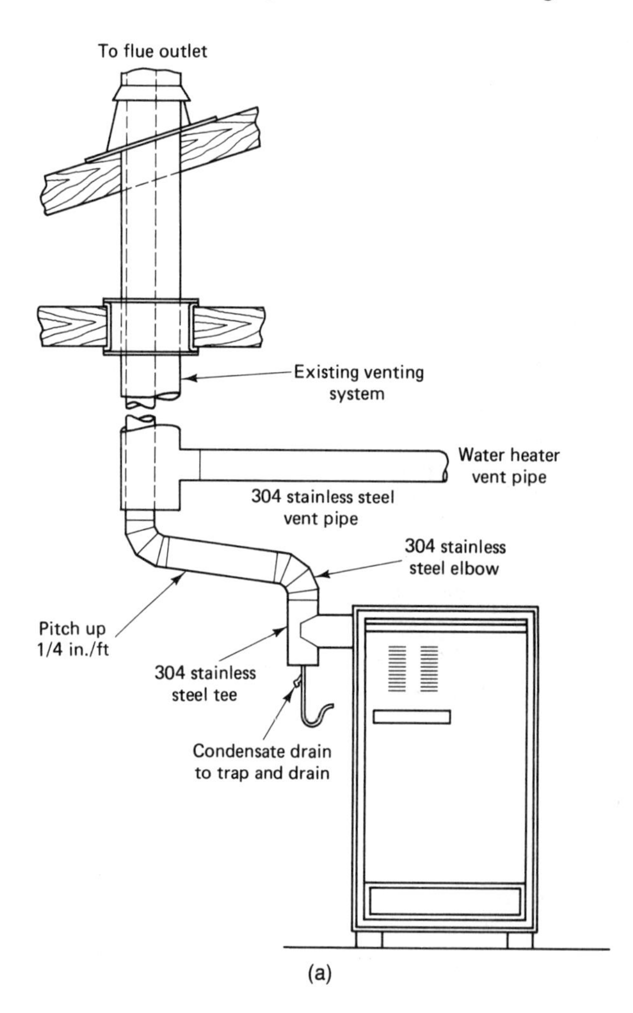

Figure 4.20 Venting and condensate trap used in gas-fired condensing appliances: (a) boiler (courtesy of Weil-McLain, a Marley Company); (b) hot air furnace (courtesy of The Williamson Company).

Since the effluents produced via the condensation of the exhaust gases must be disposed of, a condensate removal system is designed into the boiler or furnace. Most of these units vent either through an existing chimney that has been lined with stainless steel vent pipe, or through the wall of the house using stainless steel or PVC pipe. Figure 4.22 illustrates direct venting through the exterior building wall using stainless steel pipe.

The vent system on a noncondensing induced-draft system resembles those

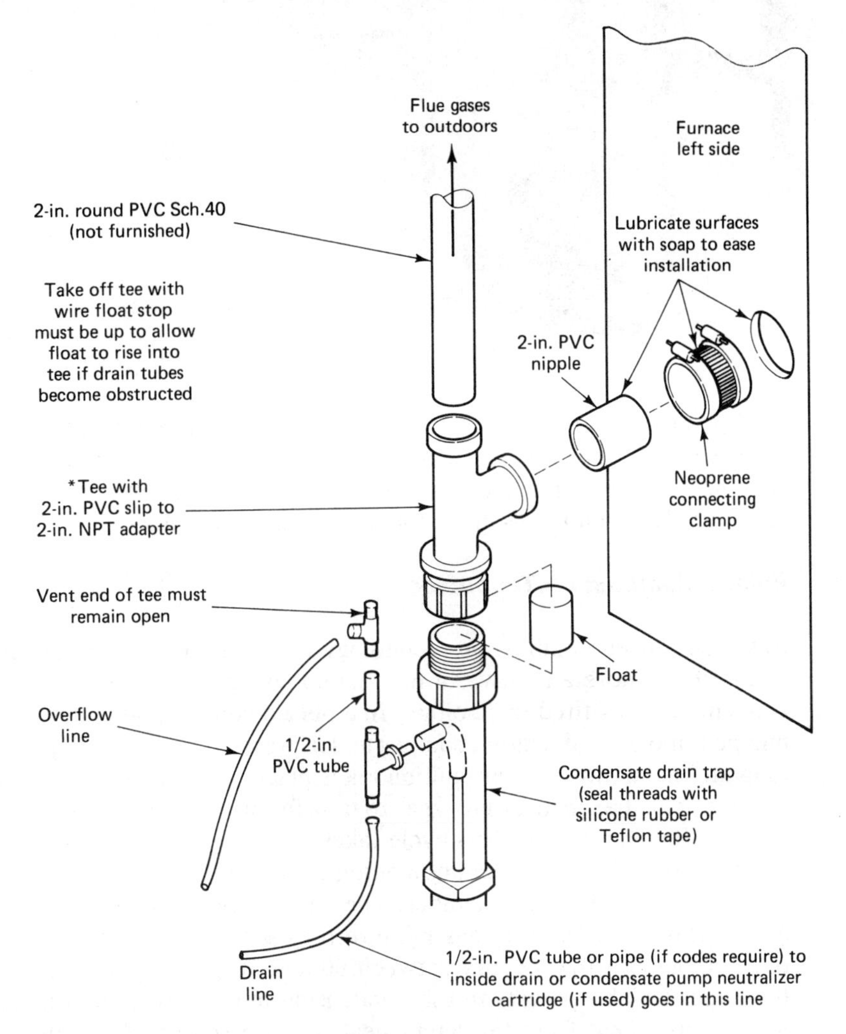

(b)

Figure 4.20 *(continued)*

Figure 4.21 Noncondensing induced-draft hot water boiler. (Courtesy of The Burnham Corporation.)

used on atmospheric combustion units, with the difference that no draft hood is used. Since the unit operates at a specified pressure created within the combustion chamber and heat transfer passages created by the blower, a draft hood is not required. This venting system is illustrated in Figure 4.23.

Pulsed Combustion Technology

Gas-fired pulsed combustion technology is a relatively recent development and is, in effect, more a relative of internal combustion diesel engine design than of conventional gas-fired technology. In a pulsed combustion boiler, air and gas are pumped into a combustion chamber under pressure and a spark plug is energized to ignite the mixture. After ignition takes place, the exhaust gases travel through a series of heat exchanger passages to transfer the heat of combustion to the boiler water. After the first ignition cycle takes place the spark plug and blower motor are deenergized; all subsequent ignition cycles are powered by residual heat and flame in the combustion chamber. The vacuum produced by the exiting exhaust gases draws in a fresh air/gas mixture for each combustion cycle. This creates ongoing ignition that occurs between 60 and 70 pulses per second. A boiler of this type, which operates in the condensing mode, gains approximately 10% of its delivered heat from the latent heat of vaporization of the flue gases and is illustrated in Figure 4.24.

The venting system used on these boilers is different from those on both atmospheric and induced-draft systems. Two ducts are connected from the boiler to the outside of the house. One duct supplies intake combustion air, the other

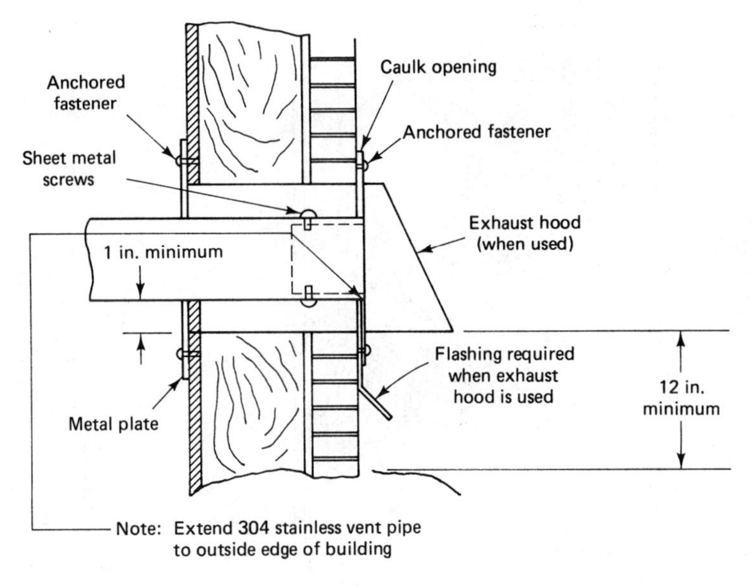

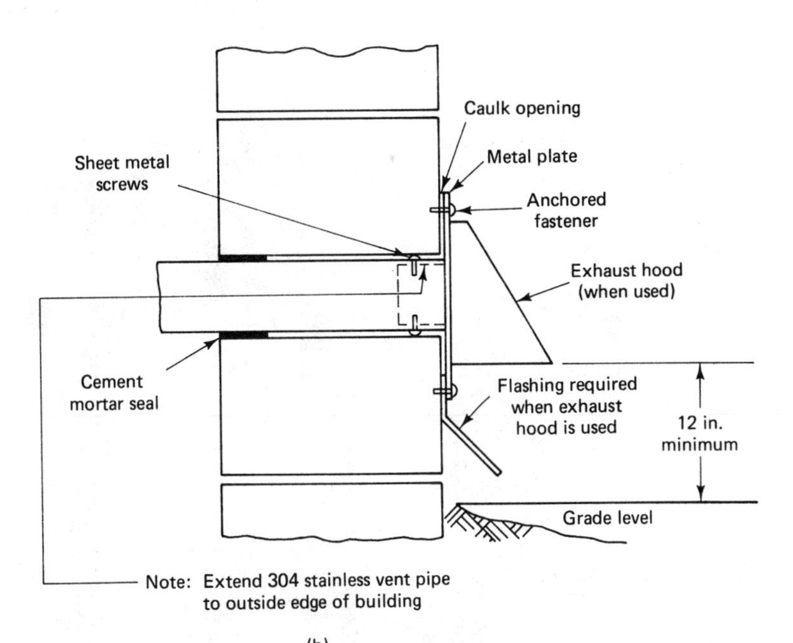

Figure 4.22 Venting an induced-draft condensing boiler through (a) a combustible wall and (b) a noncombustible wall. (Courtesy of Weil-McLain, a Marley Company.)

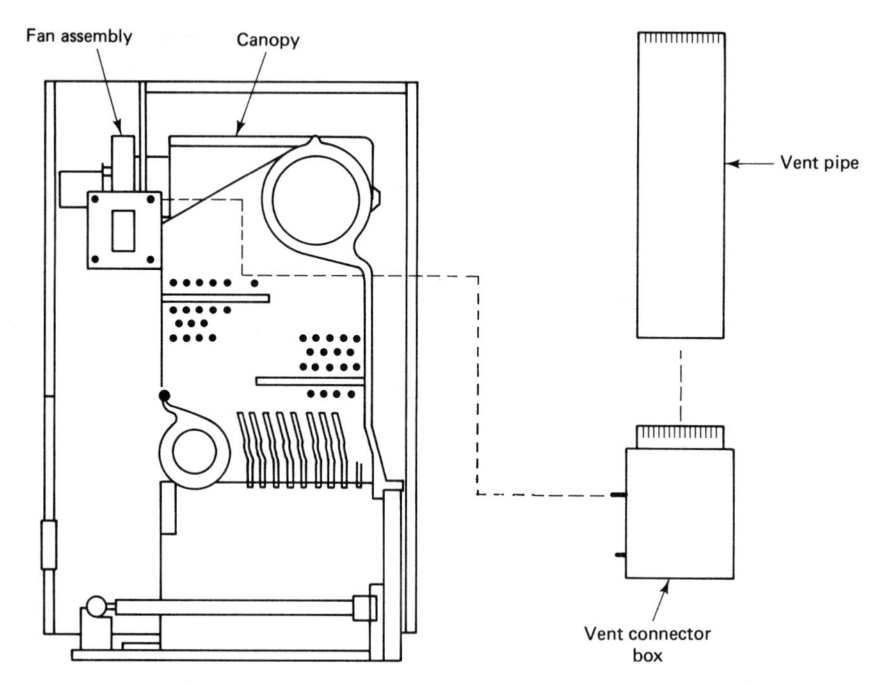

Figure 4.23 Venting configuration of a noncondensing induced-draft appliance. Note the absence of a draft hood in the induced-draft vent system. (Courtesy of The Burnham Corporation.)

Figure 4.24 Pulsed combustion condensing boiler. (Courtesy of Hydrotherm.)

serves as an exhaust pipe. In some instances mufflers may be required on the exhaust system to minimize noise from the ignition cycle.

Power Burners

Gas power burners are similar in appearance to conventional oil burners. Although most power burners are used in commercial and industrial applications, there are several units available for residential heating applications. A unit of this type is illustrated in Figure 4.25.

For help in understanding the operating sequence of these units, refer to the working drawing of the power burner illustrated in Figure 4.26. When there is a call for heat, the control mechanism begins a timed warm-up sequence that lasts approximately 40 seconds. This sequence is used to bring the burner motor up to speed and purge the burner housing. As the motor comes up to speed, a cen-

Figure 4.25 Gas power burner. These units are available in a variety of Btu outputs. Featuring high-efficiency combustion with either natural or propane gas, power burners are ideally suited for both new installations and retrofit applications. (Courtesy of The Carlin Company.)

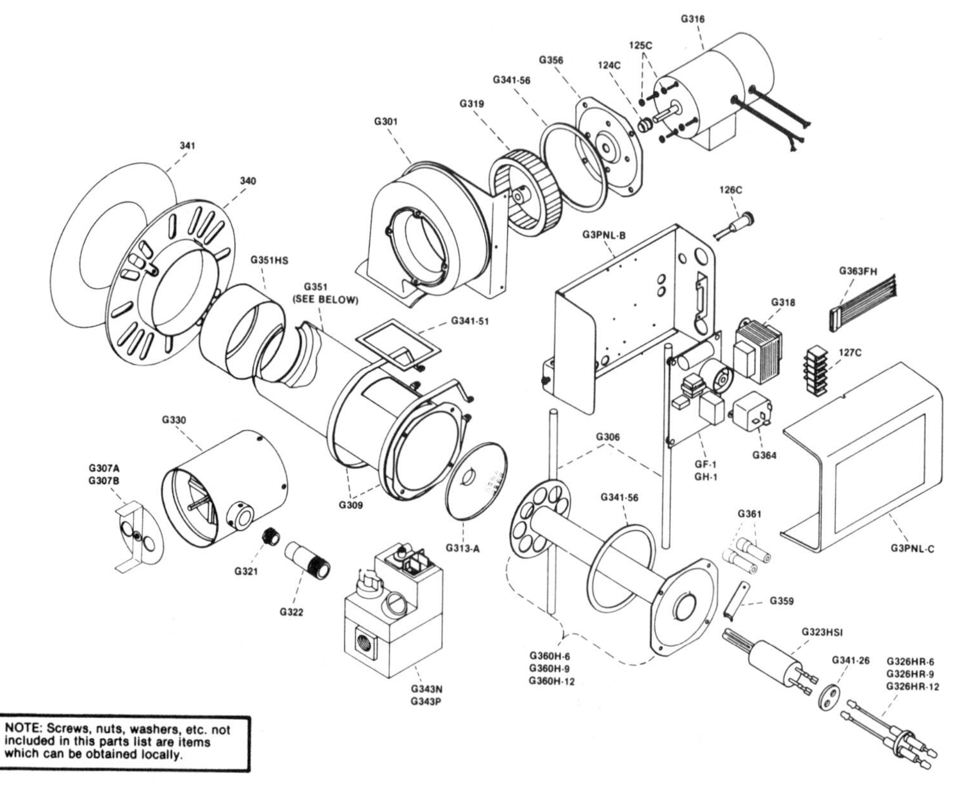

PART NUMBER	DESCRIPTION
124C	Bushing, motor shaft, nylon
125C	Washer, motor stud, nylon, 4 required
126C	Indicator light, 24-volt
127C	Terminal block
340	Mounting flange, Universal
341	Gasket for mounting flange
G301	Housing, complete w/cover plate and G341-51 gasket
G306	Pedestal legs, 3/8" OD × 12" long, 2 required
G307A	Air throttle with plugs for 'A' style air tube
G307B	Air throttle with bushings for 'B' style air tube
G309	Air tube housing clamp, 2 required
G313-A	Flameholder
G316	Burner motor, 1/50 hp, 3300 rpm, 115-volt, 60-Hz, permanent split capacitor with centrifugal switch, CCW rotation facing shaft
G318	Transformer, 40 VA rating, 115-volt primary, 24-volt secondary
G319	Blower wheel, 4" OD × 1" wide, 5/16" bore, CCW rotation facing open end
G321.063	Orifice, No. 52 drill (.063 dia.) for G3A, propane, 30,000 BTUH
G321.086	Orifice, No. 44 drill (.086 dia.) for G3B, propane, 60,000 BTUH
G321.101	Orifice, No. 38 drill (.101 dia.) for G3A, natural, 30,000 BTUH
G321.136	Orifice, No. 29 drill (.136 dia.) for G3B, natural, 60,000 BTUH
G322	Orifice nipple
G323HSI	Hot surface ignitor, Norton #201C, 120V, 5-3/8" OAL
G326HR-6	Rear ignitor assembly, overall length 7-1/2"
G326HR-9	Rear ignitor assembly, overall length 10-1/2"
G326HR-12	Rear ignitor assembly, overall length 13-1/2"
G330	Air inlet tube assembly
G341-26	Gasket, ignitor assembly

PART NUMBER	DESCRIPTION
G341-51	Gasket for air tube window
G341-56	Gasket for motor mounting ring and air tube back plate, 2 req'd
G343N,P	Gas valve, dual, 24-volt, ½" × ½", specify natural or propane
G351A-6	Air tube assembly, 3" ID flame ring, overall length approx. 10-5/16"
G351A-9	Air tube assembly, 3" ID flame ring, overall length approx. 13-5/16"
G351A-12	Air tube assembly, 3" ID flame ring, overall length approx. 16-5/16"
G351B-6	Air tube assembly, 3-3/4" ID flame ring, overall length approx. 9-7/8"
G351B-9	Air tube assembly, 3-3/4" ID flame ring, overall length approx. 12-7/8"
G351B-12	Air tube assembly, 3-3/4" ID flame ring, overall length approx. 15-7/8"
G351HS	Heat shield
G356	Motor mounting ring with G341-56 gasket
G359	Electrode hold-in
G360H-6	Ignition tube assembly with G341-56 gasket, overall length approx. 10-9/16"
G360H-9	Ignition tube assembly with G341-56 gasket, overall length approx. 13-9/16"
G360H-12	Ignition tube assembly with G341-56 gasket, overall length approx. 16-9/16"
G361	Ignition terminal boot, 2 required
G363FH	Control wire harness for Fenwal circuit board
G364	Motor relay, 24-volt coil
G3PNL-B	Control panel base
G3PNL-C	Control panel cover, less nameplate
GF-1	Control, Fenwal #05-21 hot surface ignition, w/6.8 sec. TFI
GH-1	Control, Honeywell, #S89C hot surface ignition, w/6.8 sec. TFI

Figure 4.26 Power burner component arrangement and nomenclature. (Courtesy of The Carlin Company.)

trifugally operated end switch closes the electrical circuit to the gas valve. When the gas valve is energized, the system will attempt to ignite [referred to as a **trial for ignition** (TFI)] for approximately 6 or 7 seconds. After a flame has been established, the hot surface ignition (HSI) element is electronically switched from the ignition mode to the flame-sensing mode.

These units incorporate sophisticated electronic technology to both monitor the ignition sequence and continued presence of the flame. A failure to ignite is sensed by these units within 6 or 7 seconds, resulting in a lockout of the control circuit. Should the flame be lost during the heating cycle, the unit will automatically shut down approximately 2 seconds after the flame has been lost. A double-acting draft regulator should be installed in the flue pipe, with a suitable spill switch attached to shut down the unit in the event of extended downdrafts or blocked chimney passages (see Figure 4.37 and the accompanying text for an explanation of the operation of spill switches).

Residential power burners can be used to change over oil-fired units to fire either natural or propane gas and are available in a variety of output sizes for most heating applications. The sophisticated nature of these units requires that servicing be left to qualified and experienced personnel only.

GAS SUPPLY PIPING

Although almost all packaged residential boilers, furnaces, and water heaters feature preplumbed gas valves, manifolds, and main burners, it is left to the installer to furnish the supply piping to the unit. This piping must be sized properly in order to be able to furnish a sufficient amount of gas necessary to match the manufacturer's Btu input rating of the appliance.

Only licensed technicians and those knowledgeable in gas piping procedures should perform these installations. The most commonly used materials for gas supply piping are steel, wrought iron, and ductile iron that meet applicable material and Code standards for such pipe. Also, seamless copper tubing (K or L), aluminum alloy, or steel tube may be used for supply piping as long as the gases do not contain materials that would be corrosive to these materials. Specific information relating to the standards for corrosive effects of gas on different types of metallic tubing is available in ANSI/ASTM Standard D2385 or D2420. The use of cast-iron pipe for gas line connections is prohibited.

The ability of pipe to supply sufficient quantities of gas is based on its diameter. The National Fuel Gas Code, ANSI Standard Z223.1, lists a variety of these capacities for both natural and LP gas. This standard is available from either the American Gas Association or the National Fire Protection Association.

The gas piping in the immediate vicinity of an appliance should be piped as illustrated in Figure 4.27. Note the use of a drip leg and cap in the piping circuit. This allows for trapping and removal of moisture within the gas feed lines.

All joints in the piping circuit should be sealed with the proper type of pipe

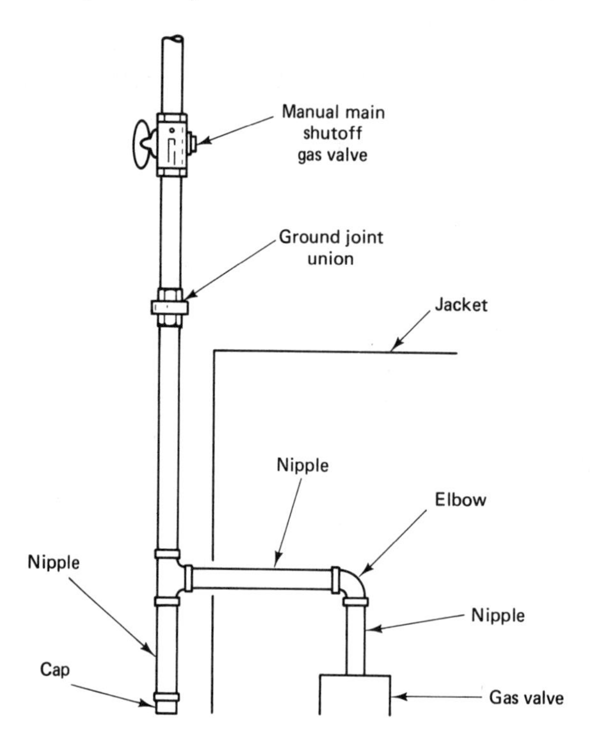

Figure 4.27 Gas supply piping to a heating appliance (typical).

joint compound. When sealing joints, the use of Teflon tape is not recommended, as it tends to break down under the influence of fuel gas and can clog small orifice and burner passages.

MISCELLANEOUS BOILER AND FURNACE CONFIGURATIONS

Many different appliance configurations are used with gas-fired heating equipment. Although some of these are not restricted to use with gas equipment, the following variations are offered to illustrate some examples of what is available. For example, warm air furnaces may be purchased in either an upflow or a counterflow design. Figure 4.28 illustrates the upflow design and Figure 4.29 shows the counterflow design. Warm air furnaces are also available in space-saving horizontal designs (Figure 4.30).

The selection of a particular furnace design is based on the requirements of the residence. For example, if the warm air ducts must be run below the level of the furnace, a counterflow design would be appropriate. If conventional ductwork is to be used, a standard upflow furnace should be specified. In locations where there is little headroom, for example, in the crawl space of a house, a horizontal furnace would have to be installed.

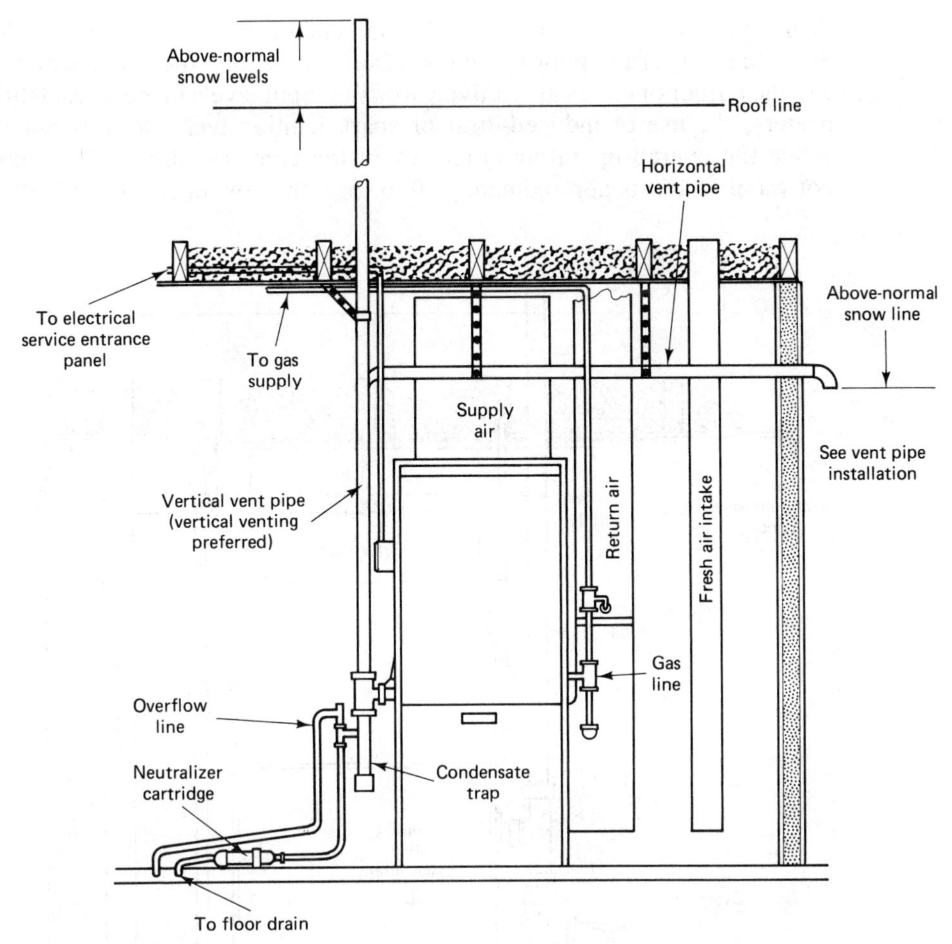

Figure 4.28 Upflow warm air furnace design. (Courtesy of The Williamson Company.)

Most gas-fired hot water and steam boilers resemble their oil-fired counterparts in overall size and appearance. Induced-draft technology has made a dramatic difference in the operating efficiency of these boilers. Figure 4.31 illustrates an induced-draft noncondensing steam boiler. Whereas the combustion efficiency of a conventional atmospheric gas-fired steam boiler is between 60 and 70%, induced-draft technology allows a steam boiler to deliver efficiencies in the range of 85%. This significantly lowers the annual operating cost of the heating system.

Little has been said in this chapter regarding gas-fired water heaters, since the operation of these units is described in Chapter 5. The conventional domestic hot water gas-fired heater has undergone little design change over the years other than improved insulation between the tank and outer jacket to minimize heat loss. Since the heat of the pilot flame contributes to heating the water within the tank and the burner operates at relatively low Btu input levels in most residential water heaters, the use of induced-draft or spark ignition technology is not required. While the overall operating efficiency of the common atmospheric combustion hot water heater is approximately 60 to 65%, the cost of raising the efficiency of

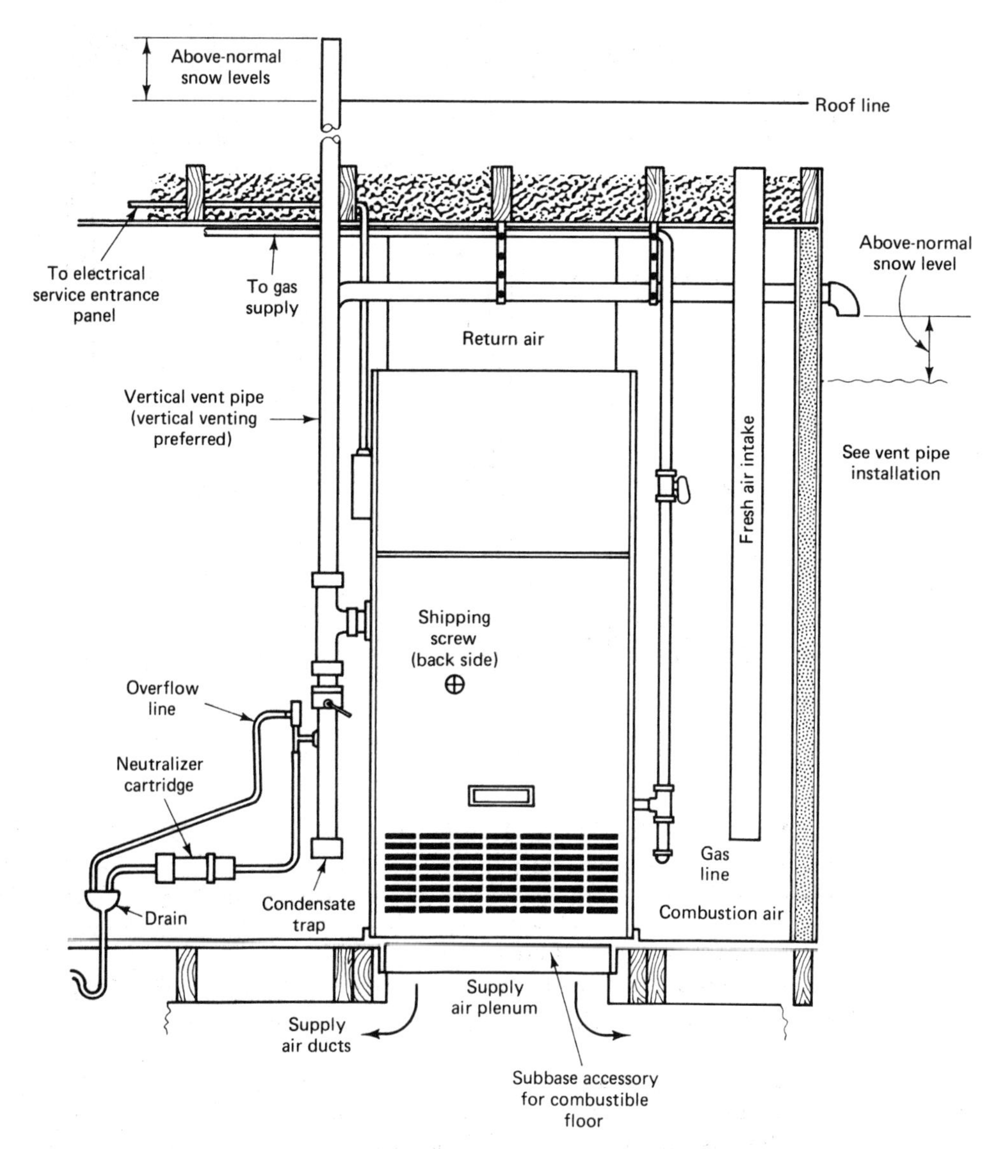

Figure 4.29 Counterflow warm air furnace. (Courtesy of The Williamson Company.)

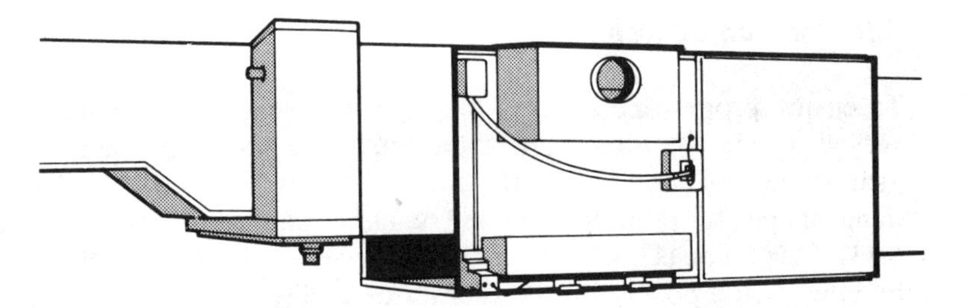

Figure 4.30 Horizontal warm air furnace. (Courtesy of The Williamson Company.)

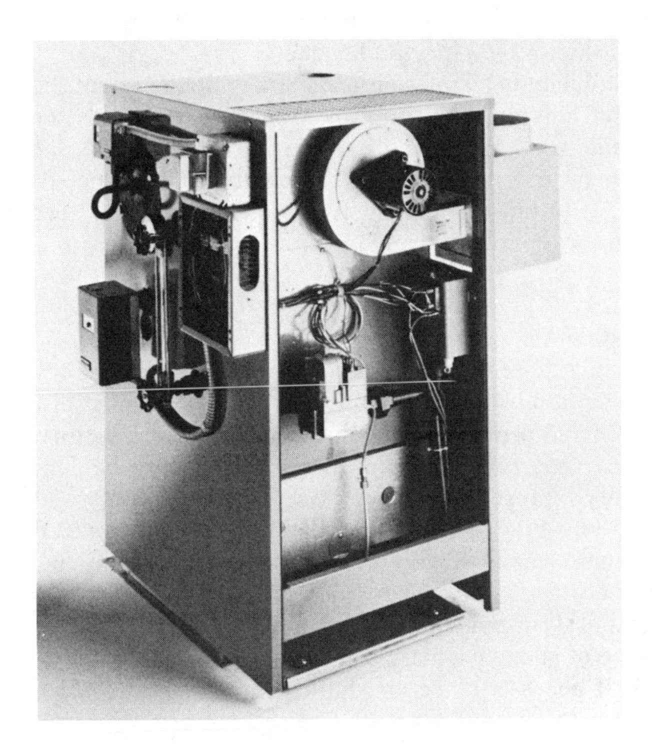

Figure 4.31 Induced-draft steam boiler. (Courtesy of The Burnham Corporation.)

the unit by the use of intermittent ignition or induced-draft technology is not very cost-effective given the pricing structure of these appliances.

MAINTENANCE PROCEDURES FOR GAS-FIRED APPLIANCES

The maintenance of gas-fired appliances involves procedures designed to ensure that the gas pilots and main burner assemblies are properly adjusted, that the flue and heat exchanger passages are kept clean and free of soot, and that the vent and gas delivery piping is in good condition.

Pilot Ignition Systems

To ensure proper operation of the pilot ignition system, the flame should be observed during the operation of the unit. The tip of the flame must be directly against the hot junction of the thermocouple (Figure 4.32.) Note the different flame shapes for pilot lights fueled by either natural or LP gas in Figures 4.12 and 4.13. If the flame is not in proper contact with the thermocouple, it can be adjusted by altering the position of the burner assembly. The upstream pilot filter should be checked at this time to make sure that it is clean. Both the pilot and thermocouple should be firmly fastened in their respective mounting brackets. Also, a check for any leaking connections should be made with a soapy water solution around all joints and tightened if necessary. Each joint should be dried off after this check to prevent corrosion of the pipe and fasteners.

The position of the pilot light in an intermittent spark ignition system should be inspected to ensure that it is properly positioned against the flame sensor (Figure 4.33). The pilot flame should surround between $\frac{3}{8}$ and $\frac{1}{2}$ in. of the sensor tip and can be adjusted by turning an adjustment screw until the proper flame position has been achieved. Consult the manufacturer's instructions for specific adjustment procedures and location of adjustment screws.

Main Burner Flames

Some main burners, especially those on induced-draft appliances, are not always adjustable, whereas those on atmospheric combustion units are almost always adjustable. If the main burners do not have an air shutter, they are factory set and cannot be field adjusted. Any problems with these burners must be handled only on the advice of factory representatives.

Prior to attempting to adjust the main burners, they should be removed from the combustion chamber and cleaned (Figure 4.34). The specific flame characteristics of main burners vary from one unit to another, depending on whether the appliance is of induced-draft or conventional combustion design.

The main burner flame of an atmospheric combustion unit should appear as illustrated in Figure 4.35. If necessary, the air shutter should be adjusted until

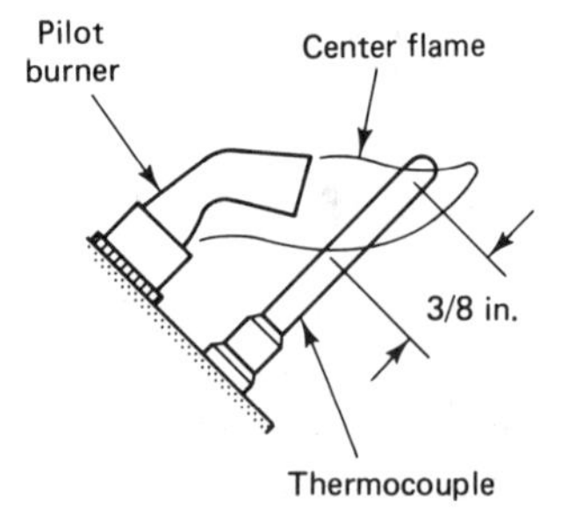

Figure 4.32 Proper flame adjustment for constant-burning pilot ignition system.

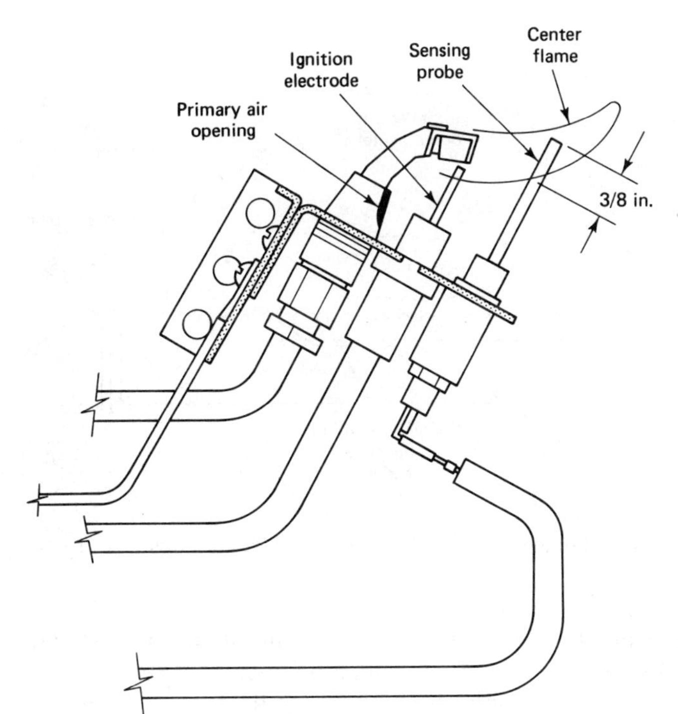

Figure 4.33 Pilot flame adjustment against flame sensor in typical intermittent spark ignition system. (Courtesy of The Burnham Corporation.)

Figure 4.34 Removal and inspection of main gas burners should be done annually to maintain correct burner operation and flame pattern. (Courtesy of Slant/Fin Corporation.)

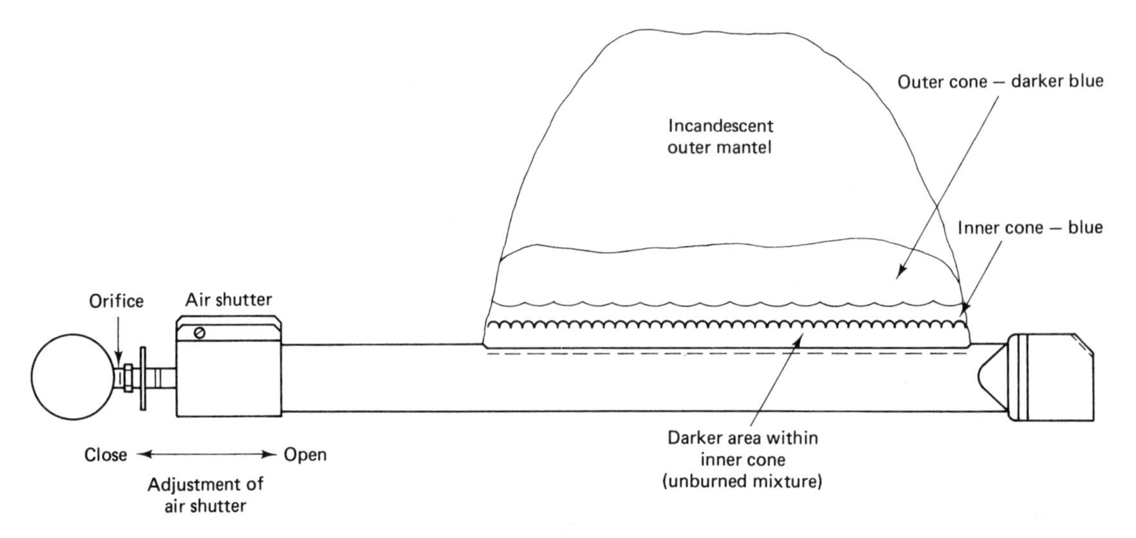

Figure 4.35 Flame pattern on typical atmospheric combustion main burner. (Courtesy of The Burnham Corporation.)

the flame pattern shows a clearly defined inner cone. When this has been achieved, the air shutter should be locked in place by tightening the locking screw.

In induced-draft units with adjustable air shutters, the flame pattern should resemble the shape illustrated in Figure 4.36. The main burner flame should be checked after the burner has been operating for approximately 10 minutes. Although occasional yellow streaking in the flame is normal, the flame should not have a constant yellow tip. Also, the flame should rest against the burner without lifting off the burner ports. The flame should not be in direct contact with the heat exchanger surfaces.

Spill (Stack) Switches

Because natural gas appliances can generate lethal amounts of carbon monoxide (CO) given the "proper" circumstances, provisions must be made in every natural gas installation to shut down the combustion process should conditions occur that

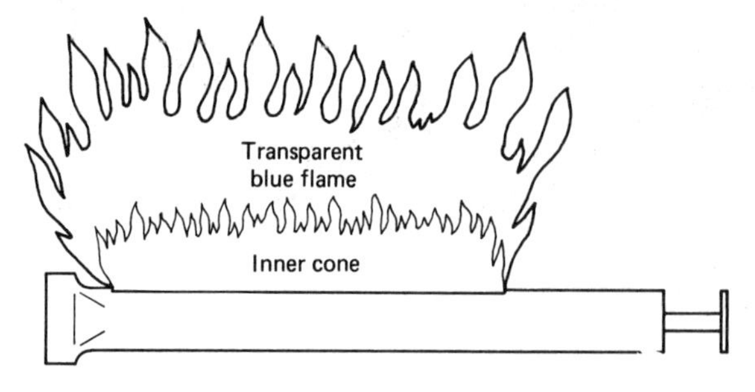

Figure 4.36 Flame pattern on typical induced-draft main burner. (Courtesy of Weil-McLain, a Marley Company.)

favor the production of the CO. Given a restricted or plugged flue passage coupled with an improperly adjusted main burner flame, CO that would ordinarily be vented to the outdoors can build up within the home. This is, of course, a deadly situation for the occupants. A **spill** or **stack switch** is a device designed to shut off all gas to the appliance in the event that the flue becomes plugged.

These switches operate in response to a rise in either temperature or pressure within the unit. A heat-sensitive spill switch is illustrated in Figure 4.37. If the flue is obstructed, heat from the vent system spills out of the draft hood, raising the temperature of the switch. Within the switch is a set of electrical contacts that will open the electrical circuit to the gas valve, closing off the supply of gas to the main burners. Pressure-sensitive switches are used in induced-draft and atmospheric appliances and respond to a rise in vent system pressure, generally caused by restrictions in the flue passages or chimney proper, which causes the electrical circuit to the gas valve assembly to open. If the gas appliance does not have a spill or pressure switch installed in the system, one can be added to ensure a safe operating environment.

Gas-Fired Furnaces

A thorough inspection of all basic components should be performed by a qualified service technician prior to the beginning of each heating season. Although the homeowner can perform many of the service procedures, it should be understood that proper training is necessary; otherwise, serious injury or death can result, due to improper and unsafe service techniques.

Gas-fired furnace service is somewhat different from that for oil-fired units due to the different nature of the combustion system. Condensing furnaces require

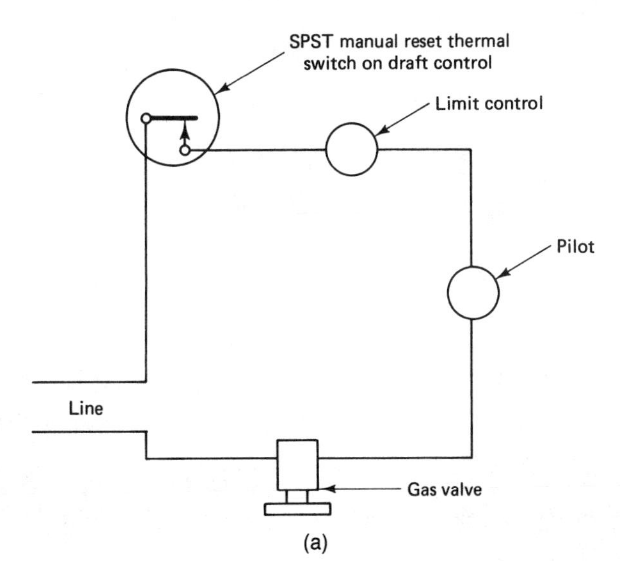

Figure 4.37 (a) Heat-sensitive spill switch installed in gas vent system. (b) The switch contacts are wired into the main gas burner circuit, deactivating the gas valve in the event of high stack temperatures, indicating an obstruction in the chimney. Induced-draft appliances use an internal switch that senses either temperature or exhaust gas pressure to perform the same function as the spill switch. (Courtesy of Field Controls Company.)

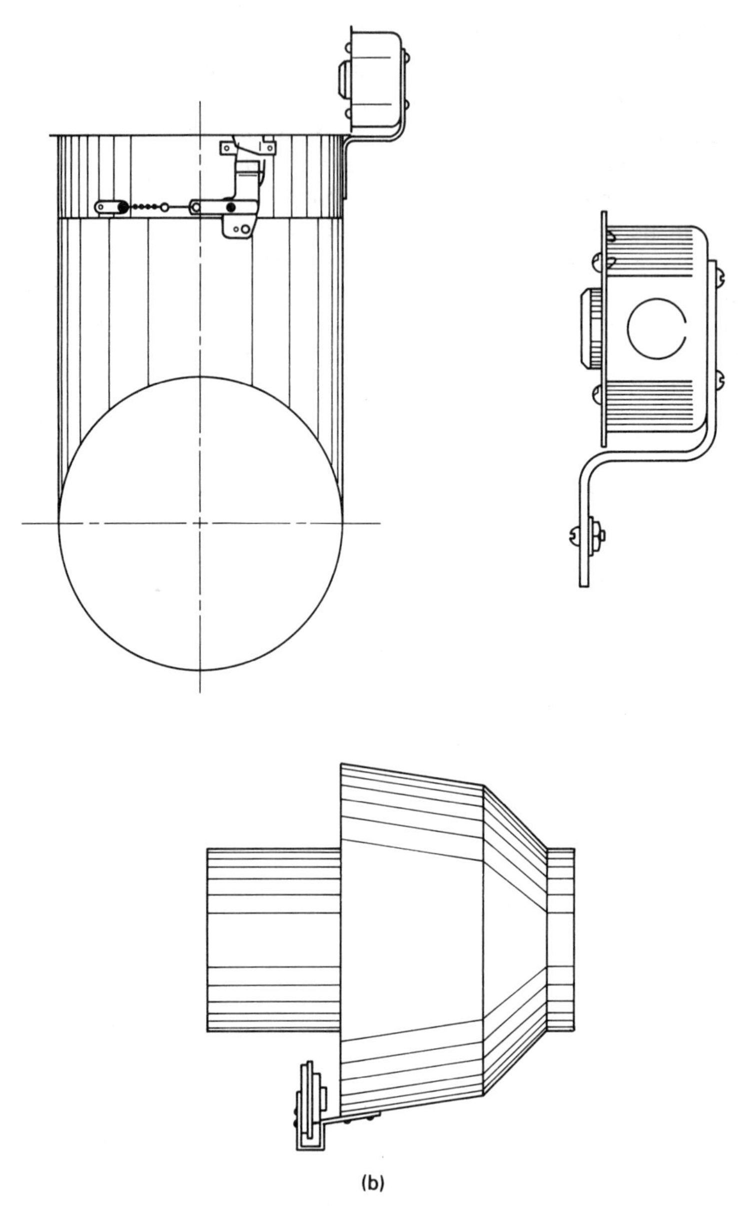

(b)

Figure 4.37 *(continued)*

special procedures due to the condensate and additional heat exchangers incorporated into these units. Condensate traps, part of the condensate removal system in all condensing appliances, must be cleaned and flushed periodically. If the furnace is equipped with a neutralizer cartridge, it should be disassembled and cleaned before each heating season. Make sure that the condensate overflow line is free of obstructions. Backflush to make sure that there are no blockages. The vent line should be securely attached to the furnace. Check the connector to make sure that it is secure.

Most condensing furnaces have both primary and secondary heat exchangers (Figure 4.38). Gas-fired furnaces that are properly installed will not normally require cleaning of the heat exchangers. However, if the filters are not changed regularly or if they fail for some reason, it might be necessary to clean the exterior surfaces of the heat exchanger. This is done by gaining access to the heat ex-

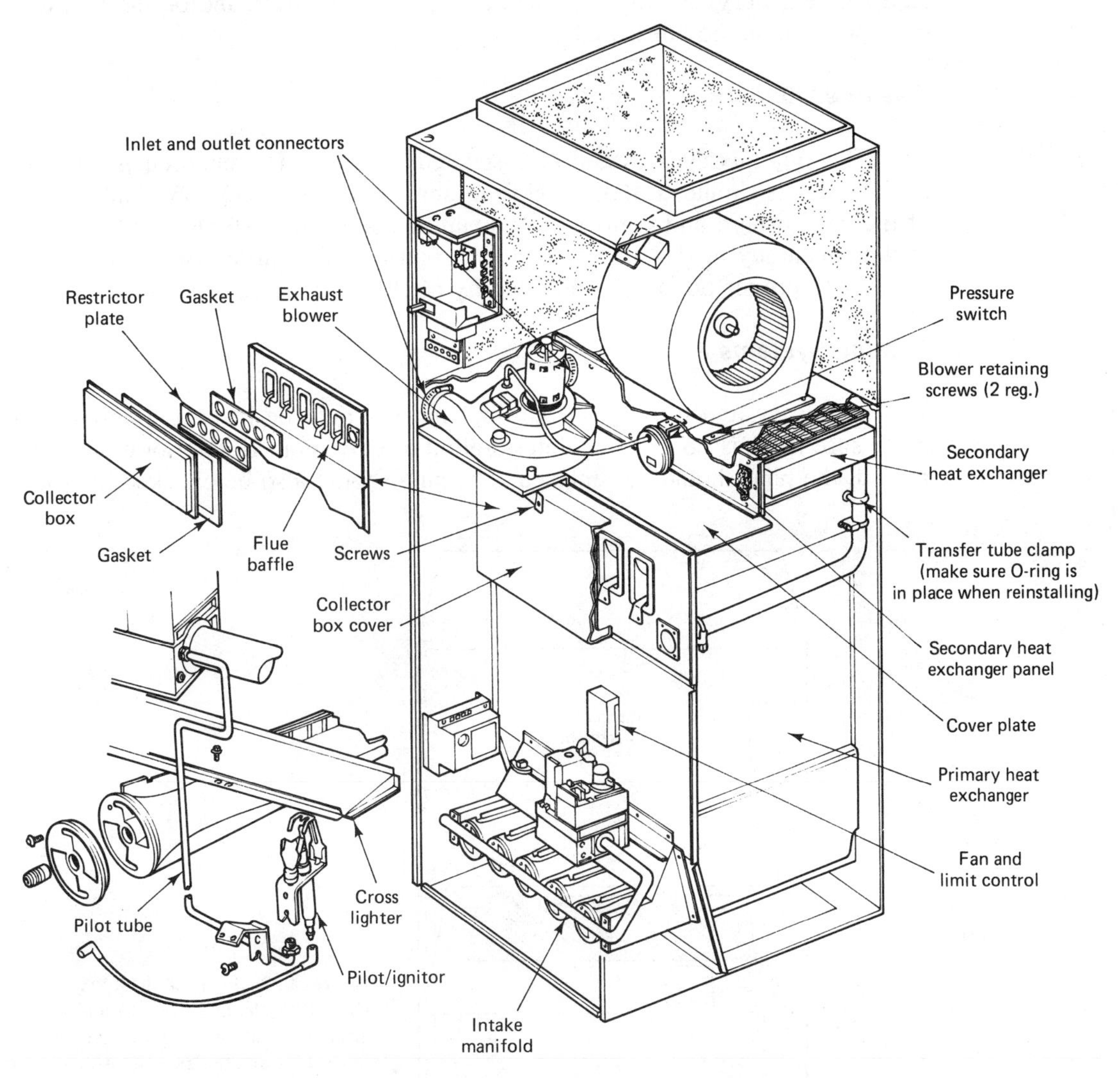

Figure 4.38 Primary and secondary heat exchanger location in condensing warm air furnace. (Courtesy of The Williamson Company.)

changer and using a stiff brush with a vacuum cleaner to remove any built-up residue on the exchanger surfaces (Figure 4.39).

If the furnace contains an electronic air cleaner and humidifier, the lint screen, humidifier pads, and after filters should be cleaned (Figure 4.40). Also, at this time the blower drive belts should be inspected and adjusted, if necessary. The humidifier pad should be cleaned at regular intervals and changed yearly. Follow the manufacturer's instructions for cleaning and replacement.

As with oil-fired units, check the belt tension on the blower pulley, and adjust if necessary. At this time the bearings on the blower motor and fan, if so equipped, should be lubricated.

Gas-Fired Boilers

All boiler flue passages should be inspected annually. The maintenance of gas-fired boilers is similar to that of oil-fired units, with the exception of adjustment of the gas burners and controls to maximize combustion efficiency. Check the cycling accuracy of the furnace controls, lubrication requirements of the circulators, and proper operation and integrity of all safety valves and gauges.

Venting Systems

The venting system of all gas-fired appliances should be inspected annually to make sure there is no external deterioration or corrosion. If deterioration is suspected, the pipe should be replaced. The pipe should also be checked internally

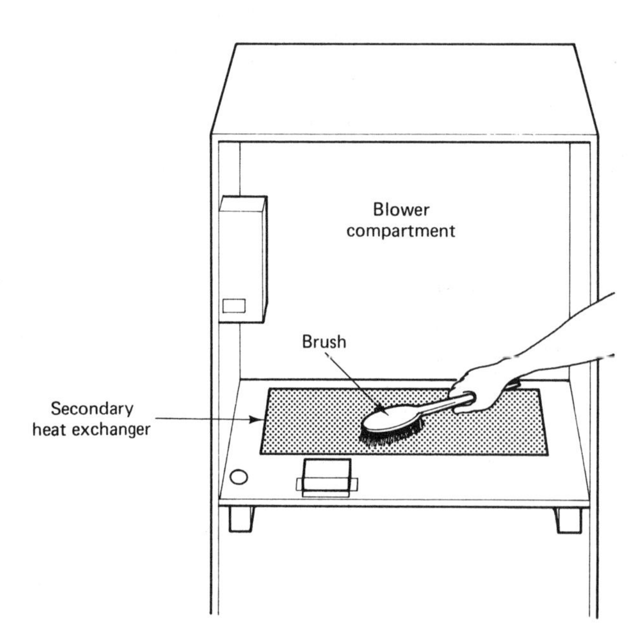

Figure 4.39 Furnace heat exchangers should be cleaned prior to and directly after the heating season to prevent loss in exchanger efficiency as well as deterioration due to formation of corrosives on the heat exchanger surfaces.

to make sure that there is no accumulation of soot or sediment. If this condition exists, the pipe should be cleaned with a stiff wire brush. At the breeching of the chimney pipe, inspect for deterioration the furnace cement or mortal holding the pipe in place in the thimble. Replace suspected joints with fresh cement or mortar. All horizontal pipe runs should be inspected for sags. Low spots in the piping can act as condensate traps which lead to pipe corrosion. If sags are found, use strapping material to lift the pipe and secure it in its proper position.

Gas Feed Lines

The main gas feed line and the supply piping in the vicinity of the boiler should be carefully inspected at the start of each heating season. Any suspect piping should be replaced. If leaks are suspected in the system, a soapy water solution should be used to inspect all joints and suspect areas.

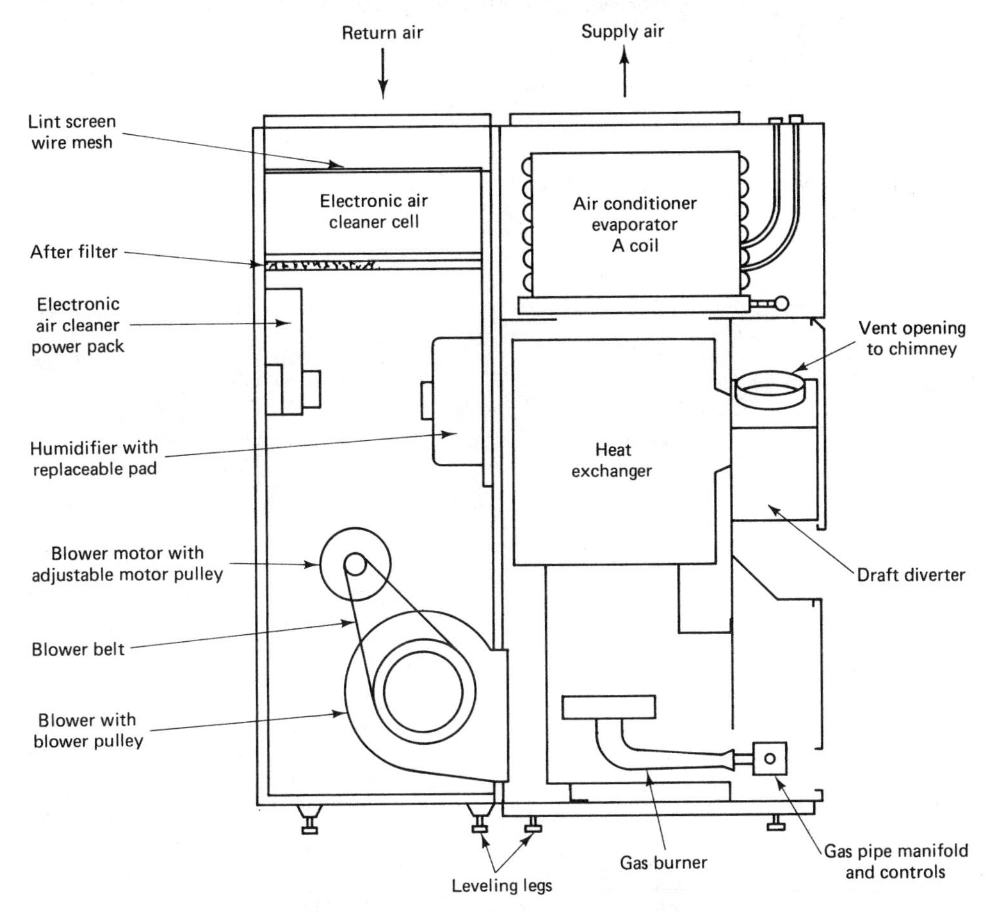

Figure 4.40 Location of electronic air cleaner and humidifier pads in typical warm air furnace. (Courtesy of The Williamson Company.)

On any new installation, or in cases where major gas components have been replaced, the operation of the complete heating system must be checked. If a smell of gas is encountered, the electrical and gas supply must be turned off promptly until the trouble has been isolated and corrected.

If the appliance uses LP gas rather than natural gas, it should be remembered that LP gas is heavier than air. As such, it will collect in the low areas or confines of the appliance room and *must* be vented before attempting to restart the appliance.

Checking gas input and pressure. The gas appliance should be checked to ensure that the gas pressure in the supply manifold is adjusted properly for both the particular appliance and the type of fuel gas used. All appliances will list the required manifold pressures for both natural and LP gas. To check the manifold gas pressure, most pressure regulators are equipped with a tapped open-

TABLE 4.1 GAS FLOW CHART (FT^3/HR)

Seconds for one revolution	Size of test dial		Seconds for one revolution	Size of test dial		Seconds for one revolution	Size of test dial	
	$\frac{1}{2}$ ft^3	1 ft^3		2 ft^3	5 ft^3		2 ft^3	5 ft^3
10	180	360	10	720	1800	62	116	290
11	164	327	11	655	1636	64	112	281
12	150	300	12	600	1500	66	109	273
13	138	277	13	555	1385	68	106	265
14	129	257	14	514	1286	70	103	257
15	120	240	15	480	1200	72	100	250
16	113	225	16	450	1125	74	97	243
17	106	212	17	424	1059	76	95	237
18	100	200	18	400	1000	78	92	231
19	95	189	19	379	947	80	90	225
20	90	180	20	360	900	82	88	220
21	86	171	21	343	857	84	86	214
22	82	164	22	327	818	86	84	209
23	78	157	23	313	783	88	82	205
24	75	150	24	300	750	90	80	200
25	72	144	25	288	720	92	78	196
26	69	138	26	277	692	94		192
27	67	133	27	267	667	96	75	188
28	64	129	28	257	643	98		184
29	62	124	29	248	621	100	72	180
30	60	120	30	240	600	102		176
31		116	31	232	581	104	69	173
32	56	113	32	225	563	106		170
33		109	33	218	545	108	67	167
34	53	106	34	212	529	110		164
35		103	35	206	514	112	64	161

ing in which to insert a U-shaped manometer. The gas pressures must be measured with both the pilot light and main burners operating. There are adjustment screws provided on most pressure regulators for the purpose of making *minor* changes in pressure. Major changes in gas pressure will generally have to be made by changing the size of the burner orifice.

Measuring natural gas input. In natural gas appliances, it is important to measure the gas input to the burners to ensure that the delivered gas will match the manufacturer's Btu rating for the boiler, furnace, or hot water heater. The Btu rating of the appliance is a function of three variables associated with natural gas: the specific gravity of the gas, its heating value per cubic foot, and the volume of gas delivered to the burners per unit of time.

A simple procedure for measuring gas input to the appliance involves the use of a gas flowchart (Table 4.1). To perform this calculation, the specific gravity

TABLE 4.1 GAS FLOW CHART (FT^3/HR) (*continued*)

Seconds for one revolution	Size of test dial		Seconds for one revolution	Size of test dial		Seconds for one revolution	Size of test dial	
	$\frac{1}{2}$ ft^3	1 ft^3		2 ft^3	5 ft^3		2 ft^3	5 ft^3
36	50	100	36	200	500	116	62	155
37		97	37	195	486	120	60	150
38	47	95	38	189	474	125		144
39		92	39	185	462	130		138
40	45	90	40	180	450	135		132
41			41	176	439	140		129
42	43	86	42	172	429	145		124
43			43	167	419	150		120
44	41	82	44	164	409	155		116
45	40	80	45	160	400	160		113
46		78	46	157	391	165		109
47	38		47	153	383	170		106
48		75	48	150	375	175		103
49			49	147	367	180		100
50	36	72	50	144	360			
51			51	141	353			
52		69	52	138	346			
53	34		53	136	340			
54		67	54	133	333			
55			55	131	327			
56	32	64	56	129	321			
57			57	126	316			
58	31	62	58	124	310			
59			59	122	305			
60	30	60	60	120	300			

of the gas along with its heating content in Btu per cubic foot must be obtained from the utility. During the time that this test is being performed, all other gas appliances should be turned off.

To calculate gas input, determine the size of the gas meter dial (either $\frac{1}{2}$, 1, 2, or 5 ft^3). While the burner is operating, one revolution of the gas dial is timed. The number of feet of gas consumed by the burners per hour is then read from the chart. This figure is then multiplied by the Btu value of the gas per cubic foot to determine the Btu input to the appliance. For example, if the size of the dial on the gas meter is $\frac{1}{2}$ ft^3, and it takes 20 seconds for the dial to complete one revolution, the burner consumes 90 ft^3 of gas per hour. Assuming a heating content of 1000 Btu/ft^3 for the gas, the input to the appliance is 90,000 Btu. If this corresponds to the nameplate rating on the unit, the system is functioning as it should. If there are any discrepancies, the gas pressure and orifice should be checked to ensure proper pressure settings and sizing and readjusted if necessary.

Combustion Efficiency Testing

The test kit used for testing the combustion efficiency of oil-fired appliances is the same as that used to take measurements on gas-fired heating equipment. Follow the instructions on the test kit. The carbon dioxide tests, stack temperature, and draft measurements are performed as described previously. The smoke test is not performed on gas-fired appliances. Combustion efficiency is then determined using the chart supplied.

5

DOMESTIC HOT WATER HEATING SYSTEMS

Domestic hot water heating represents a major energy cost that has risen dramatically during recent times. For many years, domestic hot water heating systems underwent little change in either technology or design, as there was little impetus to conserve energy. With the dramatic rises in energy prices that accompanied oil shortages in the late 1960s and 1970s, manufacturers began to look more closely at the design and operating characteristics of their units, resulting in modern domestic hot water (DHW) systems that can deliver far higher efficiencies than that of their earlier counterparts. For the most part, these devices can be expected to deliver years of relatively trouble-free economical service.

Before we undertake an examination of the various methods by which hot water can be heated, Figure 5.1 illustrates the domestic hot water system in relationship to the entire residential water system. Cold water, supplied by either a drilled well or city supply main, provides the pressure by which all the water in the home moves throughout the piping system. Normal operating pressures of residential water systems supplied by central distribution mains are in the neighborhood of 40 to 50 psi, while driven wells generally run between 30 and 50 psi, depending on the depth of the well and capacity of the water pump. Note that the cold water line splits upon entering the house, dividing into the hot and cold water service lines to all household appliances. Normal residential piping size is $\frac{3}{4}$ in. on all main service lines, with branch lines being either $\frac{3}{4}$ or $\frac{1}{2}$ in. in diameter,

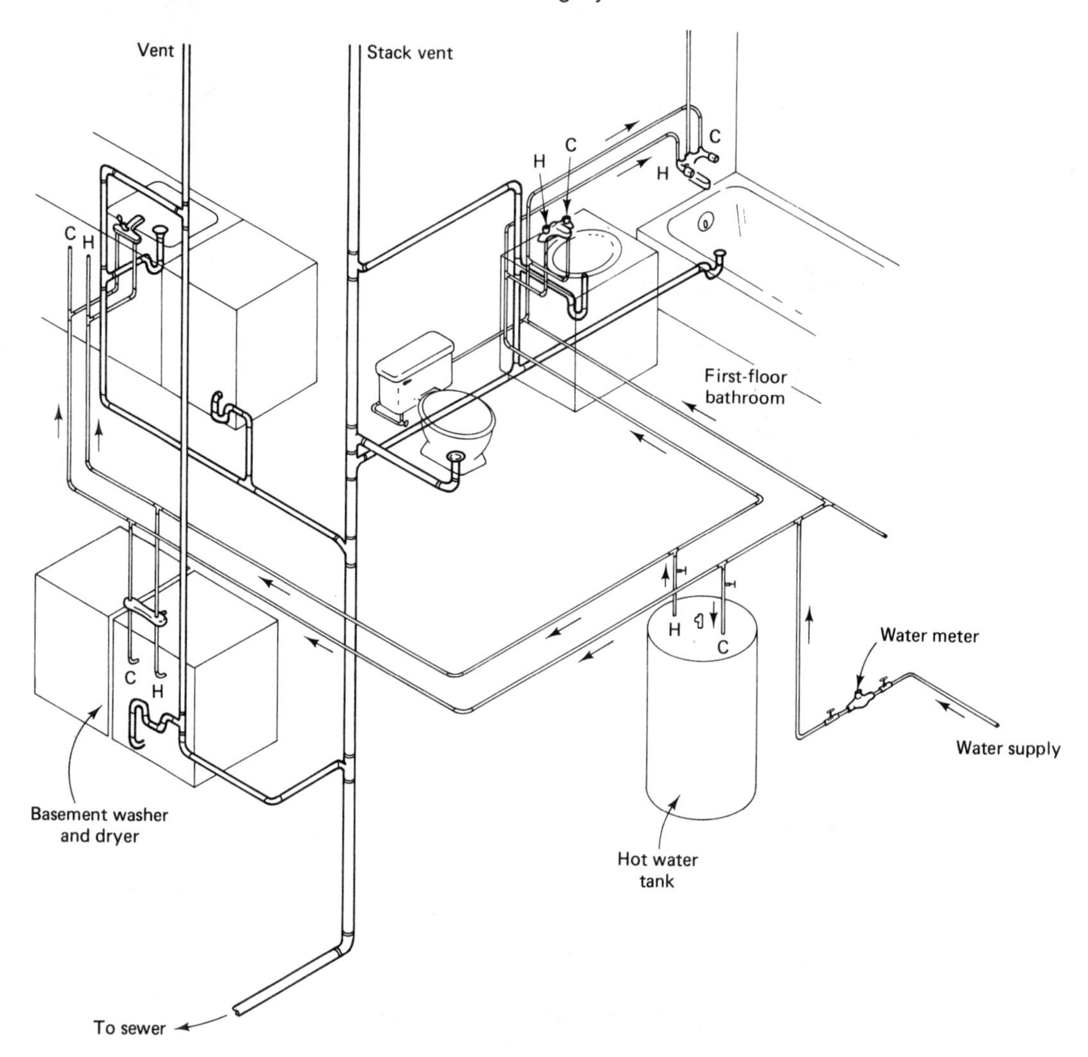

Figure 5.1 Typical residential water supply and distribution system. (From M. Greenwald and T. McHugh, *Practical Solar Energy Technology*, Prentice-Hall, Englewood Cliffs, N.J., 1985.)

depending on the requirements of the individual appliance. Normal drain line sizes are between $1\frac{1}{4}$ and $1\frac{1}{2}$ in. for sinks and tubs, and 3 to 4 in. for soil lines from all commodes. All copper tubing used in residential plumbing applications should be rated L. The other conventional residential tubing rating, M, is softer than its L counterpart and its use should be restricted to low-pressure heating system plumbing circuits. Drain lines, when plumbed in copper, are usually done with tubing rated M or with Drain-Waste-Vent (DWV) copper pipe.

Domestic hot water can be produced in a number of different ways. The most common of these are: conventional stand-alone water heaters fueled by electricity, fuel oil, or natural or LP gas, and solar domestic hot water heating systems; and hot water production within the tankless coil of a hot water or steam boiler fired by either conventional fossil or solid fuels. A discussion of each of these types of systems follows.

ELECTRIC HOT WATER HEATERS

The electric hot water heater is perhaps the most common type of water heating appliance in use today (Figure 5.2). These heaters are available in capacities of 30, 50, 66, 80, and 120 gal. The water heater should be sized according to the

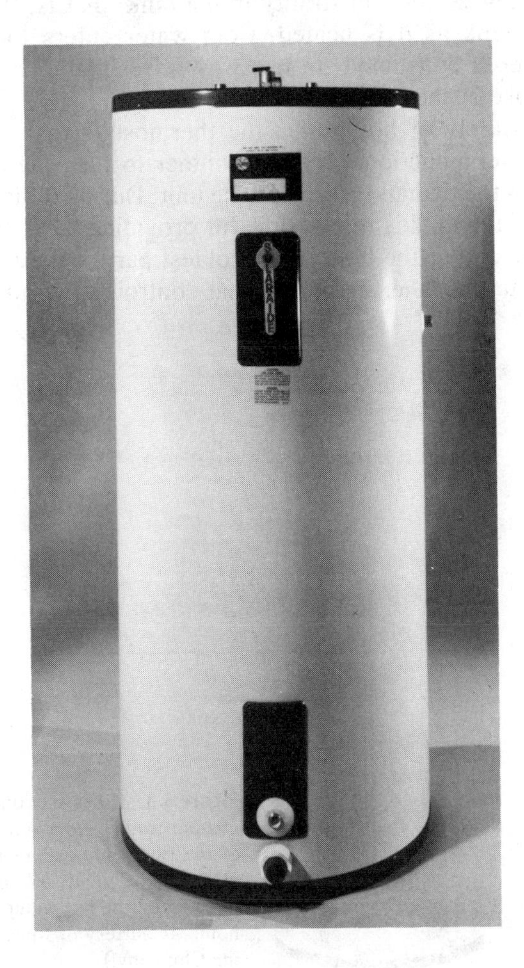

Figure 5.2 Residential electric hot water heater. (Courtesy of Rheem Manufacturing Company.)

number of people in the home and the recovery capacity of the water heater in gallons per hour that the heater can deliver. As a general sizing rule, each person in the home will consume approximately 20 gal of hot water per day. Sizing the heater in this manner would result in the purchase of an 80-gal hot water heater for a family of four people. However, if the home requires that during certain times of the day the unit will be called upon to produce 30 gal of hot water per hour, the heating elements and tank size must be sized to provide that amount of recovery.

The typical residential electric hot water heater contains two heating elements. One is located at the top of the tank, the other at the bottom (Figure 5.3). The heating elements are manufactured from a high-resistance metal that is immersed within the water of the tank. When an electrical current is passed through the elements, heat is produced that is transferred to the water surrounding the elements. The water will stratify in the tank; that is, the hot water rises to the top of the tank as it is heated. Cold water enters the bottom of the tank as the hot water is consumed. In this way a fresh supply of hot water is always available for use at the top of the heater.

To keep a ready supply of hot water, the thermostats that control each heating element work in conjunction with one another to turn the elements on selectively, depending on the demand placed on the unit. During ordinary use, the element at the bottom of the tank is responsible for providing the majority of the hot water. Since it is located at the bottom, or coldest part of the water heater, it will continue to operate until the thermostat that controls it is satisfied.

Figure 5.3 Cross section of typical residential water heater. Note the location of the two heating elements, which operate independently of one another, depending on the hot water demand in the home. (Courtesy of Rheem Manufacturing Company.)

During standby conditions, the top heating element will not come on. The upper thermostat, sensing the supply of hot water within the tank, will not energize the element until the water temperature drops to the thermostat cut-in setting. Note the temperature adjustment feature on a typical water heater electric thermostat in Figure 5.4. When hot water consumption is high, the makeup supply of cold water entering the water heater will eventually outpace the ability of the lower element to heat it sufficiently, and cold water will reach the top of the water heater. When this happens, the upper element, sometimes referred to as the quick-recovery element, will be energized to heat the water.

While the upper heating element is energized, the lower element is automatically disconnected. This switching arrangement prevents the electrical branch circuit supplying the water heater from overloading, since each heating element can draw up to 4500 W. This switching arrangement is illustrated in Figure 5.5.

Electric hot water heaters offer both advantages and disadvantages to the homeowner. They are the least expensive of the many types of water heaters available. In addition to low initial costs, they do not require an auxiliary fuel source or associated storage tanks. They are wired into a suitable outlet and are ready to provide years of trouble-free service. Also, there is very little maintenance required on these units. Only periodic replacement of the anode rods and heating elements and a monthly tank flushing keeps these systems operating reliably for many years. Where water demand is light, as in vacation homes and weekend residences, electric water heaters are the units of choice.

The major disadvantage of electric hot water heaters is based on the relatively high cost of electrical power needed to operate the units relative to lower cost fuels such as oil and natural or liquefied petroleum gas. Although much is made of electric water heaters being virtually 100% efficient in their conversion of electricity to heat, little attention is given to the overall electrical generating/supply system being perhaps 25 to 30% efficient. This figure includes thermodynamic losses that occur during the production of electricity at the power plant

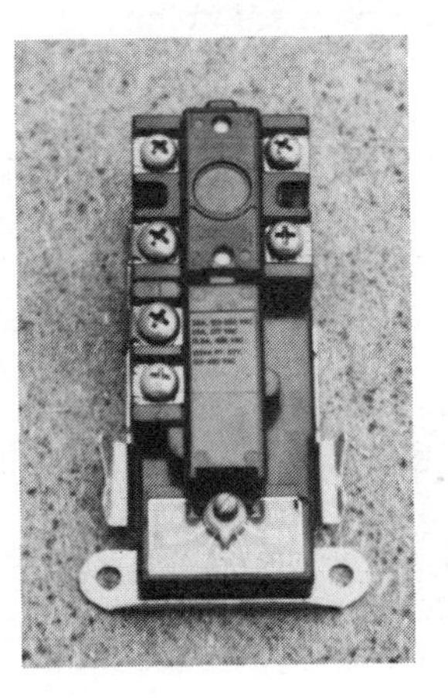

Figure 5.4 Hot water heater thermostat. The temperature is adjusted by rotating the dial to the desired temperature setting. For proper operation, these thermostats must be in direct contact with the tank shell. All power to the heater *must be turned off whenever opening the heater access covers to the thermostats.*

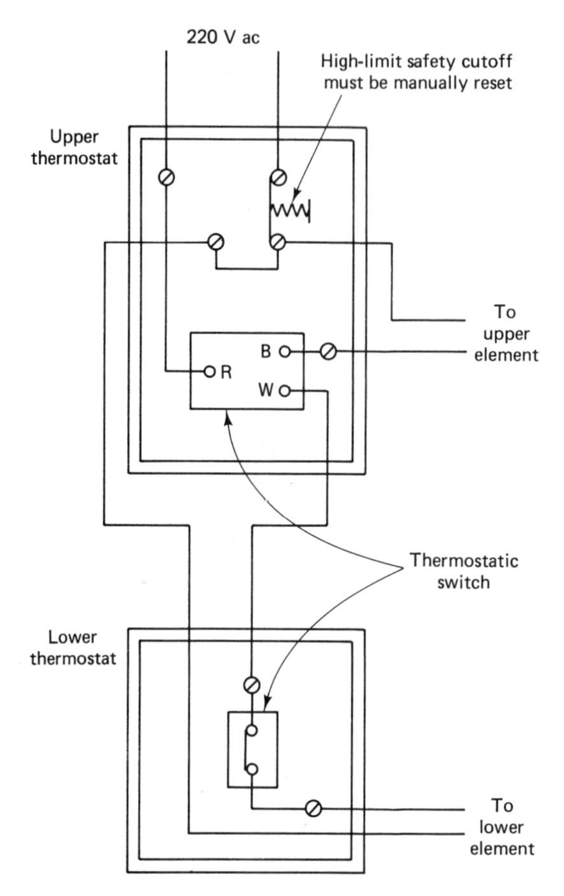

Figure 5.5 Wiring configuration of typical hot water heater thermostats. Note thermostatic switch that disconnects lower element when the upper one is energized, and vice versa. In most arrangements, the lower element provides the majority of the water heating. Should cold water temperatures activate the upper thermostat, indicating an inability of the heater to keep up with the residential demand, the upper thermostat will be energized, deactivating the lower element, until temperatures in the top of the tank satisfy the thermostat switch setting.

as well as line losses encountered during the delivery of the power from the utility grid to the residential electrical system. From an economical standpoint, the consumer pays for all the losses inherent in the utility generation and delivery system, thereby making electric water heaters perhaps the most expensive method of producing DHW in comparison to other types of water heaters.

The water heater tanks are fabricated from steel. All appropriate connections are welded onto the steel shell during initial stages of tank manufacture. After the shell has been completed, the inside of the tank shell is given a coating of porcelain enamel (glass) or concrete for corrosion protection. Concrete tank linings are superior to glass linings for corrosion protection and are also more expensive than their glass-lined counterparts. Tanks that are glass lined are equipped with anode rods made from magnesium. These rods extend into the interior of the tank and are supplied for corrosion protection. They should be replaced periodically due to electrolytic deterioration that occurs over a period of time.

Electric water heaters are available with either fiberglass or pressure-injected urethane foam insulation. Of the two options, urethane foam insulation is preferable since the resulting heat loss from foam-injected tanks is significantly less than heat losses from fiberglass-insulated tanks.

GAS-FIRED WATER HEATERS

The configuration of most gas-fired water heaters is somewhat different from that of electric water heaters (Figure 5.6). Gas-fired water heaters are set up to burn either natural or liquefied petroleum gas. Since the Btu content of LP gas is greater than that of natural gas (2200 Btu/ft^3 versus 1000 Btu/ft^3), LP water heaters are equipped with different-size gas orifices and modifications to the gas valve pressure settings to ensure that the Btu input to the water heater is the same for either type of gas used.

Note from Figure 5.6 that the combination gas control thermostat valve is located at the bottom of the heater. This valve both meters the incoming gas to the pilot light and main burners and serves to turn the gas supply on and off in response to the water temperature within the tank. The heater is equipped with a central baffled flue that dissipates the heat generated from the combustion of the gas and removes the by-products of the combustion process from within the home.

Gas-fired water heaters use a control that features a standing pilot light to ignite the main gas burner when the tank water temperature drops.* Most standing pilot systems are of the autopilot type and incorporate a pilot light that burns against a thermocouple. A thermocouple is a device composed of two dissimilar metals connected at each end. When heat is applied to one end of the thermocouple, a small electrical current is produced. This current is sufficient to operate a solenoid gas valve that serves to open and close the gas ports to the main burner assembly. A simple thermocouple is illustrated in Figure 5.7.

During normal operation, the valve assembly operates as illustrated in Figure 5.8. As long as the pilot light burns against the thermocouple, the electricity generated by the thermocouple is sufficient to energize the keeper that holds the inlet port of the gas valve open. Should a loss of pilot light occur for any reason, the keeper is deenergized, resulting in a loss of magnetism, which causes the return spring to close the gas valve (Figure 5.9). Virtually all residential water heater gas valves contain a high-limit control that will turn off the gas flow automatically should the water temperature reach a preset high limit.

Gas water heaters must be properly vented to ensure that all the by-products of combustion are removed from within the home. A typical venting configuration for most atmospheric combustion gas heaters is illustrated in Figure 5.10. Note from the illustration that a draft hood is used in the vent system. The draft hood is designed to draw dilution air from around the appliance into the vent system. This has the effect of diluting the flue gases in the chimney and lowering the flue

* See Chapter 4 for further discussion of pilot light ignition systems.

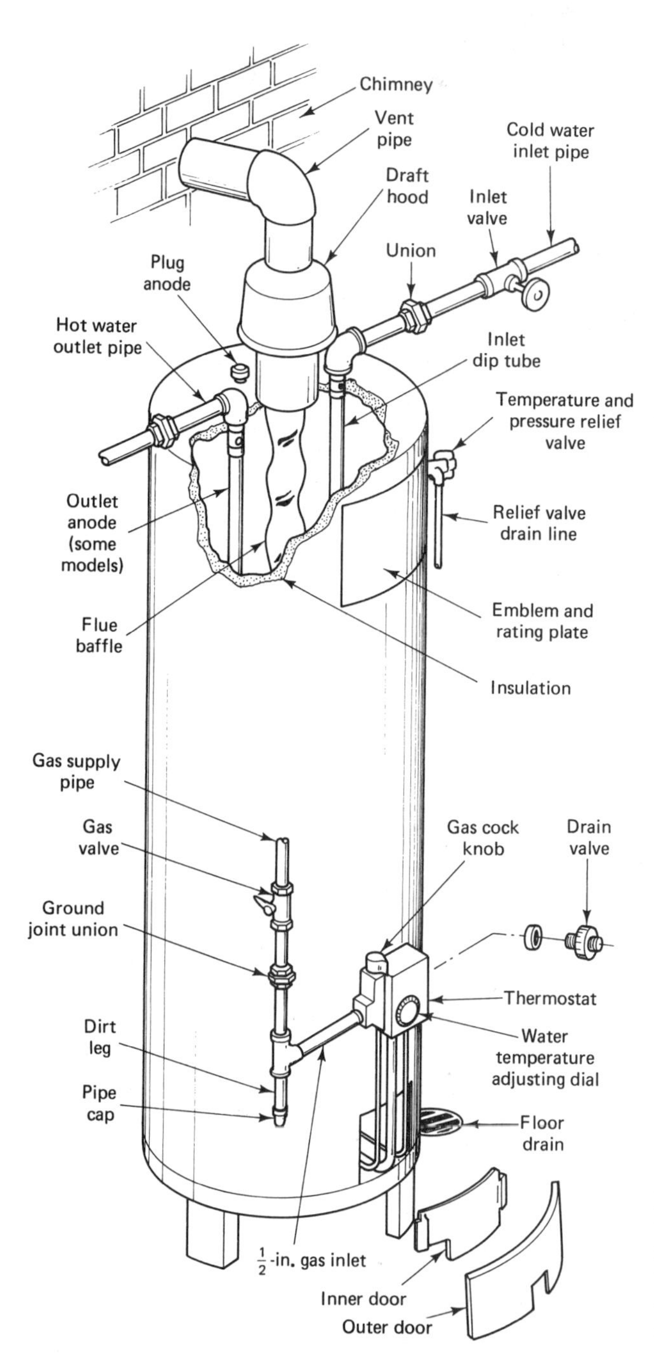

Figure 5.6 Gas-fired residential hot water heater. Note the use of an internal flue baffle to promote exhaust gas turbulence to increase heat transfer efficiency to the water within the heater. (Courtesy of A. O. Smith Corporation.)

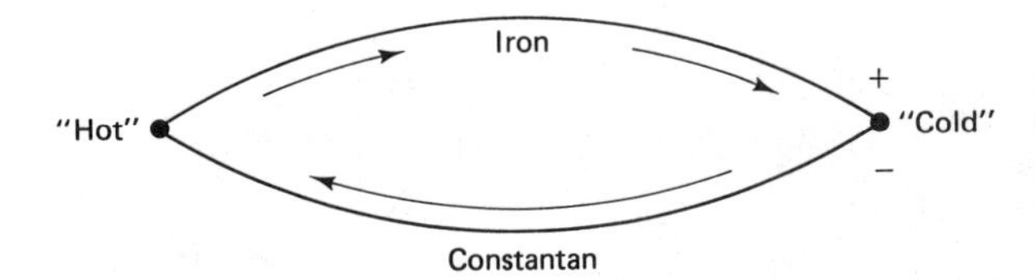

Figure 5.7 Simple thermocouple. A temperature difference across the hot and cold junctions produces a flow of electricity. (Courtesy of Robertshaw Controls Company.)

gas temperature of the exiting gases. Also, a draft hood minimizes the effects of momentary downdrafts that can occur in a chimney from time to time (Figure 5.11).

Gas-fired water heaters burn with combustion efficiencies in the range 50 to 65%. Since gas combustion is relatively clean, these units require little maintenance other than an occasional cleaning of the combustion area to prevent a buildup of dust and dirt and a check of the pilot and main burner flames. Since the construction characteristics of gas-fired water heaters are similar to their electrically powered counterparts, maintenance pertaining to anode rods and tank flushing is the same for both units.

The cost of operating gas units is less compared to electric water heaters. Of the two gas choices available, natural gas is cheaper than LP gas. Natural gas is usually the fuel of choice in cities and large surburban areas where centralized

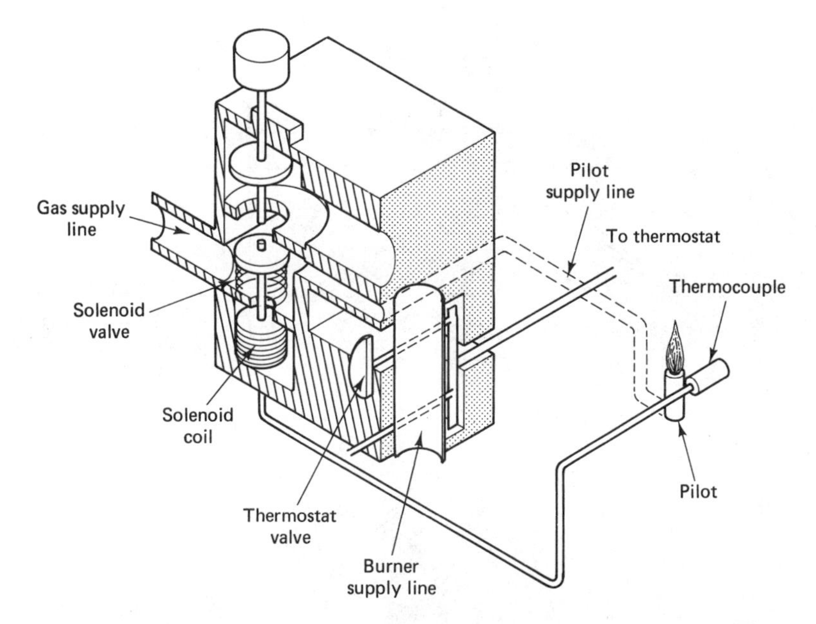

Figure 5.8 Typical pilot light/thermocouple gas ignition control system. Electricity generated by the thermocouple is sufficient to energize the keeper, allowing the gas valve to remain in the open position. (From M. Greenwald and T. McHugh, *Practical Solar Energy Technology*, Prentice-Hall, Englewood Cliffs, N.J., 1985.)

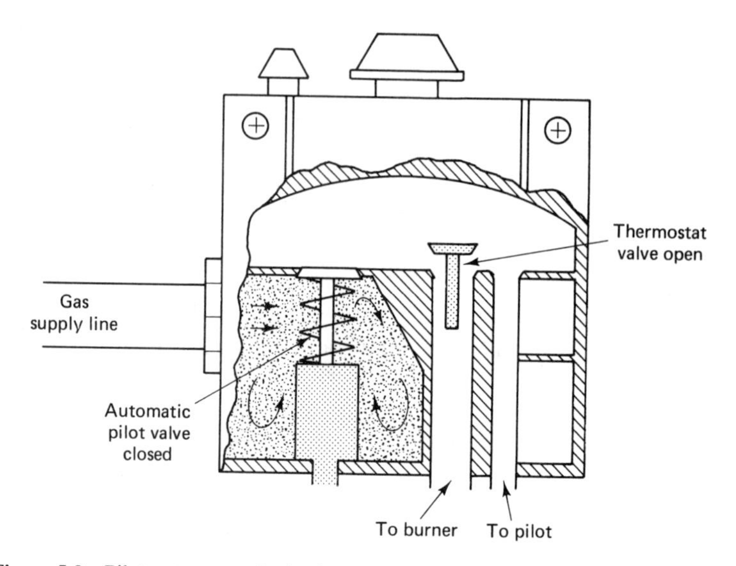

Figure 5.9 Pilot outage results in deenergizing the keeper, closing the gas valve to both the main and pilot burner assemblies. (From M. Greenwald and T. McHugh, *Practical Solar Energy Technology,* Prentice-Hall, Englewood Cliffs, N.J., 1985.)

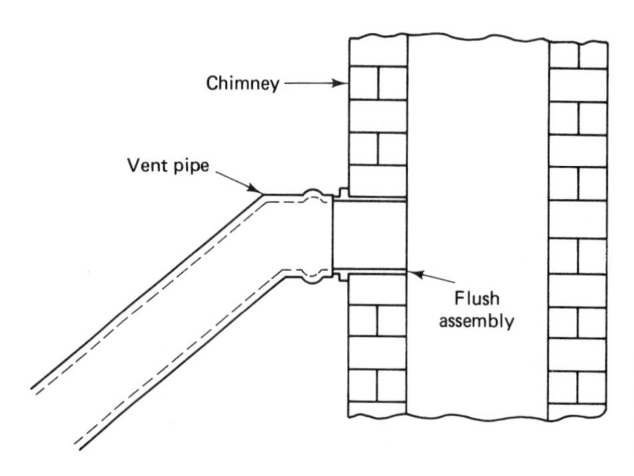

Figure 5.10 Venting configuration for atmospheric combustion appliances. The thimble should be flush with the inside of the chimney.

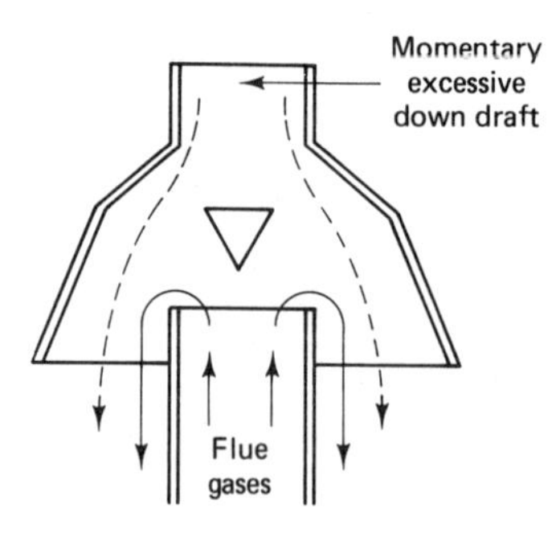

Figure 5.11 The draft hood stabilizes momentary chimney downdrafts, directing the downward air away from the combustion chamber, helping to eliminate pilot outage. (Courtesy of Robertshaw Controls Company.)

distribution facilities and underground pipelines are installed. LP gas is usually classified as a rural fuel since it requires the use of storage tanks located at the home, supplied by delivery trucks on a regular basis. These deliveries are timed according to gas consumption records of the residence.

OIL-FIRED WATER HEATERS

Oil-fired water heaters are perhaps the most efficient and cost-effective of all types of domestic hot water heaters. A typical oil-fired water heater is illustrated in Figure 5.12. A common component arrangement on these units is for placement of the oil burner at the bottom of the water heater, which is connected into a central baffled flue. The flue baffles (Figure 5.13) aid in promoting turbulence and heat transfer from the flue gases to the water within the tank.

These units feature high-speed flame retention oil burners that yield combustion efficiencies in the range 80 to 85%. A burner of this type is illustrated in Figure 5.14. The sequence of operation of a typical oil-fired water heater is as follows: Upon a drop in water temperature, the aquastat [an aquastat is an electrical switch activated by changes in water temperature (Figure 5.15)] closes the electrical circuit to the oil burner. Combustion of the oil begins and continues until the set temperature on the aquastat has been reached by the water within the tank.

Oil-fired water heaters are highly cost-effective due to the high flame temperature and efficiency of the fuel oil combustion system, as well as relatively high efficiency of heat transfer within the system. As an example of this cost-

Figure 5.12 Residential oil-fired water heater. (Courtesy of Ford Products Corporation, Valley Cottage, NY 10989.)

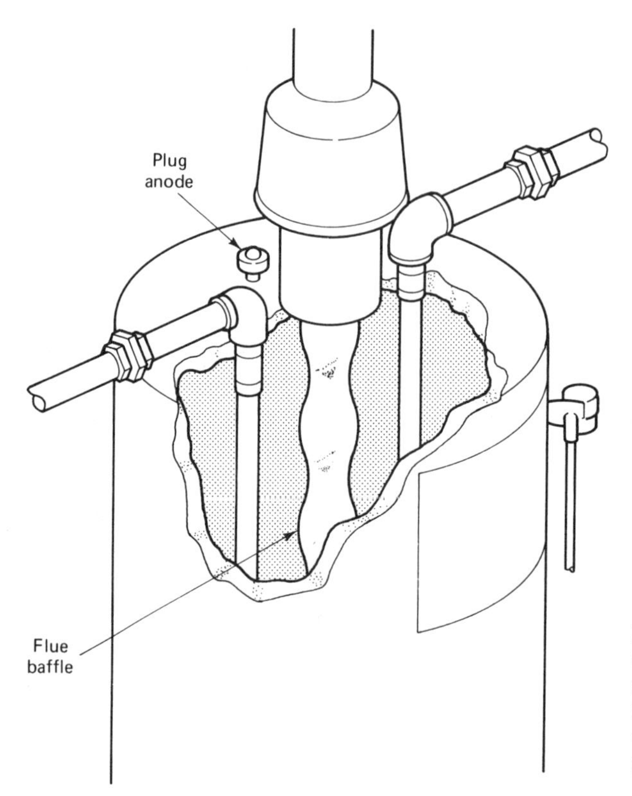

Figure 5.13 Baffles within the flue help to promote exhaust gas turbulence, increasing the efficiency of heat transfer between the flue gases and the water within the heater. (Courtesy of A. O. Smith Corporation.)

Figure 5.14 High-speed flame-retention oil burner. These units feature combustion efficiencies in the range 83 to 85%, far higher than their slow-speed counterparts of just a few years ago. (Courtesy of A. O. Smith Corporation.)

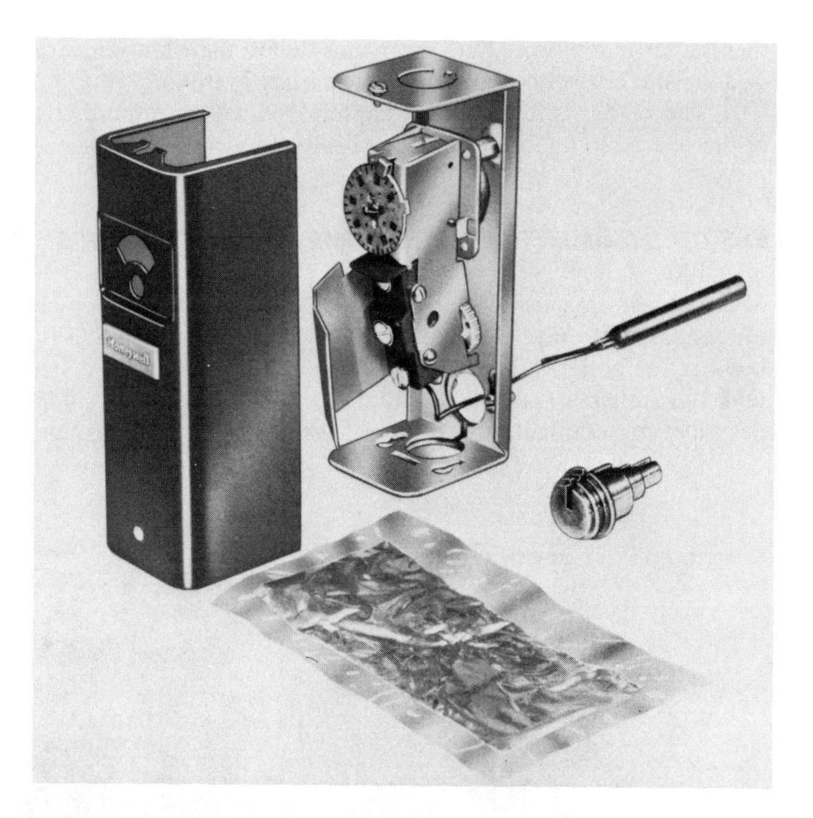

Figure 5.15 Immersion aquastat of the type used to activate the burner mechanism in a typical oil-fired water heater. The bag in the illustration contains a special conductive grease that increases heat conduction from the immersion well to the sensing bulb of the aquastat. (Courtesy of Honeywell, Inc.)

effective operation, the recovery capacity of a typical oil-fired water heater with a firing rate 0.75 gal/hr is between 100 and 120 gal of hot water per hour (rated at a temperature increase of 100°F above the incoming cold water temperature). In contrast, the standard electric or gas-fired water heater usually has a recovery rate of perhaps 30 to 35 gal/hr. Comparative cost analysis between the various systems illustrates that the oil-fired unit will produce equivalent quantities of hot water at perhaps half the cost of other fossil-fuel residential water heaters.

Oil-fired water heaters are much more expensive than either electric or gas-fired units. Oil systems require annual inspection and maintenance that should be performed by qualified personnel, procedures that are not ordinarily necessary with other types of water heaters. Also, the use of oil requires a storage tank on the premises for delivery of fuel, a situation that is not applicable with either electric or natural gas-fired units.

Given the higher initial cost and annual maintenance expenses of oil-fired water heaters compared to electric and gas appliances, oil-fired units will pay for themselves many times over in savings due to their low delivered cost per gallon of domestic hot water during their operating lifetime.

The reader is directed to Chapters 3 and 4 for a more complete discussion of the operation and maintenance of oil- and gas-fired appliances.

TANKLESS COIL (SUMMER/WINTER) DOMESTIC HOT WATER SYSTEMS

Many homes that use either hot water or steam boilers for central heating are also equipped to provide domestic hot water from a coil that is inserted into the body of the boiler. Since this type of domestic hot water system has no storage tank but rather a water coil within the boiler, the term **tankless coil** is used to describe this configuration. Figure 5.16 illustrates the position and design of a

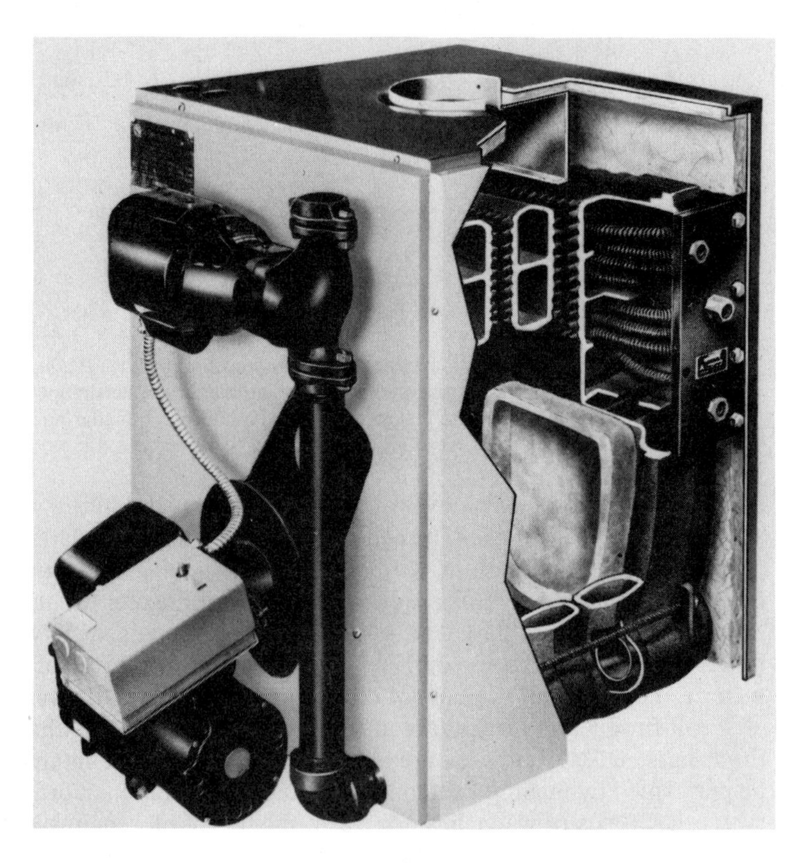

Figure 5.16 Tankless coil location in hot water boiler. Although the exact location of a coil will vary from one manufacturer to another, the coil is always placed in the top of the boiler, where it is exposed to the hottest water temperatures in the system. (Courtesy of The Burnham Corporation.)

tankless coil within a hot water heating boiler. Note from the figure that the coil is made up of tightly wound copper tubing shaped to fit in a specific boiler only, since the sizes and shapes of heating boilers vary from one manufacturer to another. These coils are available in a variety of shapes and sizes for different boilers (Figure 5.17).

Tankless coils require special aquastats that allow the boiler to maintain a specified low-limit temperature, usually between 130 and 160°F, even when the boiler is not being used to provide space heating. The boiler is on standby all year long, maintaining a preset low-limit water temperature in order to provide the residence with hot water. This type of year-round operation affords the homeowner both advantages and disadvantages. The primary advantage in this type of system is that a separate domestic hot water heater is not required: All hot water and space heating is supplied by one appliance. The main disadvantage of this system is that the boiler must maintain a minimum temperature even during the nonheating season, which can be relatively expensive, depending on the cost of the fuel and the particular boiler in the system.

During the winter months when the boiler provides space heating and cycles on and off with regular frequency, the cost of providing domestic hot water is relatively economical, featuring a system efficiency between 50 and 60%. How-

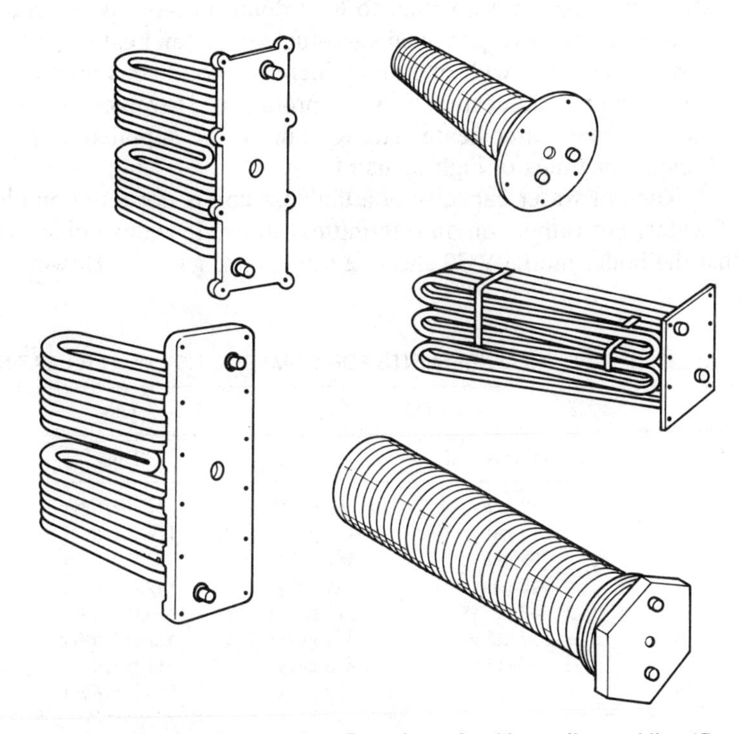

Figure 5.17 Varying shapes and configurations of tankless coil assemblies. (Courtesy of Amtrol.)

ever, during the nonheating season, the efficiency of the system drops to between 10 and 20%, since the boiler is maintaining minimum water temperature solely for residential DHW. Table 5.1 illustrates the approximate fuel consumption required for domestic hot water production on an annual basis, given a variation in family size.

Tankless coils can be installed in boilers that burn conventional fossil fuels, as well as in solid-fuel or multifuel units equipped to burn either wood or coal. In solid-fuel systems, the recovery rate of the coil will be less than that of a comparably equipped fossil-fuel boiler. Each boiler manufacturer publishes the coil ratings for a particular series of boilers. Table 5.2 illustrates the amount of hot water production available from one series of heating boilers.

One modification of the tankless coil configuration is a hybrid installation, wherein the coil (or water bottle in a solid-fuel boiler) can be used to heat the water in a conventional stand-alone domestic hot water heater. This system is illustrated in Figure 5.18. In this type of system, the coil or water bottle within the boiler draws cold water from the bottom of the conventional water heater. The water is circulated through the coil or bottle in the boiler, where it is heated and returned to the water heater. This type of system has some obvious advantages. Since solid-fuel boilers are generally on standby in order to maintain combustion, they are in a position to heat domestic hot water on a continual basis. Operating in this way, a typical solid-fuel boiler can heat up to 100 gal of domestic hot water daily and will, of course, increase wood consumption in direct relation to the amount of domestic hot water produced. However, this system also allows for an external water heater storage tank to be supplied with ample hot water, sufficient for times of high demand.

The hot water capacity of a tankless coil is not rated on a continuous draw of water, but rather, on an intermittent draw. If we examine Table 5.2, we note that the boiler model V-33 shows a rating of $4\frac{3}{4}$ gal/min. However, the hourly rate

TABLE 5.1 FUEL REQUIREMENTS FOR DOMESTIC HOT WATER HEATING

Number of people	Oil		Gas	Electric
	Tankless coil	Tank	Tank	Tank
1	90 gal/yr	62 gal/yr	87 therm/yr	1,562 kwh/yr
2	179 gal/yr	125 gal/yr	174 therm/yr	3,124 kwh/yr
3	268 gal/yr	187 gal/yr	262 therm/yr	4,686 kwh/yr
4	358 gal/yr	249 gal/yr	349 therm/yr	6,248 kwh/yr
5	447 gal/yr	311 gal/yr	436 therm/yr	7,810 kwh/yr
6	537 gal/yr	374 gal/yr	523 therm/yr	9,372 kwh/yr
7	627 gal/yr	436 gal/yr	611 therm/yr	10,934 kwh/yr
8	717 gal/yr	499 gal/yr	698 therm/yr	12,496 kwh/yr

Note: Figures are approximate.

TABLE 5.2 TANKLESS COIL RECOVERY CAPACITY

			Rating[a]			
			Steam		Water	
Boiler number	Number	NPT connection	gal/min	Δp	gal/min	Δp
V-33	222	$\frac{1}{2}$	$4\frac{1}{4}$	36.0	$4\frac{3}{4}$	42.5
V-34	226	$\frac{1}{2}$	$5\frac{1}{4}$	17.5	$5\frac{3}{4}$	20.5
V-35	232	$\frac{1}{2}$	$6\frac{1}{4}$	22.5	7	27.5
V-36	445	$\frac{3}{4}$	$7\frac{1}{4}$	19.5	$8\frac{1}{4}$	23.5
V-37	445	$\frac{3}{4}$	$7\frac{1}{4}$	19.5	$8\frac{1}{4}$	23.5
V-38	445	$\frac{3}{4}$	$7\frac{1}{4}$	19.5	$8\frac{1}{4}$	23.5

Source: The Burnham Corporation.
[a] Gal/min is based on 200°F boiler-water temperature and 40 to 140°F rise; Δp is the pressure drop in psi.

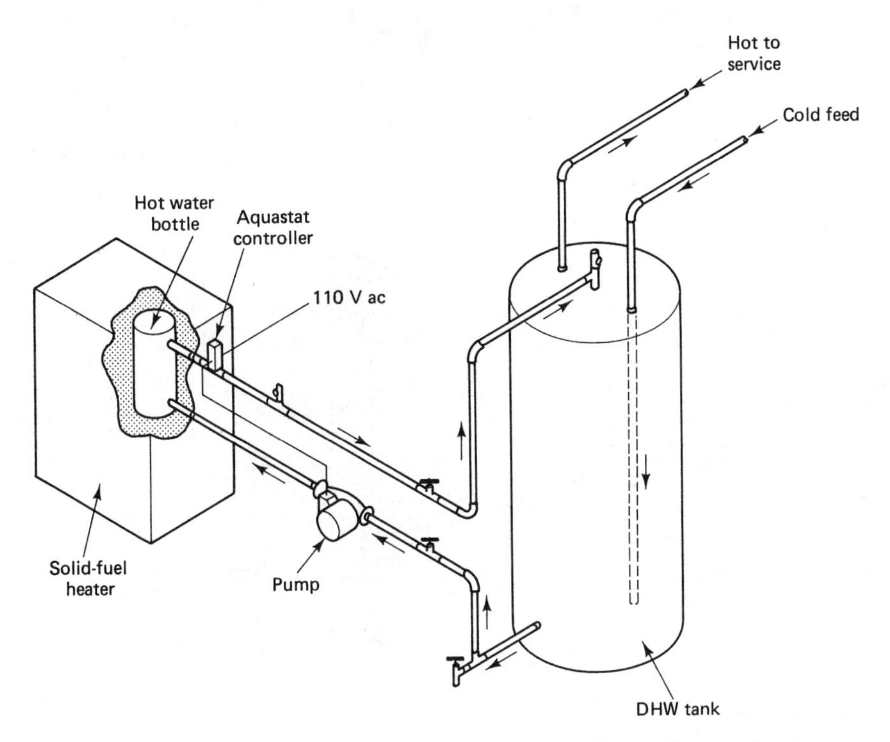

Figure 5.18 Use of a solid-fuel boiler with either a tankless coil or small water bottle to heat domestic hot water in a conventional stand-alone water heater. Note the use of a circulator and aquastat controller to keep the water moving between the water bottle and water heater. If a circulating pump is not used, a natural-convection system can be used as long as the pipe sizes are large enough (approximately $1\frac{1}{2}$ to 2 in. in diameter) to prevent overheating of the water.

of this coil is not $4\frac{3}{4} \times 60 = 285$ gal/hr as one might at first expect, but rather, approximately 40 to 50 gal/hr. This discrepancy in hourly versus intermittent draw occurs because continual draw on the coil drops the water temperature within the boiler to the point that the burner will not be able to maintain sufficient boiler-water temperature to allow the coil to produce DHW at its rated output at 100°F above the incoming water temperature (100°F is the temperature rise that is used to rate the coil recovery capacity).

Most tankless coils are easily replaced, should this become necessary, by removing the hold-down bolts and the old coil, and inserting the replacement unit. When assessing the overall performance of tankless coil heaters, the loss of off-season efficiency assumes a degree of importance in direct relation to the price of the type of fuel that is used to produce it. When oil, gas, and solid-fuel prices

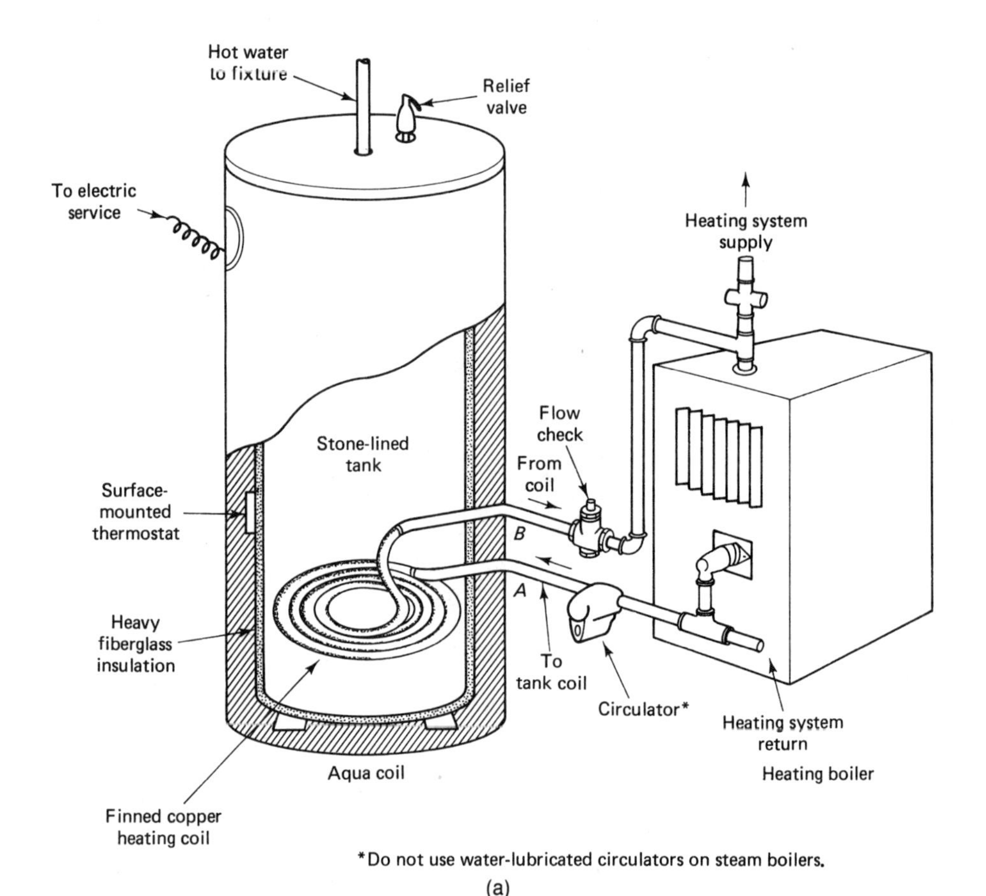

Figure 5.19 External tankless coil/aqua-booster assemblies: (a) aqua-booster coil (courtesy of Ford Products Corporation, Valley Cottage, NY 10989); (b) external hot water coil (courtesy of Weil-McLain, a Marley Company).

rise, the use of tankless coil water heaters becomes less appealing, due to the seasonal cost of fuel. One method of improving the efficiency of the tankless heater is through a modification of the piping array, which incorporates an external coil and/or storage tank assembly. Two types of external heater arrangements are illustrated in Figure 5.19. Referred to as **aqua-boosters**, or external tankless coils, these types of appliances place a tankless coil within an external storage tank, treating the coil and storage tank assembly as an independent heating zone.

This arrangement has distinct advantages over the conventional boiler tankless coil since the storage tank allows for up to 30 to 40 gal of stored hot water. Also, since the external coil assembly is treated as a separate heating zone, the boiler need not be set up to maintain minimum standby water temperatures, but rather, can sequence according to the needs of the storage tank independent of boiler-water temperature.

A second modification of this type is referred to as a **sidearm water heater** (Figure 5.20). The sidearm heater differs from the external coil in that it is usually

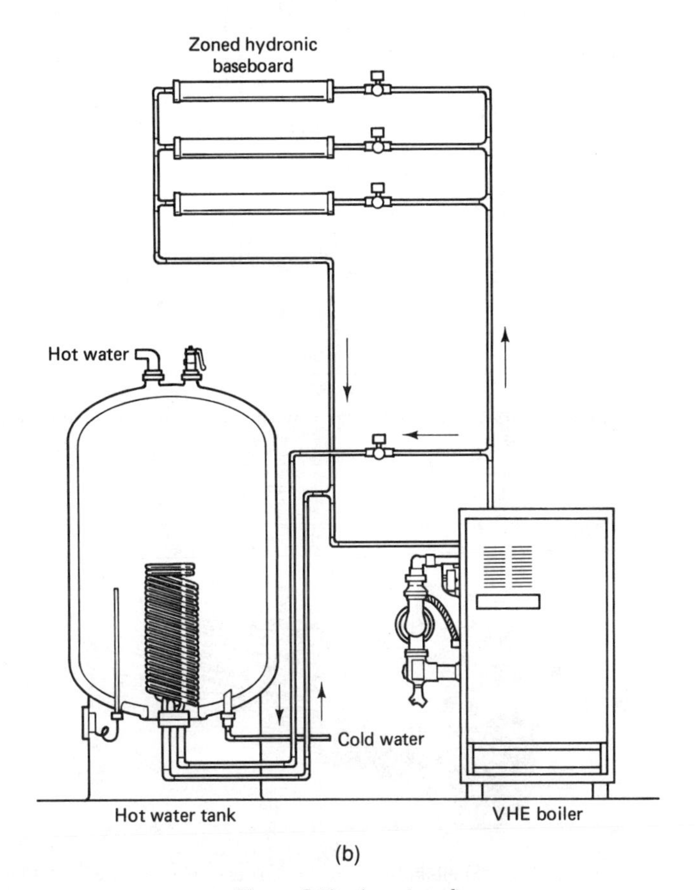

Figure 5.19 *(continued)*

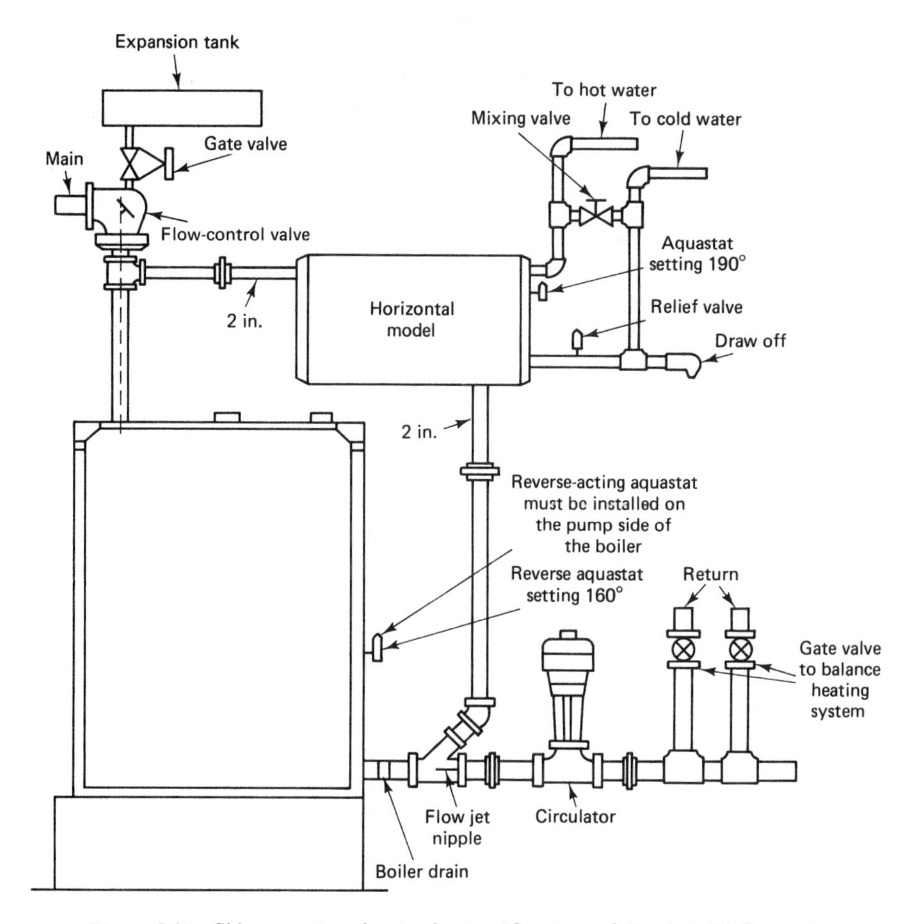

Figure 5.20 Sidearm external water heater. (Courtesy of Everhot All-Copper.)

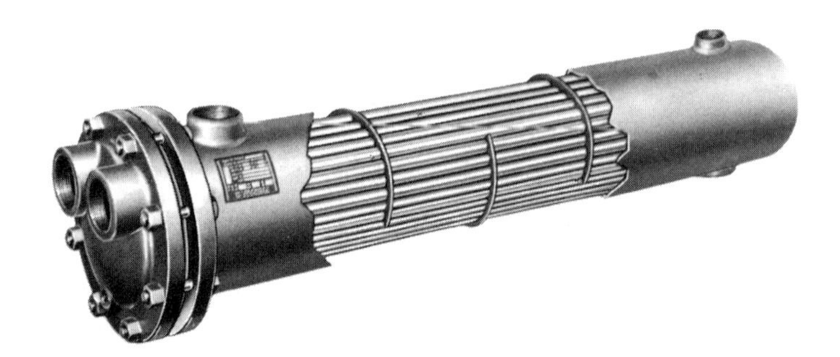

Figure 5.21 Shell-and-tube heat exchanger. (Courtesy of ITT Fluid Handling Division.)

a shell-and-tube heat exchanger rather than a storage tank with an internal coil. A typical shell-and-tube heat exchanger is illustrated in Figure 5.21. Note from the illustration that the sidearm heater is treated as a separate heating zone, similar in nature to the operation of the external tankless coil arrangement.

SOLAR DOMESTIC HOT WATER HEATERS

Solar water heaters are the most cost-effective types of solar installations. Operating on a year-round basis, these units produce domestic hot water on either a stand-alone basis or in conjunction with other types of domestic hot water heating systems. What follows is a discussion of the most common types of solar hot water heating systems, together with illustrations that focus on interfacing the solar system with the conventional domestic hot water system. There are several different types of solar DHW systems that are classified according to the method of circulation of the collecting fluid used within the system. We begin our discussion with the system components that are common to most solar domestic hot water installations.

Major Solar System Components

Solar flat-plate collectors. The liquid-cooled flat-plate collector is the most commonly used method for collecting solar energy. These units are adaptable to either domestic hot water or space heating systems, although domestic hot water constitutes the largest installed base of solar collectors in the United States.

The major components of a typical liquid-cooled flat-plate collector are illustrated in Figure 5.22. The collector consists of a frame, usually constructed

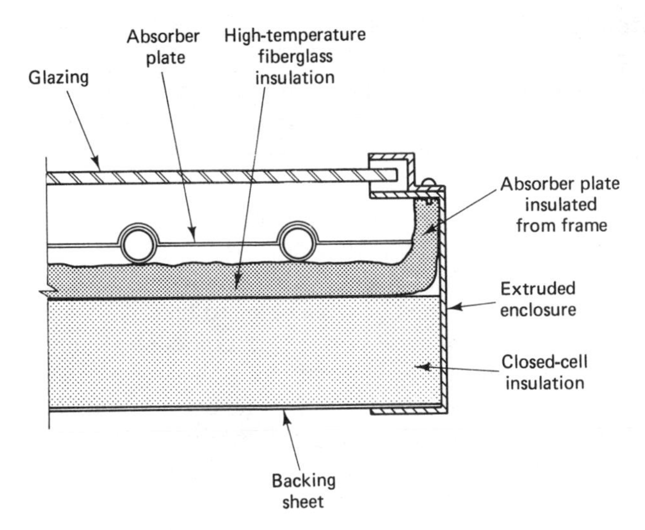

Figure 5.22 Cross section of typical liquid-cooled flat-plate solar collector.

from anodized aluminum; insulation to prevent heat losses from the absorber plate; the absorber plate, generally fabricated from copper; an exterior glazing made from tempered low-iron glass for maximum solar absorption; and stainless steel fasteners used to hold the components together. Although collectors from different manufacturers will vary in size and absorber configuration, the basic design illustrated in Figure 5.22 is fairly common.

Circulators. The circulator components that come into contact with either potable water or antifreeze solutions such as propylene glycol or silicone must be constructed from stainless steel, bronze, or brass. A typical circulating pump of this type is illustrated in Figure 5.23. In addition to stainless steel construction, these circulators are small in size, are usually water lubricated, and feature low power consumption.

Differential controllers. Differential controllers serve to turn the circulating pump on and off, depending on the temperature difference measured between the collector array and the bottom, or coldest, part of the solar storage tank. A differential controller for this purpose is illustrated in Figure 5.24. During operation, the controller compares the electrical resistance of thermistor-type temperature sensors located within the collector array and storage tank. When the temperature of the solar collectors is between 5 and 20°F warmer than the bottom of the solar storage tank, the controller will energize the circulator to begin moving collector fluid within the system. Most solar controllers incorporate advanced microprocessor circuitry into their design, which ensures high reliability and efficient long-term operations.

We now move on to a discussion of the various types of solar DHW systems available for a variety of residential installations.

Direct and Indirect Thermosiphon Systems

The simplest type of solar DHW system is classified as a **direct thermosiphon system** (Figure 5.25). The direct thermosiphon system relies on natural convection to circulate water between the solar collector(s) and the solar storage tank. This type of system is classified as a direct system because potable water is used as the collector fluid directly rather than using a synthetic heat exchange fluid. With the storage tank installed above the top of the collector, water will rise from the collector as it is heated and flow into the storage tank. Cooler water from the bottom of the storage tank will flow to the solar collector to take its place, and a circulating loop of water is established based on the natural convection of the fluid. The advantage of this system is that no external power is required for operation. The sun acts as the controller to turn the system on and off. No circulating pump is required to move the water. The direct thermosiphon system is applicable only in areas where no freezing temperatures occur.

(a)

(b)

Figure 5.23 Stainless steel circulator used in typical solar domestic hot water heating system: (a) full view; (b) sectional view. Note that the circulator is water lubricated, requiring no external oiling or maintenance. (Courtesy of Grundfos Pumps, Inc.)

Figure 5.24 Differential solar controller. The controller illustrated features digital temperature display from 32 to 212°F on collector, storage, and three auxiliary inputs. Digital displays are helpful in monitoring the performance of the solar system and useful in pinpointing malfunctions. (Courtesy of Independent Energy, Inc.)

To overcome limitations imposed by freezing temperatures, yet retain the simplicity of operation of a thermosiphon system, an **indirect thermosiphon system** is sometimes used (Figure 5.26). Note that the indirect thermosiphon system uses a solar storage tank equipped with an internal heat exchanger and antifreeze as the collector fluid. Some indirect solar designs use a charged refrigerant within a closed collector loop with insulated storage tanks located at the top of the collector assembly. When designing indirect forced-convection systems, the collector loop takes on the characteristics of a closed loop. An expansion tank has been added to the closed-loop assembly to accommodate expansion and contraction of the fluid within the collector loop during daily system operation. The main disadvantage of thermosiphon systems is the necessity of locating the solar storage tank *above* the top of the collector loop. An 80-gal solar storage tank filled with water can weigh more than 1000 lb. This weight is sometimes too great for attic or ceiling beams to support. Also, attic and ceiling installations require the presence of pressurized water lines from the home water system in these unheated areas. Even if the solar system performs satisfactorily, it is always possible for a small section of a water line to freeze or rupture, with resulting damage to the interior of the home. These factors, in addition to the simplicity of the system design and operation, should be taken into consideration when undertaking the design and installation of an indirect thermosiphon system.

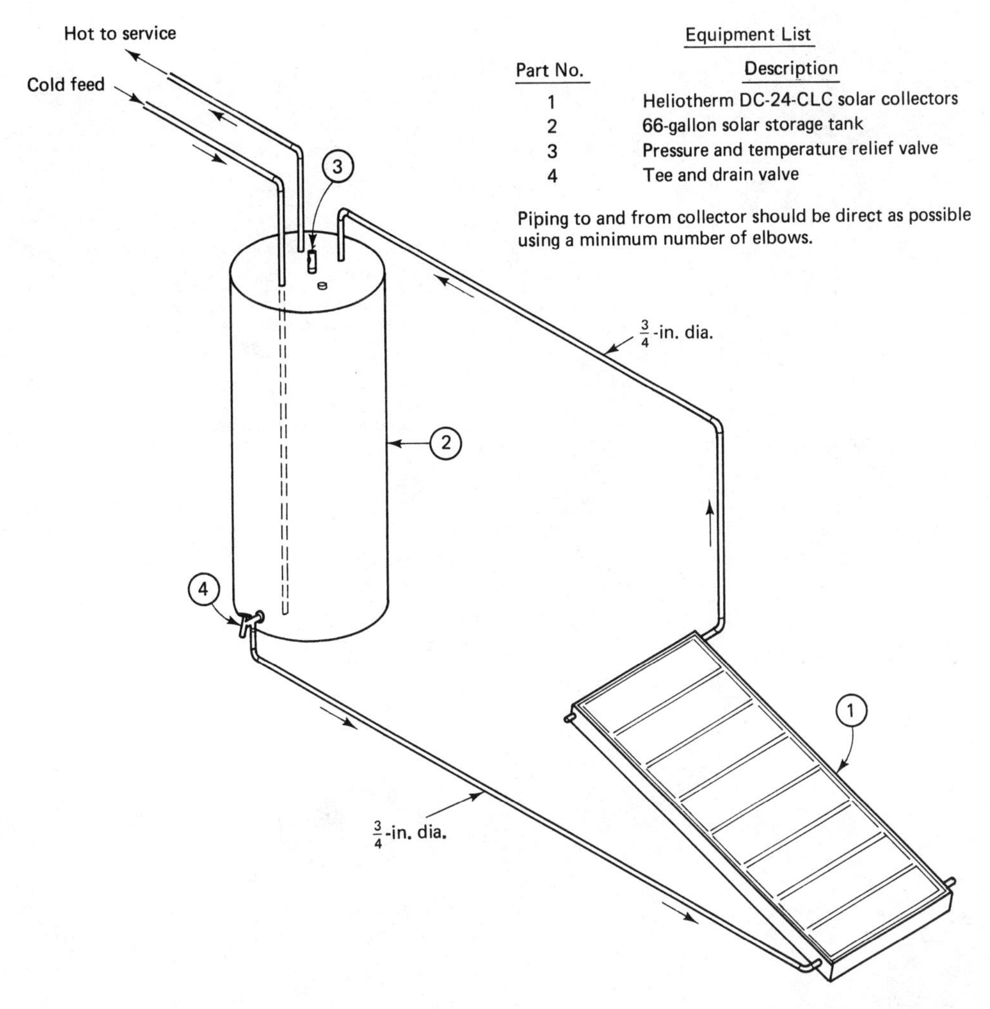

Figure 5.25 Direct thermosiphon solar system relies on varying densities of hot and cold water to establish flow from the collector to the storage tank, and back again, in a continual cycle, as long as a temperature differential exists. Note that one collector requires a minimum pipe size of $\frac{3}{4}$ in. If two collectors are used, the piping should be increased to 1 in.

Forced-Convection Direct Systems

A **forced-convection direct solar system** is similar in some respects to the direct thermosiphon system, except that a circulating pump and differential controller have been added to improve heat transfer efficiency. There are several advantages to this type of design over the direct thermosiphon configuration. Since the system utilizes a circulating pump for movement of water in the collector loop, the storage

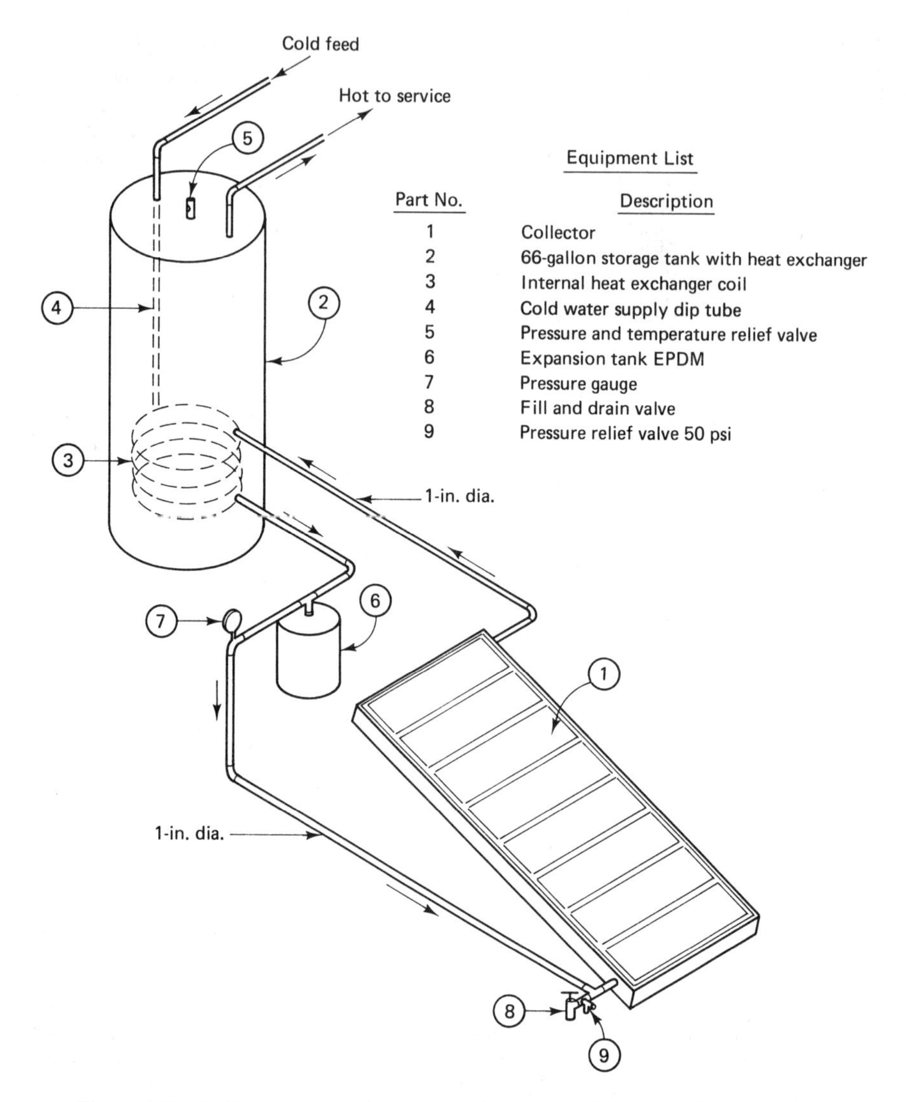

Figure 5.26 Indirect thermosiphon solar system. Note that with a closed system and internal heat exchanger in the solar storage tank, a small expansion tank has been added to the closed-loop assembly. Some indirect thermosiphon systems use a charged refrigerant closed loop and insulated storage tank above the collector, rather than antifreeze in the loop.

tank can be placed below the level of the solar collectors. Thus the normal arrangement of the solar collectors on the roof and storage tank in the basement (or lowest level in the home) is possible. Since the system uses potable water from the storage tank as the heat transfer medium, this type of system is used in warm climates only. Also, the use of residential water as the heat transfer medium requires that the circulating pump be constructed from stainless steel to ensure long-term operating reliability without fear of internal circulating pump corrosion.

Sometimes, nonfreezing capability can be added to the forced-convection direct system by installing a series of frost sensors in the controller circuitry that will automatically turn the circulating pump on when the outside temperature falls below 38°F. The constant circulation of water throughout the loop will prevent the pipes from freezing and rupturing as long as the freeze is neither too hard nor too long. This type of freeze protection is ideal in climates that experience only occasional frost and light freezes. It is not intended as foolproof freeze protection in parts of the country that experience severe winters.

Drain-Down Systems

A **drain-down solar system** is an open-loop design that circulates water from the storage tank through the collectors and back into the storage tank, using the street or pump pressure in the home to keep the collector loop filled. A circulating pump is used to move the water throughout the system (Figure 5.27).

The unique aspect of the drain-down system design incorporates the ability to isolate the collector array from the storage tank and to drain this portion of the system when outdoor temperatures fall below 38°F. Note that the system illustrated in Figure 5.27 incorporates three solenoid valves (6, 7, and 8). These valves are operated by the differential solar controller in response to temperature readings measured at the storage tank, solar collectors, and ambient frost sensors. In the drain-down mode, valves 6 and 8 close; valve 7 opens. This isolates the collector loop from the storage tank water and allows all water above the solenoid valves to drain out of the system into a convenient receptacle. For proper operation, a positive slope with no low spots must be established on all piping to and from the collectors to ensure complete drain-down. Also, it is common practice to install two frost sensors in the system. In the event of incorrect readings by one of the sensors, the system is still protected from freezing. If installed properly, these systems can be expected to perform for many years with only minor annual maintenance to the solenoid valves.

Drain-Back Systems

A **drain-back solar system** is different from a drain-down system, although the two are sometimes confused. A drain-back system uses an indirect heat exchanger along with an external water reservoir for system operation (Figure 5.28). During normal operation, the differential controller will turn on a circulating pump, which lifts potable water in the fluid reservoir and circulates it through the collector array, down into the internal heat exchanger within the solar storage tank, and back into the reservoir. Since potable water is used as the heat exchange medium, freeze protection in these systems is accomplished when the system shuts down, allowing water within the collector loop to drain back into the reservoir. Since

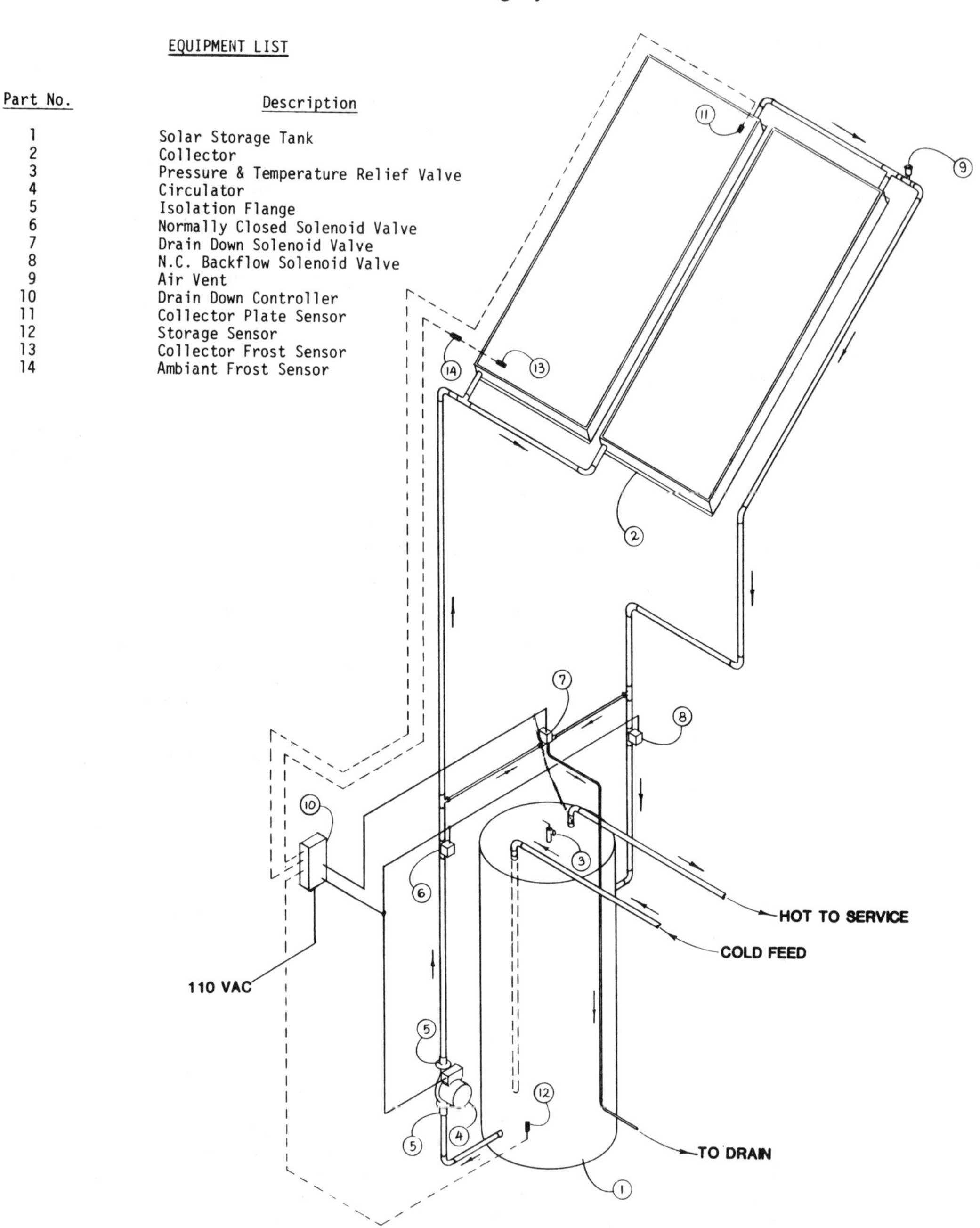

Figure 5.27 Drain-down solar system. Drain-down systems were first installed in the early and middle 1970s when the solar industry was in its infancy after the early oil shortages and energy price rises. These systems are prone to failure and freezing unless properly installed, in which case their performance is perhaps the best of all solar options.

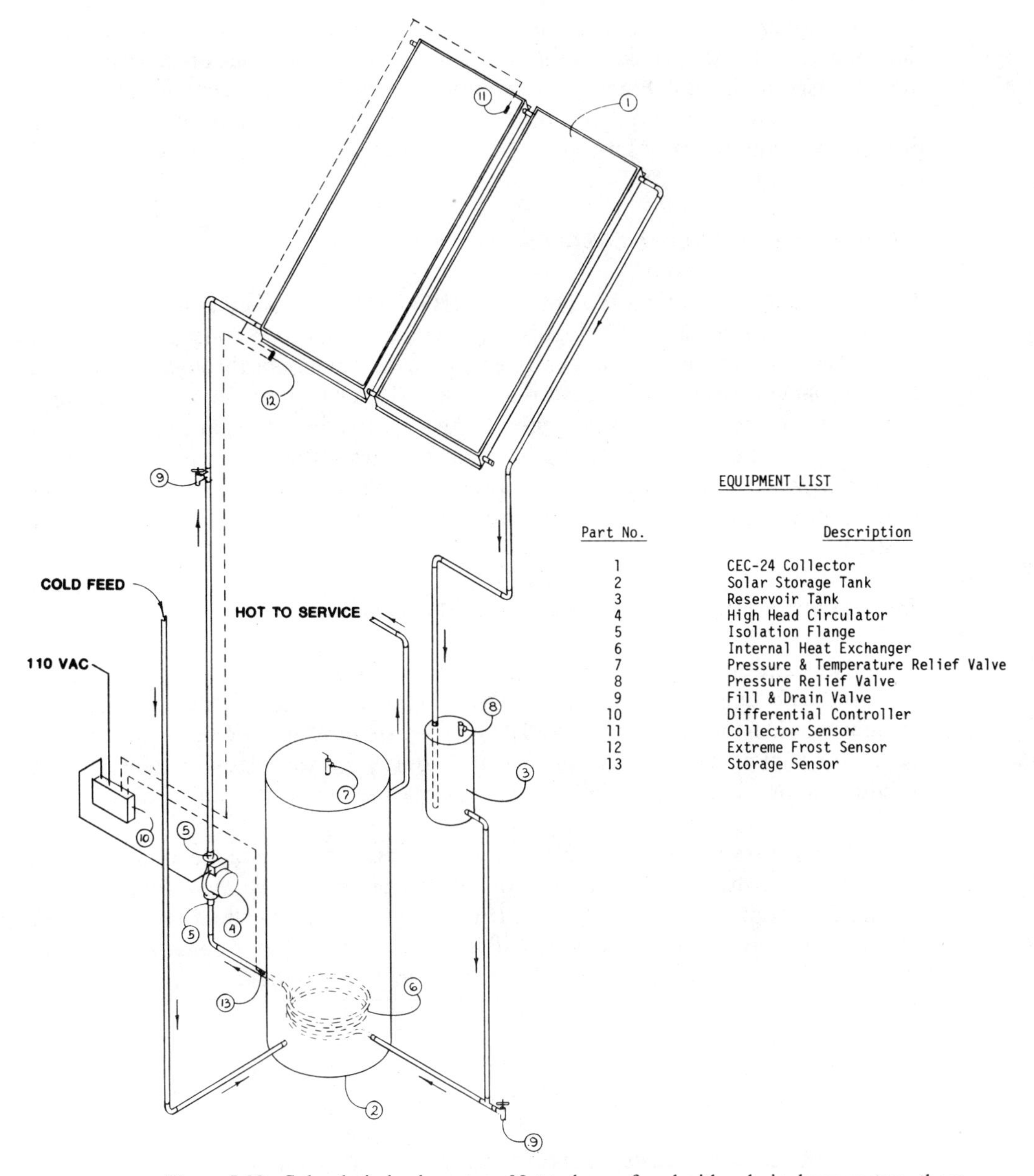

Figure 5.28 Solar drain-back system. Not to be confused with a drain-down system, these designs use a reservoir of water running through an internal heat exchanger to transfer heat from the collector array to the solar storage tank. Drain-back systems are hybrids, possessing neither the high efficiency of the drain-down system nor the freeze protection of the closed-loop antifreeze system if piping is not properly pitched.

the fluid loop is open rather than closed, an expansion tank is not required. However, since a circulating pump is required to lift water within the reservoir through the collectors, larger circulators than those found on other types of solar systems must be used to develop the required pumping pressure necessary to lift the water into the collector array. As with drain-down systems, all piping must be sloped properly to eliminate any low spots that would prevent total drainage of the circulating water.

Closed-Loop Antifreeze Systems

Perhaps the most common type of solar DHW system installed is the **closed-loop antifreeze system** illustrated in Figure 5.29. The collector loop contains a nontoxic antifreeze solution that circulates from the solar collectors through an internal heat exchanger within the solar storage tank and back up into the solar collectors. Since the collector loop contains pressurized antifreeze, an expansion tank must be used to moderate fluctuating system pressures during normal operation. Closed-loop antifreeze systems are virtually freeze-proof and offer long-term operating efficiency and reliability unmatched by any other type of solar hot water system.

Interfacing Solar with Conventional DHW Systems

Conventional bypass valving. Solar systems are usually installed in conjunction with existing conventional DHW appliances. In this design the solar system provides the majority of the DHW; the existing domestic hot water system acts as a backup to the solar system to provide for shortages in solar system production or when residential hot water demand exceeds the ability of the solar system to provide it.

When installed in conjunction with an existing stand-alone domestic hot water heater, whether electric, gas, or oil, the piping arrangement illustrated in Figure 5.30 is most often used. Note from the piping arrangement in Figure 5.30 that the incoming cold water is first diverted into the solar storage tank, where it is preheated. The solar preheated water is then fed into the incoming cold water feed line on the backup water heater. This valving system also allows the hot water from the solar storage tank to be fed directly into the hot water service line. An extra convenience with this piping arrangement allows for isolation of the solar storage tank, should this become necessary.

Tankless coil bypass valving. If the solar system is installed in a residence where domestic hot water is supplied by a tankless coil, the valving arrangement illustrated in Figure 5.31 should be used. This piping circuit offers a great deal of flexibility in channeling water from the solar system into either the tankless coil or bypassing the coil directly into the hot water service line. During the winter heating season, the solar system functions as a preheating system for

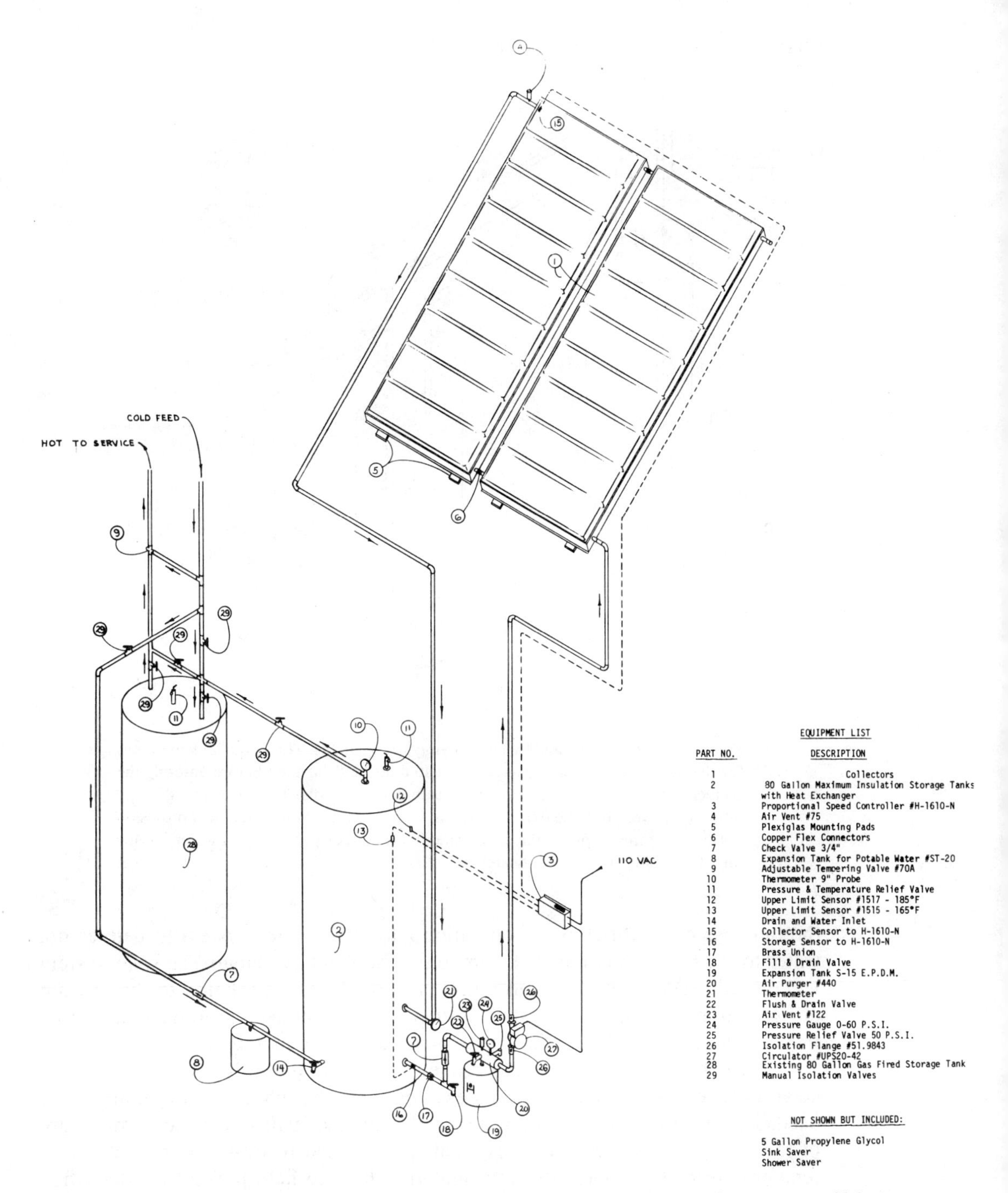

EQUIPMENT LIST

PART NO.	DESCRIPTION
1	Collectors
2	80 Gallon Maximum Insulation Storage Tanks with Heat Exchanger
3	Proportional Speed Controller #H-1610-N
4	Air Vent #75
5	Plexiglas Mounting Pads
6	Copper Flex Connectors
7	Check Valve 3/4"
8	Expansion Tank for Potable Water #ST-20
9	Adjustable Temoering Valve #70A
10	Thermometer 9" Probe
11	Pressure & Temperature Relief Valve
12	Upper Limit Sensor #1517 - 185°F
13	Upper Limit Sensor #1515 - 165°F
14	Drain and Water Inlet
15	Collector Sensor to H-1610-N
16	Storage Sensor to H-1610-N
17	Brass Union
18	Fill & Drain Valve
19	Expansion Tank S-15 E.P.D.M.
20	Air Purger #440
21	Thermometer
22	Flush & Drain Valve
23	Air Vent #122
24	Pressure Gauge 0-60 P.S.I.
25	Pressure Relief Valve 50 P.S.I.
26	Isolation Flange #51.9843
27	Circulator #UPS20-42
28	Existing 80 Gallon Gas Fired Storage Tank
29	Manual Isolation Valves

NOT SHOWN BUT INCLUDED:

5 Gallon Propylene Glycol
Sink Saver
Shower Saver

Figure 5.29 Closed-loop antifreeze solar system. These systems are the type most widely installed in the United States. Although their overall efficiency is not as high as a drain-down, they are virtually freeze-proof and require little maintenance, providing long-term reliable service and economy.

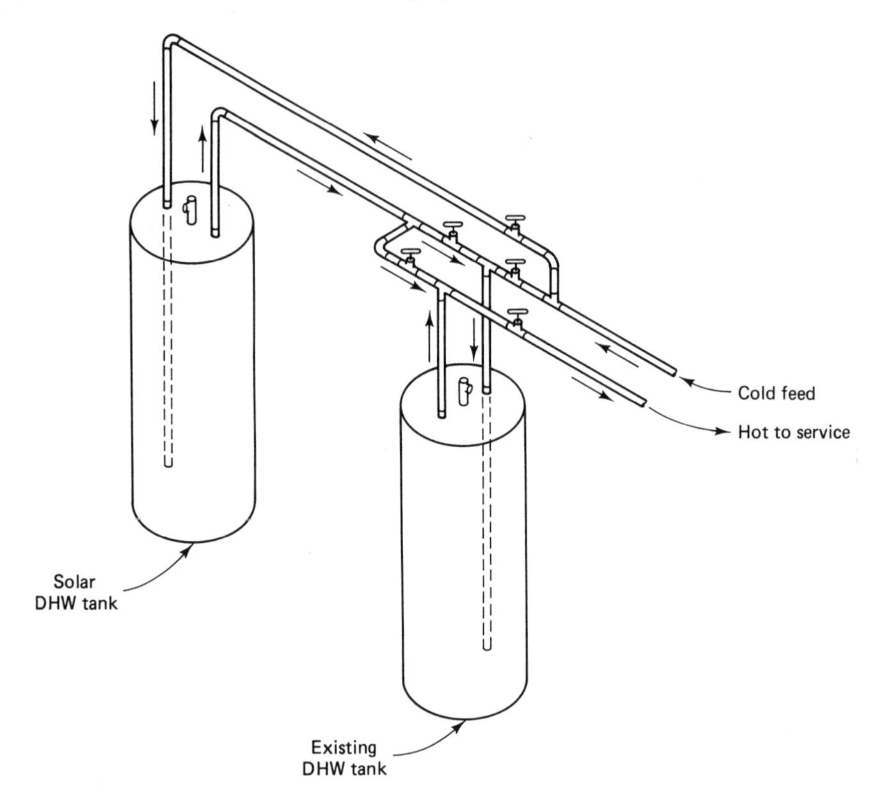

Figure 5.30 Conventional bypass valving arrangement. This arrangement allows for the incoming cold water to be preheated in the solar tank before entering the conventional backup water heater, or to bypass the solar tank and flow directly into the conventional heater. All bypass arrangements should include a provision for directing the output of the solar water heater directly into the hot water service line, bypassing the backup heater.

the tankless coil. This type of operation allows the solar system to deliver hot water to the coil in relation to the amount of sunlight available. The solar system will directly reduce the fuel consumption of the boiler in relation to the amount of hot water that the solar system (rather than the boiler) is producing. During the summer months, the boiler can be shut down completely.

During the summer it is sometimes preferable to run the output from the solar storage tank through the tankless coil even though the boiler is not being used. Although channeling the water in this manner will lower the temperature of the hot water, since part of the heat will be used to raise the internal water temperature of the boiler, this is sometimes done to help prevent condensation and corrosion in the boiler during the summer months. The reduction in water temperature from the solar storage tank when operated in this way is approximately 15 to 25°F; however, corrosion protection within the boiler is a significant

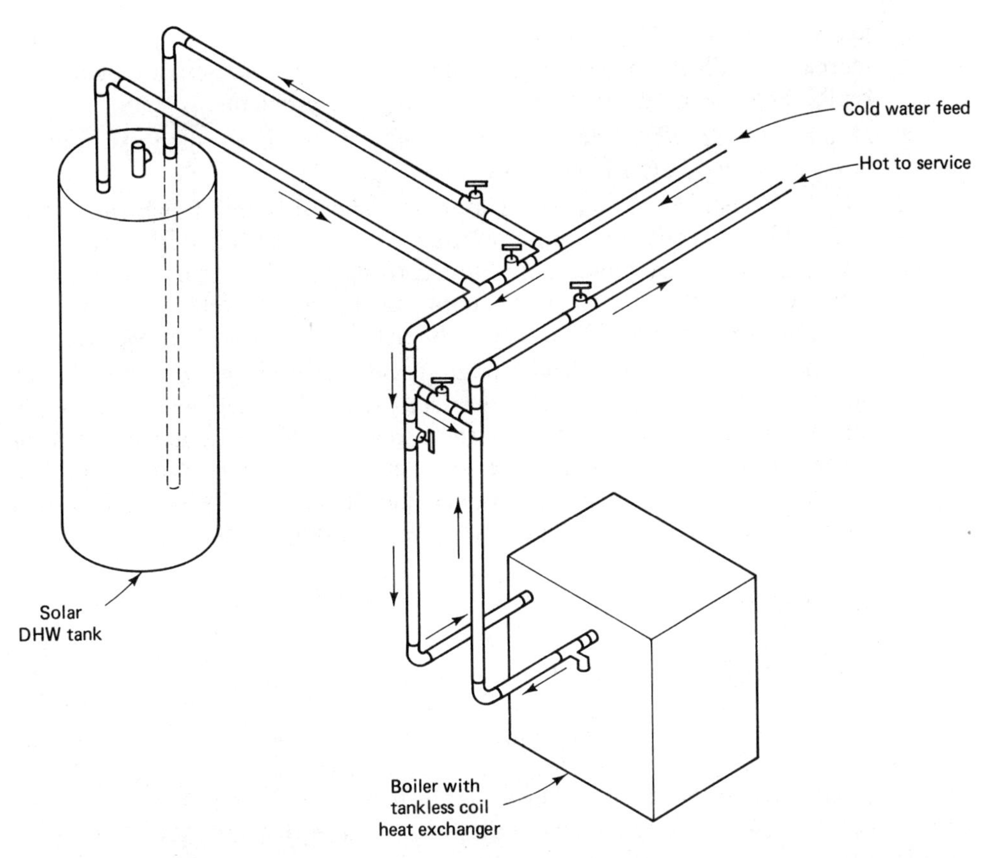

Figure 5.31 Bypass valving on tankless coil installation. Similar in configuration to the backup water heater valving arrangement, the provision for direct routing from the solar water heater to the hot water service line is important since many homeowners totally bypass the boiler tankless coil during the nonheating season when the boiler is not normally operating.

by-product of solar system output compared to the small reduction in hot water output temperature.

WATER TREATMENT

During the normal evaporation/rain cycles that are responsible for our potable water supplies, contamination of the water is an ongoing process. Several types of impurities are found in most residential drinking water, some of which require treatment. There are four basic types of impurities to be dealt with:

1. *Dissolved gases*. Two of the most common gases are carbon dioxide and

hydrogen sulfide. There are several varieties of taste and odor filters commercially available for residential water systems that use activated-charcoal replaceable cartridges that solve many of these problems.

2. *Dissolved minerals.* Minerals such as calcium and magnesium cause hardness, described later in greater detail.
3. *Suspended particulate matter.* The most common types of suspended materials are sand, clay, and iron. Filtration of these types of materials generally involves the use of replaceable filter cartridges or stationary sand filters, which utilize different grades of sand within a tank-type filter to remove the suspended particulates. This type of filter is illustrated in Figure 5.32.
4. *Microorganisms.* From a health point of view, these types of contaminants are the most dangerous. Microorganisms generally enter the water supply directly from animal or human waste disposal systems. Removal of microorganisms generally requires the installation of a chlorination system in which a chemical feed pump automatically injects the proper amount of chlorine into the water as it enters the water storage tank.

Treatment of Hard Water

Hard water is by far the most common water problem. Hard water is caused by calcium- and magnesium-based minerals dissolved in the water supply. Hardness is expressed in a term known as **grains per gallon**. To determine the exact degree of water hardness, a test kit such as that illustrated in Figure 5.33 can be used.

In addition to the test kits illustrated in Figure 5.33, many businesses that sell and install water softeners also have small bottles that they give the homeowner to return with a small sample of water. These water samples are tested to determine if hard water is present and what size of softener would be needed to correct the problem.

Hard water can create many problems, the most common of which is the familiar bathtub ring, formed by the combination of soap and minerals in the water. This scum is generally insoluble and requires extra rinsing to free the scum, or curd, from water fixtures and clothing.

Hard water is also responsible for creating major problems in residential plumbing and heating systems. Hot water lines filled with hard water tend to build up scale rapidly within the pipe. This causes restrictions within the internal diameter of the pipe, reducing both the pressure and volume of delivered water from the supply system to a significant degree. Also, high mineral residue can cause significant problems in humidifiers, water heaters, and various types of valves within the plumbing and heating system.

Water softeners operate on the principle of ion exchange. In solution, the calcium and magnesium break up into ions, charged electrical particles that combine readily with a variety of oppositely charged particles to yield the familiar undesirable results associated with hard water. In an ion exchange system, the

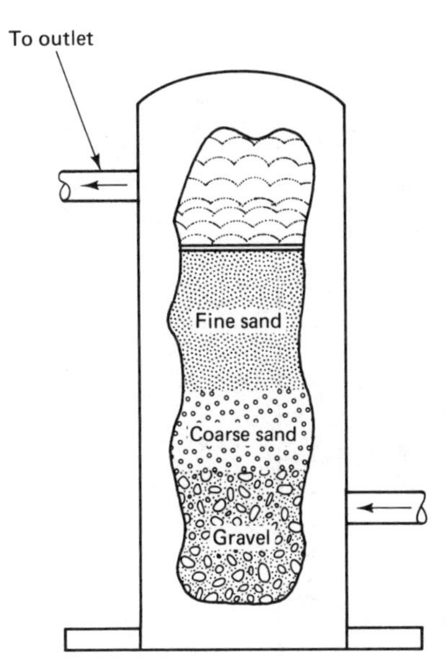

Figure 5.32 Sand filter cross section of the type used to remove sediment from the incoming water supply (typical).

Figure 5.33 Water test kit. In addition to testing for hardness, this kit can also be used to determine pH and iron content. (Courtesy of Hach Company.)

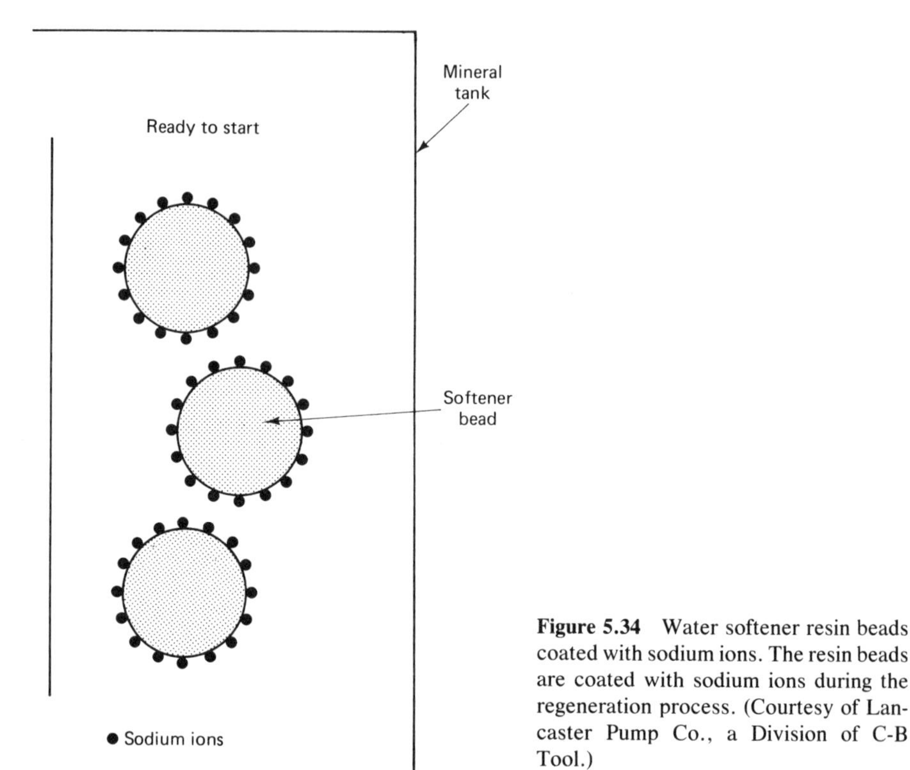

Figure 5.34 Water softener resin beads coated with sodium ions. The resin beads are coated with sodium ions during the regeneration process. (Courtesy of Lancaster Pump Co., a Division of C-B Tool.)

softener exchanges the calcium and magnesium ions in the water with sodium, or salt ions, from the softener. This ion exchange is done within the mineral tank of the water softener, which contains special resin beads that have been coated with sodium ions at the beginning of the softener cycle. This is illustrated in principle in Figure 5.34.

All water required in the residence flows through the mineral tank on its way to the various appliances in the plumbing system.* When the dissolved calcium and magnesium ions in the water come into contact with the sodium laden resin, they exchange places with the sodium ions from the resin. This process is illustrated in Figure 5.35. This ion exchange continues until the surface of the resin particles in the softener are completely coated with the calcium and magnesium ions. This condition is known as **exhaustion** and requires recharging the mineral tank, which is initiated by an automatic timer switch on the softener. Recharging the softener is known as **regeneration** and is accomplished by first

* In some instances, water softeners are connected only to the hot water lines, to eliminate what is sometimes objectionable taste, or to prevent the softening of the cold water, which is used for cooking, lawn watering, and so on.

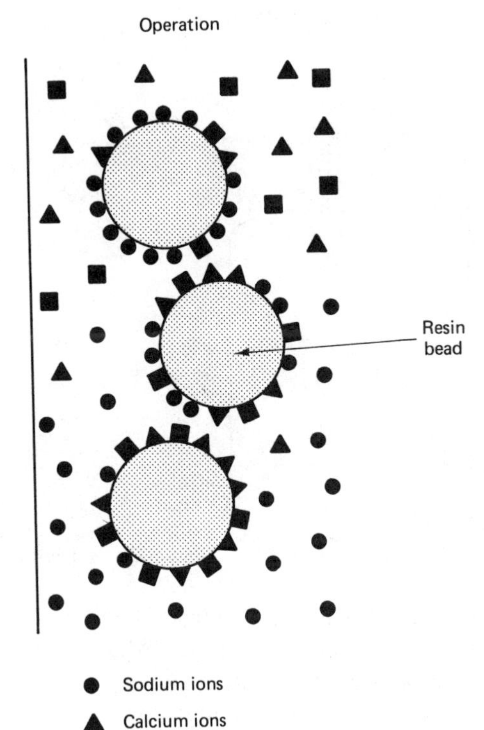

Figure 5.35 Ion exchange taking place on water softener beads. As the water flows through the softener, the calcium and magnesium ions exchange locations with the sodium ions on the softener beads. When the sodium ions have been exhausted from the softener beads, the unit must be regenerated (recharged) with salt. (Courtesy of Lancaster Pump Co., a Division of C-B Tool.)

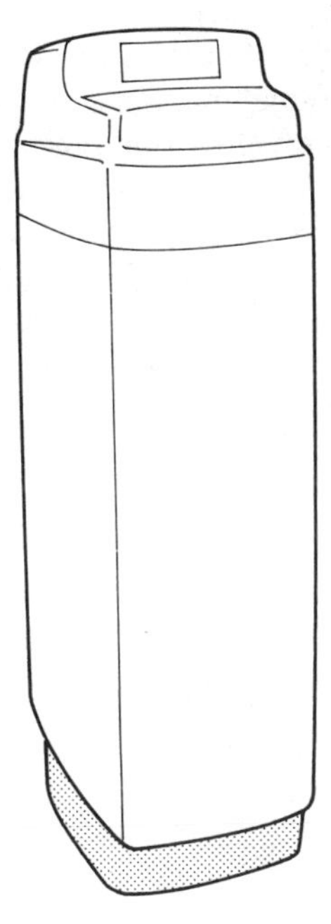

Figure 5.36 Single-unit water softener configuration (typical.)

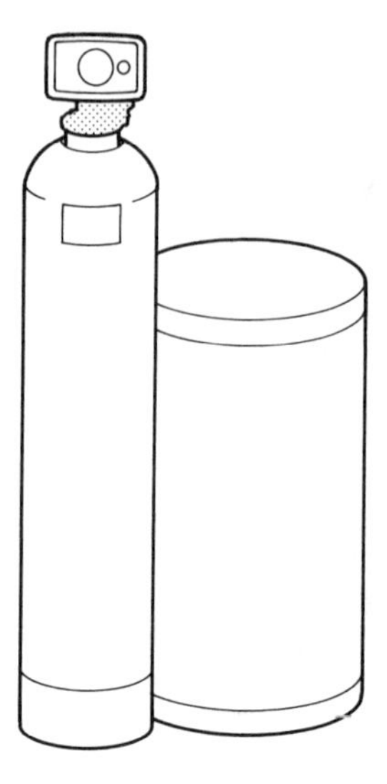

Figure 5.37 Two-tank water softener configuration. (Courtesy of Lancaster Pump Co., a Division of C-B Tool.)

backwashing the softener with salt water drawn from the brine tank. To complete the process, both the brine and mineral tank are rinsed with clean water to flush away the magnesium/calcium and excess brine residue.

Water softeners are sold in one- or two-unit configurations. The one-unit configuration uses a single cabinet that serves as the brine (salt) tank, while holding the mineral tank within. This type of design is used in installations where there is very little available floor space (Figure 5.36).

A two-unit softener configuration is illustrated in Figure 5.37. The two-tank design operates in an identical manner to the one-tank unit, with the mineral tank placed outside, and next to, the brine tank instead of inside it.

6

HEATING WITH SOLID-FUEL APPLIANCES

The popularity of solid-fuel heating was revitalized during the 1970s. Consumers saw fossil-fuel and utility rates rise to the point where the use of wood- and coal-burning stoves, furnaces, and boilers became far more economical than the use of existing heating systems. Although solid-fuel heating has always been popular in other parts of the world, the use of wood and coal in the United States waned in popularity with the availability of inexpensive fully automatic fossil-fuel heating systems. It was not until fossil fuels had become so expensive (in many cases annual fuel and utility costs had surpassed home mortgage payments by the late 1970s) that the use of wood and coal heaters spread rapidly throughout the residential market.

The modern wood- and coal-burning appliance differs markedly from units sold just 25 years ago. These changes focus primarily on aesthetics and combustion efficiency, resulting in appliances that in previous years would have been thought impossible.

In this chapter we examine the various options available for using solid fuels separately and in conjunction with fossil fuels for residential heating purposes, and we begin with an examination of the combustion characteristics of both wood and coal.

COMBUSTION CHARACTERISTICS OF SOLID FUELS

Wood

Heating value of wood. A variety of fuel woods are used in residential solid-fuel stoves, furnaces, and boilers. The heating value of wood is a function of its density. In general, hardwoods are preferable to softwoods for use in most types of solid-fuel appliances. Hardwoods are denser than softwoods and contain fewer volatiles, which when driven off during the combustion process yield undesirable by-products. The terms *hardwood* and *softwood* do not refer to the actual hardness or softness of the wood but to the classification of the tree as a leaf bearer or cone bearer. These classifications are sometimes referred to as *deciduous* (hardwoods) or *coniferous* (softwoods).

A cord of dry wood with a moisture content of about 20% contains approximately 8500 Btu/lb. Wood can be either **air dried,** in which case the wood stabilizes at the moisture content of the surrounding atmosphere (approximately 20% moisture content), or **kiln dried,** in which case the moisture content of the wood is between 12 and 15%. Moisture content detracts from the heating value of the wood since this moisture must be driven off during the combustion process; thus low-moisture-content fuel woods are desirable.

Table 6.1 illustrates the approximate Btu content per cord of fuelwood based on species. A cord of wood is a stack that measures 4 ft × 4 ft × 8 ft. Since the actual amount of wood in a cord depends on how carefully it was stacked (how much wood is in the stack opposed to spaces and holes between the logs), the figures in Table 6.1 are approximate.

Based on the number of Btu per cord given in Table 6.1, the homeowner can determine the net number of Btu per cord (assume an average appliance efficiency of 60% for conventional units and 70% for those equipped with catalytic combustion) and compare this figure with the net Btu available from other fuel sources. The prices of several fuel sources can then be compared to determine which fuels are most economical. When calculating the costs of wood heating systems, the costs for chimney repair or installation, installation of the solid-fuel unit, and annual chimney and appliance maintenance costs should be included.

To illustrate how these calculations are performed, consult Table 1.1, which lists the various Btu equivalents for a variety of common fuel sources. Let us assume that we wish to compare the use of a central heating propane furnace to a solid-fuel furnace. Knowing that propane fuel can deliver 93,000 gross Btu per gallon, and assuming an efficiency of approximately 65%, a propane furnace will deliver approximately 60,450 Btu/gal as calculated by the following formula:

$$\text{net Btu propane} = \text{gross Btu/gal} \times \text{appliance efficiency}$$

Thus,

$$\text{net Btu propane} = 93{,}000 \times 0.65 = 60{,}450 \text{ Btu/gal}$$

TABLE 6.1 HEAT VALUE OF SELECTED CORDWOODS

Species	Millions of Btu per cord[a]
Black locust	26.0
Shagbark hickory	25.5
White oak	24.5
Beech	21.0
Sugar maple	21.0
Red oak	21.0
Yellow birch	21.0
White ash	20.0
Red maple	19.0
Tamarack	19.0
Black cherry	18.5
Pitch pine	18.5
White birch	18.0
Silver maple	18.0
Norway pine	17.5
White elm	17.5
Gray birch	17.5
Hemlock	15.0
Red spruce	15.0
Balsam fir	13.5
Black willow	13.5
White pine	13.0
Basswood	12.5

[a] Btu values are approximate and based on a moisture content of 15 to 20%.

Assuming a price of $1 per gallon for propane, each dollar invested yields approximately 60,000 Btu of heating value.

To determine a net Btu value for each dollar spent on wood, we will assume a cord of mixed hardwoods that sells for $125 per cord (delivered). Using an average efficiency of 50% for the solid-fuel furnace, each cord of hardwoods will yield approximately 12,000,000 net Btu. To obtain a Btu figure per dollar, we divide 12,000,000 Btu by $125, which yields a figure of 96,000 Btu per dollar. We can see that the solid-fuel furnace yields approximately 50% more Btu per dollar than the propane-powered furnace.

When doing these calculations, it should be kept in mind that the figures do not include the cost or installation of the solid-fuel unit, nor do they take into account maintenance costs of the solid-fuel unit and chimney, which will be substantially greater than those for the propane furnace. Also, the time required to run a solid-fuel furnace is a commodity that should be given a dollar value. Since many people who use solid-fuel appliances have the time available to cut, stack, and feed the appliance, they do not normally associate a labor cost with these systems; thus this value will vary from one household to another.

Combustion cycle of wood. The combustion of wood is an evolutionary process, moving from one stage to another in three distinct phases. Wood should be stacked carefully and compactly in the firebox to aid in heat reflection from one log to another. Too much space between logs will lower the temperature of the firebox. This makes it more difficult to drive off the trapped moisture within the wood, which identifies stage 1 of combustion. As the temperature in the firebox begins to rise, moisture is driven from the wood through the open cellular structure of the ends and face of the logs.

Actual burning of the wood takes place when the temperature reaches approximately 550°F. This second stage of combustion is identified by the actual flaming of the volatile gases as they are driven from the wood.

The second stage of combustion continues until all the gases are driven off, after which what remains in the firebox is almost pure carbon, beginning the third and final combustion stage. The third-stage fire is a hot, glowing bed of coals, characterized by short red and blue flames. Third-stage combustion is virtually smokeless and very efficient.

Primary and Secondary Air. For efficient combustion to take place, specific amounts and types of combustion air patterns within the unit are required. Referred to as primary and secondary air intake patterns, these are illustrated in Figure 6.1. **Primary air**, which contains the basic oxygen supply for the combustion process, is directed over the woodpile from top to bottom. **Secondary air** is supplied to the upper areas of the firebox to aid in oxidizing the unburned volatile gases driven from the wood. Most modern solid-fuel appliances are airtight, meaning that the only air available for combustion is admitted through the air-intake dampers of the unit. This feature limits wood combustion within relatively strict parameters. This is in distinct contrast to earlier solid-fuel units, which were not airtight, making control of firebox combustion almost impossible.

Grate Configuration. One of the key components within the firebox are the steel or cast-iron grates on which the wood or coal rests. The configuration of the grates will vary from one unit to another and is dependent on whether only wood, or both wood and coal, is designed to be burned. Figure 6.2 illustrates some of the grate configurations available for solid-fuel appliances. Note that most grates contain a shaker mechanism which causes the grates to move to sift the ashes from the woodpile and deposit them below the firebox for convenient removal.

Formation of Creosote Deposits. All wood-burning appliances develop creosote as a by-product of incomplete combustion. Creosote, a complex chemical substance, is black, sticky, and shiny in appearance, and forms due to volatile gases that have not been oxidized as they are driven from the wood. These gases condense on the relatively cool surfaces of the interior of the firebox and chimney pipe. Creosote must be kept to a minimum in all appliances and chimneys since it is the chief source of chimney-fire problems. When creosote is present in suf-

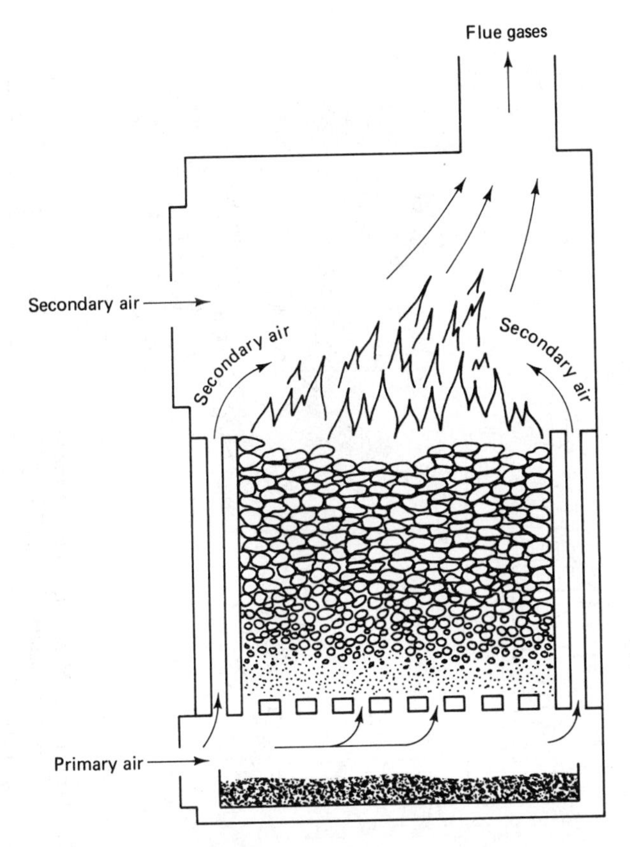

Figure 6.1 Primary and secondary combustion airflow through a solid-fuel appliance. (Adapted from John W. Bartok, *Solid Fuel Furnaces and Boilers*, Garden Way Publishing, Pownal, Vt., 1982, Fig. 7.1, p. 268. Available from Storey Communications, Inc., Pownal, VT 05261.)

ficient quantities and exposed to high temperatures within the chimney, it will ignite. If this fire is not contained, it can quickly spread to the whole house, even in instances where the chimney is in good repair. Therefore, creosote deposition should be kept to a minimum to ensure safe operating conditions. The following two guidelines will help ensure low creosote levels.

1. Use only dry hardwoods in the unit whenever possible. Hardwoods contain fewer volatile elements than do softwoods. These volatiles are the major components of creosote formation.
2. Avoid burning the appliance under restricted primary air intake whenever possible. Burning small hot fires is preferable to long-duration restricted-air burns. Restricted air fires are generally burned during the fall and spring seasons when the outdoor temperatures are around 40°F. In these instances, however, it is often less expensive to use the home heating system than the solid-fuel appliance.

Creosote Reduction Technology. Following the guidelines described above will help to ensure safe creosote levels. There are, however, two chemically based

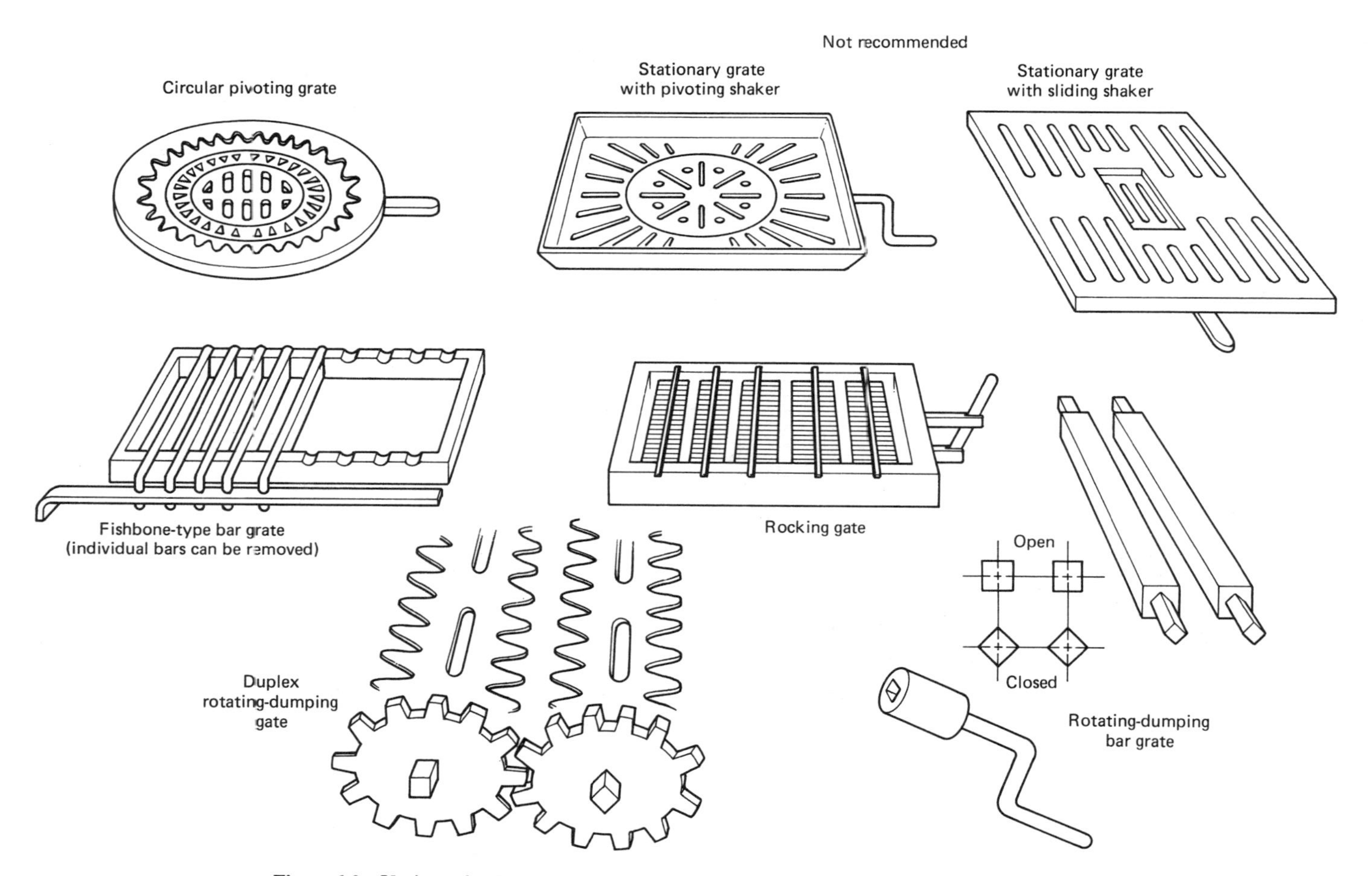

Figure 6.2 Variety of solid-fuel shaker grate configurations. (Adapted from John W. Bartok, *Solid Fuel Furnaces and Boilers*, Garden Way Publishing, Pownal, Vt., 1982. Available from Storey Communications, Inc., Pownal, VT 05261.)

Figure 6.3 Topical wood spray treatment for creosote reduction. (Courtesy of Combustion Improvers Co., Inc.)

methods in current use that aid in reducing creosote within the appliance. One method involves the use of a spray that is applied to the wood prior to loading it in the firebox. A spray of this type is available in a pressurized container (Figure 6.3). The spray acts as a catalyst to increase the temperature of the wood fire, limiting the unoxidized volatile gases available for condensation.

Perhaps the greatest advance in wood stove design is the development of the catalytic combustor. These devices operate on a principle similar to that of the automobile catalytic converter and can either be incorporated as an integral part of the appliance design or placed into the stovepipe. A cross section of a typical catalytic combustor is illustrated in Figure 6.4.

The use of a catalytic combustor will significantly reduce creosote deposits within the heating system. The combustor is made up of a heat-resistant ceramic material arranged in a honeycomb configuration. Within each honeycomb, a noble-metal catalyst is applied. When the combustor is heated to between 500 and 700°F, the catalyst achieves "light-off" status and will oxidize the wood smoke coming into contact with it well below the 1000°F normally required to burn these gases. The combustor normally operates at temperatures that range between 1220 and 1400°F. The configuration of the combustor installed within a typical radiant wood heater is illustrated in Figure 6.5.

Coal

Coal has served for many years as a major fuel for residential heating. Although its popularity waned with the availability of inexpensive oil, natural gas, and electricity, coal has once again come into widespread use with the comparative

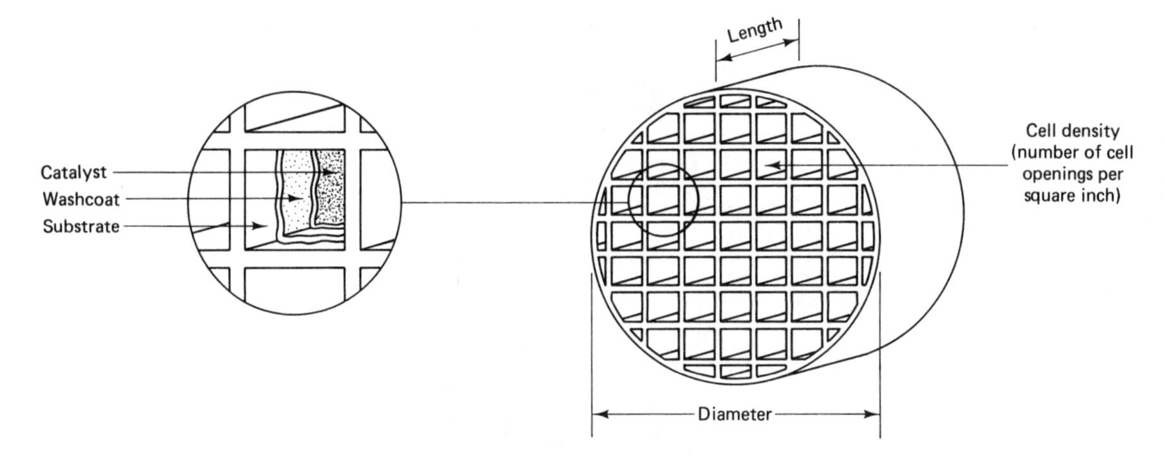

Figure 6.4 Cross section of catalytic combustor. (Courtesy of Corning Glass Works.)

price increases of these other fossil fuels. The combustion characteristics of coal are different from those of wood, requiring modifications of the appliance air intake system and grate configuration if coal, in addition to wood, is to be burned in the same unit. Although it is generally true that if an appliance is designed to burn coal, it will burn wood efficiently, the opposite is not true. Coal fires generally require less attention than do wood fires, due to the nature of combustion, although they must be properly shaken periodically if combustion is to be maintained.

Classification of coal. Coal is classified by its geological age as well as its size, based on the opening in a wire mesh screen that it will fit through. The major geological varieties of coal are lignite, bituminous, and anthracite. **Lignite** is geologically the youngest of the coals. Dark brown in color, it contains the highest moisture content of all types of coal and the least amount of energy per pound. For these reasons, lignite is not often used as a residential heating fuel in the United States, although it is popular in parts of Europe.

Bituminous coal, sometimes referred to as "soft coal," is available in several varieties. Bituminous coals are used for home heating in areas where they are readily available. Many bituminous coals contain a high percentage of sulfur, which is detectable by the odor given off during combustion.

Anthracite coal, sometimes referred to as "hard coal," is geologically the oldest of the coals. Anthracite is an excellent residential fuel. Almost pure carbon, it burns with a low blue flame and requires a minimum amount of attention during the combustion cycle. For rating purposes, coal can be considered to contain approximately 12,000 Btu/lb.

All coal contains a certain amount of fly ash, which during combustion can combine into large chunks called **clinkers**. Different coal varieties contain assorted amounts of ash that will combine at specified temperatures referred to as the *fusion temperature* of the coal. For home heating it is desirable that the coal have

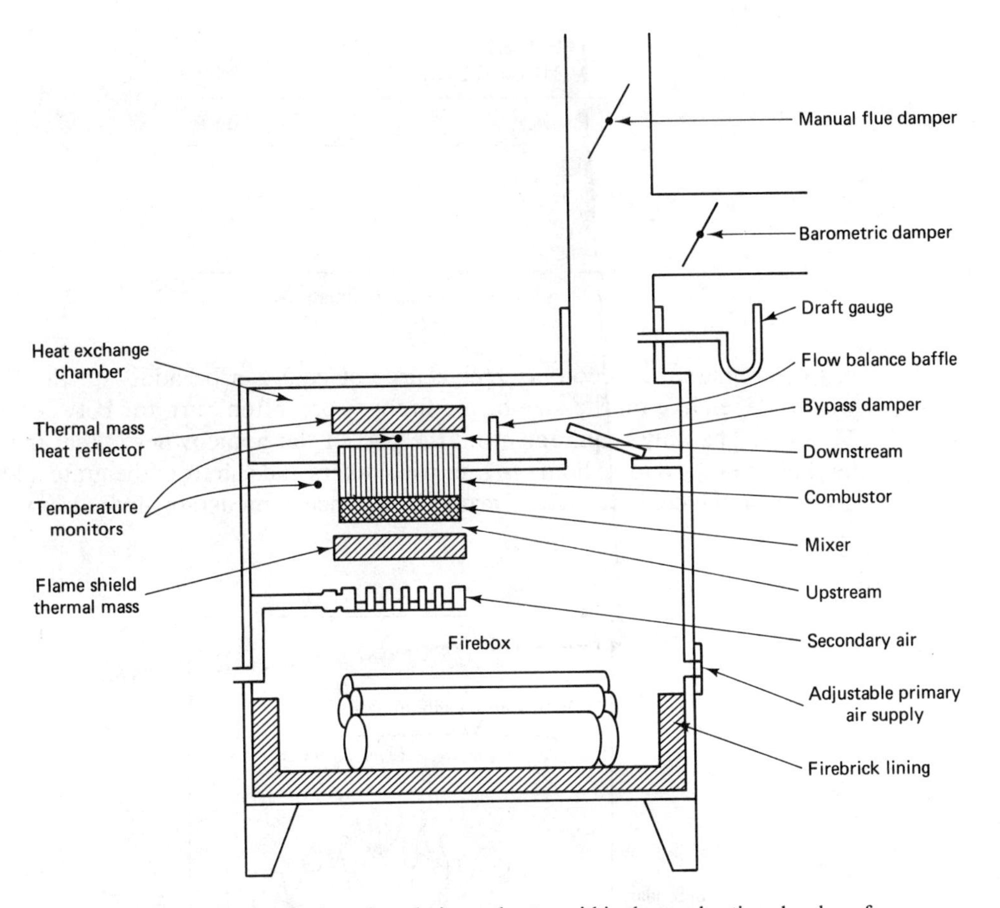

Figure 6.5 Installation of catalytic combustor within the combustion chamber of a solid-fuel appliance. Note that the combustor must be protected by a flame shield. (Courtesy of Corning Glass Works.)

the lowest ash content and highest fusion temperature possible, to keep the formation of clinkers to a minimum. Otherwise, constant attention to the fire resulting in frequent shaking of the grates to remove the clinkers from the combustion area will be needed. Both the ash content and fusion temperature of the coal are values that should be specified by the supplier prior to purchasing the coal.

The physical size of the coal chunks is important for proper operation of the appliance. Coal sizes are listed in Table 6.2. Prior to purchasing any coal, the manufacturer's specifications regarding the proper coal size to use in the particular unit should be consulted.

Combustion cycle of coal. Coal combustion temperatures are higher than those in wood fireboxes. To start a coal fire, a good wood fire must first be established; the coal is then added in a thin layer. After this initial layer of coal

TABLE 6.2 SPECIFICATIONS OF ANTHRACITE COAL

Grade	Size[a] (in.)
Pea	$\frac{1}{2}$–$\frac{3}{4}$
Nut	$\frac{3}{4}$–$1\frac{1}{2}$
Stove	$1\frac{1}{2}$–$2\frac{1}{2}$
Egg	$2\frac{1}{2}$–$3\frac{1}{2}$

[a] Sizes of coal are approximations.

has been ignited, a second, larger charge of coal can be added to the firebox, which, depending on the size of the firebox, can often burn for between 18 and 24 hours. The only requirement on the part of the homeowner is that the grates be shaken every 8 to 12 hours to remove clinkers and ash from the grates. Shaking is critical if the coal fire is to be maintained, since combustion air feeds up through

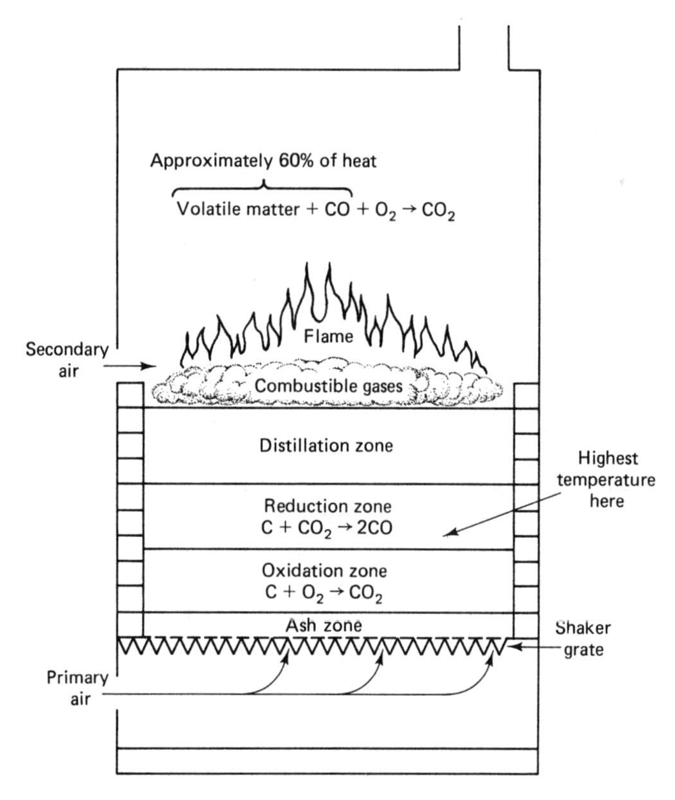

Figure 6.6 Cross section of typical coal fire. (Adapted from John W. Bartok, *Solid Fuel Furnaces and Boilers,* Garden Way Publishing, Pownal, Vt., 1982. Available from Storey Communications, Inc., Pownal, VT 05261.)

the grates rather than from above, as in a wood fire. Note the cross section of a typical coal-fired appliance in Figure 6.6.

Combustion air feeds up through the grates, where the majority of combustion takes place in the oxidation zone directly above the coal bed. Oxygen combines with the coal to produce carbon dioxide (CO_2). The CO_2 is reduced to carbon monoxide (CO) in the reduction zone. As the gases rise into the distillation zone, fresh oxygen is added to complete the oxidation of the coal.

Note from Figure 6.6 that significant amounts of CO_2 are produced during combustion. If incomplete secondary air is present, potentially lethal amounts of CO will be produced over a period of time. Efficient removal of effluents and combustion by-products via the appliance venting system is critical in all coal-burning systems. For example, the draft requirements of the average wood-burning appliance will fall between 0.03 and 0.05 in. of water column draft. In coal-burning systems, the minimum draft requirement is 0.05 in. w.c. to remove all gaseous waste products from the combustion chamber and piping. Check the draft conditions of the chimney carefully prior to placing any coal-burning appliance in operation.

With a basic understanding of the combustion characteristics of both wood and coal, we move onto a discussion of the various types of solid-fuel appliances that are available for residential heating applications.

SIZING THE STOVE FOR REALISTIC OUTPUT

Intense competition within the solid-fuel industry has led to a variety of sizing methods used to rate the output of radiant heaters. Although many of these claims are based on realistic testing and rating procedures, many are not. The ability of any solid-fuel appliance to heat a home is based primarily on the amount of wood it burns hourly. Also, the efficiency of the combustion process, coupled with the radiant surface area of the stove, contributes to its effectiveness in any specific heating situation.

The amount of wood that can be burned hourly in any solid-fuel heater *for a specified period of time* is dependent primarily on the size of the firebox. Although this may seem a bit obvious, the industry is full of claims concerning the ability of one unit or another to be rated at 150,000 Btu/hr, for example. Keeping in mind that wood contains approximately 8500 Btu/lb, it requires the combustion of over 17 lb of wood to generate 150,000 Btu. Given an approximate efficiency of 60%, a stove must burn 25 lb of wood per hour to achieve a *net* rating of 150,000 Btu. If the firebox is relatively small and can hold only 50 lb of wood, this rating is meaningless, since the heater can achieve its rated output for only 2 hours before it requires recharging. On the other hand, if an appliance is rated at 150,000 Btu/hr for a specified period of time (e.g., 6 hours), the homeowner has a realistic idea of how much heat the unit can be expected to provide.

Another fallacy related to stove rating and sizing deals with the practice on

TABLE 6.3 HEAT-CARRYING BASED ON CHIMNEY FLUE DIAMETER

Chimney flue size (nominal)	Capacity (Btu/hr)	Firing rate[a] (lb/hr)	
		Wood	Coal
6 in. round	97,000	12.1	8.1
7 in. round	131,000	16.3	10.9
8 in. × 8 in. square	150,000	18.75	12.5
8 in. round	172,000	21.5	14.3
9 in. round	218,000	27.3	18.2
8 in. × 12 in. rectangular	234,000	29.3	19.5
10 in. round	270,000	33.0	22.5
12 in. × 12 in. square	358,000	44.8	29.8

Source: Adapted from John W. Bartok, *Solid Fuel Furnaces and Boilers*, Garden Way Publishing Co., Pownal, VT, 1982. Available from Storey Communications, Pownal, VT 05261.

[a] Assume 8000 Btu/lb wood, 12,000 Btu/lb coal.

the part of some manufacturers to rate their appliances in terms of square feet, or cubic feet, of space heated. Recall from our discussion of heat loss that the construction and insulation factors vary so greatly from one house to another that this type of rating procedure is practically meaningless. In addition to construction factors, the living patterns of families vary greatly from one to another. Therefore, while one family may receive what it considers to be ample heat from a particular radiant stove, another family, used to either higher or lower temperatures, may find the same unit totally inadequate for its needs. In the final analysis, the unit under consideration should be examined carefully to determine how much fuel it can hold and what are reasonable expectations as to its efficiency based on available testing laboratory data. Note should be taken of the maximum heat capacity of the unit listed on its certification or nameplate tag. The flue collar on the stove is generally an indication of the maximum amount of heat that the unit can safely exhaust. Since the flue collar determines the diameter of the chimney in most cases, by knowing the heat-carrying capacity of various diameter chimneys, an indication of the stove rating can be determined. Table 6.3 illustrates the heat-carrying capacity of various chimney diameters, coupled with the maximum firing rates per hour using either wood or coal.

RESIDENTIAL SOLID-FUEL APPLIANCES

Residential heating with wood- and coal-burning appliances is supplied in one of several ways: by the use of room-sized radiant stoves or large radiant stoves that furnish heat to the home utilizing natural convection air currents within the residence, or by the installation of a central heating furnace or boiler operating either separately or in conjunction with an existing fossil-fuel central heating system. Whereas the initial solid-fuel appliances available in the early 1970s were primarily radiant stoves, present-day offerings include a wide variety of central heating solid- and multifuel boilers and furnaces capable of achieving high combustion efficiencies that furnish cost-effective economical central heating. Given the variety of appliances available, the radiant wood stove in all its modifications is still the most popular solid-fuel device.

Radiant Solid-Fuel Stoves

Radiant stoves range in complexity and design from simple step stoves to ornate cast-iron enameled and tiled radiant units. The term "simple" should not lead one to think of these units as inefficient or lacking in definitive features. Many of these designs have been around for years, providing reliable, efficient home heating.

The ability of any stove to heat a home is a function of three factors: the amount of wood burned per hour, which governs the total potential heat available; the efficiency of the combustion process, which is a function of the integrity of the design of the particular stove; and the amount of radiating surface available in the stove to transfer the heat generated from combustion of the wood to the interior of the home. Each of these factors is discussed in greater detail in this chapter.

Wood and coal stoves are fabricated from one of three materials: steel, cast iron, or ceramic tile. While steel and cast iron have been used for many years to build radiant wood heaters, ceramic tiles, used for many years in European tile stove designs, are relatively new to the residential home market in the United States.

There seems to be an ongoing controversy concerning the relative advantages of steel versus cast iron in radiant stove construction. Each material has its own merits. Steel allows a good deal of flexibility in stove design. It can be bent and welded in a variety of shapes. Its heat transfer rate is approximately the same as that of cast iron, depending on the thickness of the steel and cast iron used. Steel has a tendency to warp, whereas cast iron will ordinarily hold its true shape more readily. Although there are those who swear by the longevity of cast iron versus the generally less expensive steel stoves, there are literally thousands of homes in America with steel stoves that have been in use for many years with no deformation or deterioration in function or efficiency. In many instances, the choice of material resides in the application and appearance of the particular unit. Figure 6.7 illustrates a basic steel stove design that has been in existence for almost 60 years. Figure 6.8 illustrates a contemporary cast-iron enameled stove featuring a classic aesthetic design coupled with high-efficiency heating characteristics.

Figure 6.7 Radiant steel stove. This unit is capable of burning either wood or coal and features Underwriters' Laboratories approval. (Courtesy of Riteway-Dominion Manufacturing, Inc.)

Wood/coal options. Many cast-iron and steel radiant stoves are available in design configurations that enable the burning of either wood or coal with ease. In some cases, an accessory kit is necessary to enable the stove to burn coal. The configuration of the shaker grates on the stove is critical for proper stove function and determines the size of coal that will work best. The features to look for are integrity and material construction of the shaker grate assembly and the size of the ash removal drawer. Wood burning, although it does require periodic emptying of the ash drawer, generates only about one-tenth of the ash volume as a coal burner. Thus even wood/coal burners equipped with relatively large fireboxes and ash drawers will have to be emptied every day when burning coal. This volume of ash should be a consideration when examining specific appliances.

Tile and Masonry Stoves

The tile stove design has been around for many, many years. These stoves should not be confused with the tile stoves available in radiant design units. In radiant units, the tile is added as a decorative feature and to increase the thermal mass of the stove to enable it to continue to radiate heat when the fire within the unit has died down. A true tile stove design, however, is quite different from contemporary radiant stoves. The tile stove, sometimes referred to as a **Kachelofen**, is constructed entirely of similar masonry materials, all designed to expand and contract at the same rate. Figure 6.9 shows a typical Kachelofen.

In the tile stove, heat is generated by a series of small, undamped burns, resulting in intensely hot, efficient fires. Since the burns are unrestricted, little creosote is built up in the unit as the wood gases are almost completely combusted. At the present time these stoves are all custom built, although as interest in them continues to grow, they are bound to become more widely available.

Figure 6.8 Contemporary cast-iron radiant wood/coal stove. In addition to the conventional black metallic finish, these units are also available in a variety of porcelain enamel finishes. (Courtesy of Vermont Castings, Inc.)

Figure 6.9 Tile stoves not only radiate heat, but due to the amount of tile on specific stoves (which can be considerable) are capable of radiating heat long after the fire has died down. (From *Heating with Wood*, Garden Way Publishing, Pownal, Vt., p. 92. Available from Storey Communications, Inc., Pownal, VT 05261.)

Fireplace Inserts

The modern fireplace, while possessing a romantic historical background, is rather poor when it comes to efficient home heating. There are many ways to construct a fireplace, and unfortunately, most of them are inefficient from the standpoint of heat radiation and convective heat losses. Figure 6.10 illustrates the cross section of a ''Rumford''-type fireplace, designed in 1795 by Benjamin Thompson, Count von Rumford. The Rumford fireplace uses specific dimensions and relationships between the basic components of the fireplace to achieve relatively high-efficiency heat transfer. Note that the firebox is shallow and high, which allows for a great deal of heat radiation to the room. This design is contrasted with a conventional modern fireplace, designed, it would seem, to minimize everything but heat loss (Figure 6.10).*

Although the Rumford fireplace design is inherently more efficient than most modern configurations, the fireplace is still limited to efficiencies of perhaps 15 to 20% under ideal conditions. Also, once the major combustion cycle of the wood charge has been completed, most fireplaces lose more heat than they have generated into the house unless they are equipped with glass doors that prevent heat losses up the chimney.

To overcome these deficiencies, the use of fireplace inserts and fireplace stoves have become widespread. A fireplace stove is simply a radiant stove placed within or near the opening of the fireplace, with the chimney flue passage sealed tightly around the opening of the entering stovepipe. The fireplace insert is a steel or cast-iron radiant unit that is designed to fit into and be sealed within the existing fireplace opening. The fireplace stove and fireplace insert are illustrated in Figures 6.11 and 6.12.

Most inserts extend out from the front of the fireplace masonry to increase radiant surface area. The space between the physical size of the insert and the opening of the fireplace is sealed with a metal shroud that surrounds the insert appliance and is equipped with an air-tight seal against the front surface of the fireplace.

For maximum safety, a positive connection is required between the fireplace flue liner and the vent connection from the insert. This installation detail is illustrated in Figure 6.13.

The use of both fireplace inserts and stoves offers a significant increase in efficiency over the conventional fireplace design. In fact, overall operating efficiencies of inserts and fireplace stoves approach those of conventional radiant stoves. Inserts tend to cost somewhat more than an equivalent radiant solid-fuel heater, since the unit includes the sealing shroud and material necessary to install

* For a full discussion of historical and technical aspects of the Rumford fireplace, the reader is directed to *The Forgotten Art of Building a Good Fireplace* by Vrest Orton (Yankee Books, Dublin, N.H., 1969).

Figure 6.10 (a) Rumford fireplace design. This fireplace design is still perhaps the most efficient available for conventional masonry fireplaces. (b) Most modern fireplaces, due to their inefficient design, can actually account for a net heat loss when they are operating. (From John Vivian, *Wood Heat*, Rodale Press, Emmaus, Pa., 1976, p. 157.)

Figure 6.11 Fireplace stoves are conventional stoves set up on a fireplace hearth. Due to the use of the fireplace chimney, these units must have a direct connection from the stove outlet to the first layer of chimney flue tiles. Note the tee, located behind the stove, enabling the chimney to be conveniently cleaned. (Courtesy of Vermont Castings, Inc.)

Figure 6.12 Fireplace inserts have become increasingly popular, as their style has evolved into units that feature the aesthetics equal to or greater than that of the conventional masonry fireplace they replace. (Courtesy of Vermont Castings, Inc.)

a positive connection within the flue. These units are also available in steel, cast iron, and enameled finishes with or without glass-door panels supplied for viewing the fire.

We can now continue with an examination of various options available to heat the home with solid- and multifuel central heating appliances.

Solid-Fuel Central Heating Applications

Central heating applications account for an increasing number of solid-fuel appliances sold. Central heating solid- and multifuel boilers and furnaces are available in a variety of design and combustion configurations. Our examination begins with solid- and multifuel boilers for hot water and steam heating applications.

Solid-fuel boilers. Solid-fuel boilers can be installed as either stand-alone heating systems or in conjunction with an existing conventional fossil-fuel boiler as an add-on unit. Whether the boiler is installed as an add-on or stand-alone unit, the configuration of the boiler itself is the same. Prior to examining specific boiler designs, mention should be made of the certification procedures that apply to these units. The rating agency that deals with the certification of all water-filled pressure vessels is the American Society of Mechanical Engineers (ASME). Boil-

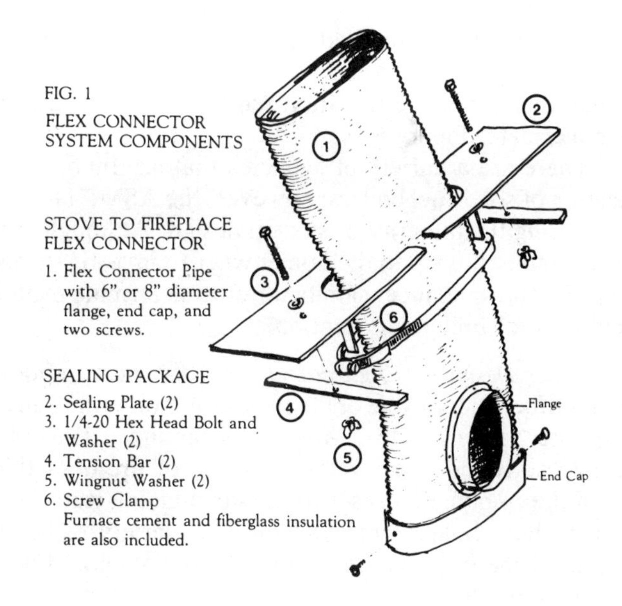

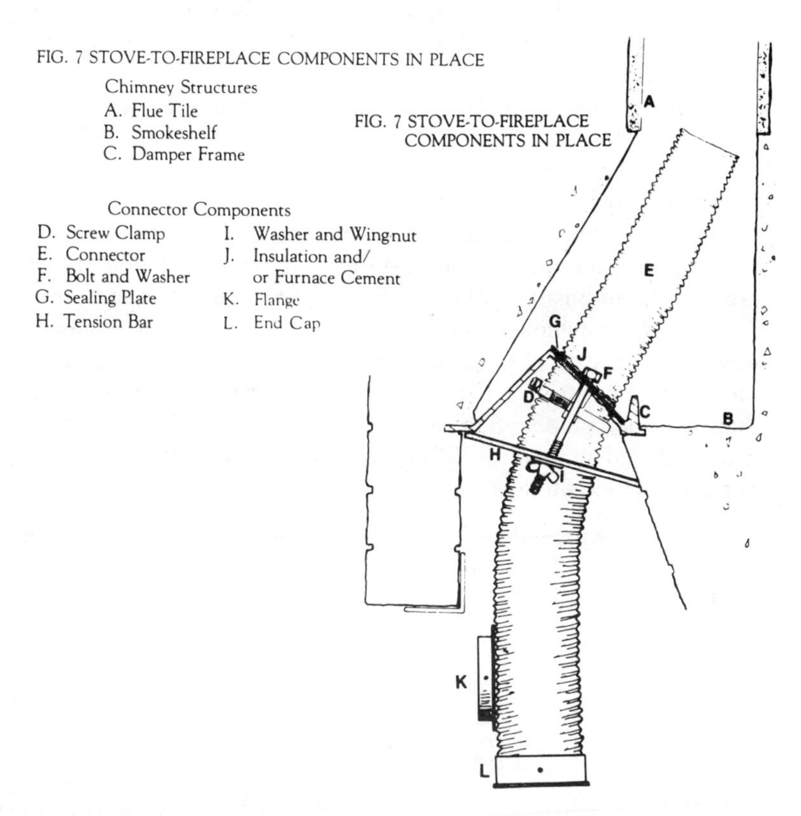

Figure 6.13 All fireplace stoves and inserts are required to have a positive flue connection running between the stove and the first line of flue tiles within the chimney. (Courtesy of Vermont Castings, Inc.)

ers that meet these certification requirements feature the certification stamp on the boilerplate (Figure 6.14).

There are a variety of agencies that are involved with the testing and certification of solid-fuel boilers; however, the ASME label is *the* label of certification for ensuring the integrity of system design. Although most other testing agencies are reputable, it is up to the homeowner to investigate all testing and certification claims by the appliance manufacturer to determine exactly what procedures have been used for boiler certification.

Boiler Design Configurations. Of the many options available in solid-fuel and multifuel boilers, one of the most common is the conventional hot water wood/coal boiler (Figure 6.15). Most units feature a firebox that is either totally or partially surrounded by water. A series of firetubes that aid in transferring the heat of the combustion gases to the surrounding water is illustrated in Figure 6.15. Most of these boilers are available in a variety of heating capacities, based on the size of the combustion chamber. Both steel and cast iron are used in boiler construction.

To obtain different heating capacities in steel boilers, the physical size of the firebox and boiler is changed; in cast-iron units, the number of sections that make up the boiler is varied.

Most solid-fuel boilers can be used as either as stand-alone units or in conjunction with a conventionally fueled boiler. Some of the installation configurations are examined later in this chapter.

The sequence of operation in these units is based on the combustion cycle of the wood. When there is either a call for heat or for the boiler to increase water temperature, additional combustion air is admitted into the firebox, which increases the intensity of the wood fire. The water temperature rises to the high-limit cutoff at which point combustion air is restricted, causing the fire to die down. Because most solid-fuel boilers do not cycle on and off as do conventional units, strict control of the water temperature within the boiler is more difficult than in other combustion systems. Control of primary combustion air can be accomplished in several different ways. An examination of some of the most popular methods follows.

Controlling Combustion Air in Solid-Fuel Boilers. The simplest type of draft-control device is the popular immersion aquastat control illustrated in Figure 6.16. In this system the control device is composed of a bimetallic element inserted

Figure 6.14 Seal of the American Society of Mechanical Engineers, commonly referred to as the ASME. This seal, patented by ASME, is an assurance of quality applied to the selection of materials and integrity of design and construction of the appliance. (Courtesy of American Society of Mechanical Engineers, reprinted from Section IV of the ASME Boiler and Pressure Vessel Code.)

Figure 6.15 Solid-fuel wood/coal boiler. (Courtesy of Royall Furnace Works.)

directly into the boiler through a tapping designated for this use. The immersion aquastat is connected to the draft door of the boiler with a connecting chain. When the boiler water cools, the aquastat rotates, tightening the chain, which opens the draft door, admitting more combustion air to the fire. Conversely, when the water temperature rises to the set point on the aquastat, it rotates in the opposite direction, loosening the chain that closes the draft door.

Induced draft has been a popular method of controlling combustion air to wood/coal appliances for many years. In this system, a fan that discharges forced air into the combustion chamber is turned on when there is a call for increased water temperature in the boiler. This rush of air is usually baffled inside the boiler so that it can be delivered above and below the grate structure simultaneously for burning both wood and coal. Induced-draft systems have the advantage of being somewhat less dependent on the chimney for maintenance of proper draft when the boiler is in the "on" cycle. This does not mean, however, that the draft requirements in the chimney can be ignored. Rather, the induced-draft fan establishes the proper operating pressure within the boiler for proper combustion. The chimney draft must still be adequate to accommodate the pressure set up by the induced-draft system as well as remove combustion by-products during low-fire conditions when the fan is not operating and the boiler is on standby. An induced-draft fan arrangement is illustrated in Figure 6.17.

A third type of draft controller used on residential solid-fuel boilers is the motorized draft controller (Figure 6.18). On a call for heat by the thermostat or control aquastat, a motorized controller is energized to open the draft door to allow increased combustion air to enter the firebox. When the boiler has reached operating temperature, a return spring is activated and closes the draft door, restricting primary combustion air.

Regardless of the method used to achieve control over primary combustion

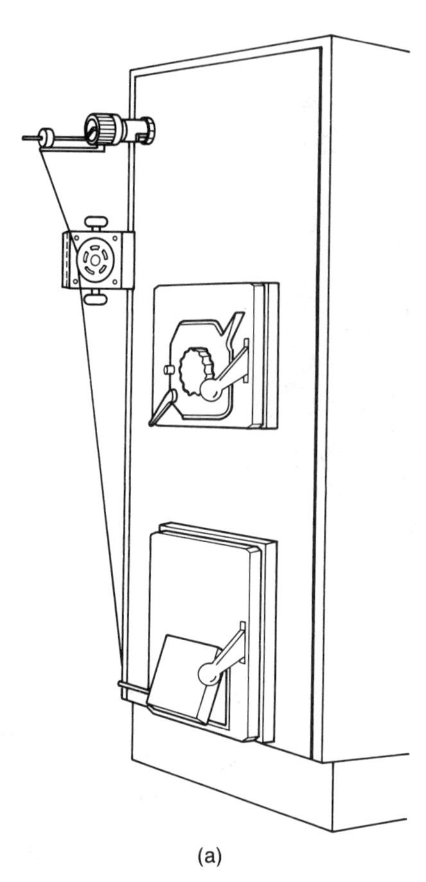
(a)

Figure 6.16 Immersion-type draft controls. Changes in boiler-water temperature act on a bimetallic element, which causes the control to rotate, raising and lowering the door that admits combustion air to the firebox. (Courtesy of Ammark Controls, Inc.)

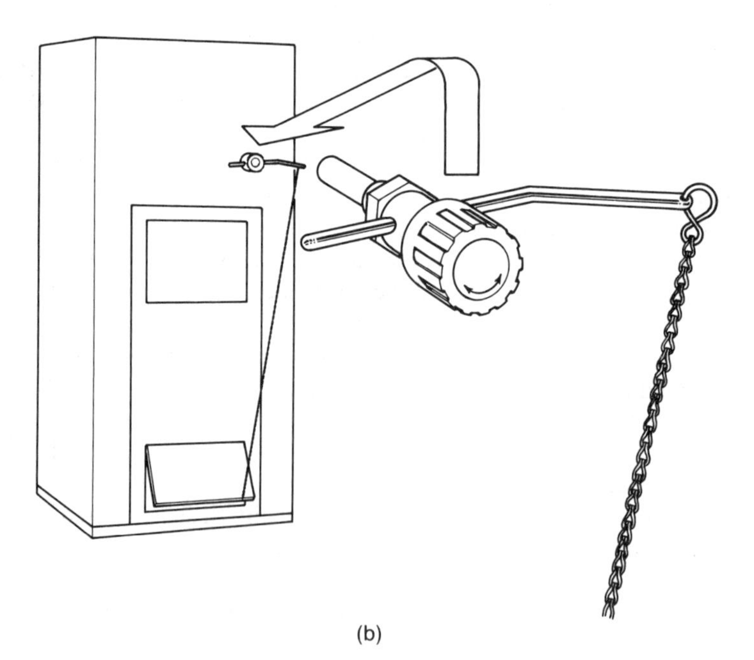
(b)

Figure 6.17 Induced-draft combustion fan. Note the flap on the left side of the fan, which is set slightly open. This setting ensures that even when the fan is not operating, sufficient combustion air will enter the firebox to prevent the fire from going out.

air entering the firebox, the sequence of operation of most units is basically the same. Almost all solid-fuel boilers employ conventional aquastats that are used to switch the circulator on when the water temperature is above the low-limit setting, assuming there is a call for heat by a room thermostat.

Overheat Controls. Within the control circuitry of most solid-fuel boilers is a provision for controlling and removing excess heat from the water if the temperature rises beyond the safe high-limit aquastat setting. This provision is necessary in all solid-fuel boilers since excessively high water temperatures can build up if either the draft control is improperly adjusted or there is no call for

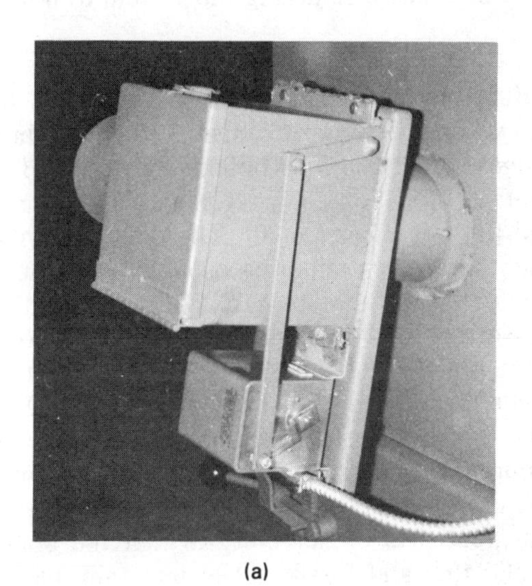

(a)

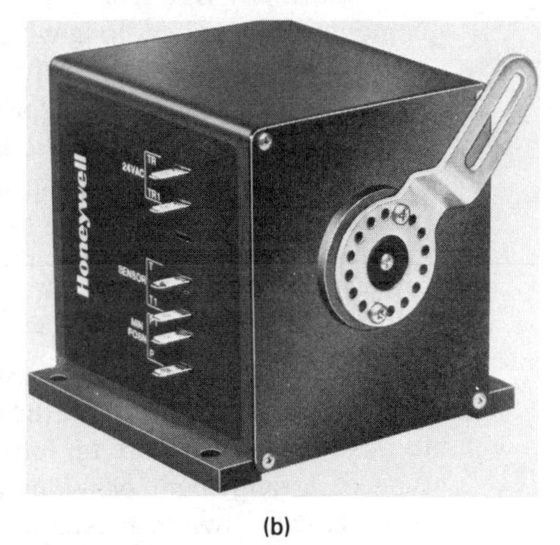

(b)

Figure 6.18 Motorized draft controls: (a) motorized control used on Eshland gasification boiler (courtesy of Eshland Enterprises); (b) Honeywell motorized controller (courtesy of Honeywell, Inc.).

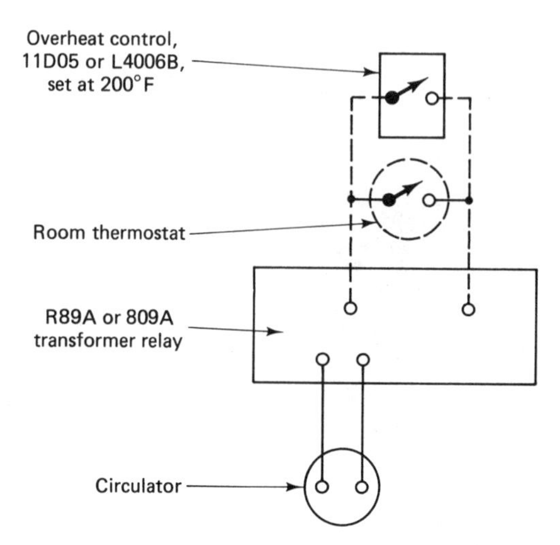

Figure 6.19 Overheat control schematic. This configuration features an immersion aquastat that closes electrical contacts on temperature rise, in essence acting as a second thermostat connected in parallel to the transformer relay.

space heating for extended periods. Most overheat protection schemes are designed to energize a zone circulator to move heated water throughout the baseboard circuit even though there is no call for heat by the room thermostat. This type of operating system is illustrated in schematic form in Figure 6.19. Note from the figure that a conventional aquastat designed to close on a temperature rise is used to initiate the overheat cycle. When the temperature drops below the aquastat setting, it will deactivate the zone circulator, returning the system to normal cycling.

Multifuel boilers. Multifuel boilers combine both a wood/coal heating combustion chamber along with a second burner mechanism for heating the boiler water with either fuel oil or fuel gas. Most multifuel boilers are designed so that the oil or gas is fired in a separate combustion chamber placed upstream of the draft stream from the wood/coal firebox (Figure 6.20). The double combustion chamber arrangement is necessary since even minor creosote deposits on the oil burner nozzle or gas orifices would render these burners totally inoperative.

Most multifuel boilers are set up for two-stage operation. The solid fuel provides first-stage heating for the boiler. The aquastat setting that controls the operation of the fossil-fuel burner mechanism is usually set 10 to 20°F below the wood aquastat controller. Should the water temperature of the boiler drop to the lower fossil-fuel setting, the burner mechanism will be activated automatically to raise the temperature up to the cutoff point on the second-stage aquastat. This type of arrangement provides a great deal of flexibility in that the boiler can automatically switch between solid fuel and fossil fuel to maintain water temperature. If the boiler is unable to keep up with the heat loss in the home due to extremely cold weather or a loss of fire, the oil or gas acts as an automatic backup to maintain even residential heating.

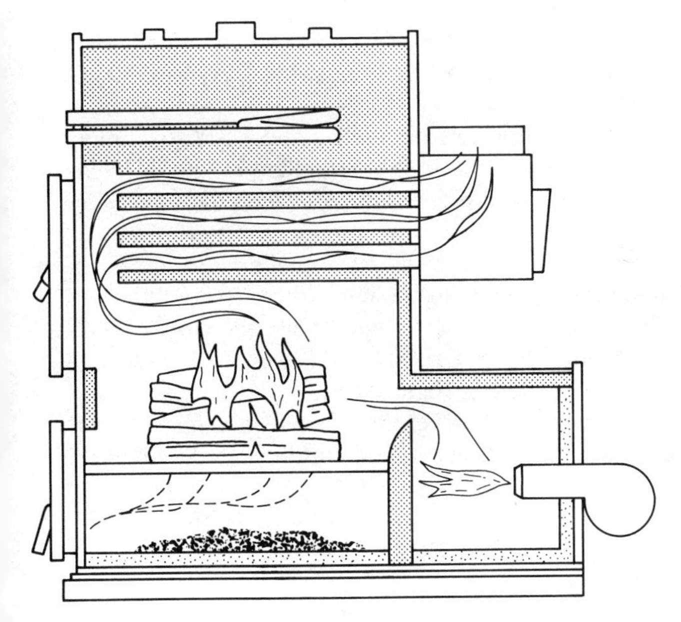

Figure 6.20 Double-combustion-chamber arrangement on multifuel boiler. Note that the oil combustion chamber is updraft of the solid-fuel firebox, preventing creosote deposition on the end cone of the burner. (Courtesy of Northland Boiler Company.)

There have always been differing opinions on whether or not one multifuel boiler is preferable to two separate units featuring a conventional fossil-fuel boiler together with an add-on solid-fuel boiler. Multifuel boilers are generally more expensive than purchasing one solid-fuel and one multifuel boiler. Some people prefer the convenience of one boiler rather than two. Two separate boilers will result in higher combustion efficiencies than one multifuel unit, since each is burning fuel within an appliance designed to maximize the combustion of a single fuel. In most multifuel boilers, however, the design must incorporate modifications that allow both solid and fossil fuels to burn in one boiler. Although the trade-offs are not necessarily significant, there are differences in the combustion efficiencies that will be achieved for each type of fuel. In the final analysis, it is up to the homeowner to determine what type of system configuration will best meet the needs of the residence.

Stoker-fired boilers. The use of stoker devices that automatically feed coal into the firebox and shake the grates to ensure proper combustion have been in use for many years. Automatic stoker systems were common in the early twentieth century in both single-family residences and multifamily and apartment buildings when coal was at the height of its popularity. After World War II, most buildings switched over to either natural gas or fuel oil, the coal boilers were either modified for oil or gas combustion or replaced entirely, and the stoker was relegated to the junk pile. The resurgence of wood and coal units has given renewed impetus to stoker-fired units. Although there are several radiant stoves on the market that utilize automatic stoker firing, the predominant use of stokers is still for central heating boilers and furnaces. In addition to coal, some of these units burn pelletized fuel. The manufacturer's specifications should be consulted to determine the suitability of the unit for pellet fuel and what sizes of coal are best suited for the particular unit.

The classic stoker system consists of a coal bin and an auger-type feed pipe. The auger rests at the bottom of the coal bin and is surrounded by an enclosing pipe through which the coal travels as it is picked up from the coal bin by the auger (Figure 6.21).

The stoker assembly is generally set up to maintain a minimum fire in the firebox and increase the feeding rate to the appliance when there is a call for heat from the room thermostat. The shaker grates are connected to the stoker mechanism, with an ashpan below the firebox that must be emptied periodically. A stoker-fired boiler of this type is illustrated in Figure 6.22.

Although there can be a significant cost differential between a unit with and without a stoker feeding mechanism, the automatic feed feature afforded by these

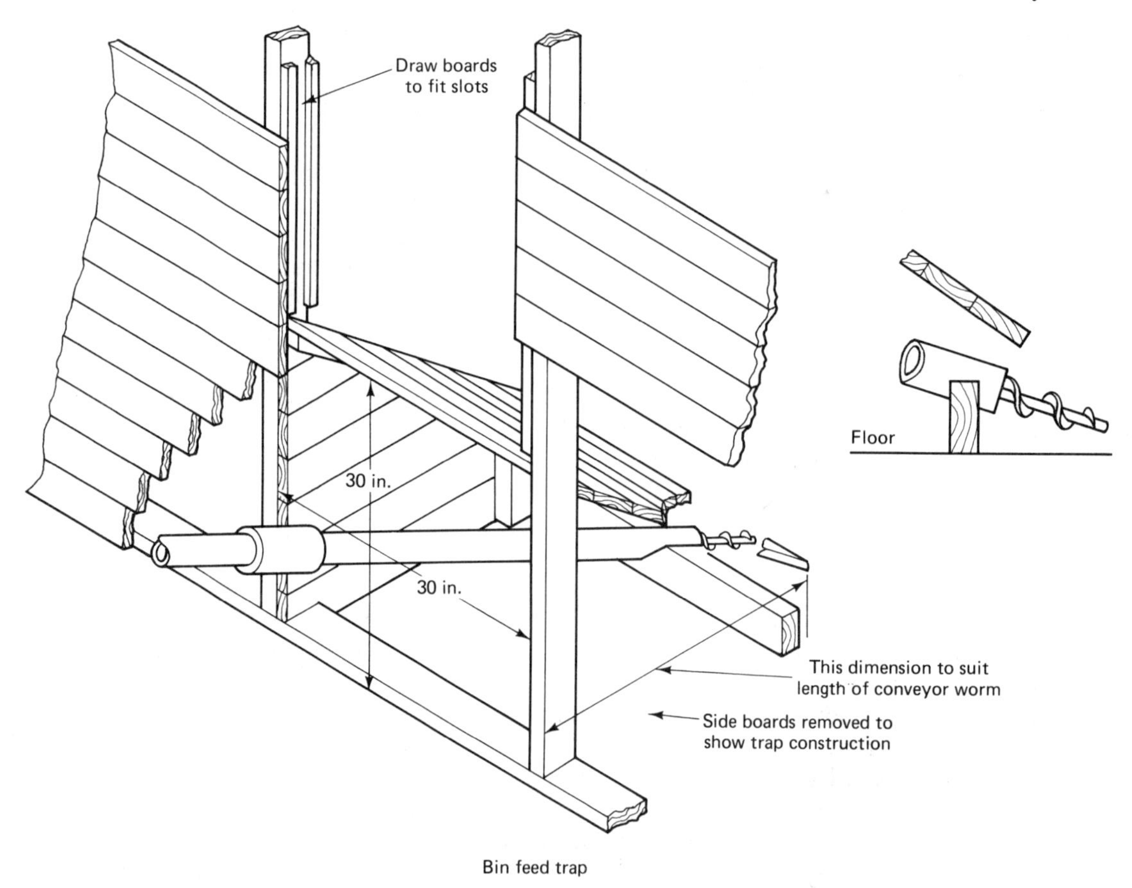

Figure 6.21 Coal auger feed and coal bin assembly. (Courtesy of Electric Furnace Man, Inc.)

(a)

(b)

Figure 6.22 Stoker-fired hot water boiler. (a) Boiler with ash door open. An optional oil burner is shown installed on the upper left portion of the boiler; the stoker motor is below the oil burner. Note the coal auger extending from the right side of the boiler. (b) Boiler rests on an airtight subbase, illustrating the stoker motor on the right and the coal combustion bin. (Courtesy of Electric Furnace Man, Inc.)

systems is usually worth the price, especially in situations where the unit must be left unattended for long periods.

Solid-fuel steam boilers. There are several solid- and multifuel boilers that are available for use in steam heating systems. Many manufacturers offer their hydronic boilers with an optional package consisting of steam trim and controls. When installed as an add-on unit, the solid-fuel steam boiler should be configured as in Figure 6.23.

Wood gasification boilers. Wood gasification combustion is a relatively new concept in solid-fuel combustion and operates differently from conventional solid-fuel appliances. In a gasification unit, an induced-draft fan supplies a large volume of air to the combustion chamber, where a charge of wood fuel is used to heat both the boiler water and a carefully designed and configured amount of high-temperature refractory material within the combustion chamber (Figure 6.24). High temperatures are achieved in these units by virtue of the combustion of the wood gases within the gasification fire tubes. When the boiler reaches operating temperature, the induced-draft fan shuts off, the air intake valve closes, and the fire within the unit goes out completely. When there is a call for heat again, or when the boiler water temperature must be raised, the induced-draft fan goes on. The high temperature within the refractory is sufficient to rekindle the fire in the combustion chamber, and the heating cycle continues. In gasification units, the boiler truly cycles on and off as in conventional fossil-fuel units. These boilers achieve high combustion efficiencies and are economical to operate on an annual basis. Also, because of the thermal mass within the unit, the boiler water temperature is maintained for long periods. It is not unusual for a gasification boiler to be off for 4 hours or more and then rekindle the fire from residual refractory heat. Gasification boilers are more expensive than their conventional counterparts, but can result in large savings over the lifetime of the unit due to reduced wood consumption and high combustion efficiency. These units are also available in multifuel configurations. The oil or gas burner acts as first-stage heat, allowing the wood fire to provide the majority of residential heating. A wood gasification boiler is illustrated in Figure 6.25 in both full view and cross section.

Solid-fuel boiler installation options. When installing a solid-fuel boiler as a stand-alone system, conventional piping procedures are used to install the unit in a series or supply loop heating system. When a solid-fuel boiler is installed as an add-on unit into an existing heating system, however, the piping considerations are somewhat different. There are two piping designs used in add-on solid-fuel boiler installations: either a series or a parallel hookup.

Series Piping. In series piping, the add-on solid-fuel boiler is installed in series with the returning water from the baseboard radiation circuit, between the backup fossil-fuel boiler and the baseboard radiation. The water returning from the baseboard heaters must first pass through the add-on solid-fuel boiler, where it is preheated, and then on through the fossil-fuel boiler and back to the baseboard radiation. This piping configuration is illustrated in Figure 6.26.

Series installations lend themselves to both single- and multizone heating systems. Although Figure 6.26 illustrates a single-zone installation, the boiler may easily be installed in a multizone system. To do this, either the zone that is used

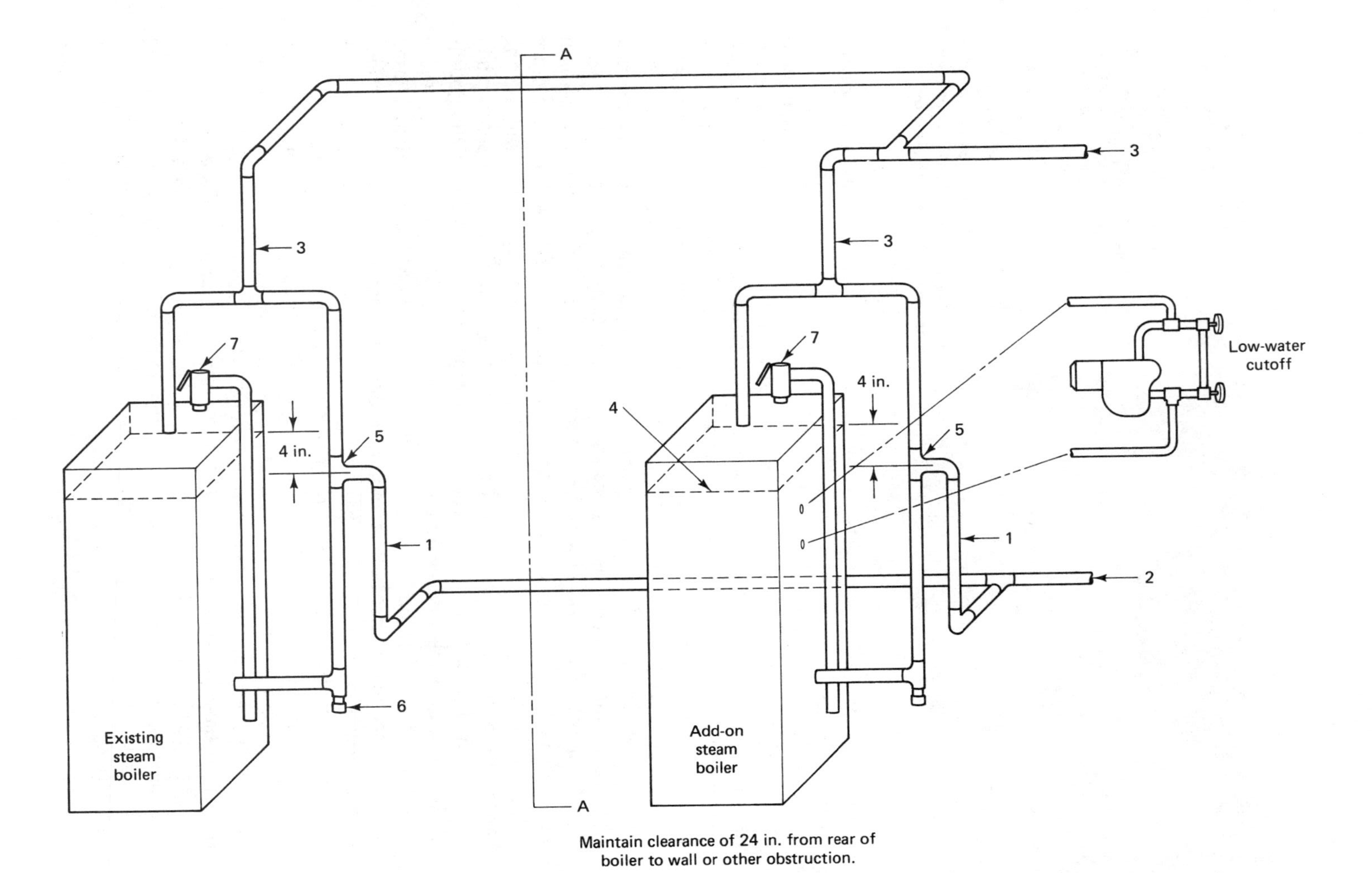

Figure 6.23 Solid-fuel steam boiler installed as an add-on to an existing conventional steam unit. (Courtesy of Northland Boiler Company.)

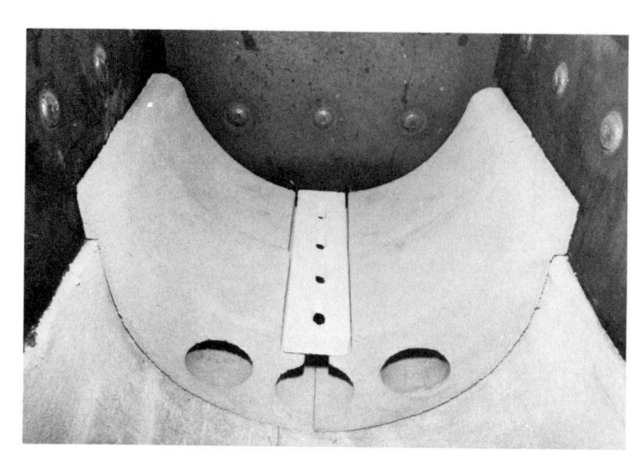

Figure 6.24 Refractory within combustion chamber of wood gasification boiler. (Courtesy of Eshland Enterprises.)

most or the one with the longest piping circuit is chosen for installation of the solid-fuel boiler.

During operation of the series add-on, on a call for heat by the zone thermostat, a zone circulator will be energized to begin moving water through the heating system and both boilers. If the solid-fuel boiler contains an induced-draft combustion system, the thermostat wiring is connected into an electrical relay that turns the induced-draft fan on, in addition to energizing the circulator relay. If the circulator is wired into the system through a low-limit aquastat to prevent water circulation below a specified temperature, the circulator will come on only after the solid-fuel boiler has reached the low-limit temperature level.

Overheat protection in a series system is achieved by wiring a high-limit aquastat into the circulator circuit to turn the circulator on if the water temperature within the boiler reaches the overheat limit. In this way the boiler will use the baseboard circuit to dump excess heat even though the thermostat setting may be satisfied. This may result in occasional overheating of the add-on zone but is preferable to possible overheating in the boiler.

For the solid-fuel boiler to provide effective space heating without activating the fossil-fuel burner mechanism, modifications are made to the aquastat settings on both boilers. An understanding of the operational aspects of the aquastat is necessary to maximize the use of the solid-fuel boiler under these circumstances. In normal operation, a fossil-fuel boiler aquastat will energize the burner if the water falls approximately 10°F below the high-limit setting if the thermostat calls for heat. If the boiler aquastat has a low-limit setting (which it will if the fossil-fuel boiler has a tankless domestic hot water coil), the burner will be activated if the water temperature falls approximately 10°F below the low-limit setting. To maximize the use of the solid-fuel boiler, the aquastat settings on the fossil-fuel boiler should be set below those on the add-on solid-fuel unit. Table 6.4 illustrates a variety of recommended settings for each unit based on individual preferences.

As the water returns from the baseboard circuit, it is heated in the solid-

(a)

Figure 6.25 Wood gasification boiler: (a) full view; (b) cross-sectional view illustrating air intake and method of operation. (Courtesy of Eshland Enterprises.)

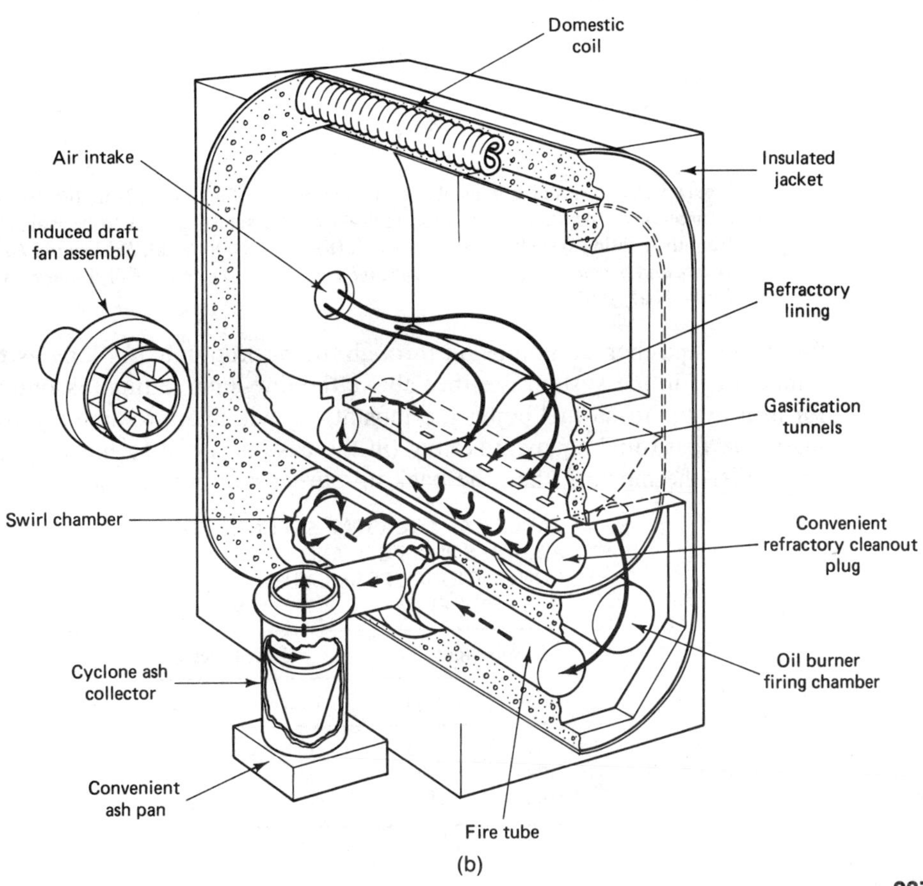

(b)

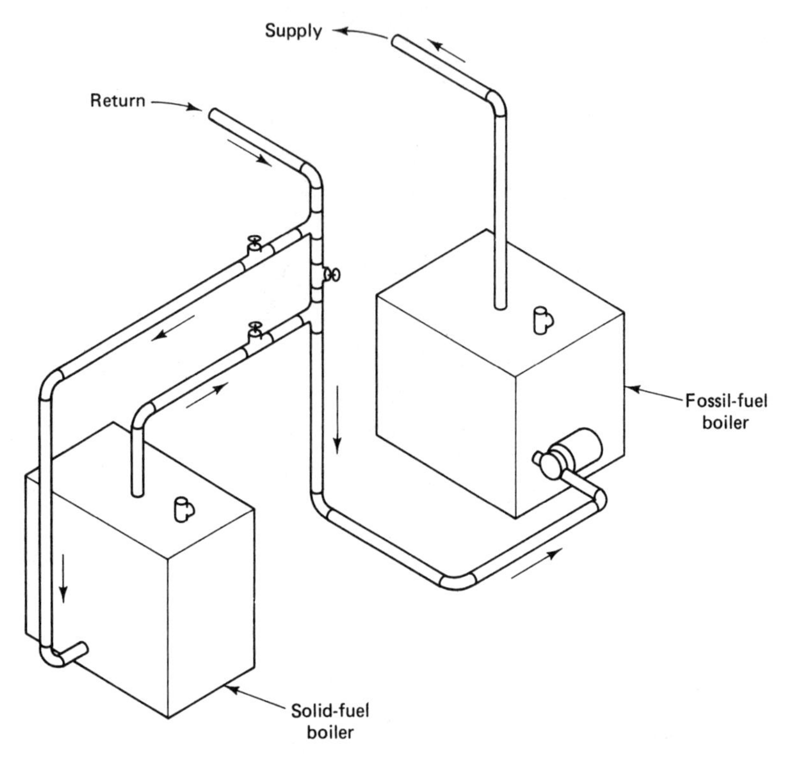

Figure 6.26 Series piping installation of add-on solid-fuel boiler. Note the use of a three-valve bypass configuration that allows the solid-fuel unit to be isolated from the heating system circuit if desired. (From M. Greenwald, *Residential Hot Water and Steam Heating: Gas, Oil, and Solid Fuels*, Prentice-Hall, Englewood Cliffs, N.J., 1987.)

fuel boiler and then continues on through the backup unit. As long as the water temperature in the system is either above the low-limit setting (assuming no call for heat by the fossil-fuel boiler thermostat), or within approximately 10°F of the high-limit fossil-fuel aquastat setting (in the case of a call for heat by the fossil-fuel boiler thermostat), then the back-up burner will not be energized. For best

TABLE 6.4 AQUASTAT SETTINGS FOR ADD-ON SOLID-FUEL AND CONVENTIONAL FOSSIL-FUEL BOILERS (°F)

Solid fuel		Fossil fuel	
Low	High	Low	High
180	200	150	170
170	190	140	160
160	180	130	150

results, the thermostat setting of the add-on boiler should be set approximately 10°F higher than the backup conventional boiler thermostat. These settings allow for true two-stage heating to take place. The solid-fuel boiler will provide first-stage heating to the extent possible. Should the heat loss in the house cause the temperature to drop to the fossil-fuel boiler thermostat setting, the burner mechanism will be started to provide additional space heating from the backup unit.

Parallel Piping. Parallel piping installations differ markedly from series piping systems. In parallel piping, the add-on solid-fuel unit circulates water between itself and the backup boiler on a continuous basis, maintaining a specified water temperature within both boilers. The parallel configuration is illustrated in Figure 6.27.

Note from the figure that the add-on unit draws water from the bottom and returns it to the top of the backup boiler. This is accomplished by the addition of a second system circulator to move water between the two boilers. This circulator is energized by an aquastat located in the solid-fuel boiler that turns on the circulator when the water temperature is high enough to move water between the two boilers without activating the fossil-fuel burner mechanism. Note also the use of flow-check valves to prevent gravity circulation between the boilers should the existing fossil-fuel boiler be at a higher temperature than the add-on unit. Figure 6.27 also illustrates an optional bypass circuit that allows gravity circulation of water to the heating system in the event of a power failure and prevents the add-on boiler from overheating during times when there is no power to operate the system circulators.

In the parallel arrangement, water will move from the backup boiler throughout the baseboard circuit. When the add-on boiler is not operating, or its water temperature is below that of the backup boiler, heating system water will bypass the add-on unit and operate from the backup fossil-fuel boiler.

Tankless Coil Installations. There are many ways in which a tankless coil in an add-on solid-fuel boiler can be plumbed into the existing hot water system. In the majority of instances, the tankless coil of the add-on boiler is piped in series with the coil of the backup boiler (Figure 6.28). Piped in this way, the domestic hot water is preheated in the solid-fuel boiler coil prior to being fed into the fossil-fuel boiler coil. This piping circuit allows the solid-fuel boiler to provide the majority of domestic hot water heating. If the hot water temperature is too cool as it leaves the solid-fuel unit, it can be raised in the fossil-fuel boiler.

Another installation design is used when the home has a stand-alone domestic hot water heater in place of a tankless coil. In this instance, the tankless coil of the add-on unit can be plumbed into the domestic hot water heater as illustrated in Figure 6.29. Note from the figure that no circulator is used in the domestic hot water circuit. Water moves through the system on an as-needed basis. However, gravity circulation will keep water moving through the tankless coil into the hot water heater as long as the solid-fuel boiler is working.

As with fossil-fuel boilers, solid-fuel units can also provide heated water to

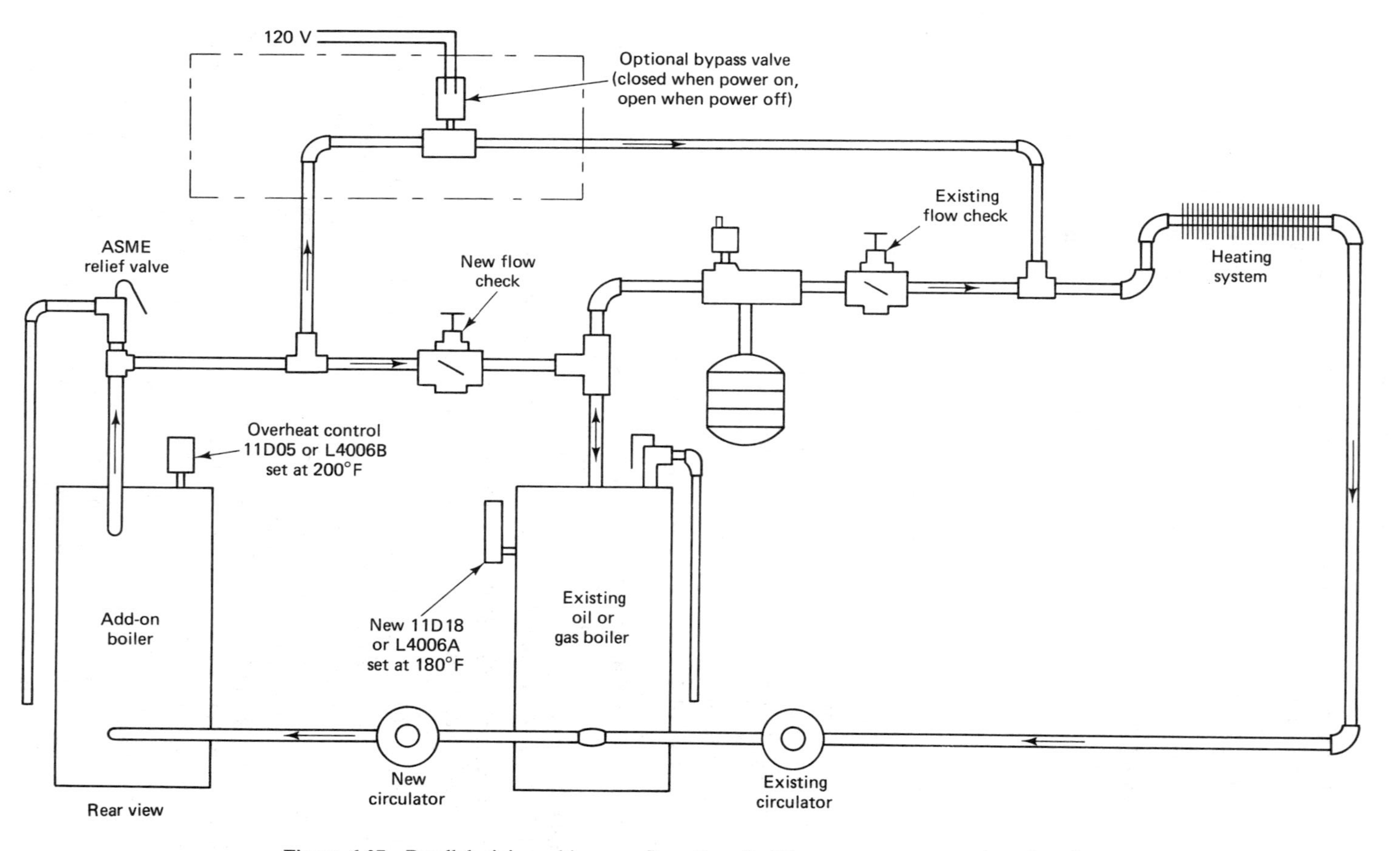

Figure 6.27 Parallel piping add-on configuration. In this arrangement, any time that the boiler water temperature in the add-on reaches the set point of the overheat control (200°F is typical), the new circulator will be energized to prevent overheating and to keep the water at the specified temperature in the existing boiler. (From M. Greenwald, *Residential Hot Water and Steam Heating: Gas, Oil, and Solid Fuels*, Prentice-Hall, Englewood Cliffs, N.J., 1987.)

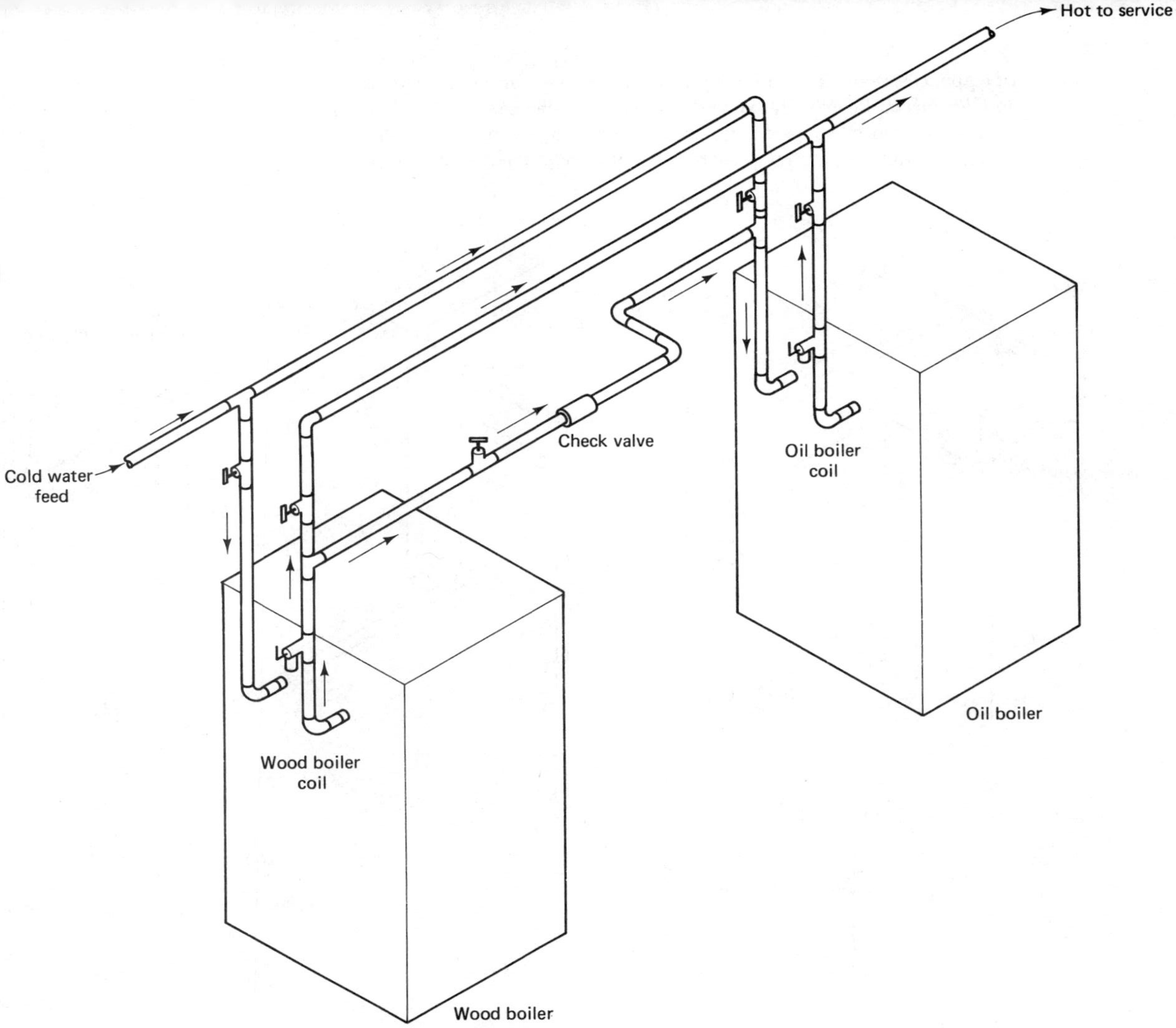

Figure 6.28 Tankless coil piping in situation where both the add-on and existing boilers have tankless water heating coils. (From M. Greenwald, *Residential Hot Water and Steam Heating: Gas, Oil, and Solid Fuels*, Prentice-Hall, Englewood Cliffs, N.J., 1987.)

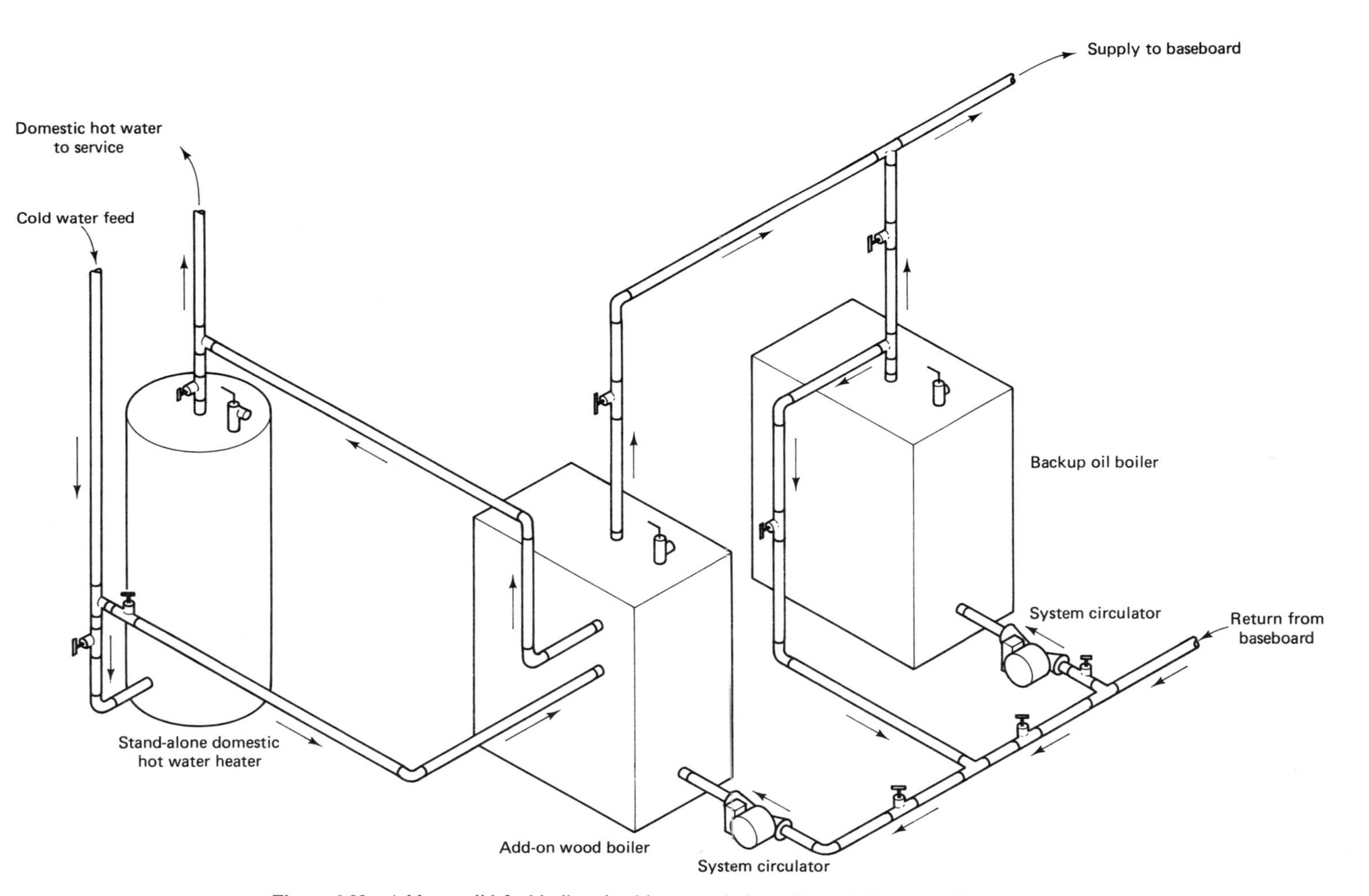

Figure 6.29 Add-on solid-fuel boiler piped into stand-alone domestic hot water heater. To improve efficiency in this type of system, a circulator can be added to the tankless coil piping circuit, controlled by an aquastat to circulate water from the coil to the storage tank. (From M. Greenwald, *Residential Hot Water and Steam Heating: Gas, Oil, and Solid Fuels*, Prentice-Hall, Englewood Cliffs, N.J., 1987.)

operate an external tankless coil. The specific domestic hot water piping configuration will depend on the existing water supply system within the home; however, most add-on solid-fuel boilers are flexible enough to permit a variety of plumbing and heating piping circuits.

Solid-fuel furnaces. Solid-fuel furnaces offer the same installation flexibility as do hot water boilers and are available in either solid fuel only or multifuel configurations. They can be installed as either stand-alone units or in conjunction with existing heating systems. A solid-fuel hot air furnace is illustrated in Figure 6.30.

The operating controls on the solid-fuel unit are essentially the same as those on conventional fossil-fuel furnaces, with the exception of the draft-control mechanism. Many solid-fuel furnaces utilize a motorized draft damper to control combustion air entering the furnace. When the temperature within the hot air plenum reaches the operating limit, the fan-limit switch activates the blower to circulate warm air throughout the heating system. A high-limit sensing switch will close the combustion air damper whenever the plenum temperature reaches the set point to prevent overheating of the furnace plenum.

Two separate combustion chambers are used: one for combustion of oil or gas, the other for combustion of wood or coal. There are different options available for control of the fossil-fuel and solid-fuel components on these units. One common type of control mechanism uses a two-stage thermostat. The first stage of the thermostat activates the solid-fuel combustion cycle, and the second stage of the thermostat energizes the oil or gas burner, placing the solid-fuel combustion cycle on hold until second-stage heating has been satisfied.

Figure 6.30 Solid-fuel hot air furnace. Note the hot air output collar on the top of the furnace plenum, which is sized for optimum hot air delivery from the unit to the warm air distribution system. Proper sizing of the distribution system is critical if the system is to work effectively. (Courtesy of Royall Furnace Works.)

When used as an add-on to an existing hot air furnace, there are several ways to install the solid-fuel furnace within the existing heating system. One installation method is a design in which the add-on furnace is placed in series with the return-air plenum, forcing the return air to travel through the add-on and then into the backup furnace, or vice versa. The original blower is used to move all air throughout the heating system and is activated by the add-on unit when the plenum temperature reaches the cut-in point. The blower can also be activated by the backup furnace in a normal operating heating cycle. This type of installation is illustrated in Figure 6.31.

Yet another method of installing a solid-fuel add-on furnace is to blow the warm air from the add-on unit directly into the existing distribution ductwork of the heating system. The add-on unit is equipped with its own blower and operating controls and taps off the return-air plenum for its air supply. This type of parallel installation is shown in Figure 6.32. Note from the figure that a back-draft damper is installed in the plenum of the fossil-fuel furnace. This prevents a reverse airflow pattern from occurring, ensuring that all air from the add-on unit circulates through the ductwork of the heating system rather than short-cycling between the add-on and backup furnace. Note too the use of an air deflector that extends into the plenum assembly from the add-on ductwork, helping to direct the air from the add-on upward, into the ductwork.

Most solid-fuel furnaces rely on induced-draft mechanisms to control wood combustion. Induced-draft technology is capable of rapidly building up the intensity of the wood fire to a level capable of providing sufficient heat for warm

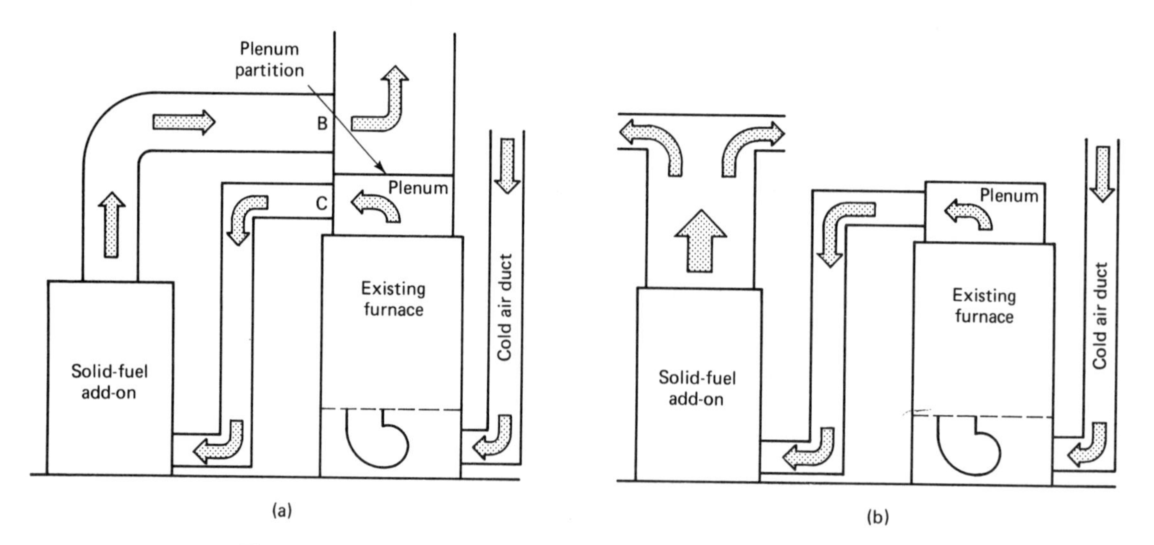

Figure 6.31 Series installation options for add-on solid-fuel hot air furnace: (a) add-on unit installed in series after the existing fossil-fuel furnace via the use of a plenum partition; (b) add-on unit installed in series with existing furnace by disconnecting existing unit from hot air plenum and replacing it with solid-fuel unit. (Courtesy of Royall Furnace Works.)

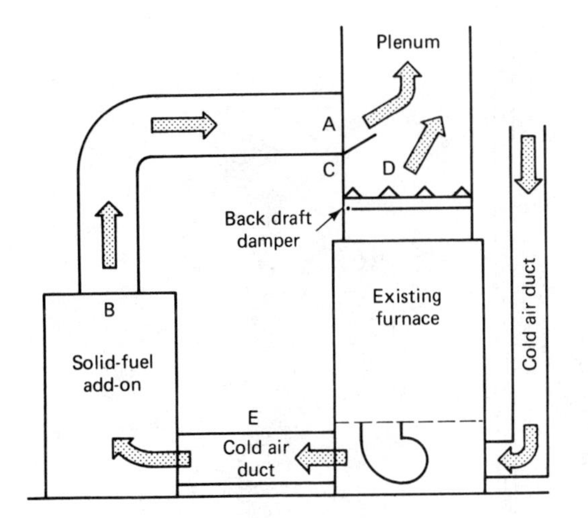

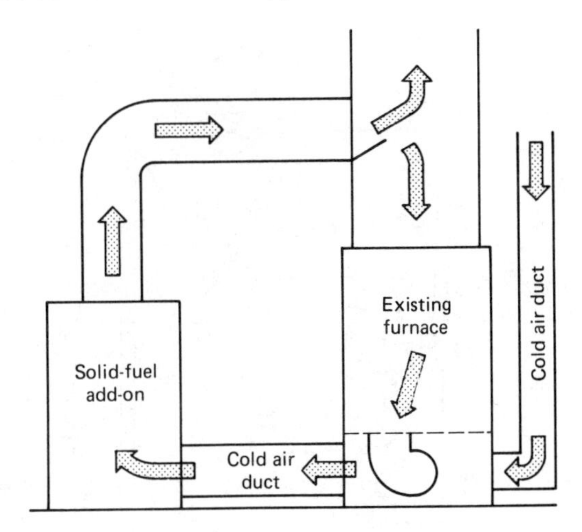

Figure 6.32 Parallel type of hot air solid-fuel add-on installation. (Courtesy of Royall Furnace Works.)

air circulation. When there is no call for heat, the furnace is on standby, maintaining a small fire within the firebox by virtue of an air-bypass setting within the induced-draft fan assembly. Many of these units are also used on a stand-alone basis, being the only source of heat in the home. In these installations, the ducting arrangement is identical to that of a conventional fossil-fuel heating system. Airflow is established around the exterior of the firebox, where it is heated and exits the top of the furnace for distribution throughout the ductwork in the home (Figure 6.33).

Under normal operating conditions, when the thermostat calls for heat, a relay switch in the thermostat circuit closes, activating the induced-draft blower. As the air temperature within the plenum rises, the system blower will turn on at the cut-in level set on the fan-limit switch (usually between 150 and 170°F). Whenever the temperature within the plenum reaches the high-limit setting in the switch, the draft blower is turned off, allowing the intensity of the fire to die down until the cut-in level is reached once again. This cycle continues until the thermostat setting has been satisfied, at which point the blower shuts off and the fire dies down to its off-cycle level.

Solid-Fuel Hot Water Heaters

Although the majority of hot water heating in solid-fuel systems is accomplished by virtue of a tankless coil installation, there are some solid-fuel domestic hot water heaters available. These units feature a variety of combustion control devices and are generally connected in series or parallel with an existing domestic hot water heater, similar to the installation configurations used in solar domestic hot water heating systems.

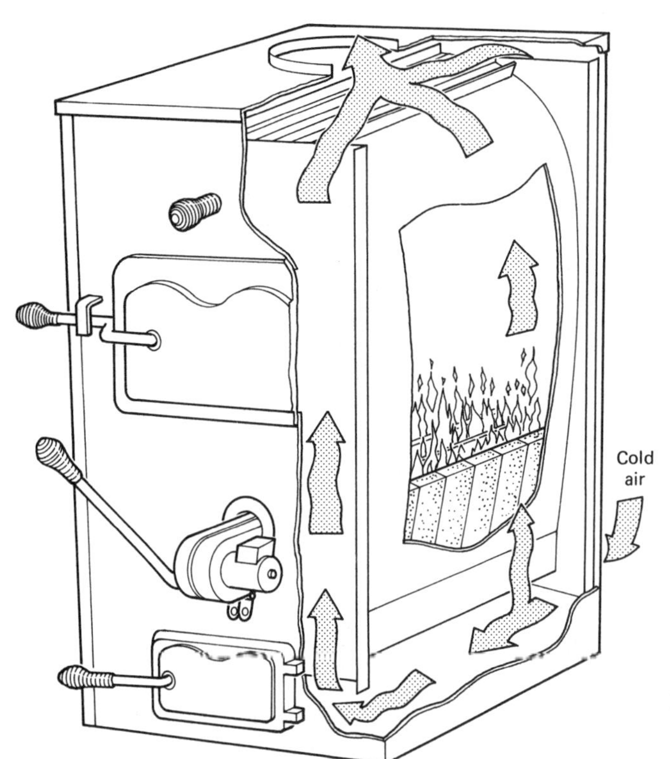

Figure 6.33 Airflow pattern around typical solid-fuel hot air furnace. (Courtesy of Royall Furnace Works.)

VENTING AND CHIMNEY REQUIREMENTS

Regardless of the type of solid-fuel heating system, the most critical aspect of the appliance installation concerns the nature and integrity of the chimney venting system, since the combustion cycle of wood leads to the formation of creosote, which can build up within the flue and chimney passages with potentially disastrous results. Two types of chimneys are available for solid-fuel appliance venting: stainless steel or masonry. The first step prior to installing any solid-fuel appliance is to determine the suitability of the present chimney for the proposed installation, if one exists. In general, the following five guidelines must be met if an existing chimney is to be used.

1. The chimney must not be used by any other appliance. A separate flue is required for each unit connected into a chimney. Under no circumstances should a solid-fuel appliance ever be connected into the same flue as any other gas, oil, or solid-fuel unit.
2. The chimney should be properly lined with a high-temperature fireclay flue liner, or stainless steel liner and insulation in the case of a chimney that has been repaired and relined.

3. There should be no loose mortar or bricks. Check the chimney to ensure that no blockages have occurred within the flue passages.
4. The chimney should be of the proper cross-sectional area to handle the maximum potential amount of heat to be produced by the solid-fuel unit. Consult Table 6.3 for determining the proper cross-sectional area of the chimney for specified heat-removal capacities.
5. In stainless steel chimney installations, make sure that the interior of the chimney is in good condition and that no corrosion exists in any of the stainless steel pipe sections. Also, it is good practice to check over the entire chimney to make sure that it was installed properly, with particular attention paid to minimum clearances and proper technique of installation.

If a new chimney is to be installed, either masonry or stainless steel is the material of choice.

Masonry Chimneys

Masonry chimneys are the most common types of chimneys. When undertaking the construction of a masonry chimney, there are certain guidelines to follow for proper installation and longevity of operation. A typical masonry chimney in cross section is shown in Figure 6.34.

Local building codes usually specify the minimum depth of footings in addition to other factors of chimney construction. All applicable codes should be checked before starting any construction. To avoid some of the common problems associated with masonry chimney construction, consult Figure 6.35.

Stainless Steel Chimneys

The use of prefabricated stainless steel chimneys has grown rapidly during recent years. This growth has mirrored that of the solid-fuel appliance business, as homeowners required a safe, quick, and convenient method for providing a chimney for stoves and central heating units.

Stainless steel chimneys can be either within or outside the house. Both interior and exterior chimney installations are illustrated in Figure 6.36. For efficient operation it is preferable to install a chimney inside the home. This allows for higher chimney temperatures, minimizing creosote buildup, and also enables heat to radiate from the chimney into the interior of the home. Exterior chimneys tend to corrode and fail more rapidly than do interior chimneys because cold outside temperatures foster continual creosote deposition in addition to a buildup of condensates such as sulfuric acid and chlorides, which may corrode even high-grade stainless steel over a sufficient period of time.

All chimneys must extend beyond the roof line a sufficient distance in order to operate safely and minimize the possibility of downdrafts and spark ignition from the vented by-products of solid-fuel combustion. The minimum height and clearances for stainless steel chimneys are illustrated in Figure 6.37.

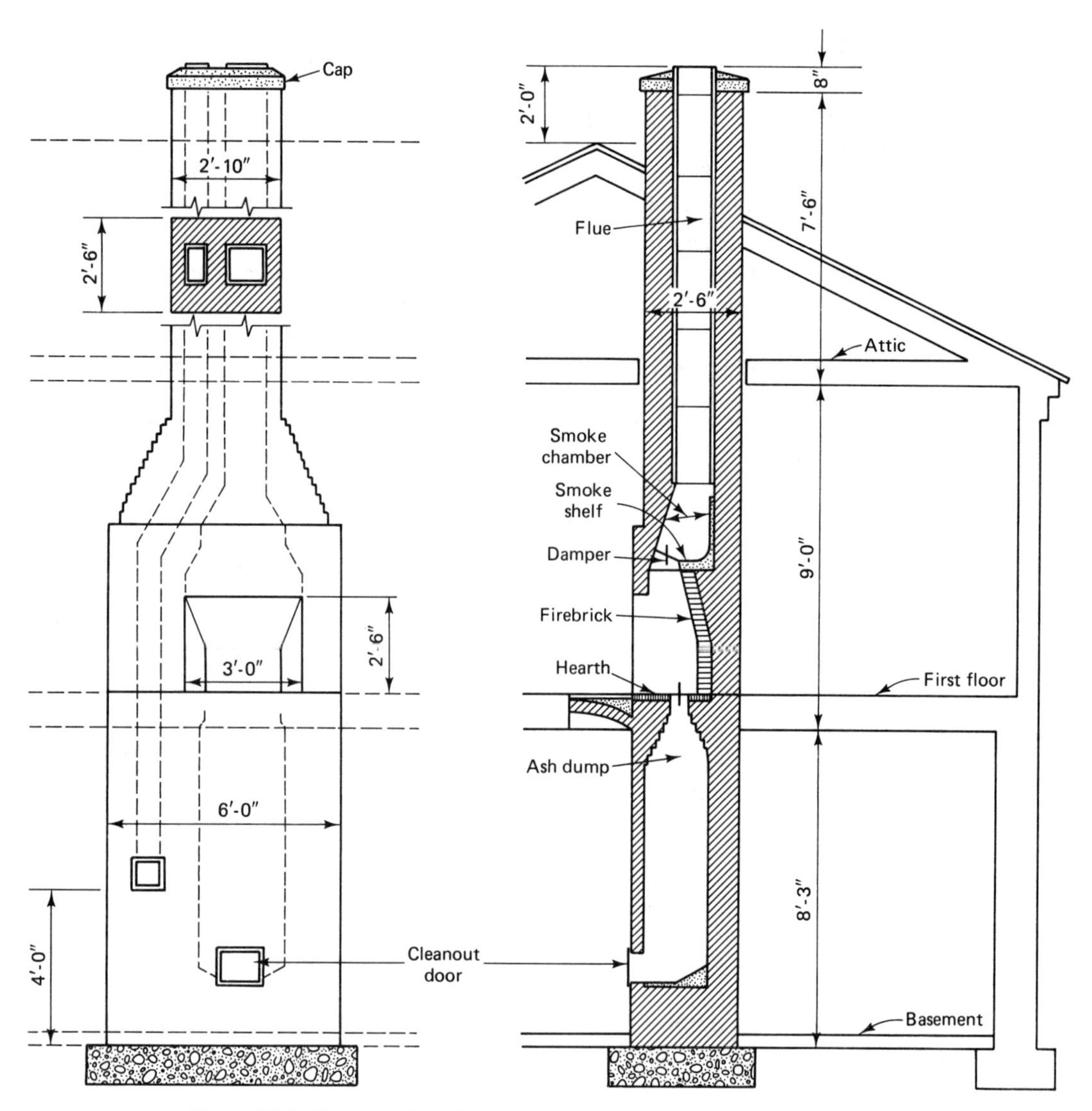

Figure 6.34 Cross section of typical masonry chimney. (Adapted from U.S. Department of Agriculture, *Farmers Bulletin,* revised 1971.)

All stainless steel chimneys undergo rigorous testing and certification procedures under the auspices of Underwriters' Laboratories in both the United States and Canada. These test procedures are periodically revised and updated to provide maximum safety and reliability. Check with the manufacturer concerning the heat rating of the chimney prior to making any purchases. Newer stainless steel chimneys are rated for high-temperature use specifically in solid-fuel appliances and so state in their respective literature.

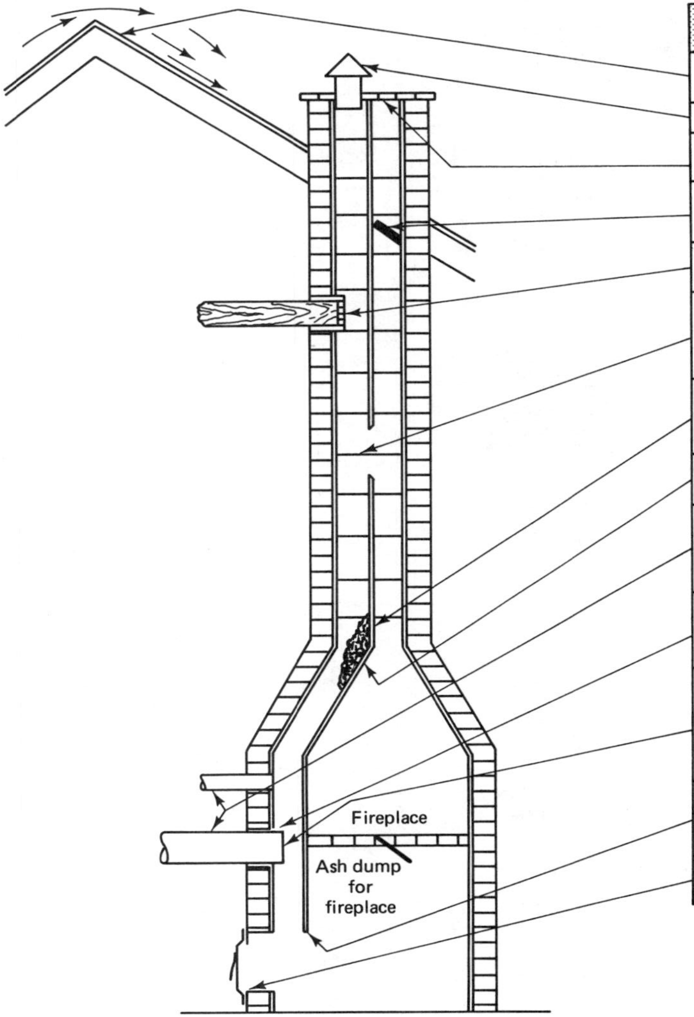

Common Chimney Troubles and Their Corrections

Troubles	Examination	Corrections
Top of chimney lower than surrounding objects	Observation	Extend chimney above all objects within 30 ft
Chimney cap or ventilator	Observation	Remove
Coping restricts opening	Observation	Make opening as large as inside of chimney
Obstruction in chimney	Can be found by light and mirror reflecting conditions in chimney	Use weight to break and dislodge
Joist projecting into chimney	Lowering a light on extension cord	Must be handled by a competent brick contractor
Break in chimney lining	Smoke test-build smudge fire blocking off other opening, watching for smoke to escape	Must be handled by a competent brick contractor
Collection of soot at narrow space in flue opening	Lower light on extension cord	Clean out with weighted brush or bag of loose gravel on end of line
Offset	Lower light on extension	Change to straight or to long offset
Two or more openings into same chimney	Found by inspection from basement	The least important opening must be closed, using some other chimney flue
Loose-seated pipe in flue opening	Smoke test	Leaks should be eliminated by cementing all pipe openings
Smoke pipe extends into chimney	Measurement of pipe from within or observation of pipe by means of a lowered light	Length of pipe must be reduced to allow end of pipe to be flush with inside of tile
Failure to extend the length of flue partition down to the floor	By inspection or smoke test	Extend partition to floor level
Loose-fitted clean-out door	Smoke test	Close all leaks with cement

Figure 6.35 Common masonry chimney problems, with some possible corrective procedures recommended. (Courtesy of R.W. Beckett, Inc.)

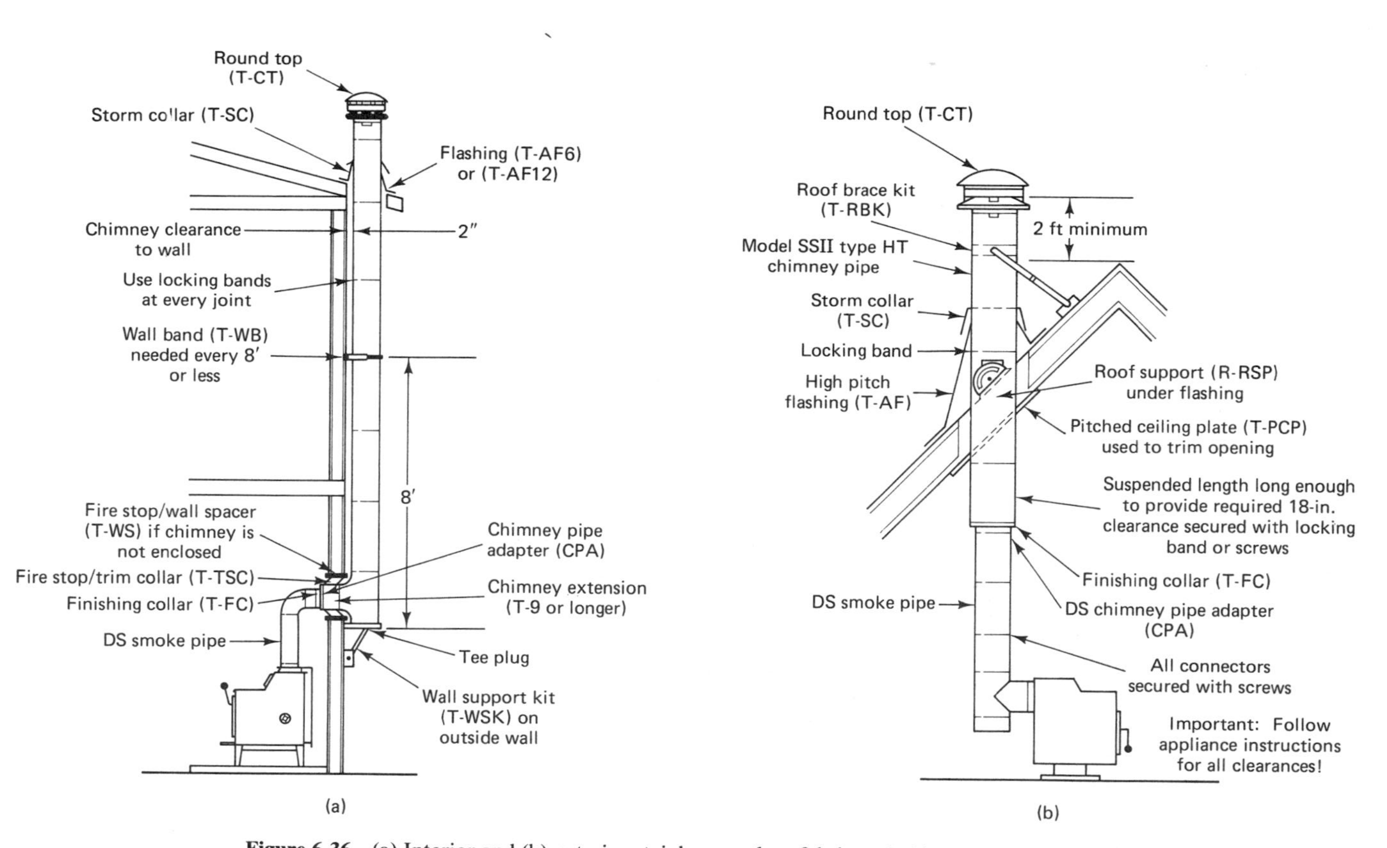

Figure 6.36 (a) Interior and (b) exterior stainless steel prefabricated chimney installations. Where possible, interior chimneys are preferred, since they operate at higher temperatures, reducing creosote deposition, and can add significant amounts of heat to the interior of a home. (Courtesy of Selkirk Metalbestos, Division of Household Manufacturing, Inc.)

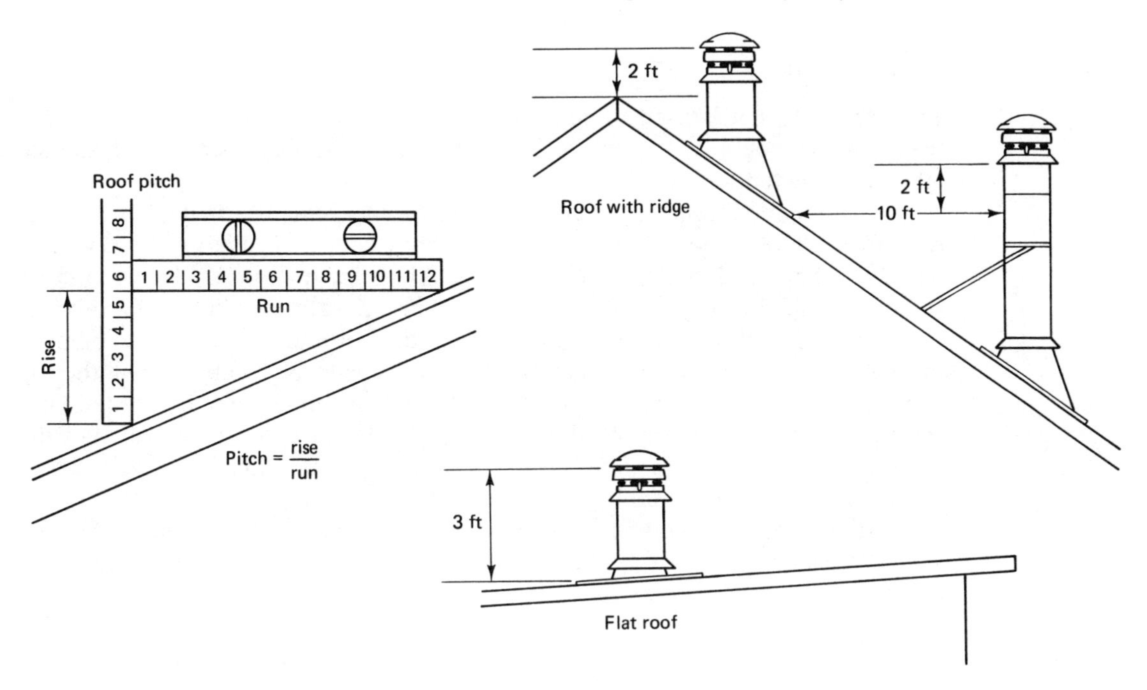

Figure 6.37 Minimum height and roof clearances of stainless steel chimneys. (Courtesy of Selkirk Metalbestos, Division of Household Manufacturing, Inc.)

Chimney Maintenance

All chimneys used in solid-fuel appliances require frequent maintenance and cleaning during the heating season. Chimneys that vent solid-fuel boilers require more maintenance than do venting furnaces or radiant stoves, since water-cooled fireboxes accelerate the buildup of creosote. In general, chimneys that vent solid-fuel boilers should be inspected twice monthly during the heating season. Those that vent hot air furnaces and stoves should be inspected and cleaned on a monthly basis.

When cleaning masonry chimneys, it is necessary to use a brush that is sized for the physical dimensions of the chimney flue liner. The use of steel brushes should be avoided in chimneys venting coal-burning units, since sparks created by the steel brushes as they are moved up and down the chimney can cause explosions due to the coal gas generated inside the chimney during the cleaning process. In this instance the use of rigid plastic brushes is recommended.

In addition to periodic cleaning, stainless steel chimneys should be oiled at the end of each heating cycle. This helps to keep the chimney resistant to any condensation that might occur during the summer months and to retard deterioration.

Vent Piping

The vent pipe that is used to connect a solid-fuel boiler, furnace, or stove into the chimney undergoes a great deal of stress during normal operation. This pipe may be subjected to temperatures in excess of 900°F.

The minimum requirements for stoves specify 24-gauge or thicker pipe. In situations where the unit will be used for burning coal exclusively, stainless steel pipe should be used instead of conventional black stove pipe. Under no circumstances should 26-gauge smoke pipe be used in any solid-fuel boiler, furnace, or stove, since it will rapidly deteriorate. A minimum of four screws should be used to secure each pipe joint, and the seams should always be locating along the top edge of the pipe. Also, the crimped ends of the pipe should face toward the unit, since this facilitates the runoff of creosote back into the unit rather than allowing it to drip or run out of the joint onto the floor. Where possible, the inside seams of each pipe section should be sealed with high-temperature furnace cement.

All vent pipes should be cleaned whenever the chimney is cleaned. The pipe should be oiled at the end of each heating season to help ensure maximum longevity.

7

RESIDENTIAL AIR CONDITIONING AND HEAT PUMP OPERATION

AIR CONDITIONING

The term *air conditioning* refers to maintaining interior temperatures at a comfortable level on a seasonal basis and encompasses both winter and summer operation. In this chapter we focus on summer air-conditioning systems and heat pumps that are used for both space heating and cooling.

Most residential air-conditioning systems and heat pumps are installed in one of three ways: as a stand-alone system; as air-conditioning and space heating units in a side-by-side or add-on design using the same air distribution network; or as an air system made up of separate room appliances. We examine the basic operating cycles and component configurations that are common to all these systems, together with major operating controls.

Moving Heat Uphill

Heat moves naturally from a warm substance to a cooler one without the need to use any external force or work being applied to the system. As long as there is a temperature difference between two objects, heat will flow from the hotter to the cooler one until both arrive at the same temperature.

To move heat from a cold area to a warmer one, work must be added to the system to move the heat "uphill." All air-conditioning systems are based on the refrigeration cycle. This cycle is used to remove heat from a cold area, for example, the inside of a refrigerator or room in the home, and deposit this heat in a warmer area, the air in the kitchen, or outdoor air. In a typical heat pump, a groundwater supply at a constant temperature would be the cold chamber; the area inside the house would be the warm chamber. To accomplish this uphill transfer of heat, an electrically driven compressor is used to circulate refrigerant fluid within a closed mechanical system. The pressure of the refrigerant varies in different parts of the system: Low-pressure refrigerant is used to absorb heat from the cold chamber in a set of evaporator coils; the heat-laden refrigerant is then raised in temperature and pressure to allow the heat to dissipate to the warm chamber in a set of condenser coils. The subtleties of the refrigeration cycle are highlighted by distinctions between various types of heat and temperature measurements that apply to all air-conditioning and heat pump systems.

Classification of Heat

There are four different classifications of heat and two types of temperature measurements referred to when describing air conditioners and heat pumps. These are:

1. *Sensible heat.* Sensible heat is heat that is added to or removed from a substance that changes the temperature of the substance but not its state. For example, when water is heated from 50°F to 90°F, it undergoes a change in sensible heat only; it remains in liquid state.
2. *Latent heat.* Latent heat is the amount of heat that is needed to change the state of a substance without changing its temperature. Changing 1 lb of water at 212°F to steam at 212°F requires 970 Btu of latent heat. Conversely, when 1 lb of ice at 32°F changes to water at 32°F, 144 Btu of latent heat is removed.
3. *Superheat.* Superheat is heat that is added to a vapor above its boiling point at specific pressure conditions. Thus, after water has been heated to steam, any additional heat that is added to the vapor raises the temperature of the vapor and superheats it. Since the vapor undergoes a temperature change only, superheat is, in effect, sensible heat.
4. *Specific heat.* The specific heat of a substance is the number of Btu necessary to change 1 lb of the substance 1°F. Thus 1 Btu is required to raise the temperature of 1 lb of water 1°F.

Temperature measurements may be made using two different techniques:

1. *Wet-bulb temperature.* The wet-bulb temperature is the lowest temperature

TABLE 7.1 BTU VALUES IN CHANGE OF STATE OF WATER

Substance	Amount	From temp. (°F)	To temp. (°F)	Temp. rise (°F)	Specific heat	Amount of heat	Type of heat	Name of change
Ice (solid)	1 lb	0°	32° =	32° at	½ Btu =	16 Btu	Sensible	(Heating)
	1 lb	32°	0° =	−32° at	½ Btu =	−16 Btu	Sensible	(Cooling)
Ice to water	1 lb	32°	32° =	0° at	144 Btu =	144 Btu	Latent	(Melting)
Water to ice	1 lb	32°	32° =	0° at	−144 Btu =	−144 Btu	Latent	(Freezing)
Water (liquid)	1 lb	32°	212° =	180° at	1 Btu =	180 Btu	Sensible	(Heating)
	1 lb	212°	32° =	−180° at	1 Btu =	−180 Btu	Sensible	(Cooling)
Water to vapor	1 lb	212°	212° =	0° at	970 Btu =	970 Btu	Latent	(Vaporization)
Vapor to water	1 lb	212°	212° =	0° at	−970 Btu =	−970 Btu	Latent	(Condensation)
Vapor (gas)	1 lb	212°	222° =	10° at	½ Btu =	5 Btu	Sensible	(Heating)
	1 lb	222°	212° =	−10° at	½ Btu =	−5 Btu	Sensible	(Cooling)

Source: The Williamson Company.

that can be indicated by a standard thermometer. Wet-bulb temperatures are obtained by placing a wet wick over the sensing bulb of the thermometer. Evaporation of water from the wick to the surrounding circulating air takes place based on the relative humidity of the air, lowering the temperature reading on the thermometer.

2. *Dry-bulb temperature*. Dry-bulb temperatures are temperature measurements taken with conventional thermometers. The difference between the wet-bulb and the dry-bulb temperature indicates the relative humidity of the surrounding air.

To illustrate the relationship of sensible, latent, super, and specific heat, consult Table 7.1, which lists the appropriate Btu values most often encountered.

THE BASIC REFRIGERATION CYCLE

The refrigeration cycle is based on the properties of substances classified as refrigerants, those materials that have boiling points well below the temperature required in the cold (evaporator) chamber. For example, R-12, a refrigerant, used in many residential air conditioners and heat pumps, boils at −21°F. This enables the material to absorb heat from its surroundings as long as they are above −21°F, cooling the area in the process and causing the refrigerant to boil as it absorbs heat. The boiling points of R-12 and R-22, two of the most widely used refrigerants, are listed in Table 7.2.

TABLE 7.2 CHARACTERISTICS OF R-12 AND R-22 REFRIGERANTS

	R-22 ($CHClF_2$)	R-12 (CCl_2F_2)
Boiling point (at 1 atm)	−41.4°F	−21.6°F
Refrigerating effect per pound	69 Btu	50 Btu
Refrigerant circulated (lb/min)	3 lb	4 lb
Freezing point	−256°F	−252°F
Flammable or explosive	No	No

Source: The Williamson Company.

Note: Standard-ton conditions; all figures approximate.

EFFECTS OF TEMPERATURE ON REFRIGERANTS

All air conditioners and heat pumps change the pressure and temperature of the refrigerant as it moves from one component of the system to another. Changes in temperature and pressure affect the boiling point of the refrigerant. As the pressure of the refrigerant is lowered, its boiling point is decreased. Conversely, as the pressure of the refrigerant increases, its boiling point increases. The temperature and pressures across the condensing and evaporator coils, as well as suction and discharge pressures at the compressor, provide a convenient means for making initial determinations as to system performance and efficiency. A temperature and pressure chart illustrating these relationships for R-12 and R-22 refrigerants is illustrated in Table 7.3.

REFRIGERANT CAPACITY AND CHARGE

Whereas heating systems are directly rated in Btu per hour delivered to the space to be heated, air-conditioning systems are rated in tons or nominal tons of unit capacity. The ton designation is based on the number of Btu of heat removed from within the conditioned area, based on 144 Btu of heat removal being required to melt 1 lb of ice (see Table 7.1). The cooling effect that is achieved by melting 1 ton of ice (remember that the heat necessary to melt the ice comes from the container or room in which the ice is located, cooling the area surrounding the ice) in a 24-hour period is

$$2000 \text{ lb} \times 144 \text{ Btu} = 288{,}000 \text{ Btu}$$

To calculate the cooling effect on an hourly basis, we find that

$$\frac{288{,}000}{24} = 12{,}000 \text{ Btu/hr}$$

TABLE 7.3 TEMPERATURE AND PRESSURE RELATIONSHIPS OF A VARIETY OF COMMERCIAL REFRIGERANTS

°F	R-12	R-13	R-22	R-500	R-502	R-717 ammonia	°F	R-12	R-13	R-22	R-500	R-502	R-717 ammonia
−100	**27.0**	7.5	**25.0**	—	**23.3**	**27.4**	16	18.4	211.9	38.7	24.2	47.8	29.4
−95	**26.4**	10.9	**24.1**	—	**22.1**	**26.8**	18	19.7	218.8	40.9	25.7	50.1	31.4
−90	**25.7**	14.2	**23.0**	—	**20.7**	**26.1**	20	21.0	225.7	43.0	27.3	52.5	33.5
−85	**25.0**	18.2	**21.7**	—	**19.0**	**25.3**	22	22.4	233.0	45.3	29.0	55.0	35.7
−80	**24.1**	22.2	**20.2**	—	**17.1**	**24.3**	24	23.9	240.3	47.6	30.7	57.5	37.9
−75	**23.0**	27.1	**18.5**	—	**15.0**	**23.2**	26	25.4	247.8	49.9	32.5	60.1	40.2
−70	**21.8**	32.0	**16.6**	—	**12.6**	**21.9**	28	26.9	255.5	52.4	34.3	62.8	42.6
−65	**20.5**	37.7	**14.4**	—	**10.0**	**20.4**	30	28.5	263.2	54.9	36.1	65.4	45.0
−60	**19.0**	43.5	**12.0**	—	**7.0**	**18.6**	32	30.1	271.3	57.5	38.0	68.3	47.6
−55	**17.3**	50.0	**9.2**	—	**3.6**	**16.6**	34	31.7	279.5	60.1	40.0	71.2	50.2
−50	**15.4**	57.0	**6.2**	—	0.0	**14.3**	36	33.4	287.8	62.8	42.0	74.1	52.9
−45	**13.3**	64.6	**2.7**	—	2.1	**11.7**	38	35.2	296.3	65.6	44.1	77.2	55.7
−40	**11.0**	72.7	0.5	**7.9**	4.3	**8.7**	40	37.0	304.9	68.5	46.2	80.2	58.6
−35	**8.4**	81.5	2.6	**4.8**	6.7	**5.4**	45	41.7	327.5	76.0	51.9	88.3	66.3
−30	**5.5**	91.0	4.9	**1.4**	9.4	**1.6**	50	46.7	351.2	84.0	57.8	96.9	74.5
−28	**4.3**	94.9	5.9	0.0	10.6	0.0	55	52.0	376.1	92.6	64.2	106.0	83.4
−26	**3.0**	98.9	6.9	0.7	11.7	0.8	60	57.7	402.3	101.6	71.0	115.6	92.9
−24	**1.6**	103.0	7.9	1.5	13.0	1.7	65	63.8	429.8	111.2	78.2	125.8	103.1
−22	**0.3**	107.3	9.0	2.3	14.2	2.6	70	70.2	458.7	121.4	85.8	136.6	114.1
−20	0.6	111.7	10.1	3.1	15.5	3.6	75	77.0	489.0	132.2	93.9	148.0	125.8
−18	1.3	116.2	11.3	4.0	16.9	4.6	80	84.2	520.8	143.6	102.5	159.9	138.3
−16	2.1	120.8	12.5	4.9	18.3	5.6	85	91.8	—	155.7	111.5	172.5	151.7
−14	2.8	125.7	13.8	5.8	19.7	6.7	90	99.8	—	168.4	121.2	185.8	165.9
−12	3.7	130.5	15.1	6.8	21.3	7.9	95	108.3	—	181.8	131.2	199.7	181.1
−10	4.5	135.4	16.5	7.8	22.8	9.0	100	117.2	—	195.9	141.9	214.4	197.2
−8	5.4	140.5	17.9	8.8	24.4	10.3	105	126.6	—	210.8	153.1	229.7	214.2
−6	6.3	145.7	19.3	9.9	26.0	11.6	110	136.4	—	226.4	164.9	245.8	232.3
−4	7.2	151.1	20.8	11.0	27.7	12.9	115	146.8	—	242.7	177.3	262.6	251.5
−2	8.2	156.5	22.4	12.1	29.5	14.3	120	157.7	—	259.9	190.3	280.3	271.7
0	9.1	162.1	24.0	13.3	31.2	15.7	125	169.1	—	277.9	203.9	298.7	293.1
2	10.2	167.9	25.6	14.5	33.1	17.2	130	181.0	—	296.8	218.2	318.0	315.0
4	11.2	173.7	27.3	15.7	35.0	18.8	135	193.5	—	316.6	233.2	338.1	335.0
6	12.3	179.8	29.1	17.0	37.0	20.4	140	206.6	—	337.3	248.8	359.1	365.0
8	13.5	185.9	30.9	18.4	39.1	22.1	145	220.6	—	358.9	265.2	381.1	390.0
10	14.6	192.1	32.8	19.8	41.1	23.8	150	234.6	—	381.5	282.3	403.9	420.0
12	15.8	198.6	34.7	21.2	43.3	25.6	155	249.9	—	405.2	300.1	427.8	450.0
14	17.1	205.2	36.7	22.7	45.5	27.5	160	265.12	—	429.8	318.7	452.6	490.0

Source: Bill Langley, *Refrigeration and Air Conditioning*, 3rd ed., Prentice-Hall, Inc., Englewood Cliffs, N.J., 1986, Table 7-1, p. 130. Reprinted by permission of Prentice-Hall, Inc.

Note: Bold figures, in. Hg vacuum; light figures, psig.

An air conditioner rated at 1 ton of capacity will remove 12,000 Btu/hr from a specific area or room. Sometimes, small air conditioners, particularly window units, are rated in Btu per hour rather than ton capacities, since these systems are relatively small in cooling capacity compared to whole-house units. After heat-gain calculations have been made, the sizing of the air conditioner or heat pump can be determined. After the unit has been sized, the refrigerant charge is determined based on the type of refrigerant used in the system. For example, Table 7.2 illustrates that 1 lb of R-22 refrigerant absorbs 69 Btu of heat. If we assume an air-conditioning unit rated at 1 ton of capacity (12,000 Btu/hr), 173 lb of refrigerant must be circulated per hour to remove 12,000 Btu of heat from the space to be cooled:

$$\frac{12{,}000}{69} = 173$$

The amount of circulating charge per minute is determined by dividing the number of pounds of refrigerant per hour by 60 and adding a small amount of refrigerant required for the tubing in the system:

R-22 refrigerant, 2-ton air-conditioning unit:

$$24{,}000 \div 69 = \frac{348 \text{ lb}}{60} = 6 \text{ lb/min} + \text{a small amount for system tubing}$$

R-12 refrigerant, 2-ton air-conditioning unit:

$$24{,}000 \div 50 = \frac{480 \text{ lb}}{60} = 8 \text{ lb/min} + \text{a small amount for tubing}$$

Note: These figures apply only to condensers with liquid receivers and are approximate values.

REFRIGERATION CYCLE DEVELOPMENT

Figure 7.1 illustrates the development of a closed-system refrigeration cycle from a simple refrigerant effect. R-22 refrigerant has been placed in a vessel located within an insulated container (Figure 7.1a). Since the vessel is vented to the air outside the container, the operating pressure of the system is zero. The refrigerant in the vessel begins to absorb heat from the inside of the insulated container, evaporating and boiling in the process. Given a sufficient quantity of refrigerant in the vessel, the temperature of the surrounding insulated container will eventually be reduced to −41°F, which is the boiling point of the R-22 refrigerant. All the refrigerant will be lost to the open air in this type of system.

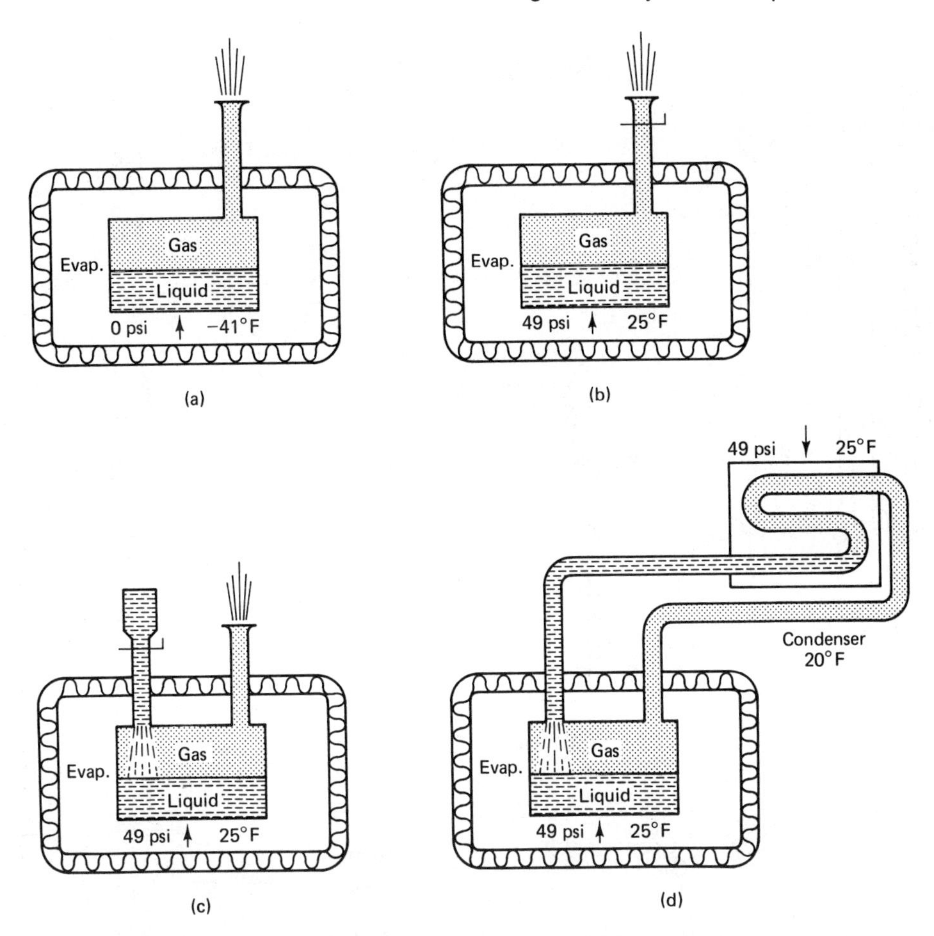

Figure 7.1 Development of the refrigeration cycle: (a) refrigerant in a closed vessel vented to the atmosphere; (b) raising the pressure and temperature of the vessel by adding a valve to the refrigerant vent line; (c) maintaining an even temperature by adding a reservoir container to the refrigerant system; (d) addition of an external condenser, closing the refrigerant loop. (Courtesy of The Williamson Company.)

Since −41°F is a temperature that is too low for most refrigeration or cooling processes, the pressure of the refrigerant vessel must be raised to obtain the desired temperature level. (See Table 7.3 for temperature/pressure relationships of R-12 and R-22.)

If we decide to cool the container to 25°F, the pressure of the R-22 refrigerant must be raised to 49 psi (Figure 7.1b). This is accomplished by adding a valve, to restrict the venting of the refrigerant to the atmosphere. This raises the pressure along with the boiling point of the refrigerant until it reaches 25°F.

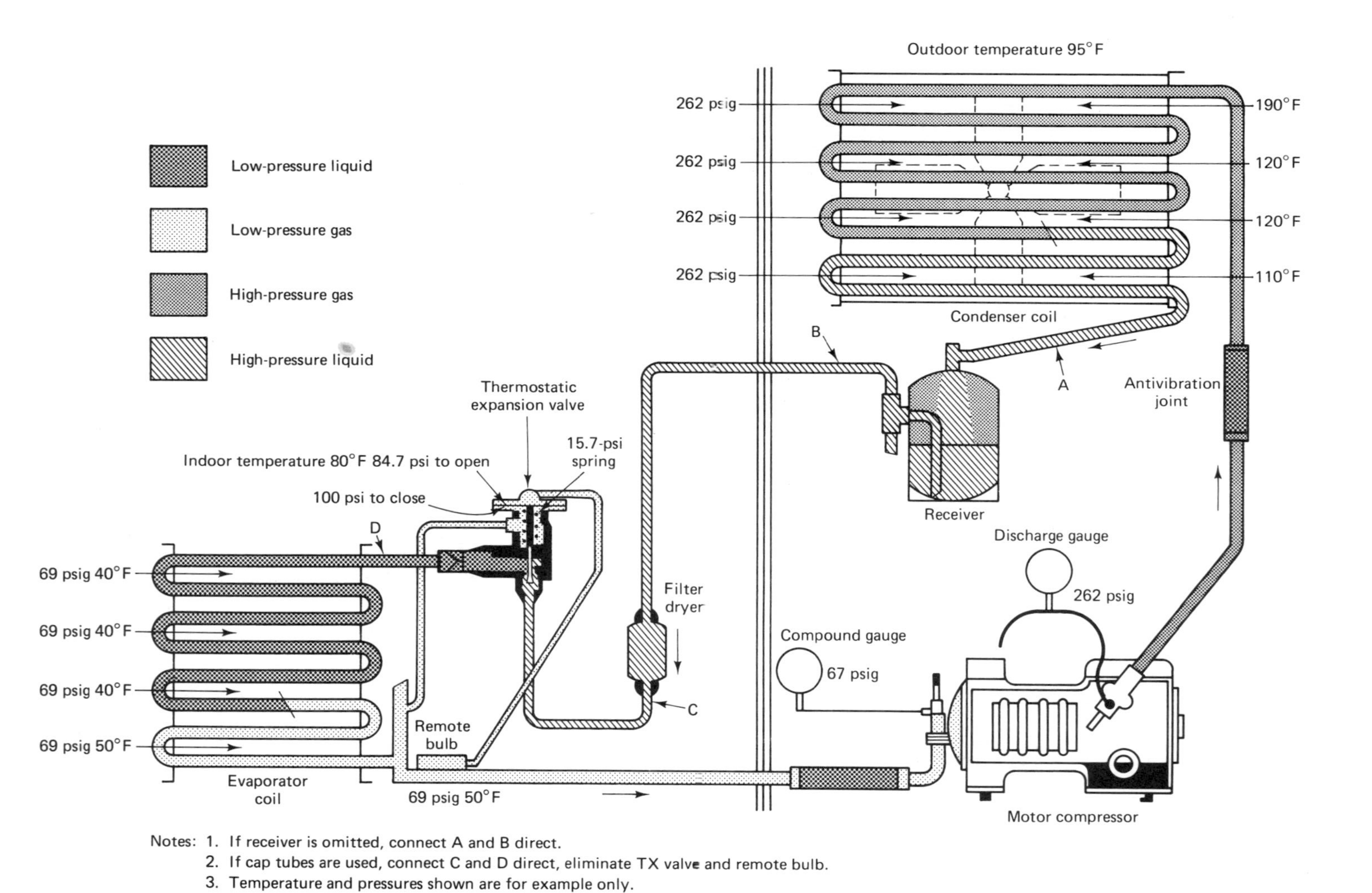

Notes: 1. If receiver is omitted, connect A and B direct.
2. If cap tubes are used, connect C and D direct, eliminate TX valve and remote bulb.
3. Temperature and pressures shown are for example only.

Figure 7.2 Typical residential air-conditioning system with operating temperatures and pressures. (Courtesy of The Williamson Company.)

If a reserve container of refrigerant is placed along with a feeding valve into the refrigerant vessel (Figure 7.1c), the flow of the refrigerant is regulated into the vessel to maintain a constant temperature of 25°F. The problem with this arrangement is that the refrigerant is wasted since no recovery system exists in the configuration as shown. By adding an external condenser (Figure 7.1d), the system will now remove the same amount of heat from the refrigerant as was picked up from the refrigerated container. As heat is dissipated from the condenser to the surrounding atmosphere, the refrigerant condenses back into a liquid and returns to the evaporator coil (refrigerant vessel) within the container to repeat the cycle within this closed system. Thus the refrigerant absorbs heat from the container via an evaporating coil where it is changed from a liquid to a gas. The gas then travels to a condenser, where it releases the absorbed heat, changing back to its liquid state.

For heat transfer to take place from the condenser to the surrounding medium (either air, water, or the ground), the condensing coils should be approximately 20°F warmer than their surroundings. To ensure that this positive temperature differential exists within the system, a compressor is used to raise the temperature and pressure of the refrigerant before it enters the condenser coils. A residential air-conditioning system is illustrated in Figure 7.2, along with typical temperatures and pressures encountered at various locations within the cooling system.

The system illustrated in Figure 7.2 is characteristic of most residential compression-cycle air conditioners and heat pumps. Although specific components will vary from one system and manufacturer to another, the basic system configuration is similar. For proper operation, the condenser coils must be properly cooled by furnishing adequate airflow or water circulation (in the case of liquid-source heat pumps) to the condenser coils. Dirty coils, which restrict convective and conductive heat transfer to the surrounding atmosphere, are one of the biggest causes of poorly operating air conditioners and heat pumps. As a general rule, 400 ft^3/min of air is required for each ton of cooling capacity. If the system compressor continually shuts off due to excessively high pressures, the coil should be checked to make sure that proper refrigerant cooling is taking place. The coils should be free of dirt and debris, and water should be able to be squirted all the way through the coils if they have been properly cleaned.

Note from Figure 7.2 that the evaporator coil low temperature is about 40°F. The coil temperature should not fall below 32°F. Under normal operating conditions, approximately 10°F of superheat is required between the inlet and outlet of the coil. Too much superheat (20 to 30°F) indicates a lack of refrigerant in the system, an insufficient quantity to cool the compressor properly. The evaporator coil should be about one-half full of refrigerant to operate properly.

To understand the operational aspects of residential air conditioners and heat pumps more fully, we examine the conventional compression refrigeration system along with its major components and control devices.

MECHANICAL COMPRESSION REFRIGERATION SYSTEMS

Most residential air conditioners and heat pumps are based on the mechanical compression system (Figure 7.3). The compressor used in most of these systems works on the reciprocating piston design in which the piston crankshaft is driven by an electric motor. The compressor is used to raise the temperature and pressure of the refrigerant high enough to allow the heat within the refrigerant to be dissipated to the environment surrounding the condensing units (air, ground, or water).

The cycle functions as follows: Liquid refrigerant at low pressure evaporates in the evaporator coils, causing the refrigerant to boil, and absorbs heat from the medium surrounding the evaporator; low pressure within the evaporator coils allows evaporation of the refrigerant to take place at low refrigerant temperatures, which ensures proper heat absorption. The low-pressure heat-laden vapor is carried by the suction line to the compressor, which compresses the refrigerant,

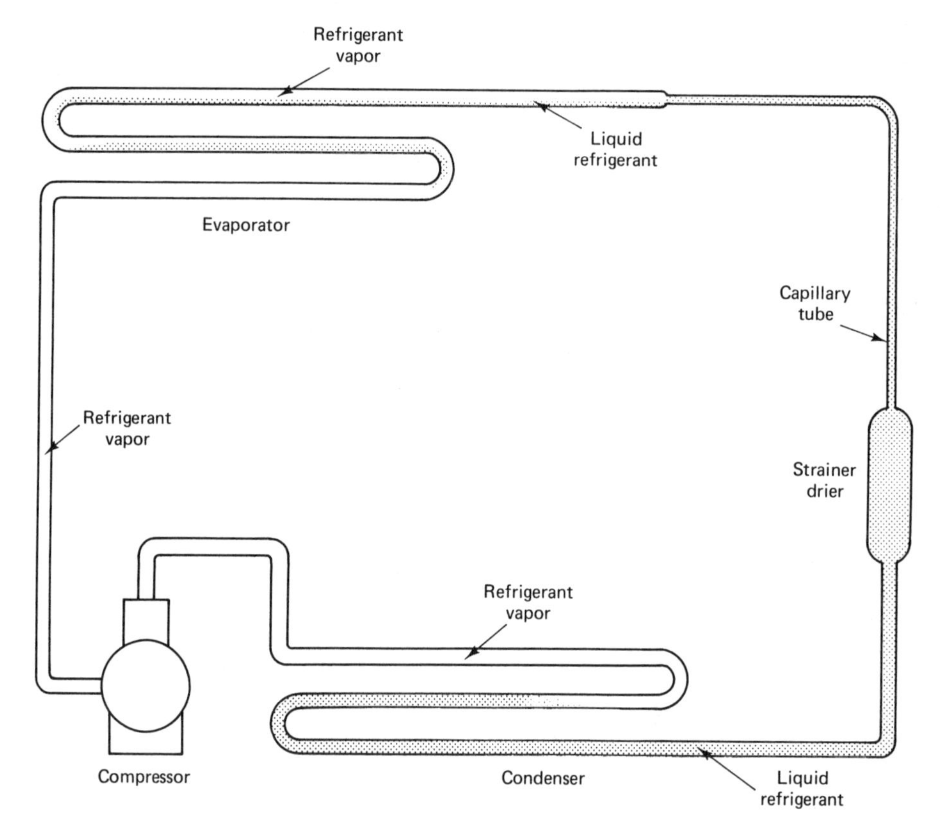

Figure 7.3 Mechanical compression refrigeration system. (From Bill Langley, *Refrigeration and Air Conditioning,* 3rd ed., Prentice-Hall, Englewood Cliffs, N.J., 1986, Fig. 2-5, p. 29. Reprinted by permission of Prentice-Hall, Inc.)

raising its temperature and discharging it into the condenser coils; in the condenser, the refrigerant releases heat to the surroundings through the coils, causing the refrigerant to condense back into a high-pressure liquid and collect in a liquid receiving tank; the liquid refrigerant is then carried into a restriction and metering device, which reduces its pressure as it enters the evaporator coils. This cycle continues in a state of balance until the room thermostat has been satisfied. The compressor in the system will cycle on and off based on refrigerant temperature and pressure during the cooling or heating cycle of the unit.

Compressors

Most air-conditioning and heat pump compressors are of the reciprocating piston type, similar in principle to the piston and valve assembly in an automobile internal combustion engine. The piston is housed in a tight-fitting cylinder and moves up on the compression stroke, compressing the refrigerant vapor, and then discharges it into the high-pressure lines leading to the condenser coils. Intake valves in the compressor meter the proper amount of refrigerant into the unit for each cycle of the system.

There are two types of compressors generally used in residential cooling applications: semihermetic and hermetic units.

Semihermetic compressors. A semihermetic compressor is illustrated in Figure 7.4. These compressors feature direct coupling of the electric motor armature to the compressor crankshaft. A common housing encloses both the motor and compressor. Most critical parts of the motor and compressor are accessible for field repairs.

Hermetic compressors. Hermetic compressors feature a housing that encloses both the electric motor and compressor whose seams have been welded to produce a hermetically sealed operating environment within (Figure 7.5). No internal repairs on these units are possible, since the shell of the unit would have to be cut open for access to operating components. The hermetic enclosure allows the compressor to operate within the atmosphere of the gaseous refrigerant, which provides proper cooling for both the motor and compressor. Hermetic compressors are most commonly found in air conditioners and heat pumps up to 5 tons in capacity.

Refrigerant Metering Devices

All refrigeration systems require a device that is capable of reducing the high-pressure liquid refrigerant as it exits the condenser to low-pressure refrigerant necessary for proper heat absorption within the evaporator coils. Two types of devices are generally used to perform this task: capillary tubes and expansion valves.

Figure 7.4 Semihermetic compressor. Semihermetic compressors are designed so that most critical components can be serviced in the field. (Courtesy of Copeland Company.)

Capillary tubes. A capillary tube is the simplest type of refrigerant flow-control device. The tube is made from small-diameter seamless copper tubing that is usually wound into a coil through which the refrigerant flows. The tube is located on the liquid line between the condensing and evaporator coils. The capillary tube acts as a restriction in the refrigerant line, allowing only a predetermined amount of liquid refrigerant to enter the evaporator. A capillary tube of this type is illustrated in Figure 7.6.

The small diameter of the tube holds back the liquid refrigerant and maintains a high operating pressure in the condenser that is required for proper heat dissipation. The refrigerant flows through the restricted-diameter tubing, changing only slightly in pressure until it reaches the last one-fourth of the tube length. At this point the refrigerant is cooled to evaporator temperature and begins to vaporize, causing a sudden drop in refrigerant pressure.

The orifice of the capillary tube is fixed, allowing a specific amount of re-

Figure 7.5 Hermetic compressor. Hermetic compressors operate within a sealed environment. Field service on these units is limited in scope, due to the sealed housing enclosing the compressor and motor assembly. (Courtesy of Copeland Company.)

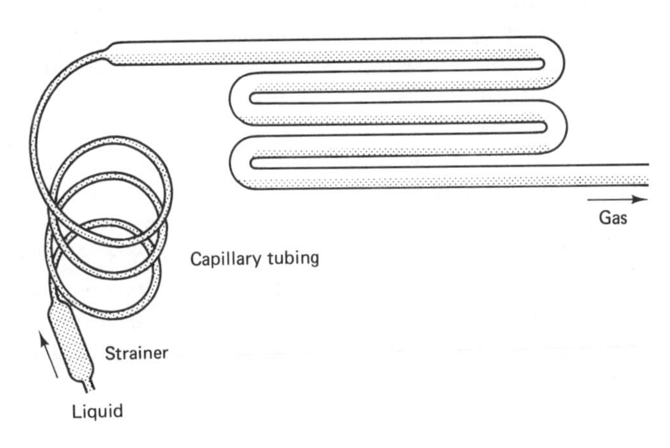

Figure 7.6 Capillary tube. (From Air-Conditioning and Refrigeration Institute, *Refrigeration and Air-Conditioning,* 2nd ed., Prentice-Hall, Englewood Cliffs, N.J., 1987, Fig. R10-13, p. 106. Reprinted by permission of Prentice-Hall, Inc.)

frigerant into the evaporator coils. The length and diameter of the capillary tubing used is a function of the type of refrigerant recommended for the unit and its cooling capacity. Because the inside diameter of the capillary tube is so small, it must be kept free of debris and wax buildup. For this reason, a filter and drier should be located at the beginning of the capillary tube assembly.

The amount of refrigerant in the system is critical for proper functioning of the capillary tube, since no liquid receiving tank is generally used in this type of system. If there is too much refrigerant in the system, high head pressures and possible flooding back of liquid refrigerant into the compressor with subsequent damage can occur. If there is too little refrigerant in the system, a resulting loss in cooling capacity will take place as well as damage to a suction-cooled compressor. Since the capillary tube is open at both ends, pressure within the system tends to equalize during the off-cycle of the compressor. This feature allows the use of low-starting-torque compressor motors.

Expansion valves. The purpose of an expansion valve in an air conditioner or heat pump is to control refrigerant flow into the evaporator coils. This enables the system to maintain constant evaporator pressure during operation. There are three types of expansion valves used in most residential cooling systems: constant-pressure expansion valves, thermostatic expansion valves, and thermal electric expansion valves. The basic operation of each type of valve is examined next in further detail.

Constant-Pressure Expansion Valves. Constant or automatic expansion valves are the simplest type of valve used to maintain constant evaporator pressure. The valve is installed in the inlet of the evaporator coil and is illustrated in Figure 7.7. The valve consists of two pressure springs; one is located above, and the other below, a valve diaphragm that is connected to a needle valve. The needle valve assembly serves to admit or restrict the flow of refrigerant through the valve. The lower spring responds to evaporator pressure; the upper valve spring has an adjustment screw to set valve pressure based on the specific operating conditions of the system, such as type of refrigerant and cooling capacity. During operation, evaporator pressure against the lower spring acts to open and close the valve during compressor operation to maintain constant pressure within the evaporator coils. When the compressor begins operation, evaporator pressure is greater than the upper spring pressure, keeping the valve closed. When the compressor has been operating for a short time, pressure in the evaporator is reduced to a point below that of the upper valve spring, causing the needle valve to open and admit refrigerant into the evaporator coils. During stable operating conditions the valve opening allows the same amount of refrigerant into the evaporator coils, which is equivalent to the pumping capacity of the system compressor. Because the valve maintains constant pressure within the coils, the evaporator pressure can be adjusted to maintain a temperature just above the freezing point of water, eliminating the formation of frost and icing on the coils.

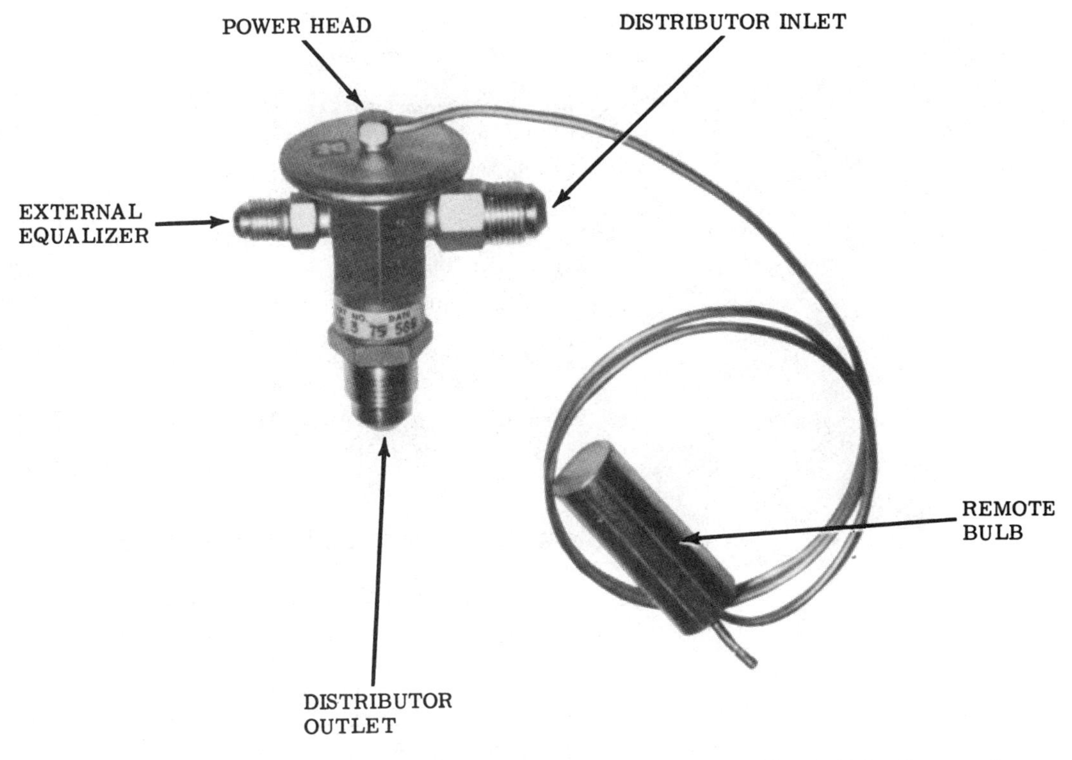

Figure 7.7 Automatic (constant) expansion valve. (Courtesy of Sporlin Valve Company.)

Thermostatic Expansion Valves. Thermostatic expansion valves are perhaps the most popular type of metering valve in current use in a variety of refrigeration and air-conditioning systems. This type of valve is designed to allow precisely the amount of refrigerant into the evaporator coils that is being vaporized and prevents liquid refrigerant from entering the compressor. A thermostatic expansion valve is illustrated in Figure 7.8.

Thermostatic valves are sensitive to three simultaneous inputs: pressure transmitted by a remote sensing bulb, pressure within the evaporator coils, and pressure exerted by a superheat spring. Pressure from the remote sensing bulb, which is equivalent to the pressure of the refrigerant as it leaves the evaporator coils, acts to move the valve into an open position, admitting liquid refrigerant into the coils. Opposing this opening force is the pressure exerted by the evaporator itself, along with the superheat spring.

During a state of operating equilibrium, the pressure imposed by the remote bulb is nearly equivalent to the pressures exerted by both the evaporator coil and superheat spring in the valve. This causes the valve to assume a stable open position, metering refrigerant into the evaporator coils and maintaining a super-

Basically, Thermostatic Expansion Valve Operation is determined by three fundamental pressures:

P_1 Bulb pressure acts on one side of the diaphragm, tends to open the valve.

P_2 Evaporator pressure acts on the opposite side, tends to close the valve.

P_3 Spring pressure – which also assists in the closing action – is applied to the pin carrier and is transmitted through push rods to a buffer plate on the evaporator side of the diaphragm.

When the valve is modulating, bulb pressure is balanced by the evaporator pressure plus the spring pressure:

$$P_1 = P_2 + P_3$$

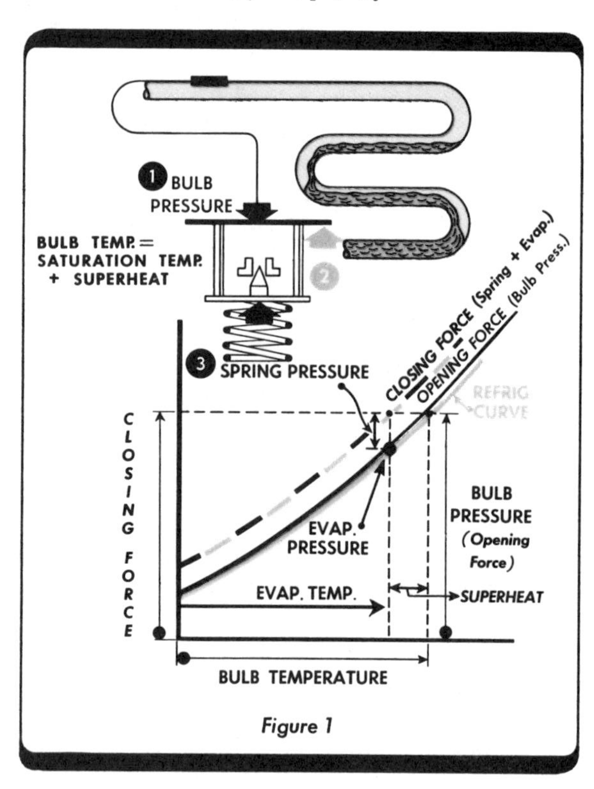

Figure 7.8 Thermostatic expansion valve. (Courtesy of Sporlin Valve Company.)

heat temperature of approximately 10°F across the evaporator coils (Figure 7.9). As the temperature of the superheated refrigerant increases, indicating a lack of refrigerant within the evaporator coils, pressure within the remote bulb assembly increases above the combined pressures of both the superheat spring and evaporator pressure. This opens the valve orifice to admit more refrigerant into the coils.

As the temperature of the refrigerant vapor decreases, pressure within the remote bulb drops to a point below that of the combined superheat spring and evaporator coil pressure, restricting or closing the valve orifice that meters refrigerant into the coils. Many thermostatic expansion valves also provide motor overload protection during times of high load conditions on the system by using a second diaphragm and metering assembly to maintain constant evaporator pressure, until the overload condition has subsided. Also, most valves use some type of bleed bypass that allows small amounts of refrigerant to bypass the valve when the unit is turned off. This equalizes system pressure for easy starting loads on the compressor motor. The bypass hole in the valve is small enough so that valve operation is not affected during normal unit cycling.

Thermal Electric Expansion Valves. Thermal electric expansion valves open and close in response to electrical currents that are produced as a result of temperatures within the evaporating coil. A thermistor temperature sensor, a solid-state device whose electrical resistance changes with temperature, is placed in an electrical circuit in series between a low-voltage power source and valve-actuating mechanism. A thermal electric expansion valve is illustrated in Figure 7.10.

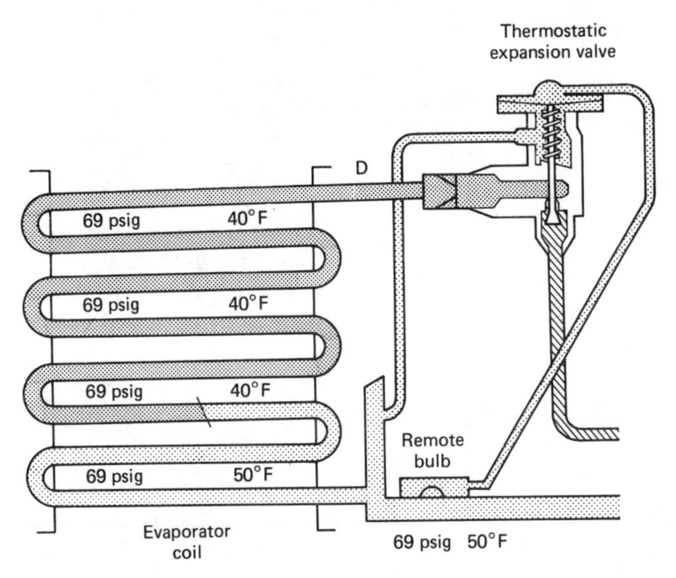

Figure 7.9 Superheat across the evaporator coils. This measurement is a good indication of the operating efficiency of the expansion valve assembly during system operation. (Courtesy of The Williamson Company.)

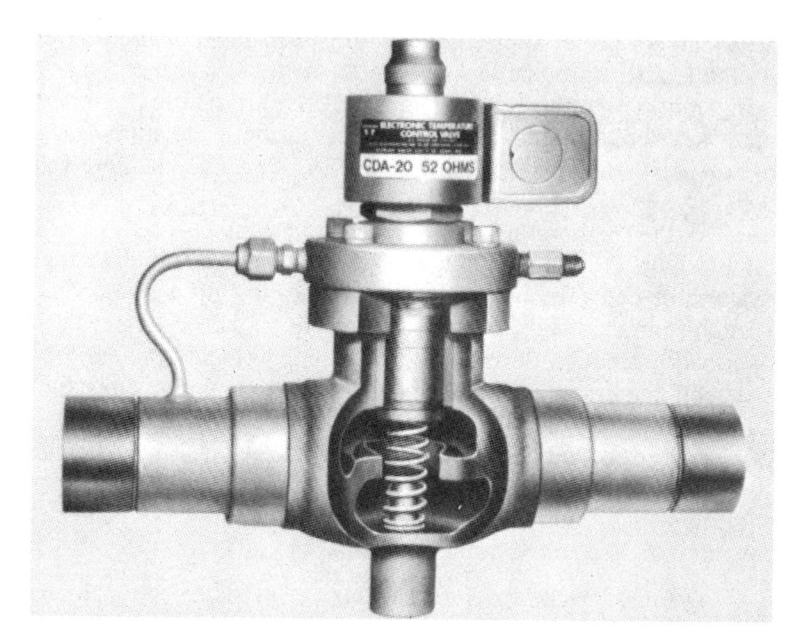

Figure 7.10 Thermal electric expansion valve. (Courtesy of Sporlin Valve Company.)

The thermistor sensor is placed in direct contact with the liquid refrigerant in the system line. When electrical current is applied to the valve, it must pass through the thermistor sensor, which will either allow full or reduced voltage flow to the valve, depending on the temperature of the refrigerant. Precise temperature monitoring allows for exact control of refrigerant pressure and levels within the evaporator coils. These valves work with many different types of refrigerant and in a variety of air-conditioning and heat pump equipment.

CENTRAL AIR-CONDITIONING SYSTEMS

Most central air-conditioning systems are installed either as a multicomponent heating/cooling system in which the air-conditioning apparatus functions in conjunction with the space heating system or as a stand-alone system or individual room units.

Warm Air Heating/Cooling Combination Systems

Conventional warm air central heating systems lend themselves to the installation and/or addition of air-conditioning equipment, since the air distribution system for the home is already in place. In some instances, existing warm air ducts may not be adequate to handle the additional requirements imposed by the installation of an air-conditioning system. Duct dimensions need to be calculated to determine if they are adequately sized for the additional air conditioner. If the combination system is to be installed as original equipment in either new construction or as the result of upgrading an older central heating unit, the duct and grill sizes can be properly engineered for the proposed installation. A typical combination warm air/air-conditioning unit is illustrated in Figure 7.11a. The unit illustrated is characteristic of those engineered for residential heating and cooling loads under 5 tons in capacity.

The evaporator coils for the air conditioner are located within the plenum assembly of the furnace. The condenser coils are usually placed on a concrete slab outside the house. The refrigerant lines are then run between the evaporator coil in the furnace plenum and the condensing unit, which also houses the com-

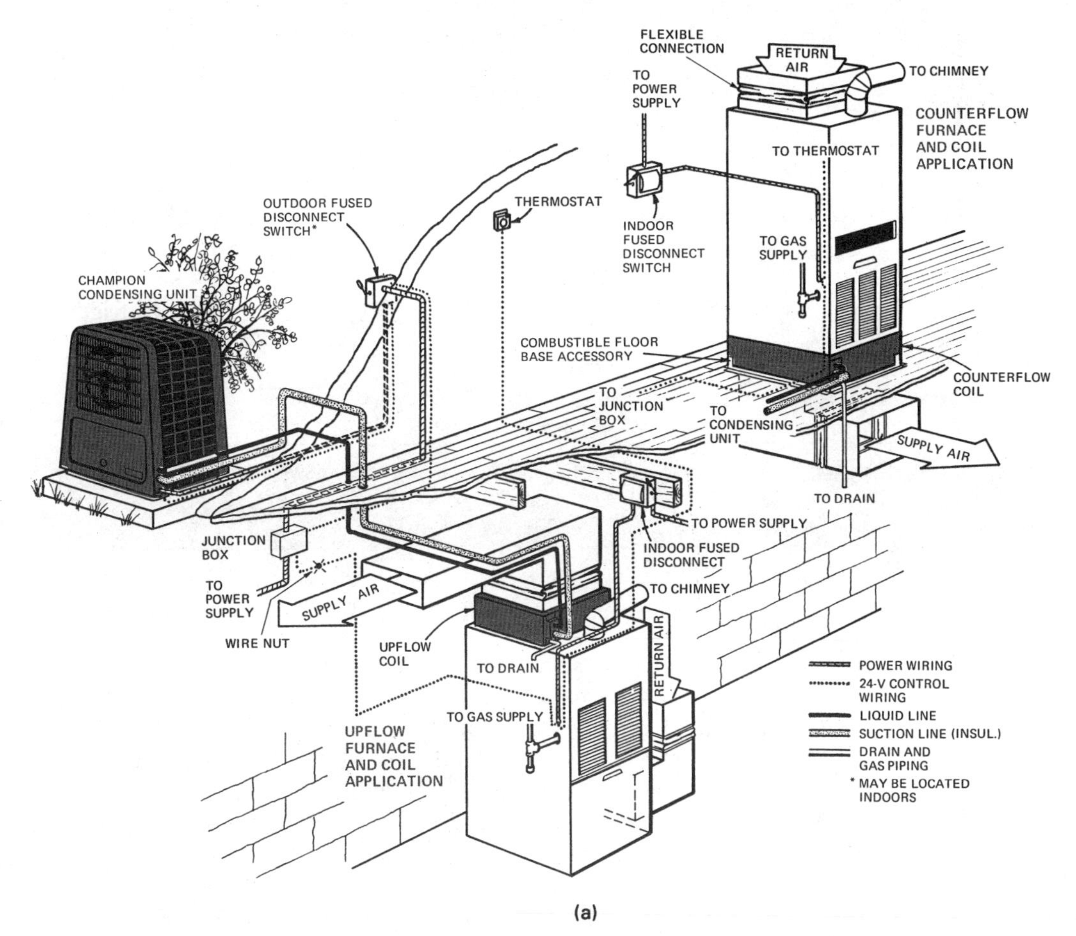

Figure 7.11 Combination warm air/air-conditioning system: (a) split system; (b) combination system. (Courtesy of York International Corp.)

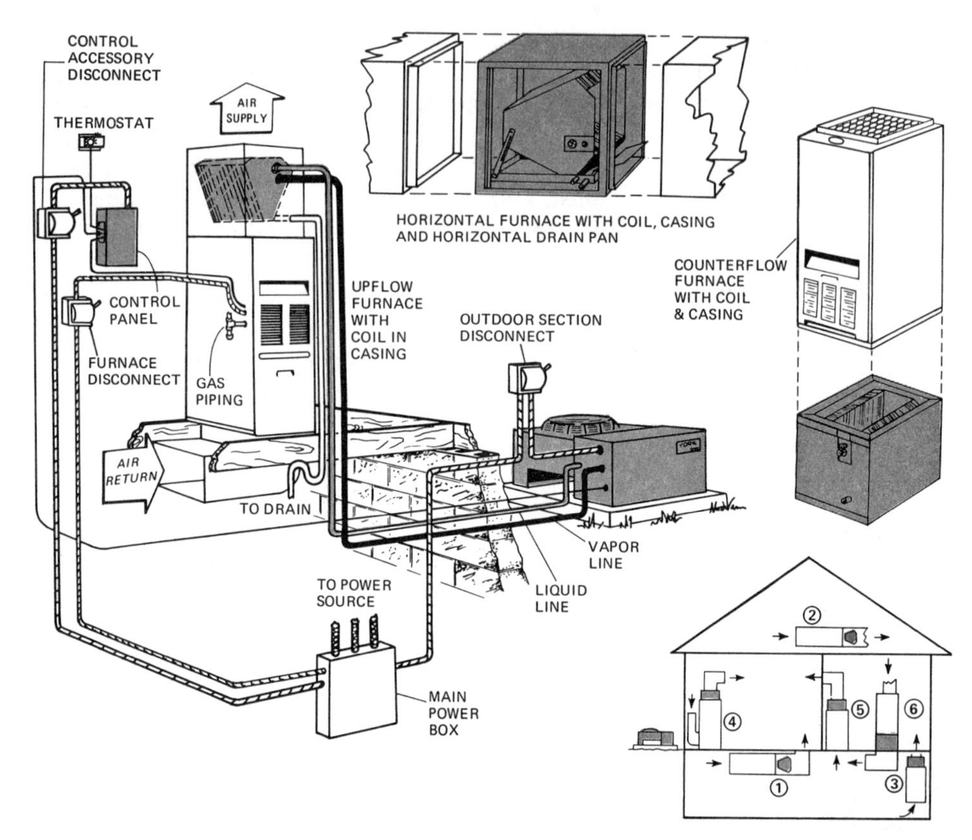

TYPICAL APPLICATIONS

①	Horizontal Suspended	③	Upflow wall-mounted	⑤	Upflow-basement or crawl space return
②	Horizontal (Attic or Crawl)	④	Upflow Plenum Return	⑥	Counterflow

(b)

Figure 7.11 *(continued)*

pressor and major control devices associated with the air conditioner (Figure 7.12). The evaporator coils are sometimes referred to as the A-coil assembly because of its shape and configuration within the furnance plenum (Figure 7.13).

Both the evaporator and condenser coils should be kept free from dirt to facilitate heat transfer in the coil assembly. Scheduled maintenance on a system with both air-conditioning and central heating capability does not add appreciably to that required in a system designed only for either heating or air conditioning.

Duct sizing. When installing an air-conditioning unit into an existing hot air central heating system, a determination must be made as to whether or not the ducting is adequate to handle the air-conditioning load or if modifications to

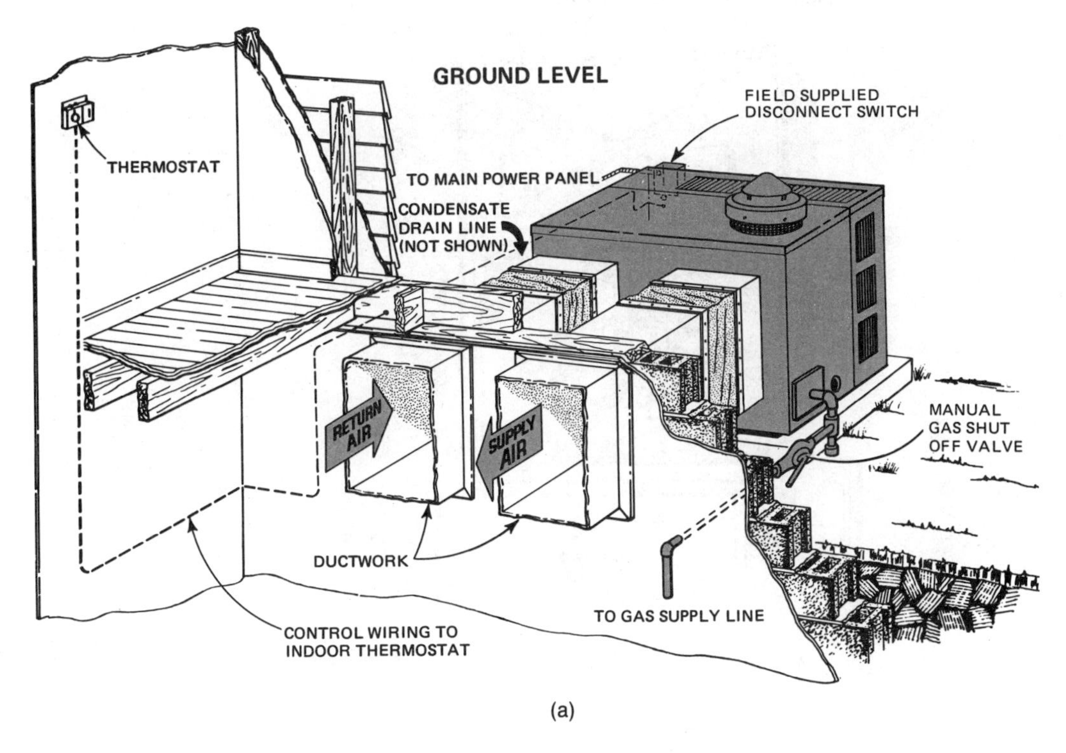

(a)

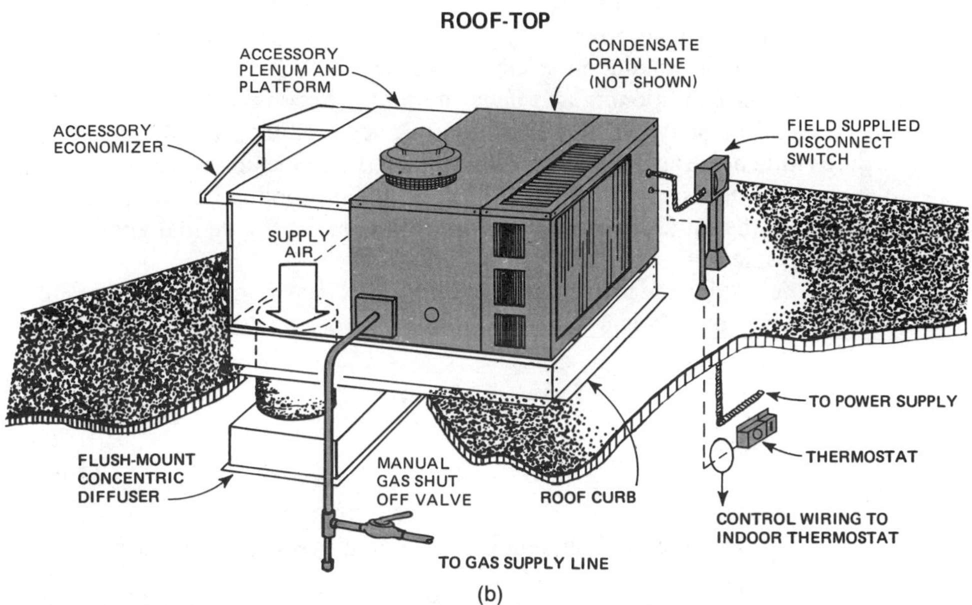

(b)

Figure 7.12 Outdoor condenser assembly, illustrating (a) ground-level and (b) rooftop mounting configurations. (Courtesy of York International Corp.)

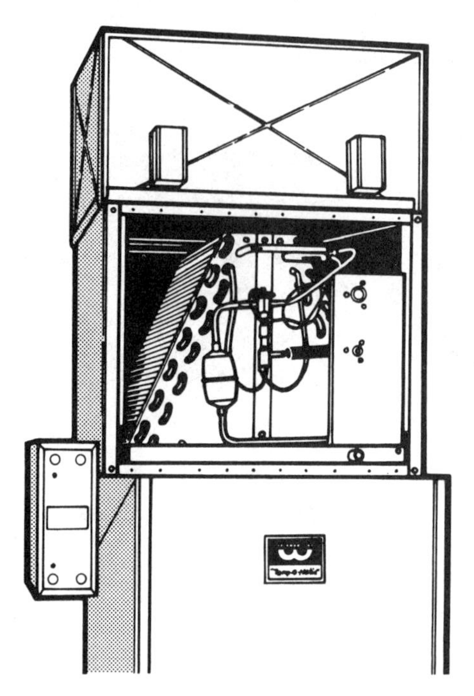

Figure 7.13 Coil assembly within a furnace plenum.

the duct system need to be made. Whereas detailed duct sizing calculations can be performed by an air-conditioning or heating engineer, some basic sizing factors are presented for making initial determinations.

Each air-conditioning and heat pump appliance is designed to operate at a specified air velocity. These velocities are calculated by the manufacturer to produce as little noise as possible while providing the proper air movement throughout the system necessary for efficient heating and/or cooling based on the size and capacity of the unit. Recommended air velocities for residential applications are listed in Table 7.4.

To determine the proper duct sizes required based on the listed air velocities, the following formula is used:

$$D = \frac{Q}{V}$$

where

D = area of duct

Q = required amount of air to be moved by the system

V = required velocity at which the air is to be moved

For example, if a particular air-conditioning system requires 1000 ft^3/min of air to be moved by the cooling system at a velocity of 750 ft/min, the duct size

TABLE 7.4 RECOMMENDED DUCT VELOCITIES IN RESIDENTIAL FORCED-AIR/AIR-CONDITIONING SYSTEMS

Designation	Recommended velocities (ft/min)	Maximum velocities (ft/min)
Outdoor air intakes[a]	500	800
Filters[a]	250	300
Heating coils[a,b]	450	500
Cooling coils[a]	450	450
Air washers[a]	500	500
Fan outlets	1000–1600	1700
Main ducts[b]	700–900	800–1200
Branch ducts[b]	600	700–1000
Branch risers[b]	500	650–800

Source: Adapted from *ASHRAE Handbook of Fundamentals,* American Society of Heating, Refrigerating, and Air-Conditioning Engineers, Atlanta, Ga., 1972, Table 6, p. 481.

[a] These velocities are for total face area, not the net free area; other velocities are for net free area.

[b] For low-velocity systems only.

is calculated as follows:

$$D = \frac{1000}{750} = 1.33 \text{ ft}^2$$

The total area of the duct should be approximately 1.3 ft^2. Any configuration or duct shape can be used that gives this overall inside surface area.

In addition to duct sizes, friction losses that occur in the ductwork as the result of the total length of the system as well as elbows and bends should be determined so that the system blower will be able to furnish the required volume of air, taking into account the frictional losses within the system. Appendix D lists the friction losses for various-size pipe and fittings. The losses of each pipe section and fitting are then added together to determine the total frictional losses that the system blower must be able to carry in addition to the basic ft^3/min requirements imposed by the size and capacity of the heating and cooling system.

In stand-alone air-conditioning units, or in combination units installed as original equipment in new construction, the duct sizes can be engineered for the unit based solely on required air capacities without regard to ducting that already exists.

Improperly sized ducts yield poor operating efficiency with even the best air conditioners and heat pumps, since improper quantities of air will not be able to deliver the rated hot or cold air to the residence. If improper duct sizing is suspected, the entire installation should be reviewed by a competent engineer to determine if and where deficiencies in system design exist and how they can best be remedied.

HEAT PUMPS

Heat pumps are refrigeration devices that function to provide both space heating and cooling to the residence. This is done by reversing the direction of refrigerant flow within the appliance. In the heat pump, the evaporator coil in the cooling cycle becomes the condensing coil in the heating mode; the condensing coil in the cooling mode becomes the evaporator coil in the space heating sequence.

Most residential heat pumps are classified according to the type of location of the evaporator and condensing coils: for example, air-to-air, water-to-air, or ground-to-air units. For purposes of illustrating the heating and cooling cycles of a residential heat pump, assume an air-to-air system.

Heat Pump Heating Cycle

A typical air-to-air heat pump operating in the heating cycle is illustrated in Figure 7.14. Heat is absorbed from the outside air and delivered indoors. Liquid refrigerant flows through the outdoor coils under low pressure and absorbs heat from the surrounding air, causing the refrigerant to vaporize. The heat-laden refrigerant vapor is drawn into the compressor, where its temperature and pressure is increased and piped into the condenser coils within the house. These coils are generally located within the furnace plenum directly exposed to circulating air

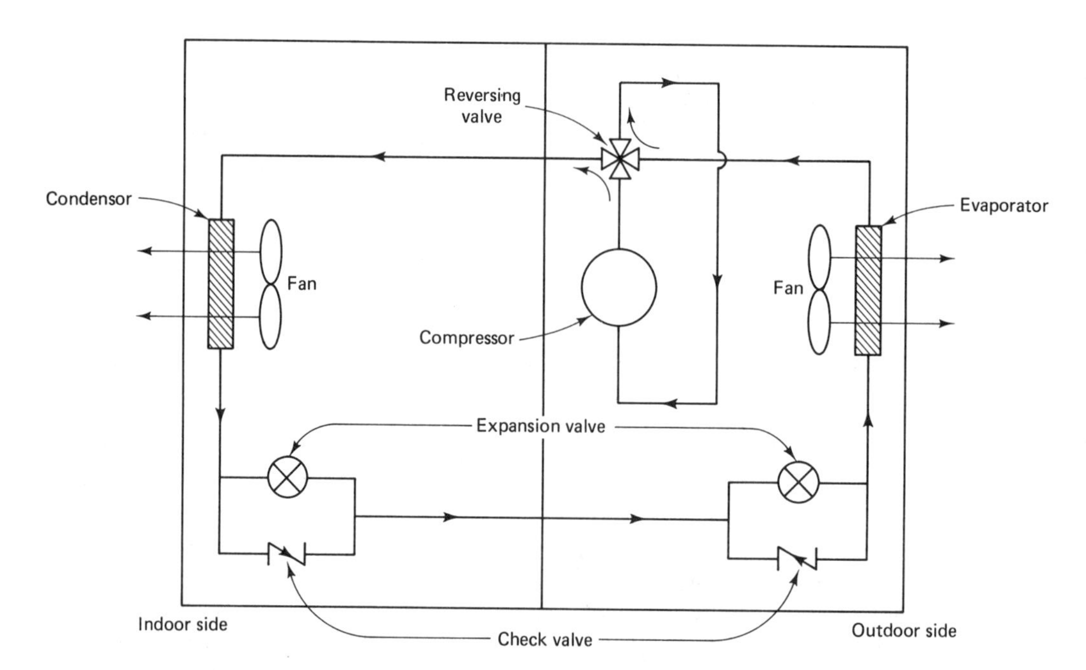

Figure 7.14 Heating cycle of air-to-air heat pump. (From M. Greenwald and T. McHugh, *Practical Solar Energy Technology*, Prentice-Hall, Englewood Cliffs, N.J., 1985.)

from within the house. Since the temperature of the condenser coils is approximately 20°F higher than that of the circulating air, heat is transferred from the refrigerant to the surrounding air. This causes the refrigerant to condense back into a liquid and return to the evaporator, where the cycle is repeated until the room thermostat has been satisfied.

Heat Pump Cooling Cycle

In the cooling mode of operation, the refrigeration flow is reversed in direction from that taken in the heating mode. Heat is removed from the interior of the residence and transferred to the outside air. During the cooling cycle, the operation of the heat pump is similar to that of a conventional air-conditioning system. The cooling cycle of the heat pump is illustrated in Figure 7.15.

To accomplish system reversal to the cooling mode, either a three- or four-way valve is used to change the direction of refrigerant flow through the evaporator and condenser coils. Note from Figure 7.15 that the direction of flow into and out of the compressor remains unchanged. In the cooling mode, the condensing coil in the residence serves as the evaporator coil, and the outdoor evaporator coil is now utilized as the condenser.

The operation of the heat pump is controlled by a special heating/cooling thermostat. The thermostat activates the reversing valve to provide the proper

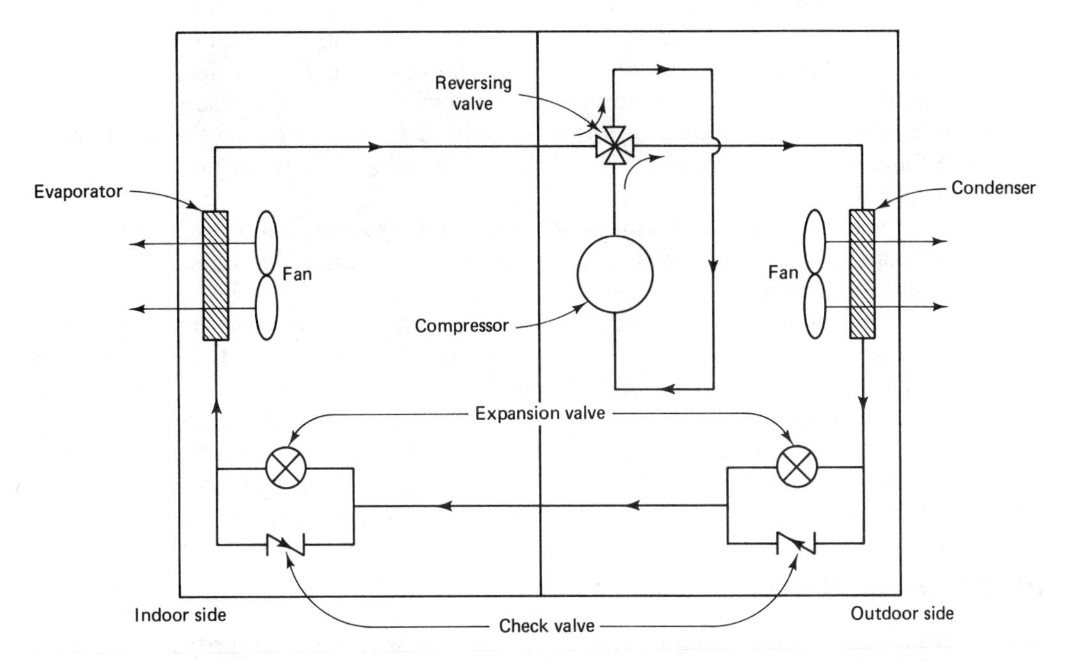

Figure 7.15 Cooling cycle of air-to-air heat pump. (From M. Greenwald and T. McHugh, *Practical Solar Energy Technology*, Prentice-Hall, Englewood Cliffs, N.J., 1985.)

direction of flow for the refrigerant, depending on whether there is a call for heating or cooling.

Heat pumps require minimum evaporator temperature levels if their performance is to be cost-effective. In air-to-air installations, heat pumps work best if outdoor temperatures do not fall below the freezing point. In situations where groundwater supplies or the earth is used as the location of the evaporator coils, year-round temperatures of about 50°F will provide very cost-effective system performance. Heat pumps are sometimes rated by their **coefficient of performance** (COP). The COP is a value that expresses the ratio of heat delivered by the heat pump versus the amount of electrical energy required to operate the system compressor. The higher the COP, the greater the overall operating efficiency of the unit, and vice versa. The COP falls off as the temperature of the evaporator location drops since there is less heat available for evaporator operation. Even in situations where adequate evaporator temperatures are available on a year-round basis, the cost of electricity must be factored into determining the cost-effectiveness of the heat pump. In areas where utility costs exceed 12 to 15 cents per kilowatthour, even high-COP heat pumps will be more expensive to operate than conventional fossil-fuel appliances for heating under most circumstances.

ROOM AIR CONDITIONERS

In a typical window air-conditioning appliance, the evaporator coils are located directly behind an air filter on the room side of the unit; the condensing coils are located outside the room within the appliance housing. A typical window air-conditioning unit configuration is illustrated in Figure 7.16. These units maintain a temperature differential of approximately 12 to 20°F between the inside room air and outdoor ambient air. Window units have a separate airflow for the condenser and evaporator coils.

Some window air conditioners function to provide cooling only; others provide space heating in addition to cooling by using either electric strip heaters or a reversible refrigeration heat pump cycle.

Annual service on window air-conditioning units involves cleaning or replacing the filters over the evaporator coils, cleaning dirt and debris from the evaporator and condenser coils and fan blades, and lubricating the motor if this is required. The fan connection on the electric motor shaft should also be checked to make sure that it is tightly set and operates properly within the fan shroud. Visible signs of rust should be inspected to make sure that no corrosion has affected the integrity of any internal tubing or components.

DETERMINING COOLING LOADS

No cooling system can perform efficiently unless it has been properly sized for a specific residence. These sizing procedures are similar to those used for determining the Btu sizing requirements for central heating systems. Cooling system

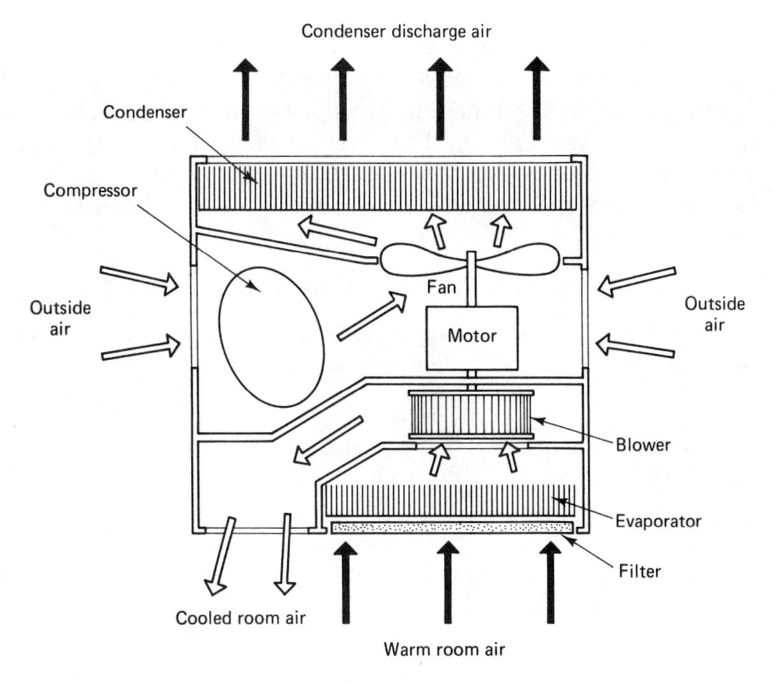

Figure 7.16 Configuration of window air conditioner. (Courtesy of ASHRAE.)

sizing techniques are referred to as heat-gain calculations and are done on either a room-by-room or a total-structure basis.

Entrance of heat into a structure during warm weather, or buildup of heat within the structure from internal heat-producing systems, can arise from several different sources. Each of these sources must be quantified when performing heat-gain calculations and generally fit into one of five basic categories:

1. Transfer of heat into the structure from the outdoors due to higher temperatures outside the home than within
2. Direct solar radiation into the home through the walls, roof, and windows
3. Build-up of heat within the home given off by the operation of electrical appliances
4. Buildup within the home of heat from people
5. Heat gain due to air infiltration

R and U Values

The *R* and *U* values that are used in heat-loss calculations are also used to determine heat gain within a residence.

***U* values.** U values indicate the amount of heat that is transferred through a building structure such as a composite wall or roof structure. U values are expressed in Btu/ft^2/°F/hr. The U values for the most common building materials are listed in Appendix A. The U value of the structure, along with the design temperature differential chosen for the specific geographical location and the total surface area of the structure, results in the heat or cooling load factor. This factor is expressed as

$$\text{heating or cooling load} = \text{area} \times \text{design temperature} \times U \text{ value}$$

For example, to calculate the heating load of a 500-ft^2 wall with a composite U value of 0.30 at a design temperature differential of 65°F, we proceed as follows:

$$\begin{aligned}\text{heating load} &= 0.30 \times 500 \times 65\\ &= (150)(65)\\ &= 9750 \text{ Btu/hr}\end{aligned}$$

***R* values.** The use of R values represents a different method for heat-load calculations. R values are units that specify the resistance of heat flow in an object or structure. R values are the mathematical reciprocals of U values:

$$R = \frac{1}{U}$$

R values for a variety of building materials are listed in Appendixes A and B, and in Table 1.3. To find the R value of any composite structure such as a wall or built-up roof, the R values of the individual materials are added together to yield the total R value. After calculating the R value of the composite structure, the U value can be calculated. To illustrate this technique, consider a typical building wall composed of the materials listed in Figure 7.17, together with their associated R values.

The U value of the wall is

$$\frac{1}{13.3} = 0.075$$

To perform a simple heat-load or heat-gain calculation, the U factor of the wall is then multiplied by the design temperature differential and the surface area of the wall. In the example above, let us assume that the wall has a total surface area of 750 ft^2 with a design temperature differential of 12°F. The cooling load for the wall is then calculated:

$$\begin{aligned}\text{cooling load} &= U \text{ value of wall} \times \text{temperature differential} \times \text{area}\\ &= (0.75)(12)(450)\\ &= 3600 \text{ Btu/hr}\end{aligned}$$

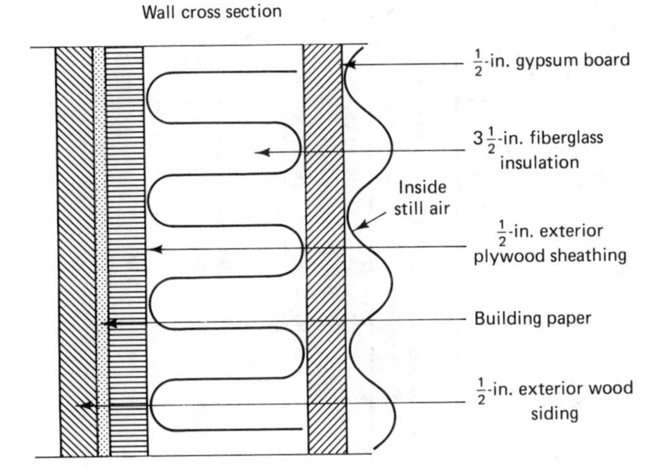

Material	R value
Inside still air	0.065
$\frac{1}{2}$-in. gypsum board (Sheetrock)	0.410
$3\frac{1}{2}$-in. fiberglass insulation	11.00
Building paper	0.060
$\frac{1}{2}$-in. plywood	0.950
$\frac{1}{2}$-in. wood siding	0.850
Total R	13.338

Figure 7.17 Structure and R-value determination for composite building wall.

Windows

Windows represent a significant source of both heat loss and heat gain in the typical home. Recent changes in design and construction have resulted in window units that feature far higher R values than those of previous units. The R and U values for window configurations most commonly found in residential applications are listed in Table 7.5.

Internal Building Heat Generation

There are many sources of heat generated within the home that must be taken into account when performing heat-gain calculations. Whereas most of this heat generation is relatively insignificant when performing heat-loss calculations, the buildup of heat within the residence during the summer months can be considerable and must therefore be accounted for when sizing any air-conditioning system. Tables 7.6 and 7.7 list some common sources of heat generation from human and mechanical sources.

TABLE 7.5 COEFFICIENTS OF TRANSMISSION (U) OF WINDOWS, SKYLIGHTS, AND LIGHT-TRANSMITTING PARTITIONS

(a) Vertical Panels (Exterior Windows, Sliding Patio Doors, and Partitions)—Flat Glass, Glass Block, and Plastic Sheet

Description	Exterior[a]		Interior
	Winter	Summer	
Flat glass			
Single glass	1.13	1.06	0.73
Insulating glass—double[b]			
$\frac{3}{16}$-in. air space	0.69	0.64	0.51
$\frac{1}{4}$-in. air space	0.65	0.61	0.49
$\frac{1}{2}$-in. air space	0.58	0.56	0.46
$\frac{1}{2}$-in. air space, low emissivity coating[c]			
Emissivity = 0.20	0.38	0.36	0.32
Emissivity = 0.40	0.45	0.44	0.38
Emissivity = 0.60	0.52	0.50	0.42
Insulating glass—triple[b]			
$\frac{1}{4}$-in. air spaces	0.47	0.45	0.38
$\frac{1}{2}$-in. air spaces	0.36	0.35	0.30
Storm windows	0.56	0.54	0.44
1–4-in. air space			
Glass blocks[d]			
6 × 6 × 4 in. thick	0.60	0.57	0.46
8 × 8 × 4 in. thick	0.56	0.54	0.44
With cavity divider	0.48	0.46	0.38
12 × 12 × 4 in. thick	0.52	0.50	0.41
With cavity divider	0.44	0.42	0.36
12 × 12 × 2 in. thick	0.60	0.57	0.46
Single plastic sheet	1.09	1.00	0.70

(b) Horizontal Panels (Skylights)—Flat Glass, Glass Block, and Plastic Bubbles

Description	Exterior[a]		Interior[e]
	Winter[f]	Summer[f]	
Flat glass			
Single glass	1.22	0.83	0.96 ·
Insulating glass—double[b]			
$\frac{3}{16}$-in. air space	0.75	0.49	0.62
$\frac{1}{4}$-in. air space	0.70	0.46	0.59
$\frac{1}{2}$-in. air space	0.66	0.44	0.56
$\frac{1}{2}$-in. air space, low emissivity coating[c]			
Emissivity = 0.20	0.46	0.31	0.39
Emissivity = 0.40	0.53	0.36	0.45
Emissivity = 0.60	0.60	0.40	0.50
Glass block[d]			
11 × 11 × 3 in. thick with cavity divider	0.53	0.35	0.44
12 × 12 × 4 in. thick with cavity divider	0.51	0.34	0.42
Plastic bubbles[g]			
Single walled	1.15	0.80	—
Double walled	0.70	0.46	—

(c) Adjustment Factors for Various Window and Sliding Patio Door Types (Multiply U Values in Parts (a) and (b) by These Factors)

Description	Single glass	Double or triple glass	Storm windows
Windows			
All glass[h]	1.00	1.00	1.00
Wood sash—80% glass	0.90	0.95	0.90
Wood sash—60% glass	0.80	0.85	0.80
Metal sash—80% glass	1.00	1.20	1.20[i]
Sliding patio doors			
Wood frame	0.95	1.00	—
Metal frame	1.00	1.10	—

Source: ASHRAE Handbook of Fundamentals, American Society of Heating, Refrigerating, and Air-Conditioning Engineers, Atlanta, Ga., 1972, Table 8, p. 370.

Note: These values are for heat transfer from air to air, Btu/hr/ft^2/°F.

[a] See part (c) for adjustment for various window and sliding patio door types.

[b] Double and triple refer to the number of lights of glass.

[c] Coating on either glass surface facing air space; all other glass surfaces uncoated.

[d] Dimensions are nominal.

[e] For heat flow up.

[f] For heat flow down.

[g] Based on area of opening, not total surface area.

[h] Refers to windows with negligible opaque area.

[i] Value becomes 1.00 when storm sash is separated from prime window by a thermal break.

Heat Transfer through Walls and Ceilings

Conventional residential construction techniques vary greatly from one geographical area to another, based on the differences in heat losses and heat gains that are anticipated. Therefore, the R and U values of walls and ceilings must be calculated individually for each building, based on the specific type of composite wall or ceiling structure encountered. A home in upstate New York might have a ceiling with an R value of 40 or above, while a nearby home built by more conventional techniques might have a ceiling R value of only 20. These two buildings will have different heat gains during the summer cooling season and thus require air-conditioning systems with different cooling capacities, even though they may be spaced only a few hundred yards apart. Consult Appendixes A and B for a list of the most applicable building materials.

Heat Loss/Gain due to Air Infiltration

Air infiltration can be a significant source of heat gain or heat loss in any building. Warm or cold air that leaks into or out of a house is replaced by new air that must be conditioned to the proper indoor temperature. An additional load is placed on the heating or cooling system in relation to the amount of fresh air that must be conditioned. It should be kept in mind, however, that air changes within a building are both normal and desired, to prevent the buildup of stale air and toxic substances normally used within the house. Table 7.8 illustrates the number of air changes taking place under average conditions in typical residences.

The heat capacity of air (C) is 0.018 Btu/ft^3/°F. Thus if a home is kept at 70°F while the outside ambient air temperature is 35°F, each cubic foot of air exchange within the home represents a heat loss of 0.63 Btu. In most circumstances it is obvious that infiltration heat losses can be considerable if air infiltration is not kept to a minimum.

Air infiltration is measured in cubic feet of crack per hour. Half the total crack length around all windows and doors in the home should be used for all calculations, since for each location where air enters a home, there must be another for it to exit. This procedure ensures that air exchange will not be counted twice. Table 7.9 lists the crack values (Q) for selected windspeeds of 15 and 20 mph in combination with both weather-stripped and nonstripped windows. Since windspeed is a primary cause of pressure differences between the inside and outside of the house, resulting in air infiltration in either direction, weather stripping will influence crack susceptibility to a significant degree.

The perimeter of a door or window is normally used as the crack length. However, if the door or window unit is not tightly sealed into the jamb, the additional perimeters of the loose seal should be added to the total crack length.

TABLE 7.6 INTERNAL BUILDING HEAT GENERATION FROM SELECTED SOURCES

(a) Heat Gain from Electric Motors (Continuous Operation[a]) (Btu/hr)

Nameplate[b] or brake horsepower	Full-load motor efficiency (%)	Location of equipment with respect to conditioned space or airstream[c]		
		Motor driven machines in $\frac{hp \times 2545}{\%\ eff.}$	Motor out–driven machine in hp × 2545	Motor in–driven machine out $\frac{hp \times 2545\ (1 - \%\ eff.)}{\% = eff.}$
$\frac{1}{20}$	40	320	130	190
$\frac{1}{12}$	49	430	210	220
$\frac{1}{8}$	55	580	320	260
$\frac{1}{6}$	60	710	430	280
$\frac{1}{4}$	64	1,000	640	360
$\frac{1}{3}$	66	1,290	850	440
$\frac{1}{2}$	70	1,820	1,280	540
$\frac{3}{4}$	72	2,680	1,930	750
1	79	3,220	2,540	680
$1\frac{1}{2}$	80	4,770	3,820	950
2	80	6,380	5,100	1,280
3	81	9,450	7,650	1,800
5	82	15,600	12,800	2,800
$7\frac{1}{2}$	85	22,500	19,100	3,400
10	85	30,000	25,500	4,500
15	86	44,500	38,200	6,300
20	87	58,500	51,000	7,500
25	88	72,400	63,600	8,800
30	89	85,800	76,400	9,400
40	89	115,000	102,000	13,000
50	89	143,000	127,000	16,000
60	89	172,000	153,000	19,000
75	90	212,000	191,000	21,000
100	90	284,000	255,000	29,000
125	90	354,000	318,000	36,000
150	91	420,000	382,000	38,000
200	91	560,000	510,000	50,000
250	91	700,000	636,000	64,000

(b) Recommended Rate of Heat Gain from Commercial Cooking Appliances Located in the Air-Conditioned Area

Appliance	Capacity	Overall dim. (inches) width × depth × height	Miscellaneous data (dimensions in inches)	Manufacturer's input rating		Probable max. hourly input (Btu/hr)	Recommended rate of heat gain (Btu/hr)			
				Boiler hp or watts	Btu/hr		Without hood: Sensible	Without hood: Latent	Without hood: Total	With hood[a]: All sensible
Gas-burning, counter type										
Broiler-griddle		31 × 20 × 18			36,000	18,000	11,700	6,300	18,000	3,600
Coffee brewer			With *warm* position		5,500	2,500	1,750	750	2,500	500
per burner			With storage tank							
Water heater burner					11,000	5,000	3,850	1,650	5,500	1,100
Coffee urn	3 gal	12-in. dia.			10,000	5,000	3,500	1,500	5,000	1,000
	5 gal	14-in. dia.			15,000	7,500	5,250	2,250	7,500	1,500
	8-gal. twin	25-in. wide			20,000	10,000	7,000	3,000	10,000	2,000
Deep fat fryer	15 lb fat	14 × 21 × 15			30,000	15,000	7,500	7,500	15,000	3,000
Dry food warmer per ft² of top					1,400	700	560	140	700	140
Griddle, frying per ft² of top					15,000	7,500	4,900	2,600	7,500	1,500
Short-order stove, per burner			Open grates		10,000	5,000	3,200	1,800	5,000	1,000
Steam table per ft² of top					2,500	1,250	750	500	1,250	250
Toaster, continuous	360 slices/hr	19 × 16 × 30	2 slices wide		12,000	6,000	3,600	2,400	6,000	1,200
	720 slices/hr	24 × 16 × 30	4 slices wide		20,000	10,000	6,000	4,000	10,000	2,000

Source: ASHRAE Handbook of Fundamentals, American Society of Heating, Refrigerating, and Air-Conditioning Engineers, Atlanta, Ga., 1972, Table 30, p. 417.

[a] For intermittent operation, an appropriate usage factor should be used, preferably measured.

[b] If motors are overloaded and amount of overloading is unknown, multiply the heat-gain factors by the following maximum service factors.

Horsepower:	$\frac{1}{20}$–$\frac{1}{8}$	$\frac{1}{6}$–$\frac{1}{3}$	$\frac{1}{2}$–$\frac{3}{4}$	1	$1\frac{1}{2}$–2	3–250
Ac open type	1.4	1.35	1.25	1.25	1.20	1.15
Dc open type	—	—	—	1.15	1.15	1.15

No overload is allowable with enclosed motors.

[c] For a fan or pump in air-conditioned space, exhausting air, and pumping fluid to outside the space, use the values in the last column.

TABLE 7.6 INTERNAL BUILDING HEAT GENERATION FROM SELECTED SOURCES (*continued*)
(b) Recommended Rate of Heat Gain from Commercial Cooking Appliances Located in the Air-Conditioned Area

Appliance	Capacity	Overall dim. (inches) width × depth × height	Miscellaneous data (dimensions in inches)	Manufacturer's input rating		Probable max. hourly input (Btu/hr)	Recommended rate of heat gain (Btu/hr)			
				Boiler hp or watts	Btu/hr		Without hood			With hood [a]
							Sensible	Latent	Total	All sensible
Gas-burning, floor-mounted type										
Broiler, unit		24 × 26 grid	Same burner heats oven		70,000	35,000	Exhaust hood required	Exhaust hood required	Exhaust hood required	7,000
Deep fat fryer	32 lb fat		14-in. kettle		65,000	32,500				6,500
	56 lb fat		18-in. kettle		100,000	50,000				10,000
Oven, deck, per ft^2 of hearth area			Same for 7 and 12 high decks		4,000	2,000				400
Oven, roasting		32 × 32 × 60	Two ovens—24 × 28 × 15		80,000	40,000				8,000
Range, heavy duty		32 × 42 × 33								
Top section			32 wide × 39 deep		64,000	32,000				6,400
Oven			25 × 28 × 15		40,000	20,000				4,000
Range, jr., heavy duty		31 × 35 × 33								
Top section			31 wide × 32 deep		45,000	22,500				4,500
Oven			24 × 28 × 15		35,000	17,500				3,500
Range, restaurant type										
Per 2-burner sect.			12 wide × 28 deep		24,000	12,000				2,400
Per oven			24 × 22 × 14		30,000	15,000				3,000
Per broiler-griddle			24 wide × 26 deep		35,000	17,500				3,500

Electric, counter type										
Coffee brewer										
Per burner				625	2,130	1,000	770	230	1,000	340
Per warmer				160	545	300	230	70	300	90
Automatic	240 cups/hr	27 × 21 × 22	4-burner + water htr.	5,000	17,000	8,500	6,500	2,000	8,500	1,700
Coffee urn	3 gal			2,000	6,800	3,400	2,550	850	3,400	1,000
	5 gal			3,000	10,200	5,100	3,850	1,250	5,100	1,600
	8-gal twin			4,000	13,600	6,800	5,200	1,600	6,800	2,100
Deep fat fryer	14 lb fat	13 × 22 × 10		5,500	18,750	9,400	2,800	6,600	9,400	3,000
	21 lb fat	16 × 22 × 10		8,000	27,300	13,700	4,100	9,600	13,700	4,300
Dry food warmer, per ft^2 of top				240	820	400	320	80	400	130
Egg boiler	2 cups	10 × 13 × 25		1,100	3,750	1,900	1,140	700	1,900	600
Griddle, frying, per ft^2 of top				2,700	9,200	4,600	3,000	1,600	4,600	1,500
Griddle-grill		18 × 20 × 13	Grid, 200 in.^{2}	6,000	20,400	10,200	6,600	3,600	10,200	3,200
Hotplate		18 × 20 × 13	2 heating units	5,200	17,700	8,900	5,300	3,600	8,900	2,800
Roaster		18 × 20 × 13		1,650	5,620	2,800	1,700	1,100	2,800	900
Roll warmer		18 × 20 × 13		1,650	5,620	2,800	2,600	200	2,800	900
Toaster, continuous	360 slices/hr	15 × 15 × 28	2 slices wide	2,200	7,500	3,700	1,960	1,740	3,700	1,200
	720 slices/hr	20 × 15 × 28	4 slices wide	3,000	10,200	5,100	2,700	2,400	5,100	1,600
Toaster, pop-up	4 slices	12 × 11 × 9		2,540	8,350	4,200	2,230	1,970	4,200	1,300
Waffle iron		18 × 20 × 13	2 grids	1,650	5,620	2,800	1,680	1,120	2,800	900
Electric, floor-mounted type										
Broiler							Exhaust hood required	Exhaust hood required	Exhaust hood required	
No oven			23 wide × 25 deep grid	12,000	40,900	20,500				6,500
With oven			23 × 27 × 12 oven	18,000	61,400	30,700				9,800
Deep fat fryer	28 lb fat	20 × 38 × 36	14 wide × 15 deep kettle	12,000	40,900	20,500				6,500
	60 lb fat	24 × 36 × 36	20 wide × 20 deep kettle	18,000	61,400	30,700				9,800
Oven, baking, per ft^2 of hearth			Compartment 8 in. high	500	1,700	850				270
Oven, roasting, per ft^2 of hearth			Compartment 12 in. high	900	3,070	1,500				490
Range, heavy duty		36 × 36 × 36								
Top section				15,000	51,100	25,600				8,200
Oven				6,000	20,400	10,200				3,200
Range, medium duty		30 × 32 × 36								
Top section				8,000	27,300	13,600				4,300
Oven				3,600	12,300	6,200				1,900
Range, light duty		30 × 29 × 36								
Top section				6,600	22,500	11,200				3,600
Oven				3,000	10,200	5,100				1,600

TABLE 7.6 INTERNAL BUILDING HEAT GENERATION FROM SELECTED SOURCES (*continued*)
(b) Recommended Rate of Heat Gain from Commercial Cooking Appliances Located in the Air-Conditioned Area

Appliance	Capacity	Overall dim. (inches) width × depth × height	Miscellaneous data (dimensions in inches)	Manufacturer's input rating		Probable max. hourly input (Btu/hr)	Recommended rate of heat gain (Btu/hr)			
							Without hood			With hood [a]
				Boiler hp or watts	Btu/hr		Sensible	Latent	Total	All sensible
Steam heated										
Coffee urn	3 gal			0.2	6,600	3,300	2,180	1,120	3,300	1,000
	5 gal			0.3	10,000	5,000	3,300	1,700	5,000	1,600
	8-gal twin			0.4	13,200	6,600	4,350	2,250	6,600	2,100
Steam table per ft^2 of top			With insets	0.05	1,650	825	500	325	825	260
Bain marie per ft^2 of top			Open tank	0.10	3,300	1,650	825	825	1,650	520
Oyster steamer				0.5	16,500	8,250	5,000	3,250	8,250	2,600
Steam kettles per gal capacity			Jacketed type	0.06	2,000	1,000	600	400	1,000	320
Compartment steamer per compartment		24 × 25 × 12 compartment	Floor mounted	1.2	40,000	20,000	12,000	8,000	20,000	6,400
Compartment steamer	3 pans 12 × 20 × $2\frac{1}{2}$		Single counter unit	0.5	16,500	8,250	5,000	3,250	8,250	2,600
Plate warmer per ft^3				0.05	1,650	825	550	275	825	260

Note: Heat gain from cooking appliances located in the conditioned area (but not included in the table) should be estimated as follows:

1. Obtain *probable maximum hourly input* by multiplying the manufacturer's hourly input rating by the usage factor of 0.50.
2. If appliances are installed without an exhaust hood, the estimated latent heat gain is 34% of the *probable maximum hourly input* and the sensible heat gain is 66%.
3. If appliances are installed under an effective exhaust hood, the estimated heat gain is all sensible heat.

[a] For poorly designed or undersized exhaust systems the heat gains in this column should be doubled and half of the increase assumed as latent heat.

(c) Rate of Heat Gain from Miscellaneous Appliances

Appliance	Miscellaneous data	Manufacturer's rating		Recommended rate of heat gain (Btu/hr)		
		Watts	Btu/hr	Sensible	Latent	Total
Electrical appliances						
Hair dryer	Blower type	1580	5400	2300	400	2700
	Helmet type	705	2400	1870	330	2200
Permanent wave machine	60 heaters @ 25 W, 36 in normal use	1500	5000	850	150	1000
Neon sign, per linear ft of tube	$\frac{1}{2}$ in dia			30		30
	$\frac{7}{8}$ in. dia			60		60
Sterilizer, instrument		1100	3750	650	1200	1850

Appliance	Miscellaneous data	Manufacturer's rating		Recommended rate of heat gain (Btu/hr)		
		Watts	Btu/hr	Sensible	Latent	Total
Gas-burning appliances						
Lab burners						
Bunsen	$\frac{7}{16}$ in. barrel		3000	1680	420	2100
Fishtail	$1\frac{1}{2}$ in. wide		5000	2800	700	3500
Meeker	1 in. diameter		6000	3360	840	4200
Gas light, per burner	Mantle type		2000	1800	200	2000
Cigar lighter	Continuous flame		2500	900	100	1000

TABLE 7.7 HEAT INPUT FROM A VARIETY OF HUMAN ACTIVITIES[a]

Degree of activity	Typical application	Total heat adults, male (Btu/hr)	Total heat adjusted[b] (Btu/hr)	Sensible heat (Btu/hr)	Latent heat (Btu/hr)
Seated at rest	Theater—Matinee	390	330	225	105
	Theater—Evening	390	350	245	105
Seated, very light work	Offices, hotels, apartments	450	400	245	155
Moderately active office work	Offices, hotels, apartments	475	450	250	200
Standing, light work; or walking slowly	Department store, retail store, dime store	550	450	250	200
Walking; seated Standing; walking slowly	Drugstore, bank	550	500	250	250
Sedentary work	Restaurant[c]	490	550	275	275
Light bench work	Factory	800	750	275	475
Moderate dancing	Dance hall	900	850	305	545
Walking 3 mph; moderately heavy work	Factory	1000	1000	375	625
Bowling[d] Heavy work	Bowling alley Factory	1500	1450	580	870

Source: ASHRAE Handbook of Fundamentals, American Society of Heating, Refrigerating, and Air-Conditioning Engineers, Atlanta, Ga., 1972, Table 29, p. 416.

[a] Tabulated values are based on 75°F room dry-bulb temperature. For 80°F room dry-bulb, the total heat remains the same, but the sensible heat values should be decreased by approximately 20% and the latent heat values increased accordingly.

[b] *Adjusted total heat gain* is based on normal percentage of men, women, and children for the application listed, with the postulate that the gain from an adult female is 85% of that for an adult male, and that the gain from a child is 75% of that for an adult male.

[c] Adjusted total heat value for *sedentary work, restaurant,* includes 60 Btu/hr for food per person (30 Btu sensible and 30 Btu latent).

[d] For *bowling* figure one person per alley actually bowling, and all others as sitting (400 Btu/hr) or standing (550 Btu/hr).

TABLE 7.8 AIR CHANGES IN TYPICAL RESIDENCE

Type of room or building	Number of air changes taking place per hour
Rooms with no windows or exterior doors	$\frac{1}{2}$
Rooms with windows or exterior doors on one side	1
Rooms with windows or exterior doors on two sides	$1\frac{1}{2}$
Rooms with windows or exterior doors on three sides	2
Entrance halls	2

Source: ASHRAE Handbook of Fundamentals, American Society of Heating, Refrigerating, and Air-Conditioning Engineers, Atlanta, Ga., 1981, Table 2, p. 22.8.

Note: For rooms with weather-stripped windows or with storm sash, use two-thirds of these values.

TABLE 7.9 AIR INFILTRATION *Q* VALUES (ft^3/hr-ft)

	Wind speed	
	15 mph	20 mph
Weather-stripped	25	35
Non–weather-stripped	40	60

Source: ASHRAE Handbook of Fundamentals, American Society of Heating, Refrigerating, and Air-Conditioning Engineers, Atlanta, Ga., 1981, Table 3, p. 22.9.

Note: For wood double-hung windows.

Design Temperature Values

Design temperatures and degree-days for the majority of geographical areas in the United States are listed in Appendixes E and H. Although cooling loads vary greatly from one section of the country to another, it is generally acceptable to use a temperature differential of 12 to 15°F when calculating the cooling load of a typical home. An interior temperature of 75°F is the standard when performing heat-gain calculations for air-conditioner sizing. The norm for performing heat-loss calculations for sizing heating systems is 68 to 70°F.

*Total Heat-Gain Load Calculation Procedure**

To illustrate how some of these concepts are applied when doing heat-gain calculations, let us perform a very simple heat-gain analysis that focuses on determining the cooling load for air-conditioning purposes based on the heat-gain factors just mentioned. For purposes of this calculation, consider a single-story residence with an overall base measurement of 30 ft × 50 ft with window and door areas as illustrated in Figure 7.18. The house is on an unheated foundation with an unheated attic that runs the full length of the structure, located in New York City. The design temperature differential for air-conditioning purposes is 25°F. To proceed, each of the heat-gain/loss sources must be calculated as follows:

1. *Walls*. A composite wall structure as illustrated in Figure 7.18 with a total

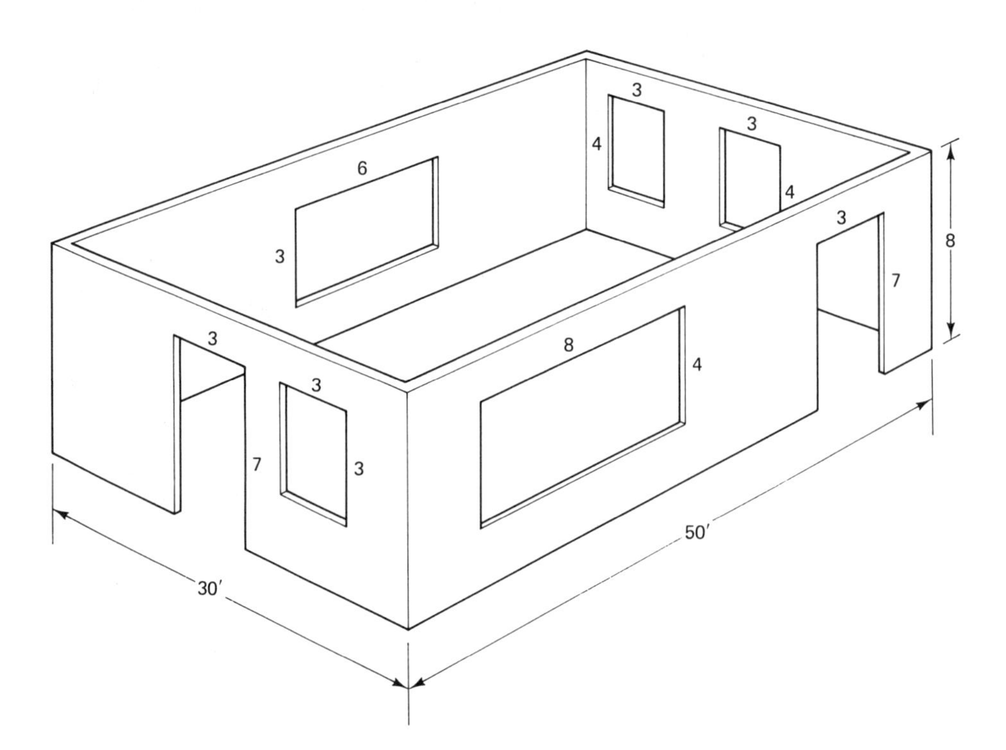

Figure 7.18 Sample home used as a basis for undertaking simple heat-gain calculations.

* The following example is a shortened version of heat-gain calculation procedures for illustration purposes. The individual factors necessary for a total heat-gain calculation include additional factors, such as determination of heat gained through passive solar radiation through building windows. For a comprehensive discussion of heat-loss and heat-gain calculations, the reader is directed to the *ASHRAE Handbook of Fundamentals*, published by the American Society of Heating, Refrigerating, and Air-Conditioning Engineers, Atlanta, Ga.

R value of 13.33 is assumed for the house. The net surface area of the wall must be calculated to determine the overall heat gain. The total area of the four walls equals the gross wall surface area minus the surface area of the windows and doors:

$$\begin{aligned}\text{net wall area} &= 30 \text{ ft} \times 8 \text{ ft } (2) + 50 \text{ ft} \times 8 \text{ ft } (2)\text{—windows and doors} \\ &= 480 \text{ ft} + 800 \text{ ft—windows and doors} \\ &= 1280 - 125 \\ &= 1155 \text{ ft}^2 \text{ net wall area}\end{aligned}$$

The U value of the wall, given an R value of 13.33, is

$$U_{\text{wall}} = \frac{1}{13.33} = 0.07$$

The heat loss of the total wall area in the home is

$$U_{\text{wall}} \times \text{surface area of the wall} \times \text{temperature differential:}$$

$$0.07 \times 1155 \times 25 = 2021 \text{ Btu/hr heat gain}$$

$$\text{net window and door area (from above)} = 125 \text{ ft}^2$$

$$\text{net ceiling area} = \text{length} \times \text{width of house} = 1500 \text{ ft}^2$$

$$\text{net floor area} = \text{length} \times \text{width of house} = 1500 \text{ ft}^2$$

2. *Windows*. All the windows in the house are assumed to be double-glazed storm units with insulating glass and $\frac{1}{2}$ ft of air space. The U value for these units (taken from Table 7.5) is 0.65. The design temperature differential for our calculations is 25°F (assuming a maximum load outdoor temperature of 100°F). Disregarding incoming solar radiation (which can be considerable and should be determined in a full-length calculation), the heat gain through the windows is

$$U_{\text{window}}(0.65) \times 25 \times 125 = 2031 \text{ Btu/hr}$$

3. *Ceiling*. Assume a composite ceiling structure as follows:

$$\begin{aligned}\tfrac{1}{2} \text{ in. of gypsum board} &= R\text{-}0.410 \\ 6 \text{ in. of fiberglass insulation} &= R\text{-}19.0 \\ \text{total } R \text{ value} &= 19.41 \\ U \text{ value for the above} &= \frac{1}{19.41} = 0.05\end{aligned}$$

For purposes of this example, the maximum heating load will take place with an attic temperature, due to solar heat gain, of 100°F. [During the winter, the attic temperature can be assumed to be $\frac{1}{2}$ of the difference between the indoor and outdoor temperature (thus with an indoor temperature of 70°F and an outdoor temperature of 0°F, the attic will be approximately 35°F).] For a more complete description of net gain through a roof, see Appendix F.

The heat gain through the ceiling in this instance is

$$U_{\text{ceiling}}(0.05) \times 1500 \times 25 = 1875 \text{ Btu/hr}$$

4. *Floor.** Since the temperature of the floor over a slab or unheated basement is usually lower than the design temperature (75°F) called for in air conditioning, no significant heat gain can be expected from this source.

5. *Miscellaneous Heat Gain.* From Tables 7.6 and 7.7 assume the following parameters for this home, accompanied by the relevant heat-gain figures:

3 people, moderately active work	1350 Btu/hr
Miscellaneous appliances†	2500 Btu/hr

6. *Air Infiltration.* The total perimeter area of the doors and windows in the sample house is 122 linear feet. Heat gain due to air infiltration is determined by the formula

$$\text{Btu/hr heat gain} = \text{crack length} \times Q \times T \times C$$

where

$$Q = \text{determined from Table 7.9}$$

$$T = \text{temperature differential}$$

$$C = 0.018 \text{ heat capacity of air constant}$$

In this example we assume a design temperature differential of 25°F, with a Q value of 25, derived at a wind speed of 15 mph, considered average for the location of the residence:

$$122 \times 25 \times 25 \times 0.018 = 1372.5 \text{ Btu/hr heat gain due to air infiltration}$$

7. *Total Heat-Gain Determination.* The total heat gain in this simple ex-

* Although slab floors in direct contact with the earth may be neglected for purposes of determining the cooling load of a residence, these slabs can be responsible for considerable heat loss during the winter. Uninsulated, unheated floor slabs are not recommended in climates where the annual number of heating degree-days exceeds 4000, since heat loss from the perimeter of this type of slab can create considerable downdrafts from windows and exterior walls. For a full discussion of the calculation of heat loss from both heated and unheated slabs, see *ASHRAE Guide to Fundamentals*, available from American Society of Heating, Refrigerating, and Air-Conditioning Engineers, Inc., 1791 Tullie Circle NE, Atlanta, GA 30329.

† Although the appliances listed are commercial, these figures are offered as approximations of typical residential conditions.

ample is calculated by adding up all the individual heat-gain figures determined:

Heat gain from:	Net gain (Btu/hr)
Walls	2,021
Ceiling	1,875
Windows	2,031
Floor	0
Interior heat gain	3,850
Air infiltration	1,372
Total Btu heat gain	11,149

Given that 1 ton of air conditioning is equivalent to removing 12,000 Btu/hr from a confined space, the residence in this example could be adequately handled by a 1-ton air conditioner.

Although the calculations above are highly simplified, they serve to illustrate the basic procedures that are involved in performing heat-gain and heat-loss calculations.

8

SMALL-SCALE HYDROELECTRIC GENERATION

The use of water to power a variety of mechanical devices can be traced back many thousands of years. The two most common uses of water-driven machines were for grinding grain and pumping water. Waterwheels were adapted to provide power for more diverse applications, such as running sawmills and assorted industrial machinery. With the development of coal, oil, steam, and electrical power, most water-driven installations were abandoned. Water-driven machines have undergone design changes, which have increased their efficiency: from the old paddle-wheel type of mill to modern hydraulic turbines incorporating high operating efficiencies. It was not until dramatic price increases in fossil fuels and utility rates took effect that the idea of resurrecting abandoned hydro facilities and the development of new ones began to take hold. A variety of old water-powered facilities, from grain-grinding units to small municipal hydroelectric facilities, were purchased and rehabilitated with an eye towards producing low-cost hydroelectric power on a decentralized basis.

There is a distinction between general-purpose hydro facilities and "small-scale" hydro facilities. This book focuses on small-scale units that are suitable, in both generating output and economic viability, for use in residential applications, as opposed to larger, high-head commercial and industrial projects.

APPLICATIONS AND BASIC REQUIREMENTS

Small-scale facilities are advantageous in locations where the residential property contains a small stream or river that can be channeled through a turbine to generate electricity for consumption in the home. There are many technical, legal, and economic factors that must be considered prior to undertaking this type of installation.

Perhaps the most important factor associated with small-scale hydro facilities is the economic feasibility of the project. Small-scale hydro systems are expensive. The most advantageous case is where an older existing hydro facility can be reconstructed by the homeowner, saving considerable expense over the alternative of building a facility from scratch. Even when existing sites can be rebuilt, the cost of rehabilitation can be significant: It is not unusual to spend more than $20,000 to rebuild an existing small hydroelectric plant. In the case of new construction, the cost will be even higher.

More subtle consideration associated with these plants concerns the legal and environmental approvals that must be obtained prior to site approval in most instances. Usually, the local Department of Environmental Conservation (DEC) must be consulted to determine if there are any restrictions concerning the water rights of the proposed facility. If a dam is to be constructed, it could affect water both upstream and downstream from the facility, and specific figures regarding dam capacity, generating capacity, and all construction aspects of the installation will be carefully scrutinized. Often, the DEC will work in conjunction with the local building inspector and utility company officials to develop a comprehensive outline for construction. Depending on the complexity of the approval procedure, the homeowner may be required to engage legal counsel to work through the various aspects of this approval process.

THE BASIC SMALL-SCALE HYDROELECTRIC FACILITY

Most residential hydro facilities generate low electrical output (between 1 and 25 kW of generating power). The low power definition is not fixed, but rather depends on the potential power available in the water source as well as the size of the generator to be installed. A typical small-scale residential hydroelectric facility is illustrated in Figure 8.1. Referring to the figure, the following definitions and explanations of terms relative to these facilities are appropriate. The source of the water (1) may be either a river, stream, or lake with adequate storage capacity. Water storage should be determined during the low-water season rather than using high-water measurements. If the water source experiences significant dry periods or times of restricted water flow during the year on a predictable basis, the site is probably not a good one for further consideration.

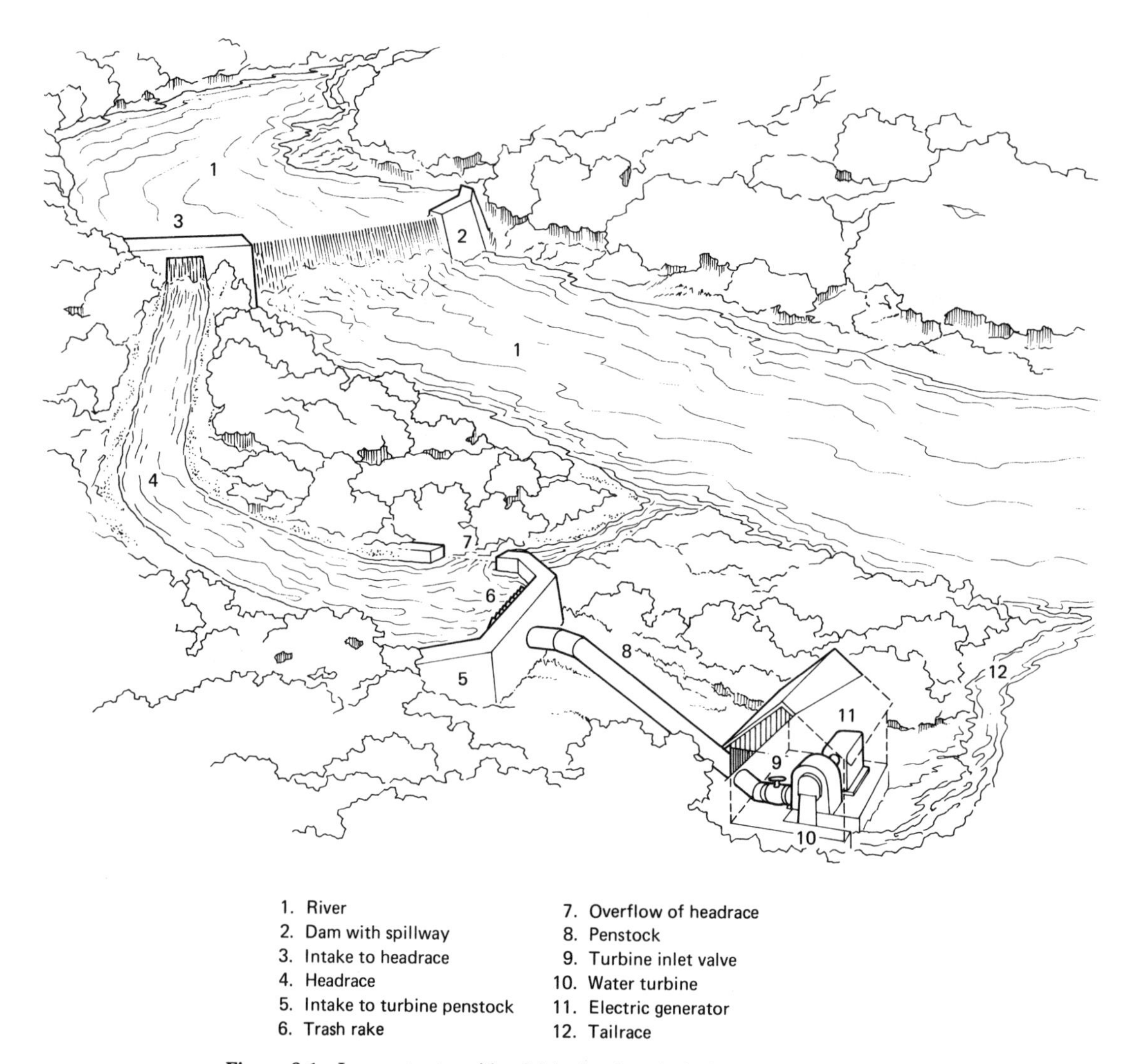

Figure 8.1 Low-output residential hydroelectric facility. The principal operational components of the system, while not identical, are typical of most systems. (Adapted from Carol Hupping Stoner, ed., *Producing Your Own Power,* Rodale Press, Emmaus, Pa., © 1974, Fig. 13, p. 69. Permission granted by Rodale Press, Inc., Emmaus, PA 18049.)

The dam (2) may be constructed from a variety of materials, which will be examined later in more detail. Note that the dam contains a spillway that allows excess water from the reservoir to escape from the reservoir and that is sized to handle the maximum amount of water that can be expected in the source. Both the headrace (4) and the tailrace (12) are open-channel waterways, designed to deliver the requisite amounts of water to the system at relatively low velocities. Prior to entering the penstock (8), which is a pipe that delivers water to the turbine, the water approaches the penstock intake area (5), which contains a trash rake (6). The trash rake is designed to protect the turbine and prevent debris from entering the penstock. The rake should be cleaned every day. Water travels down the penstock (8), where it is metered to the water turbine (10) through a turbine inlet valve (9). The turbine is connected to the electrical generator through a belt or shaft which couples the turbine runner to the generator or alternator armature. Water exits the turbine assembly and powerhouse through a tailrace, which returns it to the stream.

Perhaps the most important term discussed in a hydroelectric facility is **head**. "Head" refers to the total height of water, which is measured from the dam or reservoir to the level of the tailwater. There are two types of head in hydro facilities: artificial and natural. **Artificial head** refers to the height of the water measured from the dam or reservoir to the turbine. **Natural head** refers to the vertical height of water available when it follows its natural course in the streambed. The location of the powerhouse can take advantage of either artificial or natural head. Figure 8.2 illustrates a powerhouse location in which only artificial head is utilized by the powerhouse, located directly below and in close proximity to the dam.

A powerhouse location that utilizes both artificial and natural head is illustrated in Figure 8.3. The remote location of the powerhouse will usually be more expensive than using a location that is directly adjacent to the dam. However, in instances where a rapid drop-off of water occurs as it travels downstream, the

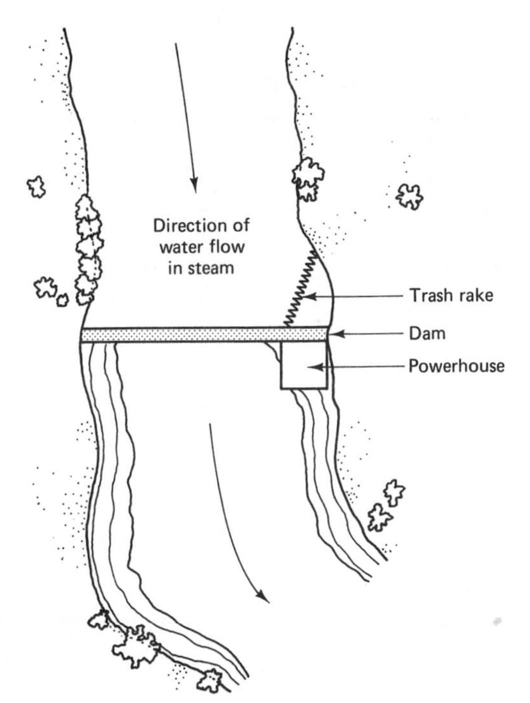

Figure 8.2 Powerhouse location utilizing artificial head only.

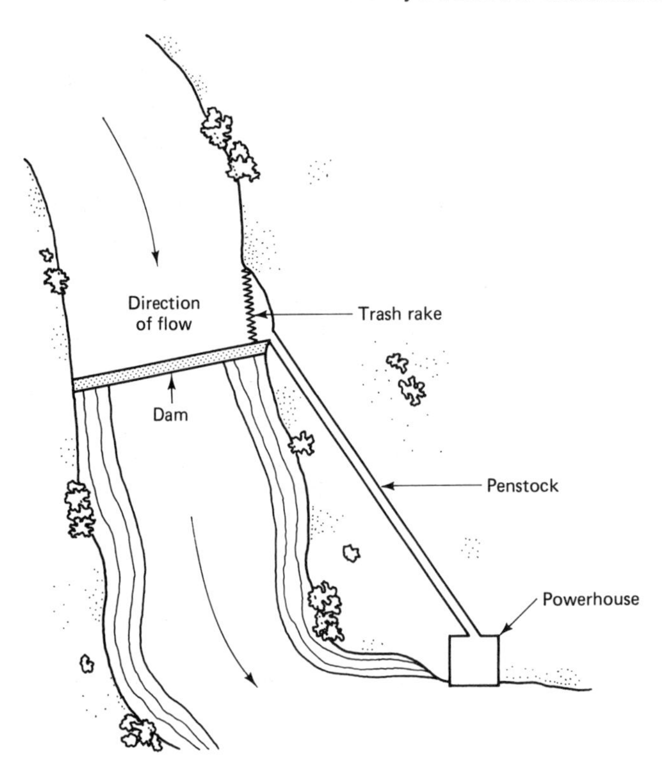

Figure 8.3 Powerhouse location that takes advantage of both natural and artificial head.

additional cost of the remote installation may well be justified by virtue of the additional generating power available due to higher head available downstream.

SMALL-SCALE HYDRO DAMS

Small-scale hydro facilities use dams made from four types of materials: wood, stone, concrete, and earth. Each of these dams types is illustrated in cross section in Figure 8.4.

Dams should be located at the narrowest part of the stream or pond to facilitate construction whenever possible. If this is not practical, any natural restriction in the stream or pond might also be a suitable location for a dam.

Any type of dam must be properly engineered to prevent erosion of the dam as well as provide adequate strength for water storage. Earthen dams (Figure 8.4d) are perhaps the weakest of the four types of dams illustrated. Generally constructed where masonry or concrete materials are not available, the dam must be anchored so that it is not undermined by seepage of water between the bottom of the dam and the earth above it. Earthen dams are not adequate in instances where the water moves quickly, since erosion will undermine the structure in

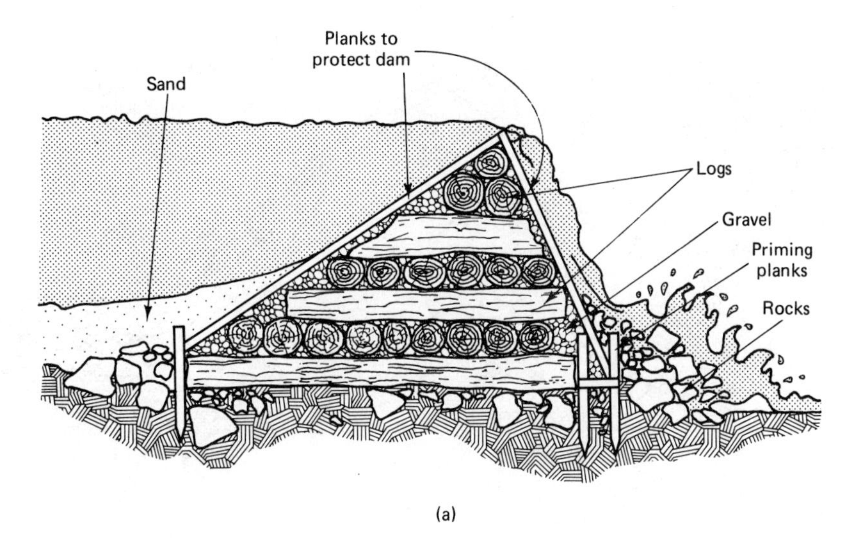

(a)

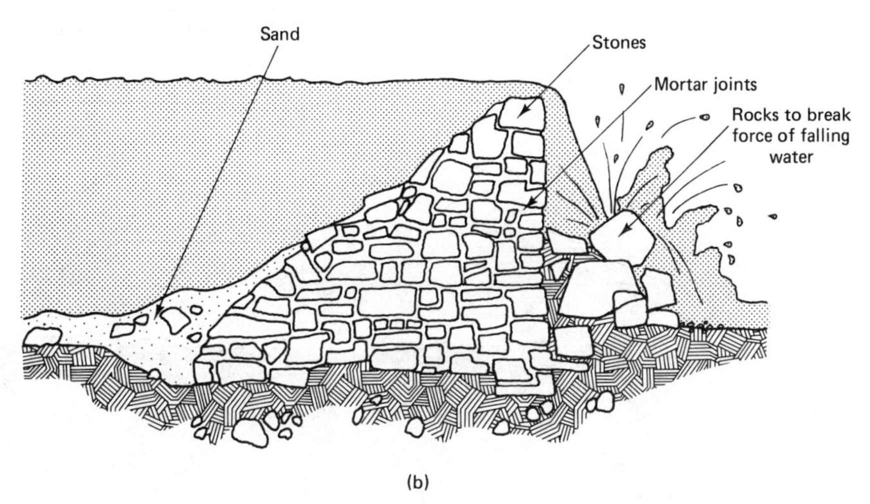

(b)

Figure 8.4 Typical small-scale hydroelectric dam construction techniques: (a) wooden dam; (b) stone dam; (c) concrete dam; (d) earthen dam. (Adapted from Peter Clegg, *Energy for the Home,* Garden Way Publishing, Pownal, Vt., pp. 146–147. Available from Storey Communications, Pownal, VT 05261.)

relatively short order. When making earthen dams, only heavy soil or clay soils that resist water should be used.

Timber dams (Figure 8.4a) are economical in areas where lumber is plentiful. Either oak or chestnut is the best lumber to use for these dams, although most dense hardwoods will be adequate. The timber members should be approximately 6 in. in diameter and the cross members spiked at 2- to 3-ft intervals. Stones are

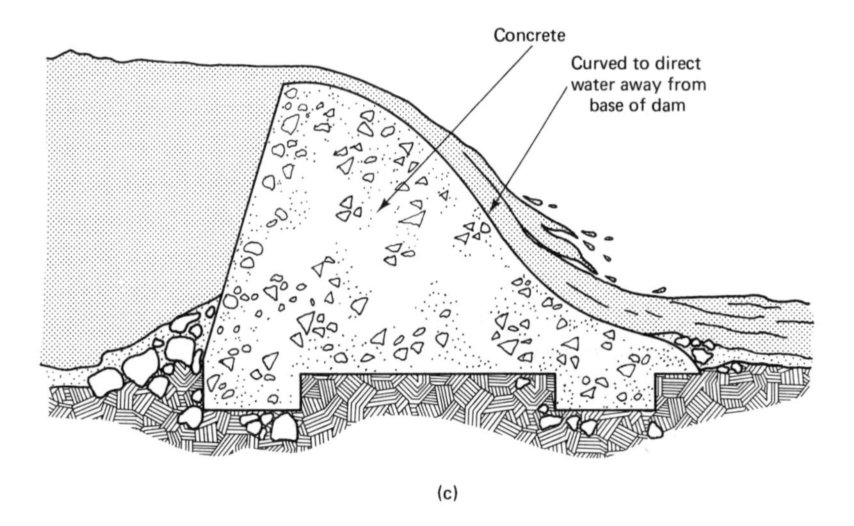

(c)

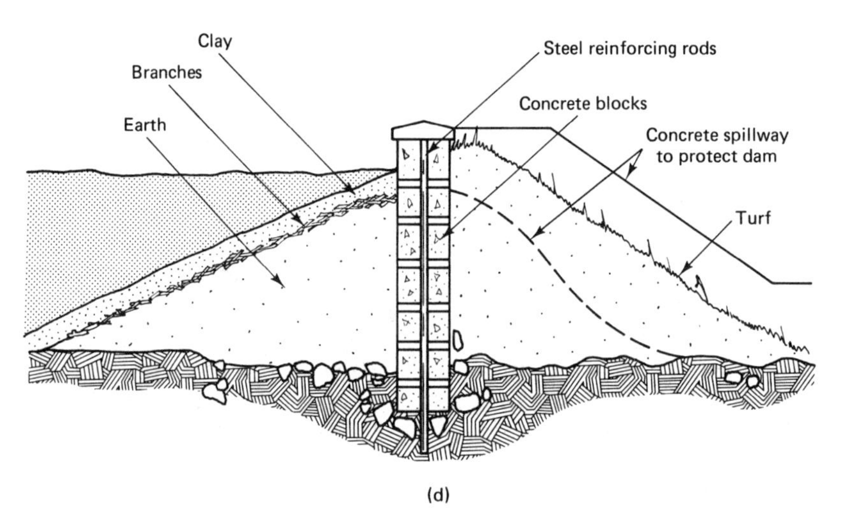

(d)

Figure 8.4 *(continued)*

used to fill the spaces between the timbers and the dam and should be sealed with planks on both the upstream and downstream sides. Space between the planks should be sealed with clay to prevent water seepage. The dam is anchored into the stream using a series of planks that are driven into the streambed and fastened to the framing lumber in the dam. Note the use of stone on the downstream side of the dam, which breaks the force of the spill water, preventing erosion of the heel of the dam.

The use of masonry materials in dam construction is preferable to other options given the availability of both materials and funds. Building a stone or concrete dam (Figure 8.4b and c) requires a knowledge of local soil conditions and an engineering knowledge of masonry construction techniques and stress analysis. If the proposed dam will be higher than 6 to 8 ft, a structural engineer should be consulted either to verify the integrity of the design of the dam or to design one that will meet all expected stresses. A stone dam (Figure 8.4b) should be bonded with concrete and anchored on a solid footing or foundation in the stream, which will prevent the structure from shifting and eventually collapsing. As a general rule, the base of the dam should be equal to its height. Rocks and crushed stone should be placed at the heel of the dam to prevent spillwater erosion.

A concrete dam also requires a firm footing for stability. The base of the dam should be about $1\frac{1}{2}$ times its height, and the apron or top of the dam should be arched as illustrated to help break the force of the spill water.

DETERMINING AVAILABLE ENERGY FOR POTENTIAL SITES

The procedure for evaluating small-scale hydro sites involves the qualification of two factors: volume of water flow and available head. Each of these factors is examined below.

Measuring Flow Rates in Small Water Sources

There are three ways to measure the volume of water flow in small sources, depending on the size of the source and the available materials for measuring purposes. These are the bucket method, the weir method, and the float method.

Measuring water flow using the bucket method. The bucket method is the easiest to undertake given the right conditions. If the stream is small, and if it is estimated that the flow rate is less than 1 ft³/sec, a small dam can be built, diverting the entire flow of the stream through a pipe into a standard 5-gal bucket. The formula to determine the volume of water flow in cubic feet per second is as follows:

$$\text{volume (ft}^3\text{/sec)} = \frac{\text{volume of bucket (ft}^3\text{)}}{\text{time required to fill bucket (sec)}}$$

Therefore, if it takes 4 seconds to fill a bucket with a volume of 6 ft³, the flow is

$$\text{flow (ft}^3\text{/sec)} = \frac{6}{4} = 1.5 \text{ ft}^3\text{/sec}$$

If the stream or water output from the source is too large for the bucket method, either a weir will need to be constructed or a float measurement is taken.

Calculating stream flow with a weir. A **weir** is a small rectangular notch of known dimensions cut into a wooden dam (Figure 8.5). By measuring the height of water behind and above the bottom of the weir, its flow rate can be calculated from a weir table. Since the weir is a temporary structure, it can be constructed from tongue-and-groove planks and sealed with clay or plastic sealing compound on the dam or upstream side of the weir. To ensure accurate measurements, all water flow from the stream or pond must go through the weir; therefore, it should be firmly anchored and sealed on the bottom of the stream or river bed by the use of anchor planking set into the streambed.

A rectangular notch should be cut into the center of the weir, with the edges of the opening sawed on a slant to produce a sharp edge upstream. This edge prevents the wooden planking from breaking down under the action of the running water. The dimensions of the weir notch are important. The length of the notch

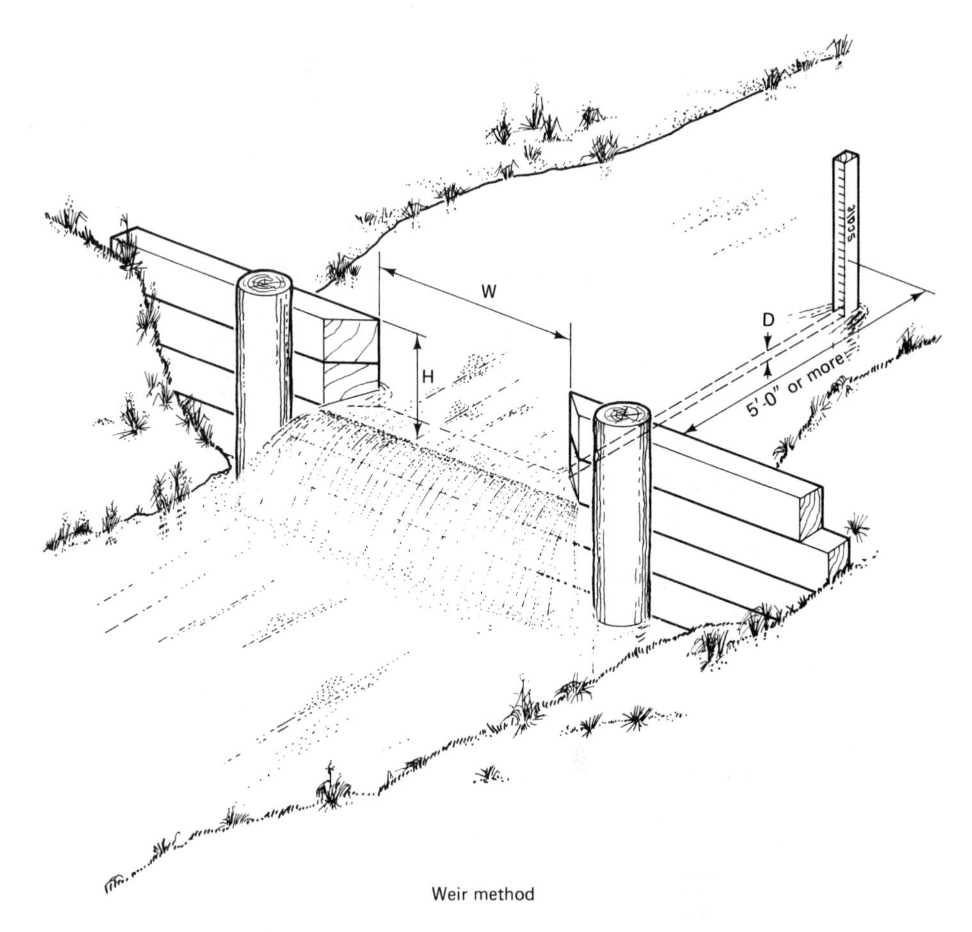

Figure 8.5 Configuration of a weir, used to measure the volume of water flow in a small stream.

should not be less than three times the height of the opening. The bottom of the notch should be about 1 ft above the water surface below the dam and perfectly level. Approximately 6 ft upstream from the weir, a stake is driven into the streambed, with its top level with the bottom notch of the weir. After the weir has been completed, the water in the pond will rise, due to the dam, and achieve a relatively stable flow through the weir. At this time, the height of the water above the stake should be measured. With this figure, the volume of water flow through the weir can be determined from the figures in Table 8.1, which illustrate the volume of flow in cubic feet per minute per inch of weir notch. Multiplying this figure by the width of the notch gives the total stream flow in cubic feet per minute.

To illustrate this procedure, let us assume that the depth of the water over the stake in the stream is 5.9 in. and that the total length of the weir notch is 48 in. The total volume of the stream flow is therefore

$$\text{volume} = 5.9 \times 48 = 283 \text{ ft}^3\text{/min of stream flow}$$

TABLE 8.1 WEIR TABLE

Inches depth over Stake D		$\frac{1}{8}$ inch	$\frac{1}{4}$ inch	$\frac{3}{8}$ inch	$\frac{1}{2}$ inch	$\frac{5}{8}$ inch	$\frac{3}{4}$ inch	$\frac{7}{8}$ inch
1 inch ...	.40	.47	.55	.65	.74	.83	.93	1.03
2 "	1.14	1.24	1.36	1.47	1.59	1.71	1.83	1.96
3 "	2.09	2.23	2.36	2.50	2.63	2.78	2.92	3.07
4 "	3.22	3.37	3.52	3.68	3.83	3.99	4.16	4.32
5 "	4.50	4.67	4.84	5.01	5.18	5.36	5.54	5.72
6 "	5.90	6.09	6.28	6.47	6.65	6.85	7.05	7.25
7 "	7.44	7.64	7.84	8.05	8.25	8.45	8.66	8.86
8 "	9.10	9.31	9.52	9.74	9.96	10.18	10.40	10.62
9 "	10.86	11.08	11.31	11.54	11.77	12.00	12.23	12.47
10 "	12.71	12.95	13.19	13.43	13.67	13.93	14.16	14.42
11 "	14.67	14.92	15.18	15.43	15.67	15.96	16.20	16.46
12 "	16.73	16.99	17.26	17.52	17.78	18.05	18.32	18.58
13 "	18.87	19.14	19.42	19.69	19.97	20.24	20.52	20.80
14 "	21.09	21.37	21.65	21.94	22.22	22.51	22.70	23.08
15 "	23.38	23.67	23.97	24.26	24.56	24.86	25.16	25.46
16 "	25.76	26.06	26.36	26.66	26.97	27.27	27.58	27.89
17 "	28.20	28.51	28.82	29.14	29.45	29.76	30.08	30.39
18 "	30.70	31.02	31.34	31.66	31.98	32.31	32.63	32.96
19 "	33.29	33.61	33.94	34.27	34.60	34.94	35.27	35.60
20 "	35.94	36.27	36.60	36.94	37.28	37.62	37.96	38.31
21 "	38.65	39.00	39.34	39.69	40.04	40.39	40.73	41.09
22 "	41.43	41.78	42.13	42.49	42.84	43.20	43.56	43.92
23 "	44.28	44.64	45.00	45.38	45.71	46.08	46.43	46.81
24 "	47.18	47.55	47.91	48.28	48.65	49.02	49.93	49.76

Source: Courtesy of Kvaerner Hydro Power Inc./Leffel Turbines.

Measuring water flow using the float method. When using the float method for flow calculations, the cross-sectional area of the streambed is measured and multiplied by the velocity of water flow in the stream, which is shown by a floating bottle or piece of wood. The cross-sectional area of the stream is determined by taking a series of measurements across the stream to obtain the average depth and multiplying this figure by the width of the stream where the measurements were taken (Figure 8.6).

A 30-ft length is marked out along the stream with two stakes. Some small pieces of wood or other small floats are thrown into the stream and timed to determine how long it takes the float to travel the 30-ft distance. This test should be repeated several times to ensure the accuracy of the results. To determine the total stream flow, the velocity of float travel is multiplied by the cross-sectional area of the stream. For example, let us assume that a series of five readings across a stream give depth readings of 2 ft, 1 ft 6 in., 1 ft 6 in., 2 ft 0 in., and 2 ft 6 in., resulting in an average depth of 1 ft 9 in. If the width of the stream at this point was 12 ft, the cross-sectional area of the stream at this point is

$$12 \times 1.9 = 22.8 \text{ ft}^2$$

The velocity of the stream based on the results of the float experiments can now be determined. Assume that the float travel averaged 30 ft in 45 seconds. The formula to determine the velocity of flow rate in the pond is

flow rate = length of measured course of float × 60 sec × 0.8 (coefficient to obtain average flow between the top and bottom of the stream) divided by the length of time to run the course

Thus

$$\text{velocity} = \frac{30 \times 60 \times 0.8}{45} = \frac{1440}{45} = 32 \text{ ft/min}$$

To determine the flow in cubic feet per minute, the velocity in feet per minute

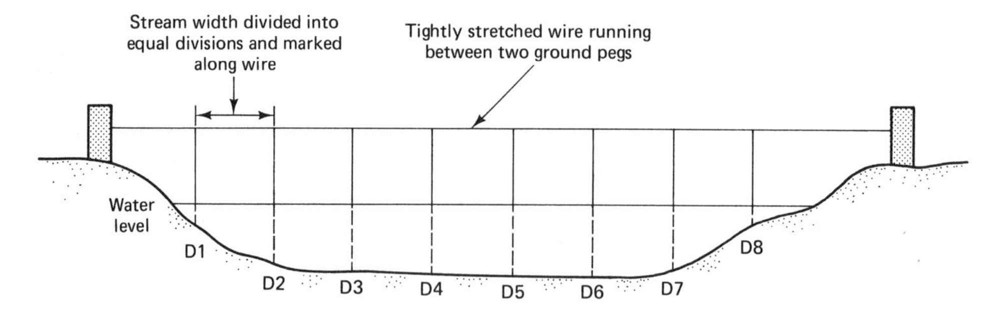

Figure 8.6 Method of determining the cross-sectional area of a small stream.

is multiplied by the cross-sectional area of the stream:

$$\text{flow rate} = \text{velocity} \times \text{cross-sectional area} = 32 \times 22.8$$
$$= 729 \text{ ft}^3/\text{min}$$

This figure, 729 ft^3/min, represents the flow of water in the water source and should not be confused with the available amount of water that can be delivered to a turbine and converted into electrical power. After the flow rate has been calculated, the total head is determined. From this figure the frictional losses associated with delivering water from the dam to the powerhouse are subtracted to yield the "net head" available for power generation. The net head multiplied by the volume of available water determines the net horsepower available at the powerhouse. This figure can then be used to size the turbine and generator and to define the overall parameters in which a system will operate.

Head Determination

Most residential hydroelectric systems are low-head applications: those that operate with less than 50 ft of head. Most residential systems are, in fact, well below the 50-ft head height.

There are two methods used to calculate the head of the system, which is defined as the difference in height between the headwater (water output level behind the dam) and the tailwater (water output from the turbine). The easiest method to take head measurements is to use a surveyor's level and a stadia rod (scale) (Figure 8.7).

If a surveyor's level is not available, a more tedious procedure can be used that involves making a series of small height measurements using a carpenter's level and a 2 in. × 6 in. × 10 ft board, from the headwater down to the tailwater. The individual measurements are then added together as illustrated in Figure 8.8.

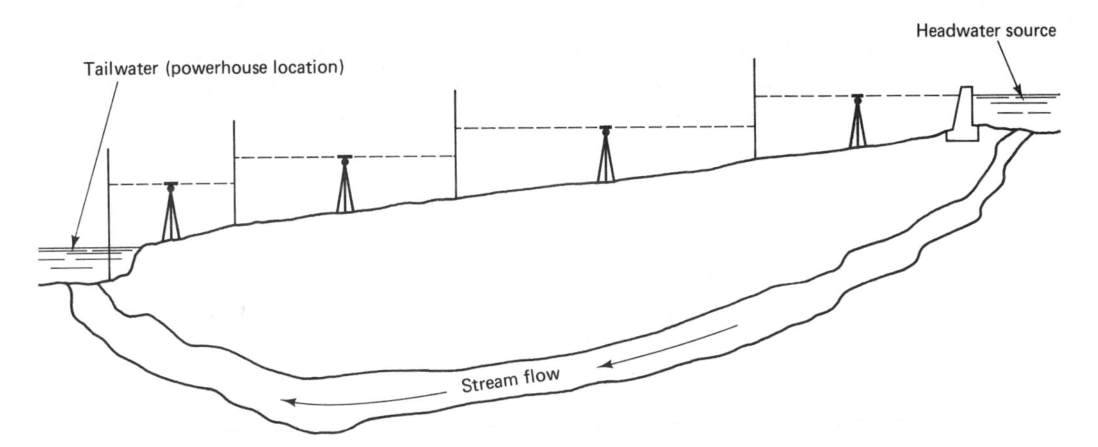

Figure 8.7 Measuring water source head using a surveyor's level.

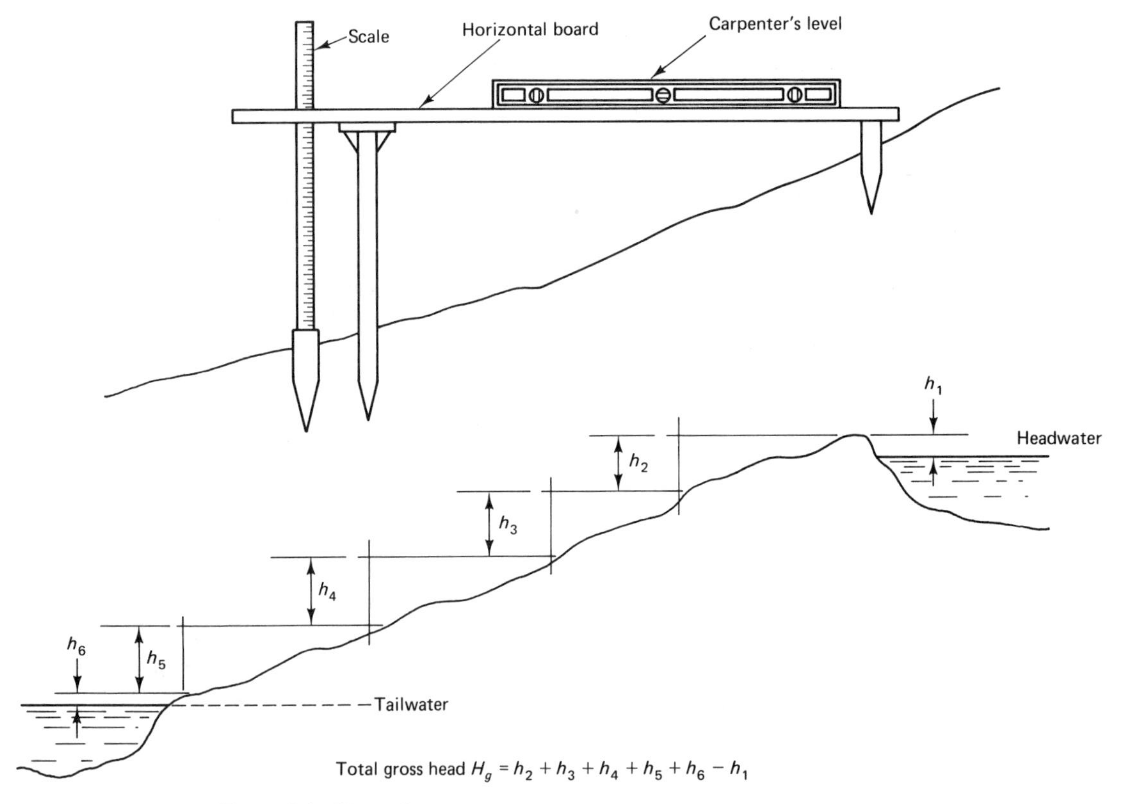

Figure 8.8 Measuring water source head with a carpenter's level. (Adapted from Carol Hupping Stoner, ed., *Producing Your Own Power,* Rodale Press, Emmaus, Pa., © 1974, Fig. 9, p. 65. Permission granted by Rodale Press, Inc., Emmaus, PA 18049.)

Head losses. After the gross head measurements have been calculated, allowance must be made for the various losses that will occur in the water delivery system due to friction, pipe bends, and other irregularities in the delivery system. Sizing of the delivery pipe (penstock) is critical for proper and efficient system operation. If the pipe diameter is too small, an otherwise well-designed system will operate poorly. If the pipe run is too long, excessive frictional losses will result. The balance of proper pipe size, length of run, number of bends, and type of pipe requires a good deal of thought before final selections can be made. The loss in head as a function of the length of the pipe run and the type of pipe used in the system is shown in Figure 8.9.

After selections have been made, and the head loss has been calculated, the net head of the system can be determined by subtracting head losses from the gross head figure. Once the net head figure is derived, the net power output of the system can be determined.

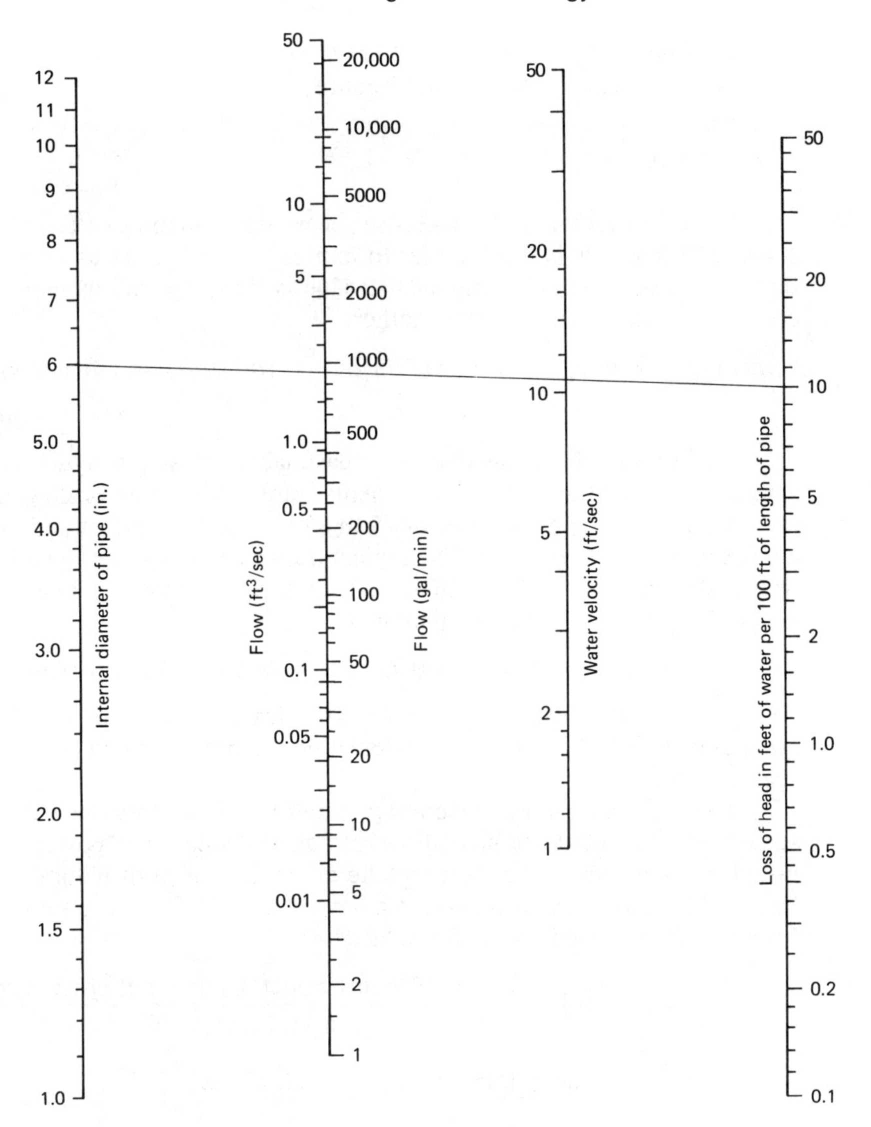

Figure 8.9 Water flow and head loss. (Adapted from Carol Hupping Stoner, ed., *Producing Your Own Power,* Rodale Press, Emmaus, Pa., © 1974, p. 73. Permission granted by Rodale Press, Inc., Emmaus, PA 18049.)

Calculation of Net Power Output

Net power output of any hydroelectric generator is composed of the following factors:

1. Efficiency of the water turbine
2. Efficiency of the generator or alternator

3. Efficiency of the transmission system (e.g., belts)
4. Water flow, in cubic feet per minute
5. Net head of system
6. Constant factor

Efficiency factors 1, 2, and 3 describe what is sometimes referred to as **overall system efficiency**. It is often easier to express these factors as one figure rather than the three separate components. This is done by multiplying the separate component efficiencies by one another:

$$(\text{generator efficiency} \times \text{turbine efficiency} \times \text{transmission efficiency}) = \text{overall efficiency}$$

To perform these calculations, reasonable efficiency figures must be assigned to each of the system components. Most alternators or dc generators operate within an efficiency range of 75 to 80%. Small-scale hydraulic turbines are approximately 80% efficient. The typical transmission system, whether it be belt or gear driven, is about 95% efficient. Therefore, to obtain the overall efficiency, we perform the following calculation:

$$\text{overall efficiency} = 0.75 \times 0.80 \times 95 = 0.57 \text{ efficiency}$$

Although this figure may vary from one installation to another, it falls within acceptable limits of accuracy for calculating the power output of a typical small-scale system.

To determine the net generating output based on previous calculations, assume that a potential site has a flow rate of 80 ft^3/min. The powerhouse is to be installed remote from the dam to take advantage of both natural and artificial head, which has been calculated at a net value of 35 ft. The net power available from the site is based on the formula

$$\text{power (kW)} = \frac{\text{flow (ft}^3\text{/min)} \times \text{net head} \times \text{net efficiency}}{\text{constant factor (708)}}$$

Substituting our figures in the calculations above, we find that

$$\text{power} = \frac{80 \times 35 \times 0.57}{708} = 2.25 \text{ kW}$$

After net power has been determined, an energy survey of the residence should be conducted to determine the percentage of the electrical needs of the home that will be provided by the hydro facility. This survey is similar to the one described for wind generator sizing (see Chapter 9). After the loads have been determined, an analysis is made to determine how the power is used during the day and at what times this load will peak. A daily or monthly electrical consumption figure can be calculated, and from this, what the hydro facility will be able to supply and what will need to be made up from either the local utility system or backup generator.

If the output of the hydro system is small (less than 1 kW/hr), the use of a battery storage system would be practical. Connecting the output of the generator to a small resistive heating load such as a hot water heater or space heating baseboard circuit would also be practical in this instance.

The methods of interfacing the hydro facility with existing power in the home are basically the same as for wind generator installations: a separate dc wiring system might be most adequate or a synchronous inverter for cogeneration and feedback into the grid might be the system of choice if the site can produce a meaningful amount of excess power.

After the site has been thoroughly investigated and power calculations have been performed, the last stage prior to construction is the selection of the turbine to be used. There are several types of water turbines available for small-scale residential use.

CLASSIFICATION OF WATER TURBINES

A variety of water turbines are available designed for specific flow rates and head pressures. The type of water turbine most people are familiar with is the old-style waterwheel.

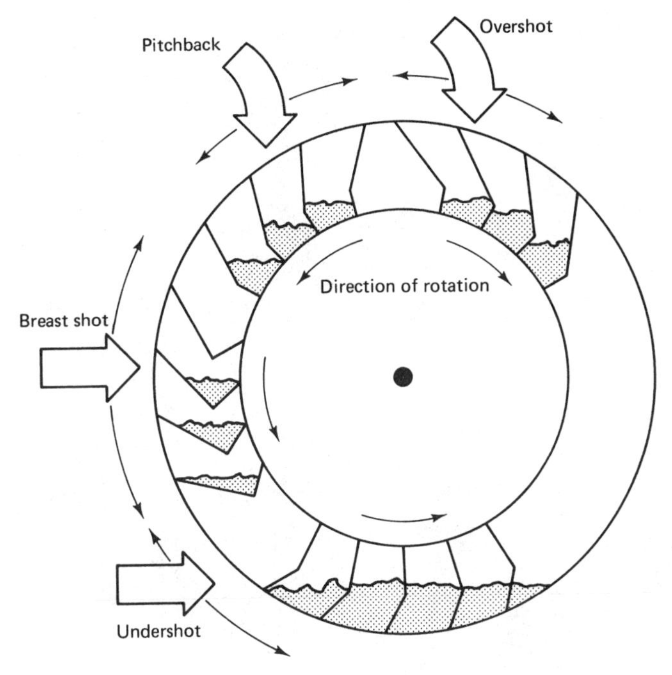

Figure 8.10 Waterwheel classifications. (Adapted from Peter Clegg, *Energy for the Home,* Garden Way Publishing, Pownal, Vt., p. 135. Available from Storey Communications, Pownal, VT 05261.)

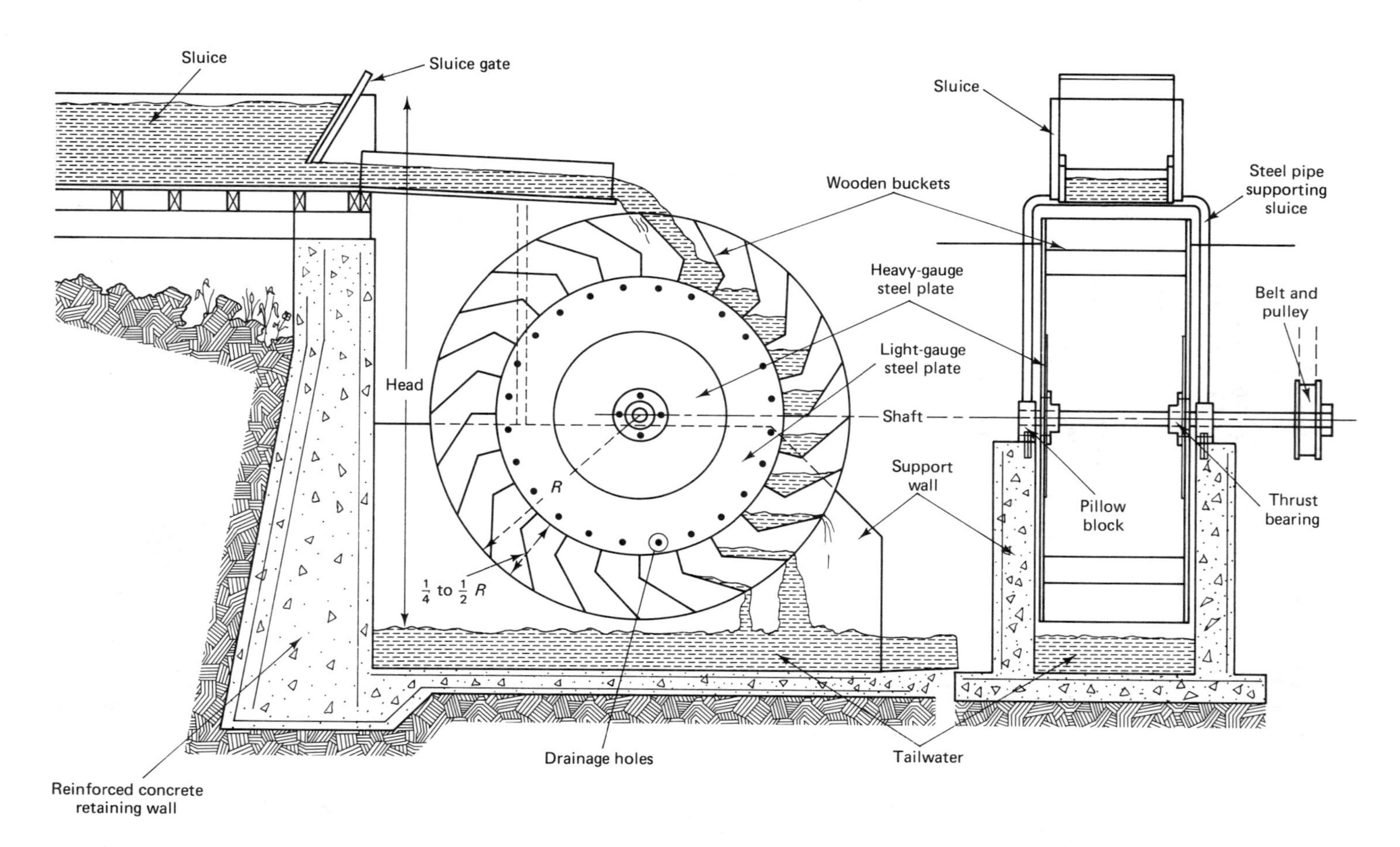

Figure 8.11 Operation of overshot waterwheel (typical). (Adapted from Peter Clegg, *Energy for the Home*, Garden Way Publishing, Pownal, Vt., p. 136. Available from Storey Communications, Pownal, VT 05261.)

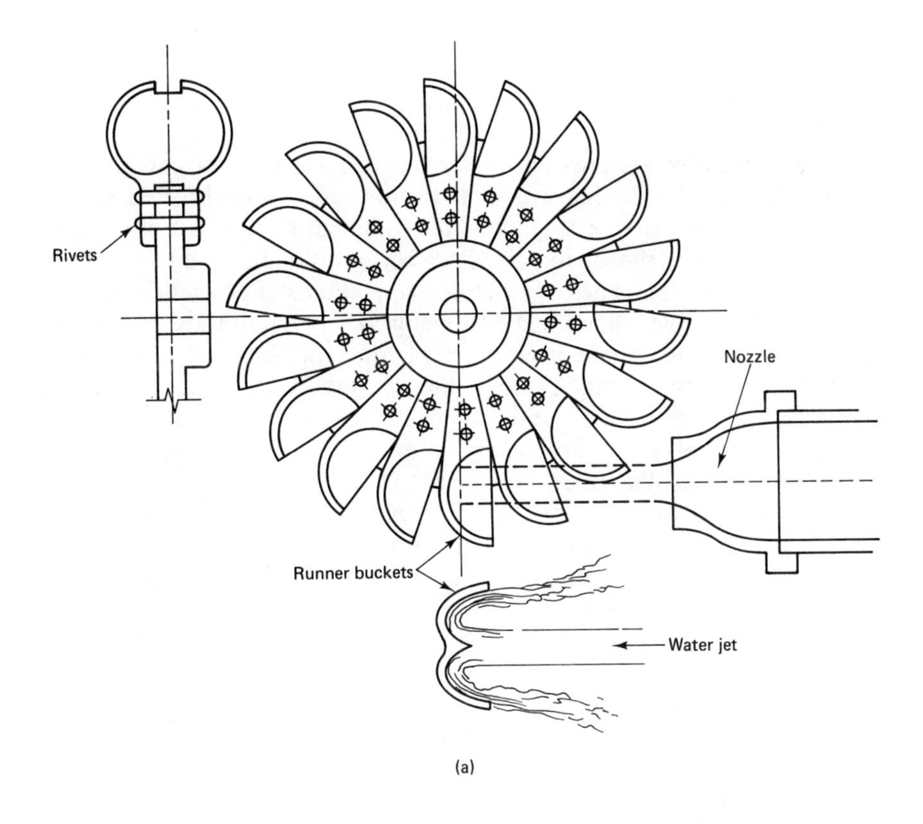

(a)

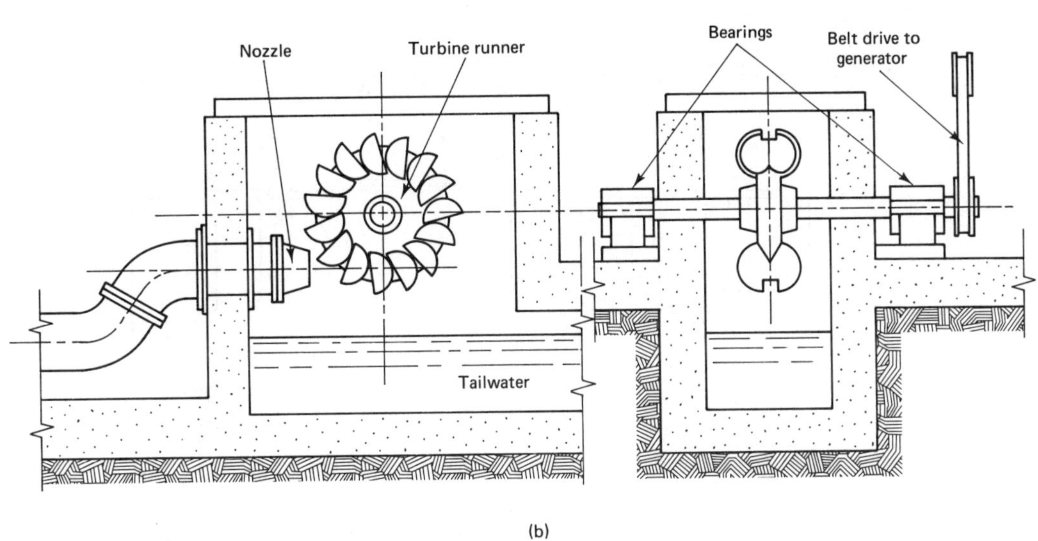

(b)

Figure 8.12 Impulse-type turbine, commonly referred to as a Pelton wheel: (a) principle of operation; (b) small impulse turbine operating in concrete housing. (Adapted from Carol Hupping Stoner, ed., *Producing Your Own Power,* Rodale Press, Emmaus, Pa., © 1974, Fig. 27, p. 83. Permission granted by Rodale Press, Inc., Emmaus, PA 18049.)

Waterwheels

The basic waterwheel operates under one of four types of conditions, based on the location of the driving water in relation to the wheel: overshot, undershot, pitchback, or breast shot (Figure 8.10). Although the earliest wheels were of the undershot design, the overshot wheel became dominant up to the time of the industrial revolution. The overshot design is the most efficient driving method of the four alternatives illustrated. All waterwheels require a good deal of head to operate efficiently. These wheels are picturesque, but their usefulness for generating electricity is limited due to their relatively slow rotation. It is possible, however, to gear up the rotational speed of the system to run a small generator. Most waterwheels require relatively little maintenance. The operation of a typical overshot wheel is illustrated in Figure 8.11.

Water is delivered to the wheel from the penstock and metered by a sluice gate that regulates the flow to drive the wheel efficiently and not empty out the drive bucket until it is almost in contact with the tailwater. This allows for maximum gravity drive of the wheel. The power available from the wheel is determined by both the diameter and width of the wheel, as well as the volume of water available from the stream. Since the rotation of the wheel is relatively slow (between 2 and 10 revolutions per minute), the gearing that is required to bring the system up to speed for most electrical generating requirements results in high transmission losses that make the wheel unsuitable for this task in most instances.

There are two types of turbines that are used in most all residential hydroelectric installations: impulse and reaction turbines.

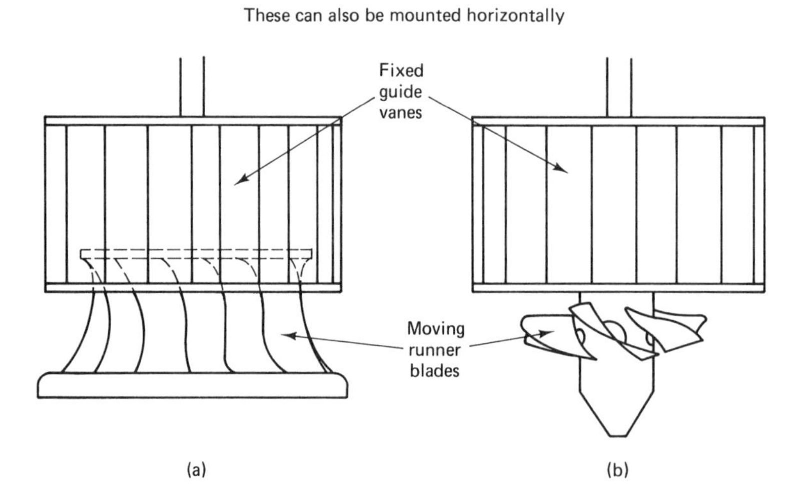

Figure 8.13 Reaction-type turbines: (a) propeller type; (b) Francis type. (Courtesy of Kvaerner Hydro Power Inc./Leffel Turbines.)

Impulse Turbines

Impulse turbines are used primarily in high-head installations, since their operation depends on the velocity of the water that is delivered to the turbine. An impulse turbine, sometimes referred to as a **Pelton wheel**, is illustrated in Figure 8.12. The impulse turbine illustrated is a double-bucket arrangement. Water is directed into the buckets from a nozzle. As the water hits the bucket, its direction is reversed and it imparts energy to the turbine wheel in the opposite direction, causing the

Figure 8.14 Commercially available residential hydroelectric turbine. (Courtesy of Kvaerner Hydro Power Inc./Leffel Turbines.)

TABLE 8.2 SIZING TABLE OF THE TYPE USED IN TYPICAL RESIDENTIAL INSTALLATIONS THAT MATCH SYSTEM OUTPUT TO AVAILABLE HEAD AND FLOW[a]

Electrical capacity	Head (ft)	Style[b]	Water (ft^3/min)	Electrical capacity	Head (ft)	Style[b]	Water (ft^3/min)
½ kW or 500 W	8[c]	HL	104	5 kW or 5000 W	8	OT	760
	9[c]	HJ	92		9	OT	600
	10[c]	HJ	82		10	LR	590
	11[c]	F	74		11	LR	535
	12[c]	F	68		12	LR	490
1 kW or 1000 W	8[c]	IL	190		13	JP	470
	9[c]	HL	175		14	JP	435
	10[c]	HL	155		15	JP	400
	11	HL	140		16	JP	365
	12	HJ	127		17	JP	340
	13	HJ	118		18	JP	330
	14	HJ	110		19	JP	320
	15	HJ	105		20	JP	315
	16	HJ	100		21	HL	300
	17	HJ	94		22	HL	290
	18	HJ	90		23	HL	285
	19	F	84		24	HL	275
	20	F	80		25	HL	260
	21	F	76				
	22	F	74				
	23	F	72				
	24	F	70				
	25	F	68				
2 kW or 2000 W	8[c]	JP	330	7½ kW or 7500 W	11	OT	800
	9[c]	JP	290		12	OT	740
	10[c]	JP	260		13	LR	680
	11[c]	HL	245		14	LR	630
	12[c]	HL	225		15	LR	590
	13[c]	HL	215		16	JP	550
	14	HL	190		17	JP	515

	15	HL	178		18	JP	490
	16	HL	166		19	JP	480
	17	HL	156		20	JP	450
	18	HJ	153		21	JP	430
	19	HJ	148		22	JP	410
	20	HJ	140		23	JP	400
	21	HJ	133		24	JP	390
	22	HJ	127		25	JP	380
	23	HJ	120				
	24	F	116				
	25	F	110				
3 kW or 3000 W	8	LR	470	10 kW or 10,000 W	12	OT	980
	9	JP	415		13	OT	900
	10	JP	370		14	OT	840
	11	JP	340		15	OT	780
	12	JP	310		16	LR	715
	13	JP	280		17	LR	670
	14	JP	260		18	LR	650
	15	IL	250		19	JP	610
	16	IL	240		20	JP	580
	17	HL	225		21	JP	550
	18	HL	210		22	JP	525
	19	HL	200		23	JP	500
	20	HL	190		24	JP	490
	21	HL	180		25	JP	480
	22	HL	170				
	23	HL	165				
	24	HJ	162				
	25	HJ	158				

Source: Kvaerner Hydro Power Inc./Leffel Turbines.

[a] Standard rating for alternating-current units is three-phase 60 Hz and either 120 or 240 or 480 V. These Hoppes hydroelectric units may also be furnished for 50-Hz current.

[b] Style refers to various sizes of Hoppes hydroelectric units.

[c] Furnished in direct current only.

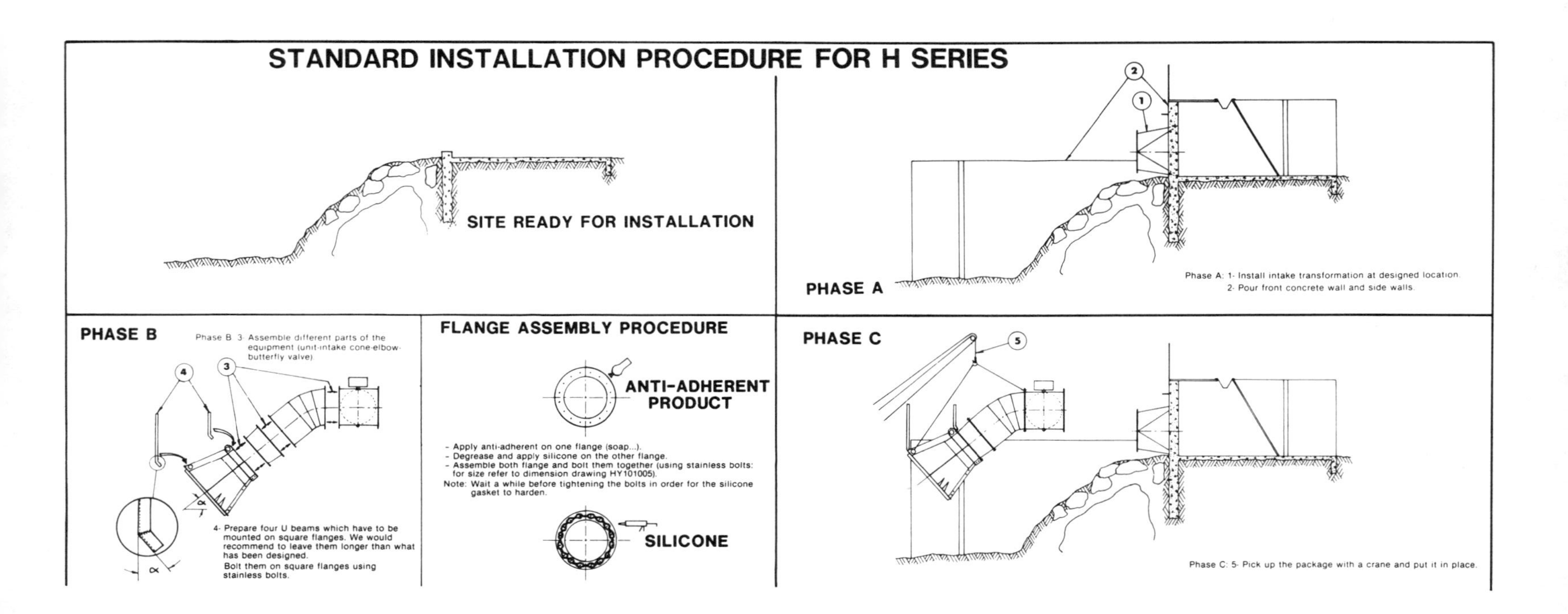

STANDARD INSTALLATION PROCEDURE FOR H SERIES
SITE READY FOR INSTALLATION
PHASE A
Phase A: 1- Install intake transformation at designed location.
2- Pour front concrete wall and side walls.
PHASE B
Phase B: 3- Assemble different parts of the equipment (unit-intake cone-elbow-butterfly valve).
4- Prepare four U beams which have to be mounted on square flanges. We would recommend to leave them longer than what has been designed.
Bolt them on square flanges using stainless bolts.
FLANGE ASSEMBLY PROCEDURE
ANTI-ADHERENT PRODUCT
- Apply anti-adherent on one flange (soap...).
- Degrease and apply silicone on the other flange.
- Assemble both flange and bolt them together (using stainless bolts: for size refer to dimension drawing HY101005).
Note: Wait a while before tightening the bolts in order for the silicone gasket to harden.
SILICONE
PHASE C
Phase C: 5- Pick up the package with a crane and put it in place.

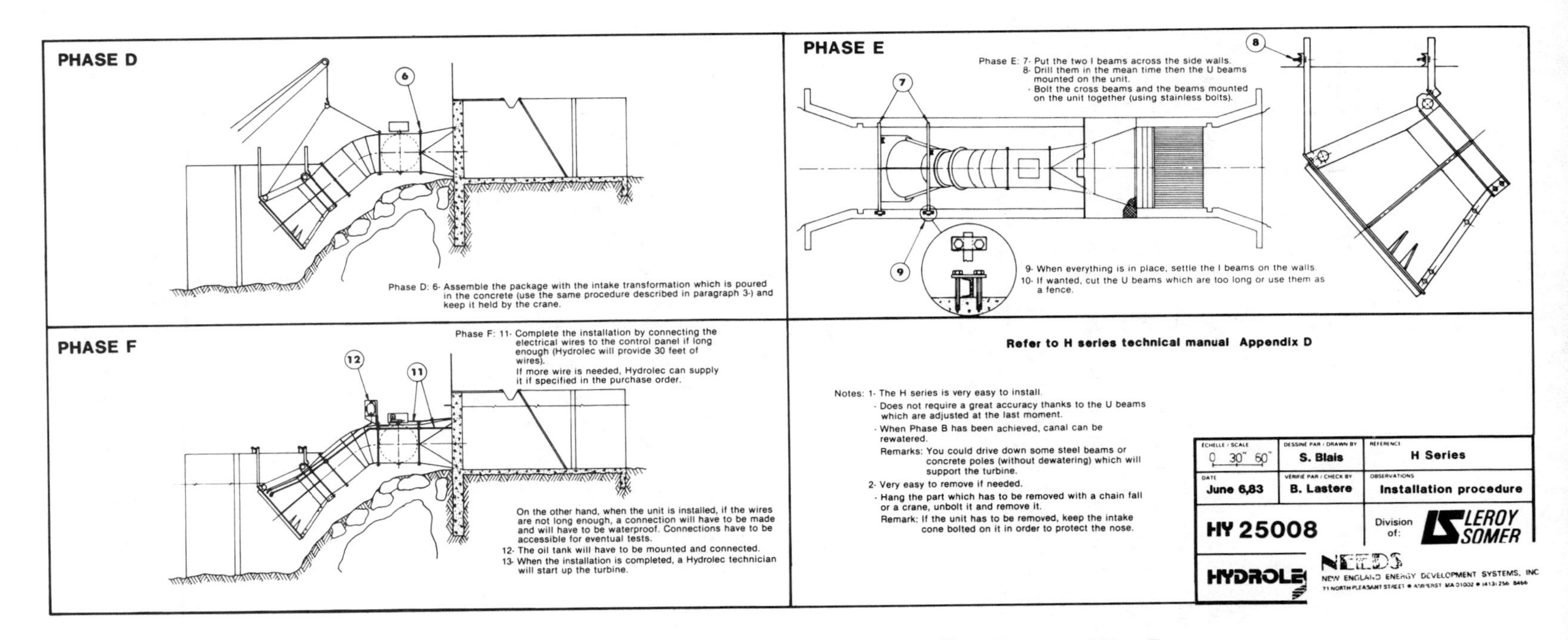

Figure 8.15 Site and installation plan for residential hydro facility. (Courtesy of New England Energy Development Systems, Inc.)

wheel to rotate. Since the velocity of water striking the bucket determines the operational characteristics of the wheel, and velocity is directly proportional to the head of the water source, impulse turbines require high water heads.

Reaction Turbines

A **reaction turbine** uses the weight of the water to rotate the turbine. Reaction turbines are either of the propeller type (Figure 8.13b) or the Francis type (Figure 8.13a). The reaction wheel turbine rotates as the water falls through a duct or pipe feed system onto the turbine wheel, which is confined within. Reaction turbines are used primarily in residential installations, since they operate well over a wide range of conditions and develop sufficiently high speed for direct coupling to a generator or alternator.

Commercially available residential hydro units can be sized by using tables that are prepared for specific turbines and generating plants. Table 8.2 is a table of this type, and the turbine that is described is illustrated in Figure 8.14.

SITE PLANS

After all the specifics of the installation have been determined, a site plan is drawn up that contains specific installation and design features of the system. Some plans, such as the one illustrated in Figure 8.15, illustrate not only the site details, but a step-by-step plan for installing the system.

Several firms produce a variety of residential-sized hydroelectric turbines and generators that will provide efficient, reliable power for virtually all sites that meet the prerequisites of adequate head and water flow. If the home has a water source that meets these requirements, these systems provide a highly reliable, cost-effective source of electrical power.

9

RESIDENTIAL WIND ENERGY SYSTEMS

HISTORICAL DEVELOPMENT OF WIND POWER

The use of wind for providing power dates back many thousands of years. The earliest recorded use of wind energy systems can be traced to ancient Persia, where windmills were used to both grind grain and pump water (Figure 9.1). These two uses have constituted the most widespread applications of wind-powered devices throughout history. In many parts of the world, water pumping windmills continue to be a familiar sight.

Although there have been many different design changes and innovations directed to wind energy systems over the years, the greatest impact on machine configuration took place beginning in the early twentieth century, when wind machines began to be used for generating electricity.

The wind machine with which most people are familiar is the American water pumping farm windmill (Figure 9.2). These windmills took on their present configuration during the nineteenth century. Made of galvanized steel, the water pumper is very reliable and durable (many are still in serviceable operating condition after 100 years of continuous use). The multivane rotor gives the unit very high starting torque in low wind speeds, which makes them ideal for pumping applications. However, the large number of rotor blades are aerodynamically inefficient at high wind speeds, when the large number of vanes tend to present

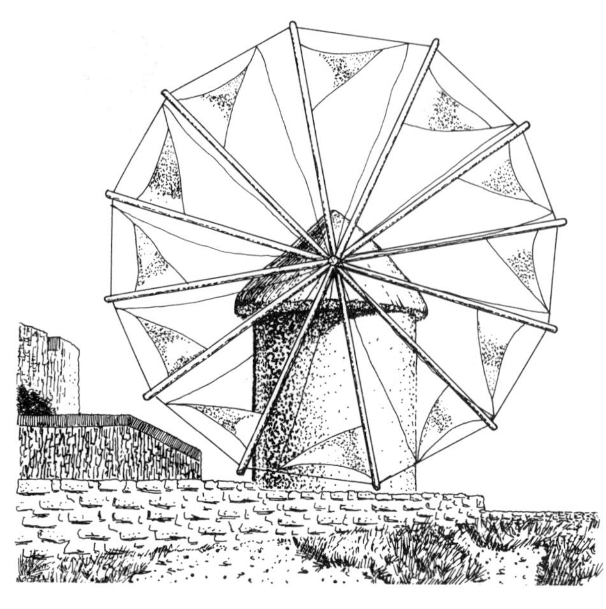

Figure 9.1 Design of early Persian windmill. The basic design of this machine has gone unchanged for centuries. Wind machines of this type are still in use today, providing power for a variety of mechanical tasks.

Figure 9.2 American water-pumping farm windmill. These machines provide the high torque in low wind speeds that is necessary for water-pumping applications.

a solid face to the wind. During these wind speeds, the rotor wheel can be pushed into the tower, resulting in total destruction of the machine.

The familiar design of the Dutch windmill lends itself to applications that also focus on grinding and water pumping tasks. Like its American water pumping counterpart, the Dutch windmill, graceful in appearance and design, is capable of providing significant horsepower to accomplish a variety of mechanical tasks; however, the machine is aerodynamically inefficient in high wind speeds and is therefore limited to low-rotational-speed, high-torque applications.

The dawn of the twentieth century, marked by the widespread use of electrical power, gave rise to a new generation of wind energy devices. Whereas the cities were the first to experience the benefits of centralized generation and distribution of electricity, the rural areas and farms depended on gasoline generators and battery storage systems to supply the electrical power necessary for their needs. These early systems were often sized at 32 V and provided only minimal power for radios and other small appliances. It was in the rural market that the earliest wind generators found their widest use. They were produced in 32-V output sizes to match the existing small home generators and wiring systems. A typical wind generator of this type is illustrated in Figure 9.3. Note from the illustration that this machine features a three-bladed rotor design. The blades are aerodynamically designed and balanced to be highly efficient in a variety of wind regimes, and capable of producing relatively high rpm rates in winds up to approximately 40 mph with automatic feathering devices for overspeed protection.

The use of aerodynamically shaped blades allows machines to develop very high tip speed ratios,* which are necessary for the production of electrical power. These units are either of the direct-drive or gear-driven configuration (depending on whether the generator is directly attached to the rotor or driven by a gear train). These generator designs are discussed in more detail later.

First-generation wind machines were used primarily for battery charging and are operated in the following manner. When the wind blows, the power from the generator is directed through a control box and used to charge storage batteries, supply power to appliances, or both if the machine is generating enough power. During windless periods, the appliances are operated from the battery bank, which is recharged by the wind generator when there is sufficient wind.

These units were highly reliable and were often sold with extended warranties (some were warranted for life). For the most part they lived up to their expectations. Battery storage systems used on these generators are designed for deep cycling and last for many years if properly maintained.

Toward the middle of the twentieth century, 220/110-V power became the normal household service, replacing the outdated 32-V systems. Wind generator manufacturers adapted to this change and began to sell their units in the 110-V configuration. By this time, however, the Rural Electrification Administration

* **Tip speed ratio** refers to the speed of the blade at its tip in relation to the speed of the wind stream through the blade. For example, a blade with a tip speed of 50 mph in a wind stream of 10 mph has a tip speed ratio of 5:1.

Figure 9.3 Electrical wind generator, circa 1930.

had brought centralized utility power to most rural areas. With the coming of centralized power, the major market for wind generators disappeared.

The American wind energy industry went into a long period of dormancy, which extended from post–World War II years to the mid-1970s, by which time most companies that had been producing wind generators were out of business. Those companies that stayed in business produced few machines or had diversified product lines that enabled them to survive the downturn in demand for their wind generators. It was not until the oil and gas disruptions of the late 1960s and early 1970s that wind energy was again examined as one possible alternative to expensive oil-, gas-, and utility-produced electrical power.

Wind energy, like many alternate energy sources, is highly dependent for its success on the price structure of the fossil fuels and energy rate structures that it replaces. Federal tax incentives coupled with oil and gas prices that rose exponentially during the 1970s and early 1980s helped the industry regain a partial foothold in the market in which it was once dominant. Whether the industry can survive without tax incentives and stable fossil- and solid-fuel prices is questionable. The economics of wind-generated electricity can be beneficial, depending on alternative fuel price structures.

SITE ANALYSIS

Before installing any wind system, a detailed site analysis must be made to determine the suitability of the location. This site analysis consists of the following components:

1. Visual inspection of the site for topographical considerations
2. Calculation of prevailing wind speed
3. Determination of power availability at the site

Figure 9.4 illustrates the availability of potential energy from the wind within the continental United States.

Visual Inspection and Topographical Considerations

Proper siting of a wind system is critical for efficient system functioning. The machine must be located in an area where it has free unobstructed access to the wind.

As a general rule, the wind generator should be situated at least 50 ft higher than any structure within 500 ft of the generator. This assures that there will not be any turbulence from nearby trees or buildings which might otherwise obstruct the wind. If there are buildings that are near the generator site, the 500-ft minimum should be sufficient to minimize wind turbulence.

Height is a significant factor in overall generator performance: The higher up you go, the greater the availability of wind. If possible, the machine should be placed on a tower with a minimum height of 100 ft, for optimum performance. Although it may not always be possible to erect a 100-ft tower, this height will almost always optimize the performance of the machine for maximum winds available at a particular site.

Other site factors that affect wind generator performance relate to topographic considerations, such as nearby hills and valleys. For example, machines placed on hilltops without nearby obstructions are almost certain to perform well. However, if a location affords a valley between two hills or mountains, a wind channel is often established in the valley, and the machine might perform better in the valley than on the hilltop. This can only be determined by measuring the wind speeds at each location.

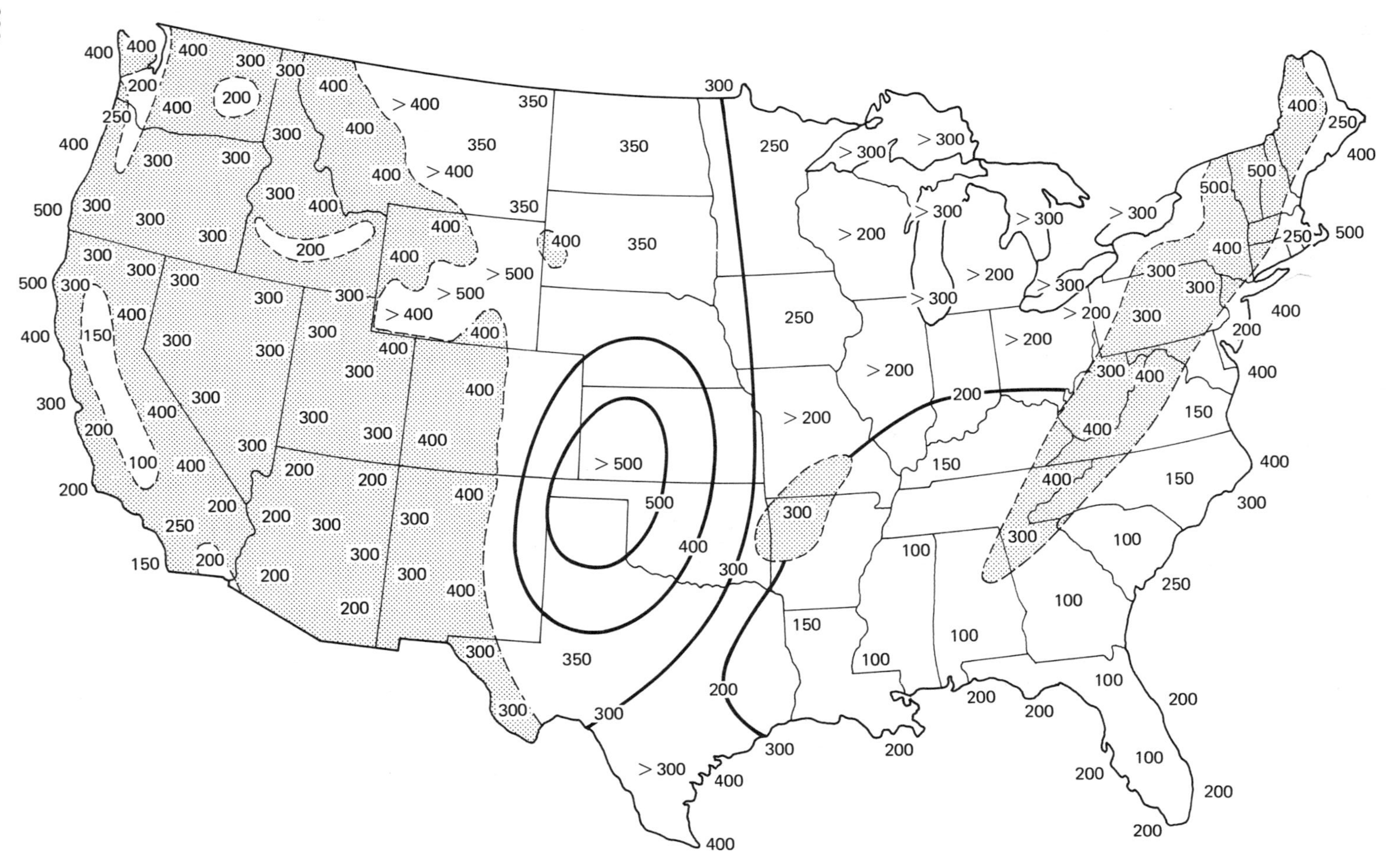

Figure 9.4 Annual wind power estimates for the continental United States. Figures listed are in watts/square meter (W/m^2). (Courtesy of Omnion Power Engineering Corp.)

Measuring Wind Speeds

There are many ways to measure wind speeds. The first procedure involves gathering any available data from the U.S. Weather Service measured as close as possible to the proposed site.

The National Climatic Center in Asheville, North Carolina, has computerized records of wind data for approximately 1000 locations throughout the United States. To take advantage of this service, the closest locality for which data are available is requested, and the Center will send an analysis of data for a nominal fee.

Local area airports are another source of wind data. The airports may often have a logged record of wind speeds available for the entire year and can be a valuable source of information regarding local wind and climatic conditions. Also, local radio and television stations can sometimes provide local wind speed information abstracted from news programs and Weather Service information files.

Although wind speed data may be available for a location as close as 1 mile to the proposed wind generator site, prevailing winds are very site specific and as such must still be measured at the exact location of the proposed generator. Variations of wind speeds between two locations 1 mile apart may be significant enough to determine whether a proposed wind system is or is not going to be cost-effective. Local wind speed information is helpful in establishing the parameters under which the machine may be expected to operate, but specific energy potential must be calculated at the site.

Site analysis focuses on two types of winds: prevailing winds and energy winds. Prevailing winds are those that blow on the average of 4 to 6 hr/day, for between 4 and 6 days per week. Energy winds are those winds over 12 mph. These categories are not mutually exclusive. In some areas, energy winds might be prevailing winds. However, it is important to quantify the energy winds available at a particular location. There is relatively little energy available in wind speeds below 12 mph. If a proposed location has prevailing winds between 8 and 10 mph, it does not have sufficient energy winds available to make the installation cost-effective. Conversely, if the prevailing winds at a location are 12 to 16 mph, it can be assumed that the site is worthy of further consideration.

Wind is not only variable on a day-to-day basis but on a seasonal basis as well. Most detailed site analyses should be taken over the period of a year to determine generator output accurately.

The key to reliable site analysis data is to take wind speed measurements as often as possible. An **anemometer** is the device used to measure wind speed and is available in a variety of configurations. The simplest type of anemometer is the hand-held unit (Figure 9.5). Although these units are useful for measuring the wind speed at any given time, they are not suitable for long-term detailed site analysis studies, since readings must be taken manually throughout the day and night.

Most anemometers used for long-term site analysis studies are micropro-

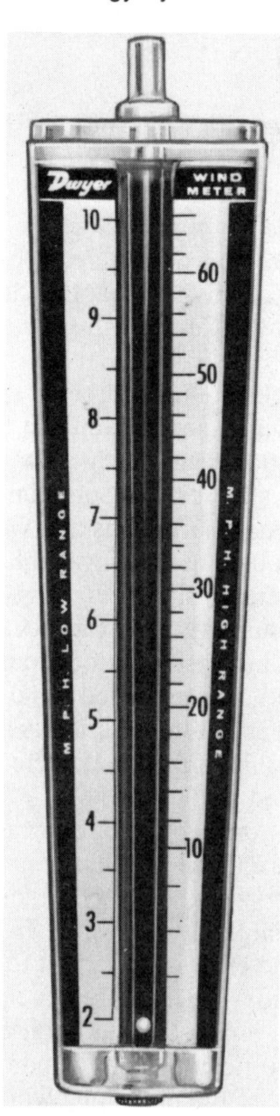

Figure 9.5 Hand-held anemometer, useful for taking spot wind speed measurements. (Courtesy of Dwyer Instruments, Inc.)

cessor driven and range in complexity from simple data-gathering capability to those units that can store and interpret data and transmit it to remote computer terminals for further analysis. One type of logging anemometer, which stores data for up to one year, is illustrated in Figure 9.6. This unit is battery driven and works by measuring the total amount of wind that has blown past the anemometer cups. Each given quantity of wind is stored as a pulse in the memory system of the unit. By dividing the total number of seconds that the anemometer is in operation by the total number of pulses measured, average wind speed can be obtained over the time during which the measurements have been taken. Although this type of instrument is more accurate than a hand-held unit, it only averages

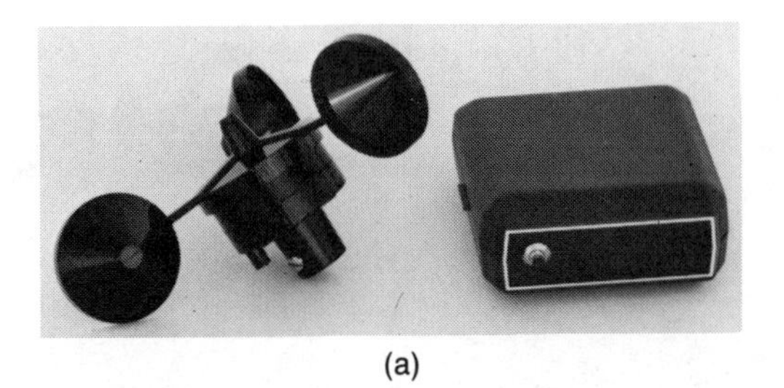

(a)

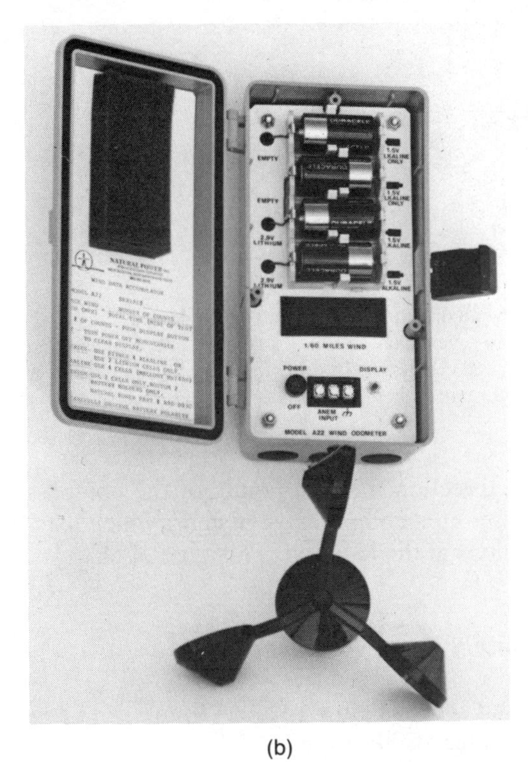

(b)

Figure 9.6 Logging anemometer: (a) indoor model, operating on 120 V ac, featuring updated wind speeds every 2 seconds shown on digital display; (b) outdoor model, rated for harsh, exposed conditions. This unit runs on alkaline or lithium batteries for 3 years of unattended operation. (Courtesy of Natural Power, Inc.)

the wind speed at the location over a given period. Wind generators are highly sensitive to gusts of wind that can generate significant quantities of electrical power, even though they may last for only hours or minutes. These energy winds would not show up on an averaging type of anemometer, but will only be averaged in with times of the day when little or no wind is available. This leads the observer to the conclusion that there is less power available at the site than there actually is.

To remedy this shortcoming, a more sophisticated anemometer might be used. The anemometer illustrated in Figure 9.7 is capable of measuring the actual wind speeds at the location in discrete categories (0 to 2 mph, 2 to 4 mph, 4 to 6

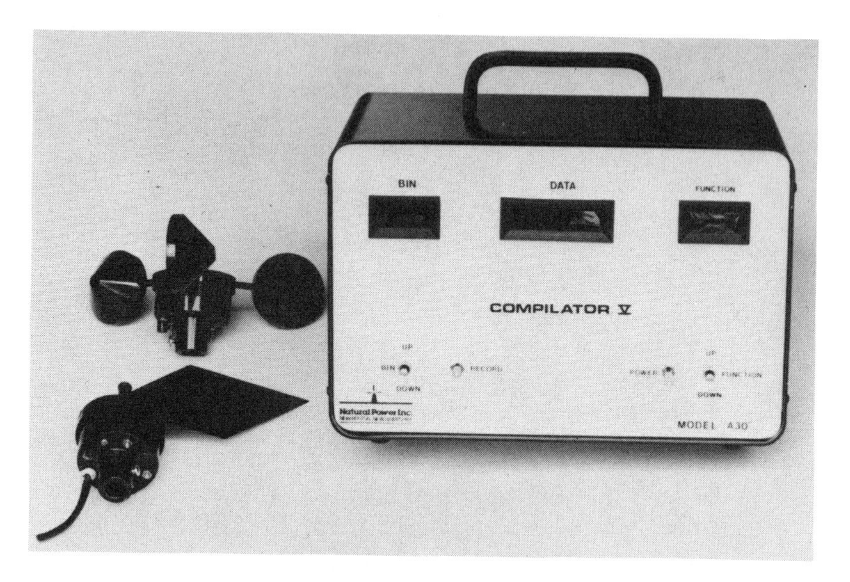

Figure 9.7 Wind speed/wind direction compilator. These devices are capable of receiving inputs from up to seven anemometer heads, and feature adjustable sampling periods and wind speed categories. Information from this unit can be transferred to computer casette tapes for further processing and analysis. (Courtesy of Natural Power, Inc.)

mph, etc.), as well as wind direction from 16 points of the compass. With this information it is possible to set up a comprehensive chart which lists the various wind speeds and wind directions at the location. This type of chart, called a **wind rose**, is illustrated in Figure 9.8.

Determining Power Availability

When analyzing data, all wind speeds below 10 and 12 mph can be ignored since relatively little energy is available at these speeds. Winds above 35 to 40 mph can also be ignored, since most wind generators are designed to feather or shut down at these speeds to avoid potential rotor damage. Concentrating on the useful energy winds can give a reasonable idea of potential energy available at the location.

Yet another type of wind energy measuring device is one that allows the anemometer to be programmed to the output characteristics of almost any wind generator output curve and displays the theoretical output of a particular machine under wind conditions at the specific site.

Once all available data have been gathered, it is interpreted to determine wind generator electrical output and the economic benefits to be derived from the proposed installation.

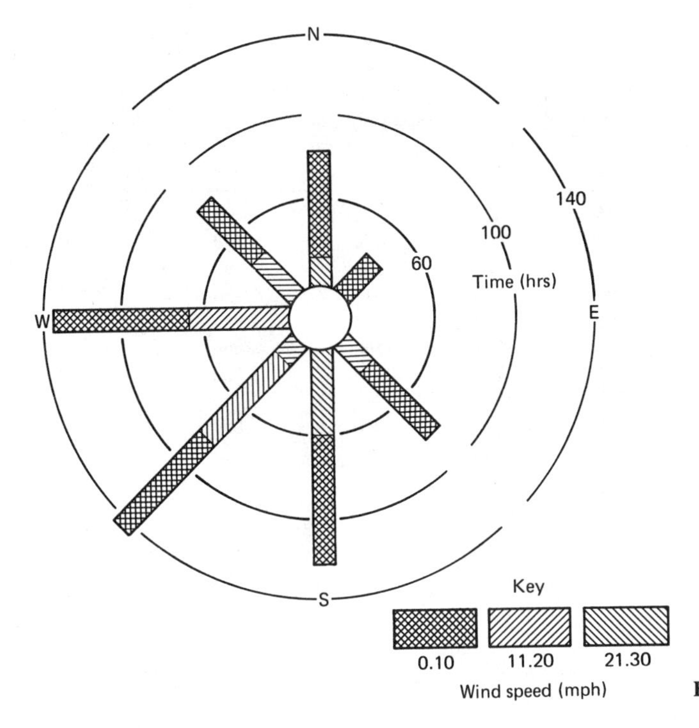

Figure 9.8 Wind rose.

DETERMINING GENERATOR OUTPUT

The potential energy available from the wind is a function of both available wind speed and the diameter of the propellers of the wind generator. Table 9.1 lists the available wind-generated energy, in watts of electrical output, based on a variety of wind speeds and propeller diameters.

Table 9.1 presents some interesting phenomena regarding wind speeds and blade diameter relationships. Two rules govern available electrical output from wind generators: the law of the square of the diameter of rotor blades and the law of the cube regarding wind speed. Each of these phenomena is now examined in more detail.

Blade Diameter versus Generator Output

Table 9.1 illustrates the relationship between blade diameter and generator output. For example, the output of a 6-ft-diameter propeller at 5 mph is approximately 5 W, while that of a 12-ft-diameter propeller at the same wind speed is approximately 21 W. Note that this relationship exists at a variety of wind speeds for both-size propellers: The power available at twice the diameter of a machine is approximately four times that of the smaller diameter. Thus the available power of a wind

TABLE 9.1 WINDMILL OUTPUT (W) AS A FUNCTION OF WIND SPEED AND PROPELLER DIAMETER

Propeller diameter (ft)	Wind velocity (mph)					
	5	10	15	20	25	30
2	0.6	5	16	38	73	130
4	2	19	64	150	300	520
6	5	42	140	340	660	1,150
8	10	75	260	610	1,180	2,020
10	15	120	400	950	1,840	3,180
12	21	170	540	1360	2,660	4,600
14	29	230	735	1850	3,620	6,250
16	40	300	1040	2440	4,740	8,150
18	51	375	1320	3060	6,000	10,350
20	60	475	1600	3600	7,360	12,760
22	73	580	1940	4350	8,900	15,420
24	86	685	2300	5180	10,650	18,380

Source: Adapted from Carol Hupping Stoner, ed., *Producing Your Own Power*, Rodale Press, Emmaus, Pa., © 1974, Table 1, p. 17. Permission granted by Rodale Press, Inc., Emmaus, PA 18049.

generator is proportional to the square of the diameter: If you double the size of the propellers, power output increases by a factor of approximately 4.

Law of the Cube

Note from Table 9.1 that power availability increases with increasing wind speed. However, the relationship is not a simple geometrical progression but increases with the cube of the wind speed: If you double the wind speed, the available power increases by a factor of 8. Although the theoretical output of a wind generator should increase by a factor of 8 when doubling wind speed, most systems fall short of this ideal, due to the inefficiencies of the blades, generator, and line losses associated with the installation.

Most residential wind systems use blade diameters between 12 and 24 ft. The majority of these systems are in the 12- to 14-ft category. From Table 9.1 we see that with wind speeds below 10 mph, a machine equipped with a 14-ft-diameter propeller generates relatively little electrical power. However, between 15 and 20 mph, available power increases significantly. For most installations, 12-mph prevailing winds should be considered the minimum required for a proposed wind electric system to be cost-effective.

MATCHING POWER REQUIREMENTS WITH GENERATOR CAPACITY

The underlying principle that applies to the installation of any alternative energy technology is to install the smallest system necessary to obtain the maximum benefits. This principle makes the installation highly cost-effective while spending the least amount of money. Sizing the system involves first calculating the power requirements of the residence. After this has been determined, and if the site analysis indicates a reasonable amount of power potential at the location, a variety of machines can be considered based on what the homeowner can afford and which unit will yield the most cost-effective installation.

Determining Residential Power Consumption

There are two methods that can be used to determine the expected or actual power consumption of a residence: either actual meter readings on a month-by-month basis or an examination of all the appliances in the home along with their power consumption multiplied by their actual or expected usage each month. Once residential power consumption has been calculated, the output of the wind generator based on the available wind speeds is compared to consumption to determine the energy contribution of the wind generator on a month-by-month basis.

Table 9.2 lists many household electrical appliances together with their estimated energy consumption in kilowatthours (kWh) on a yearly basis. Although most appliances operate continuously throughout the year, some, such as room air conditioners, do not. Therefore, it is a good idea to single out those appliances, such as heaters, room air conditioners, and window fans, that are seasonal in nature and then calculate the monthly power consumption.

If the wind generator will be installed in an existing residence, the easiest way to determine energy consumption is to examine the utility bills for the preceding year, which will give an accurate month-by-month indication of electrical consumption.

After the power consumption of the residence has been calculated, it is good practice to examine the electric bill, together with a list of all appliances in the home to determine where and how the electrical consumption can be decreased. The best way to decrease utility costs is to consume less. Although this may seem a bit obvious, the benefits are perhaps more subtle. For example, electric hot water heaters are expensive to operate. By switching from an electric to an oil-fired water heater, or solar-assisted water heater, monthly power consumption can be decreased significantly. Tactics such as these enable the installation of a smaller wind generator, resulting in less money spent per month on electricity, and a smaller, less expensive, and more efficient wind generator installation.

Matching Machine Output to Household Requirements

Each wind generator manufacturer publishes an output curve for each of their machines, which charts the electrical output of the unit based on a variety of wind speeds. Although actual output may vary from published figures (output curves have been known to be a bit optimistic), they are an indication of *approximate* generating capacity that can be expected from a particular machine. Figure 9.9 is a theoretical output curve for a 110-V ac generation equipped with field regulation.

TABLE 9.2 ENERGY REQUIREMENTS OF COMMON HOUSEHOLD APPLIANCES

	Average wattage	Est. kWh annually		Average wattage	Est. kWh annually
Air cleaner	50	200	Radio/record player	100	100
Air conditioner (room)	1,500	1,250	Range with oven	12,500	1,200
Blender	375	15	Range without oven	12,500	1,250
Broiler	1,250	100	Refrigerator		
Carving knife	100	5	12 ft^3	250	750
Clock	2	15	Frostless 12 ft^3	350	1,200
Clothes dryer	4,500	1,000	Refrigerator/freezer		
Coffee maker	900	100	14 ft^3	350	1,200
Deep fryer	1,500	75	Frostless 14 ft^3	600	1,800
Dehumidifier	240	375	Roaster	1,250	200
Dishwasher	1,200	350	Sandwich grill	1,000	25
Fan			Sewing machine	75	10
Circulating	100	45	Shaver	15–25	2
Whole house	300–500	300	Sunlamp	275	15
Window	200	175	Television		
Freezer 15 ft^3	350	2,000	Black and white		
Frostless 15 ft^3	450	1,750	Tube	150	350
Frying pan	1,200	150	Solid state	50	125
Hair dryer	175	15	Color		
Heater (portable)	1,000–2,500	150	Tube	300	700
Heating pad	35–100	10	Solid state	200	450
Heat lamp	250	15	Toaster	1,000	35
Hot plate	1,250	75	Toothbrush	7–10	1
Humidifier	175	175	Trash compactor	400–500	50
Iron	1,000	150	Vacuum cleaner	600	50
Mixer	125	15	Video recorder	35	50
Oven (microwave)	1,500	200	Waffle iron	1,000–1,200	20
Radio	75	75–100	Washing machine	500	100
			Waste disposer	400–600	30
			Water heater	4,500	5,000

Source: M. Greenwald and T. McHugh, *Practical Solar Energy Technology*, Prentice-Hall, Inc., Englewood Cliffs, N.J., 1985, Table 10.3, p. 227.

Note: Figures are approximations and may vary from one manufacturer to the other.

To expand this analysis further, Table 9.3 lists the average monthly output, in kilowatthours, of a variety of wind generators based on their factory rating. Using this information, a family with a monthly electrical consumption of 500 kWh would get approximately 40%, or 200 kWh, of electrical power from a wind generator installation, assuming 12-month prevailing wind speeds of 14 mph. Using a utility cost of 0.12 cent per kilowatthour, the generator in this instance would save the family approximately $288 per year, out of a total utility bill of $720 per year, as follows:

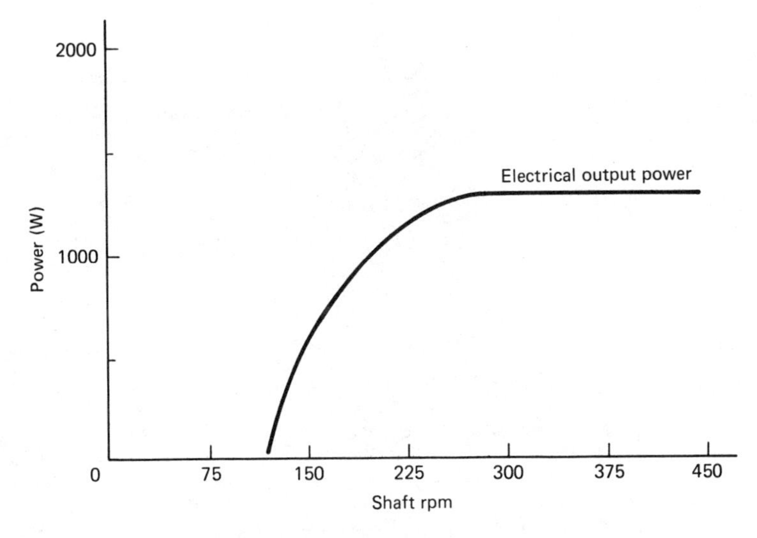

Figure 9.9 Wind generator output curve.

$$\text{yearly electrical consumption} = 500 \text{ kWh} \times 12 \text{ months}$$
$$= 6000 \text{ kWh/yr}$$

$$6000 \text{ kWh/yr} \times 0.12 \text{ cent/kWh} = \$720\text{/yr}$$

$$40\% \text{ of } 6000 \text{ kWh} = 0.40 \times 6000 = 2400 \text{ kWh}$$

$$\text{amount saved} = 0.12 \text{ cent/kWh} \times 2400 \text{ kWh} = \$288$$

To perform a return-on-investment (ROI) calculation, we must determine the costs of the following:

1. Total generator costs
2. Cost of the tower
3. Installation of electrical wiring
4. Cost of the storage system
5. Installation costs of generator, including site preparation, concrete, miscellaneous equipment

For purposes of illustration, we can assume the following costs of the proposed installation:

Cost of generator:	\$4000
Cost of tower:	\$1500
Wire installation:	\$350
Storage system:	\$2500
Installation of generator:	\$500
Total installed cost of system:	\$8850

TABLE 9.3 AVERAGE MONTHLY WIND GENERATOR OUTPUT (kWh) BASED ON WIND SPEED AND GENERATOR RATING

Nominal output rating of generator (W)	Average monthly wind speed (mph)					
	6	8	10	12	14	16
50	1.5	3	5	7	9	10
100	3	5	8	11	13	15
250	6	12	18	24	29	32
500	12	24	35	46	55	62
1,000	22	45	65	86	104	120
2,000	40	80	120	160	200	235
4,000	75	150	230	310	390	460
6,000	115	230	350	470	590	710
8,000	150	300	450	600	750	900
10,000	185	370	550	730	910	1090
12,000	215	430	650	870	1090	1310

Source: Adapted from Carol Hupping Stoner, ed., *Producing Your Own Power*, Rodale Press, Emmaus, Pa., © 1974, Table 2, p. 21. Permission granted by Rodale Press, Inc., Emmaus, PA 18049.

The ROI calculation is based on

$$\text{return on investment} = \frac{\text{first-year energy savings}}{\text{net system purchase price}}$$

Thus

$$\frac{\$288}{\$8850} = 0.035\% \text{ return on investment}$$

It can be seen from the calculations above that a 3.4% return on investment is not a very strong reason on which to base the decision to install the system. There are many other ways in which to invest the equivalent amount of money that would yield higher returns than installing a wind electric generating system. At the calculated rate of avoided electrical cost, the proposed system would take 50 years to pay for itself, not including annual maintenance and repair costs.

However, if the system costs approximately $10,000 to install, with an annual savings of $750 in electrical costs, the ROI would now be

$$\text{ROI} = \frac{\$750}{\$10{,}000} = 7\tfrac{1}{2}\%$$

A $7\frac{1}{2}\%$ return on investment makes the proposed installation highly cost-effective and worth considering.

It is usually advantageous to include escalating utility costs as well as a reasonable estimate of yearly maintenance expenses on both the wind generator and storage or conversion system to make the ROI calculations more accurate.

With an understanding of the basics of the economics involved in these installations, our examination moves onto design specifications of wind generators, support structures, and energy-conversion devices.

WIND MACHINE AND TOWER DESIGN

Wind generators and support towers, despite their seemingly uncomplicated and graceful appearance, are complicated engineering structures. Designed to operate in a variety of environments, they must withstand wind speeds in excess of 100 mph with very low mean time between failures (MTBF). They must do so in an economic climate that is competitive with utility-generated electrical power. We begin our discussion with the various generator designs available for residential applications.

Wind Generator Design

Although the early wind electric systems were almost always built around the dc generator, the modern wind machine is also available in an alternator version (similar in principle to the automobile alternator). Figures 9.10 and 9.11 illustrate the dc generator and alternator designs, respectively.

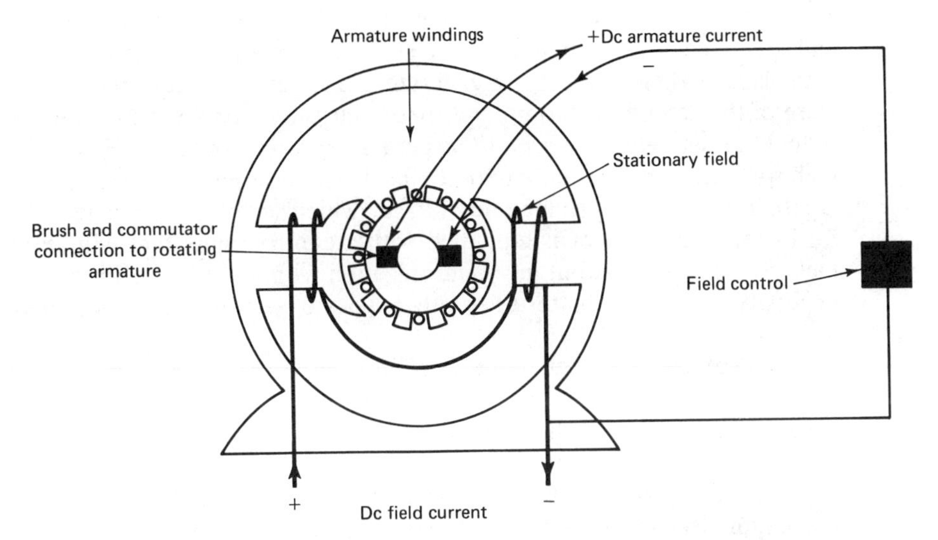

Figure 9.10 Dc generator basic wiring and configuration. Note that the field control circuit runs from the negative brush terminal to the negative field winding. As the rotational speed of the generator increases, so does the field current, increasing the electrical output of the generator. The field control circuit, normally located in the wind generator control panel, regulates the output of the generator, maintaining it within the safety limits of the unit.

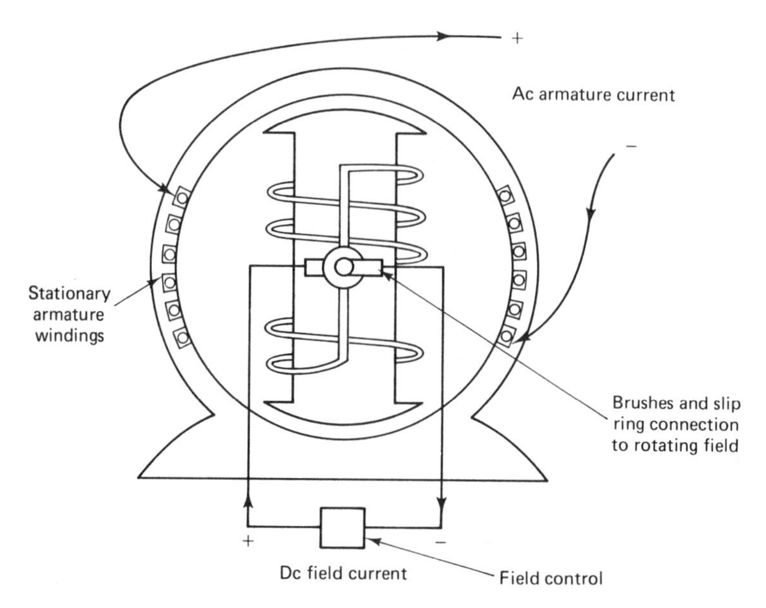

Figure 9.11 Alternator design configuration. Note the connection of the field control circuit to the rotating field circuit for alternator control.

There are advantages and disadvantages to both designs. Alternators are more efficient at producing electricity than are generators; they are also lighter and less expensive to manufacture. There are, in addition, no armature brushes to replace as there are with dc generators. However, most of the early machines were of the generator design, and these units are capable of maximum electrical output at relatively low revolutions per minute (rpm), whereas alternators require high rpm rates for maximum output. On alternator-equipped machines, the output is either rectified to produce dc voltage, which enables the machine to be used for battery charging, or is equipped with a control panel that will automatically lock the alternator output in synchronization with the utility grid, eliminating the necessity for storage batteries or other types of energy-conversion devices.

Direct and indirect drive. Wind generators are built in either the direct or indirect drive configuration. In the direct-drive unit, the blades of the machine are attached directly to the armature or alternator shaft. No gears are used in this system to step up the rotational speed of the generator or alternator shaft. For this reason, direct-drive units use alternators or generators that are designed for maximum electrical output at relatively low rotational speeds.

Indirect-drive units use gears, belts, pulleys, or transmission devices to increase the rotational speed of the blade shaft to a minimum speed required for efficient electrical output.

Due to the different rotational speed requirements of generators versus alternators, most alternators are belt or gear driven, whereas the direct-drive configuration is usually found with dc generators. These categories are not exclusive; there are direct-drive alternators available which have been engineered specifically for this type of application.

Direct-drive machines have a definite advantage over their belt- or gear-driven counterparts, in that they usually achieve maximum output at relatively slow rotational speeds (between 350 and 400 rpm). Low rotational speeds increase the life expectancy of the unit: The more slowly a device rotates, the less wear and tear it experiences, hence the longer its mean time between failures (MTBF).

The output of most wind generators is usually designed to peak at wind speeds between 25 and 30 mph. In practice, machine manufacturers rate their systems at maximum output, *which occurs at a specified wind speed, generally in the neighborhood of 22+ mph.* During normal operation, wind speeds of this magnitude are encountered only a small percentage of the time. Therefore, what is most critical for efficient system operation is the output of the generator at the wind speeds that will most often be encountered at the specific site. Generator and output curves should be analyzed very carefully prior to making any decision.

Horizontal and vertical axis designs. Wind machines are classified according to the axis of rotation of the blades: either horizontal or vertical, using the ground as a plane of reference. The horizontal-axis machine is by far the most common type of design (Figure 9.12). The vertical-axis machine (Figure 9.13) has been experimented with for many years, but was not investigated in depth with regard to electrical generation until recently. Horizontal-axis propeller-driven machines are slightly more efficient than their vertical-axis counterparts when rating machine overall output. Most vertical-axis designs that use aerodynamically designed blades are not self-starting and must be motored up to operate when the machine senses energy winds. Horizontal-axis propeller-driven units are self-starting.

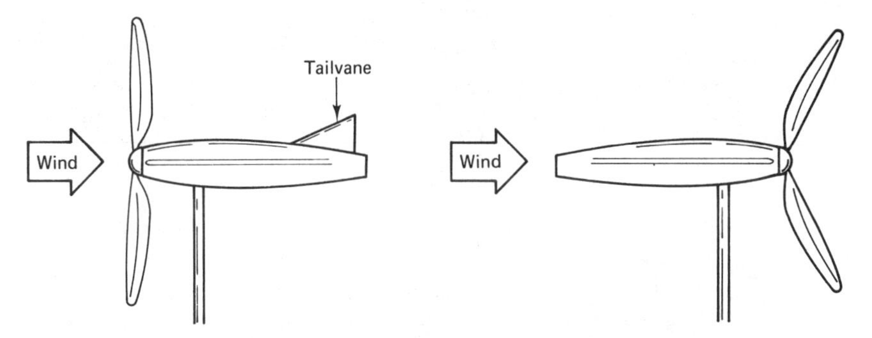

Figure 9.12 Horizontal-axis wind electric generators (typical). Horizontal-axis machines have the longest track record for successful use in electrical generating applications, dating back to the early twentieth century. These units feature self-starting rotors. The upwind design requires a tail vane to orient the blades into the wind, while the downwind design does not.

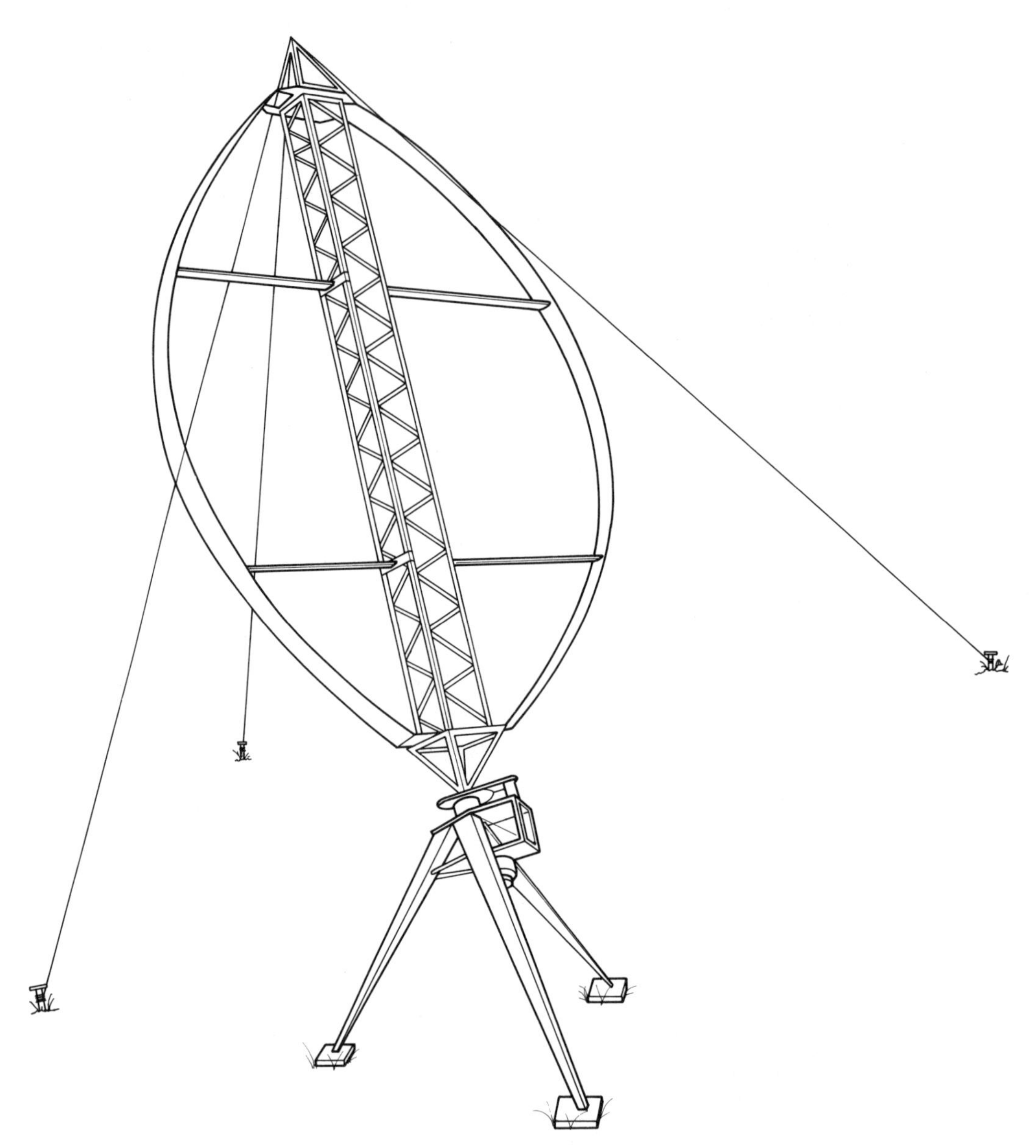

Figure 9.13 Vertical-axis wind generator design (Darrieus). Although this basic wind energy design dates back to the early twentieth century, applying it to the generation of electricity is a relatively new concept. Note that the generator is housed at the base of the system. Although these units are omnidirectional, requiring no orientation into the wind, they are not self-starting and require either auxiliary horizontal-axis starters, such as small Savonius rotors, or must be motored up to operating speed when energy winds are present.

Another difference between these two designs is the method in which the generator is turned into the wind. Vertical-axis machines are omnidirectional; that is, they respond to winds from any direction without the need to orient the blades of the machine in order to begin rotating. Horizontal-axis machines must be oriented into the wind. This is usually accomplished by using a tail vane, which will keep the machine perpendicular to the direction of wind flow (Figure 9.14).

Tail-vane-oriented windmills are upwind designs: The blades are upwind of the approaching wind stream and supporting tower; the downwind design (see Figure 9.12) uses the drag effect on the blades to orient the machine into the wind. The propellers are placed downwind of the tower. Although both designs have been used successfully, the upwind design has been more popular, since it also allows the machine to be completely turned out of the wind in excessively high wind speeds. This is accomplished by turning a crank on the tower, which turns the tail vane parallel rather than perpendicular to the propellers (Figure 9.15).

Maneuvering the tail vane in this way has proven to be highly effective in shutting down the generator in high winds which could otherwise destroy the machine. Different wind generators use different designs to prevent the rotor from overspeeding. If the rotor turns out of control, the machine can be destroyed due to excessive centrifugal forces built up by the blades.

To understand the principles of overspeed control, it is necessary to be familiar with the aerodynamic forces that act on a typical windmill blade.

Aerodynamic forces acting on windmill blades. There are two primary aerodynamic forces acting on a typical propeller blade: lift and drag (Figure 9.16). The principle of **lift** accounts for the fact that airplanes fly. Note from the illus-

Figure 9.14 Tail vane orientation of horizontal-axis wind generator, with tail vane straight out, maintaining position of blades perpendicular to the wind stream for generating purposes.

Figure 9.15 Tail folded, maintaining blade position of generator parallel to the incoming wind stream. In this position, the generator blades will turn only six or eight revolutions per minute, in essence keeping the unit shut down.

tration in Figure 9.16 that as the wind moves across the aerodynamically shaped blade, it splits into two directions. Wind moving across the top of the blade must travel farther than that moving below it. A low-pressure or suction area forms on the top side of the blade. A high-pressure area forms below the blade, pushing it upward. The **drag** effect on the blade results from the resistance of the blade in the wind stream and opposes the lifting force, which pushes the blade backward. In a properly designed windmill blade the lift forces are greater than the drag forces, causing the blade to move upward. On a typical horizontal-axis wind generator, the lift forces cause the blades to rotate about their axis on the generator or alternator shaft, generating electricity in the process.

To maximize lift forces, the blade is angled into the wind. This angle is referred to as the **angle of attack** and is designed into either the rotor of the machine or the blade itself. When the blade is mounted on the machine, it presents the angle to the wind that is necessary to maximize aerodynamic lift and begin the blade rotating in slow wind speeds.

Aerodynamic drag is a principle familiar to anyone who has ever stuck a hand out of the window of a moving car. The wind forces acting on the hand, pushing it backward, are drag forces. The typical wind anemometer operates on aerodynamic drag: The wind pushes the anemometer cups around a vertical axis. Where drag is used to rotate a closed flat shape rather than an aerodynamic propeller, the drag device can never move faster than the wind stream moving through it. It is this principle that makes the typical anemometer a highly reliable and accurate measuring device.

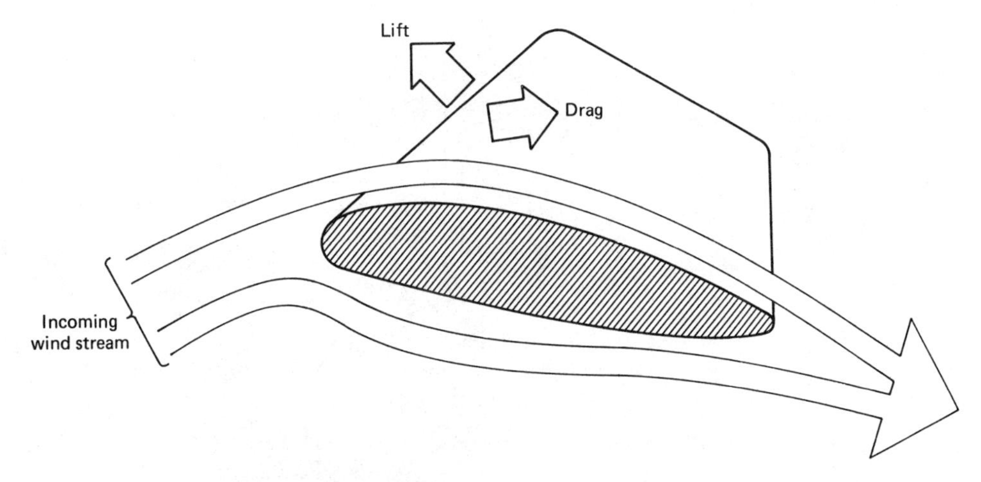

Figure 9.16 Lift and drag forces acting on an aerodynamically configured windmill blade. Note that the drag force is perpendicular and in opposition to the lift forces acting on the blade.

Overspeed control mechanisms. Residential wind generators are designed for optimum performance in wind environments ranging from 12 mph to approximately 40 mph. Although there is a great deal of potential energy in wind speeds greater than 40 mph, the destructive force at these wind speeds usually convinces most owners to turn off their units when these conditions are expected. This procedure does not mean, however, that the machine cannot operate safely in this wind environment. Overspeed control devices are designed into the machines to keep potential rotational speeds at safe levels in all wind speeds encountered in the operating environment.

Overspeed control is usually accomplished by devices that activate to feather or partially rotate the blades, which increases their angle of attack to the point where the blades lose their aerodynamic lift. This causes them to stall (much the same as an airplane will stall if it attempts to climb too quickly or at too steep an angle). In addition to feathering mechanisms, some machines are designed to lift the blades parallel rather than perpendicular to the wind stream (this procedure is sometimes referred to as "helicoptering" the blades), or to automatically fold the tail vane parallel to the blades when the wind speed increases to the feathering point. These design configurations are examined below in more detail.

Flyball Governors. The original and perhaps the least complicated method of overspeed rotor control is the flyball governor (Figure 9.17). The flyball governor is elegantly simple in design and operation. As rotor speed increases, the flyball weights begin to "fly out" in response to increasing centrifugal forces. This causes the gearing within the rotor to rotate the blade shafts, feathering the blades and causing the blades to lose lift, thus slowing the rotation of the rotor. Should the rotor rpm increase to the point where the flyballs swing out their

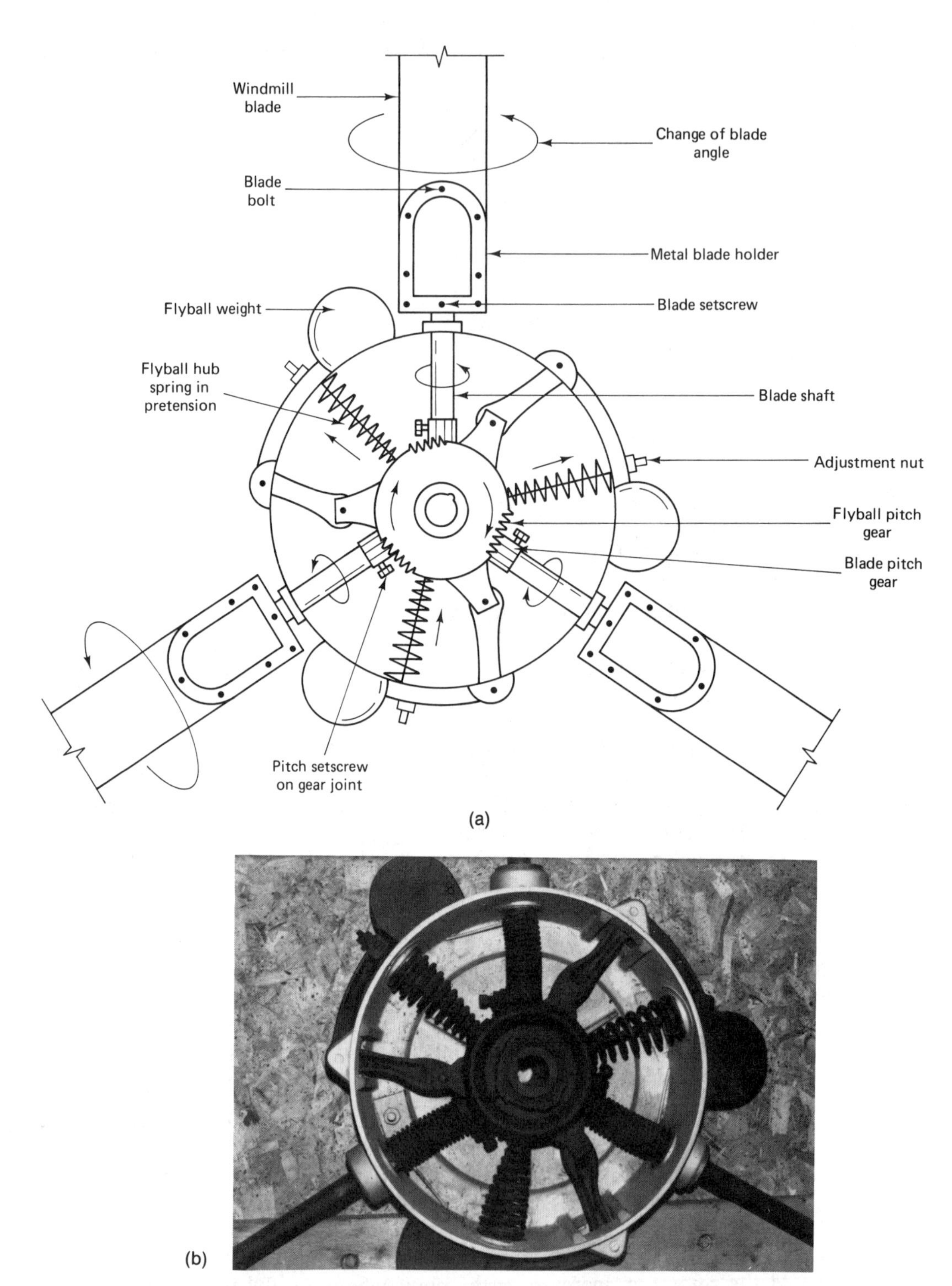

Figure 9.17 Flyball governor. Increasing rotational speed causes the weights to fly out due to increasing centrifugal forces, rotating the windmill blade. This rotation changes the angle of attack on the blade, decreasing its lift. If the flyball weights move out sufficiently, the blades will stall, stopping rotation of the wind generator until the unit resets as the wind speed decreases. This action is fully automatic and virtually maintenance free.

maximum amount, blade feathering is such that they will go into an aerodynamic stall, causing the rotor to stop turning. The flyball governor is highly reliable and requires little maintenance. It is usually constructed of cast steel or cast iron and weighs approximately 100 lb. The large mass of the rotor does make it less than ideally responsive to changing wind speeds, which is why the centrifugal governor was developed.

Centrifugal Governors. The centrifugal governor (Figure 9.18) uses the centrifugal force of the blades (rather than the flyball weights) to feather the rotor. Constructed from stainless steel and cast iron, these rotors are light in weight, reliable, and sensitive to changing wind gusts. As the speed of the rotor increases, the blades move out on their shafts due to centrifugal force. The blades are bolted to the shaft by a rotating ball joint and secured with a heavy-duty spring. The shaft rotates in response to blade movement, causing the angle of attack of the blades to change based on their distance of travel along the shaft. In high winds blade travel results in a feathered stall. As the wind speed dies down, spring tension brings the blades back down on the shaft, restoring the proper angle of attack of the blades, which enables the rotor to turn normally.

Rotor and Tail Vane Feathering. Some machines are designed to lift the entire rotor parallel to the wind stream, rather than feathering the individual pro-

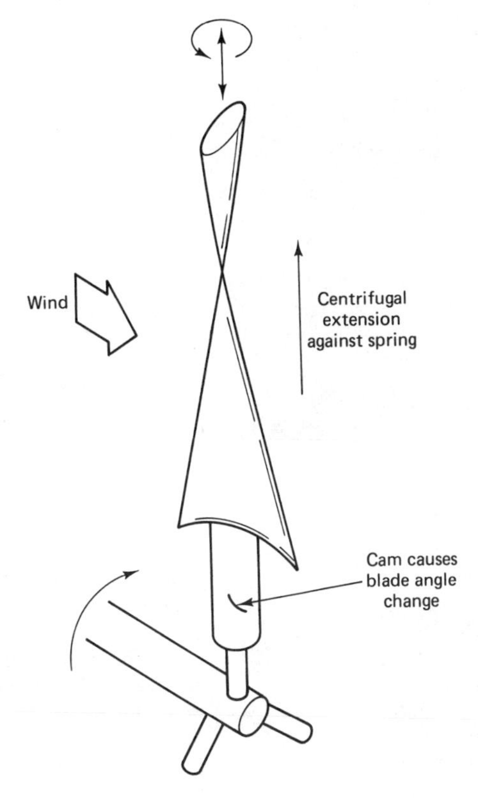

Figure 9.18 Centrifugal governor. This system uses the weight of the blades as the governing mechanism. As wind speed increases, the blades move out on the spring-loaded, cam-guided shaft, causing them to rotate and eventually stall, given the proper wind speeds. This governing device is lighter in weight than the flyball mechanism, making the unit more responsive to wind gusts and varying wind speeds.

pellers. This process, sometimes referred to as "helicoptering" the blades, is illustrated in Figure 9.19. Another variation of this principle is to furl the tail vane automatically as rotor speed increases. This is accomplished by placing the rotor off center, which creates a moment of rotation about the machine, causing the tail vane to begin to fall in toward the rotor assembly as the machine rpm increases.

All the methods described for overspeed control are reliable and time proven. The devices described here are not the only ones available for overspeed control. Some machines use electrical or mechanical braking devices; others use

Figure 9.19 Rotor feathering mechanism, which rotates the rotor and blade assembly parallel to the incoming wind stream. Note that in this system, the feathering mechanism achieves overspeed control by changing the axis of rotation in reference to the ground from horizontal to vertical. (Courtesy of Northern Power Systems, Inc.)

hydraulic mechanisms within the blade design to feather the blade tips. The feathering configurations most often encountered in typical residential wind generators, however, are those we have just examined.

Support Structures

All wind generators must be adequately supported on a structure strong enough to withstand both varying weather conditions as well as the stresses and forces exerted on it by the operation of the wind system. Most residential generators

Figure 9.19 *(continued)*

are installed on steel towers. Some, however, are installed on telephone poles, which can be hinged to allow the machine to be raised up or down, allowing service and maintenance to be performed on the machine at ground level.

Self-supporting towers. The self-supporting tower is the most common type of structure used for wind generators. Made from galvanized steel, these towers are classified as open truss designs, using a crisscross tensioned lattice assembly to maintain the integrity of the tower (Figure 9.20).

Self-supporting towers have many advantages over other types of structures: They are easily assembled and can be raised piece by piece; platforms are easily built on top of the tower that afford a convenient place to perform routine maintenance and repair procedures; and they are highly corrosion resistant, being fabricated from hot-dipped galvanized steel and galvanized fasteners. Most towers have a ladder built into one side made up of pegs or conventional rungs. The steps on most ladders can easily be removed to help prevent unauthorized access by strangers or young children (removing the bottom 10 ft of the ladder usually accomplishes this).

Figure 9.20 Four-post self-supporting steel tower.

An industrial-quality safety belt *must* be worn at all times when working on any windmill tower. These belts are designed to prevent accidental falls and have saved many lives. Failure to use a safety belt can be fatal.

Most self-supporting towers are bolted into steel angle-iron supports that are embedded in the ground, resting on either a concrete footing or pads located below the frostline.

Self-supporting towers are of either the three- or four-post design. Both three- and four-post designs are strong and durable and can be used interchangeably.

Normal maintenance on this type of tower involves touching up any areas of the tower where the galvanizing might scratch or wear away. Tower bolts should be checked periodically to make sure that they are tight (sometimes the bolts loosen up or crack due to machine and/or wind vibration). Also, the tower should be electrically grounded, using a ground wire and an approved ground rod. Since windmill towers are likely lightning targets, the ground rod is mandatory. The ground wire should be installed vertically from the tower directly to the ground rod without any sharp bends or curves and should extend 8 ft into the ground. If the tower is placed on a concrete pad, the ground rod should be located adjacent to the pad in direct contact with the earth.

Guyed towers. Some windmill towers use guy wires for support (Figure 9.21). Guyed towers are slender in design and feature a series of tensioned wires securely fastened to ground anchors. The tension on the wires is critical for proper support of the tower and should be checked periodically. Only heavy-duty aircraft-grade cable and turnbuckles should be used in securing these towers.

Guyed towers are less expensive than their self-supporting counterparts, although they can be used with equal confidence in most residential installations. These towers are also constructed from galvanized structural steel and fastening devices and should provide years of reliable, trouble-free service.

Pole support structures. The use of telephone-type poles for supporting wind generators is rarely recommended by machine manufacturers, although given the proper pole and installation procedure, they offer a less-expensive alternative to factory-built towers.

Most generators on poles fall into one or two categories: either guyed poles with fixed mounts or guyed poles with hinged mounts. In the fixed-mounted installation, pole spikes are used for climbing the pole, similar to those used on utility poles. Maintenance on any wind generator installed in this way is awkward, but possible.

To facilitate servicing the machine, some of these installations feature a hinged arrangement at the base of the pole that enables the pole to be raised or lowered from the ground using a heavy-duty winch or vehicle. This type of arrangement is illustrated in Figure 9.22. Most hinged poles are custom made and

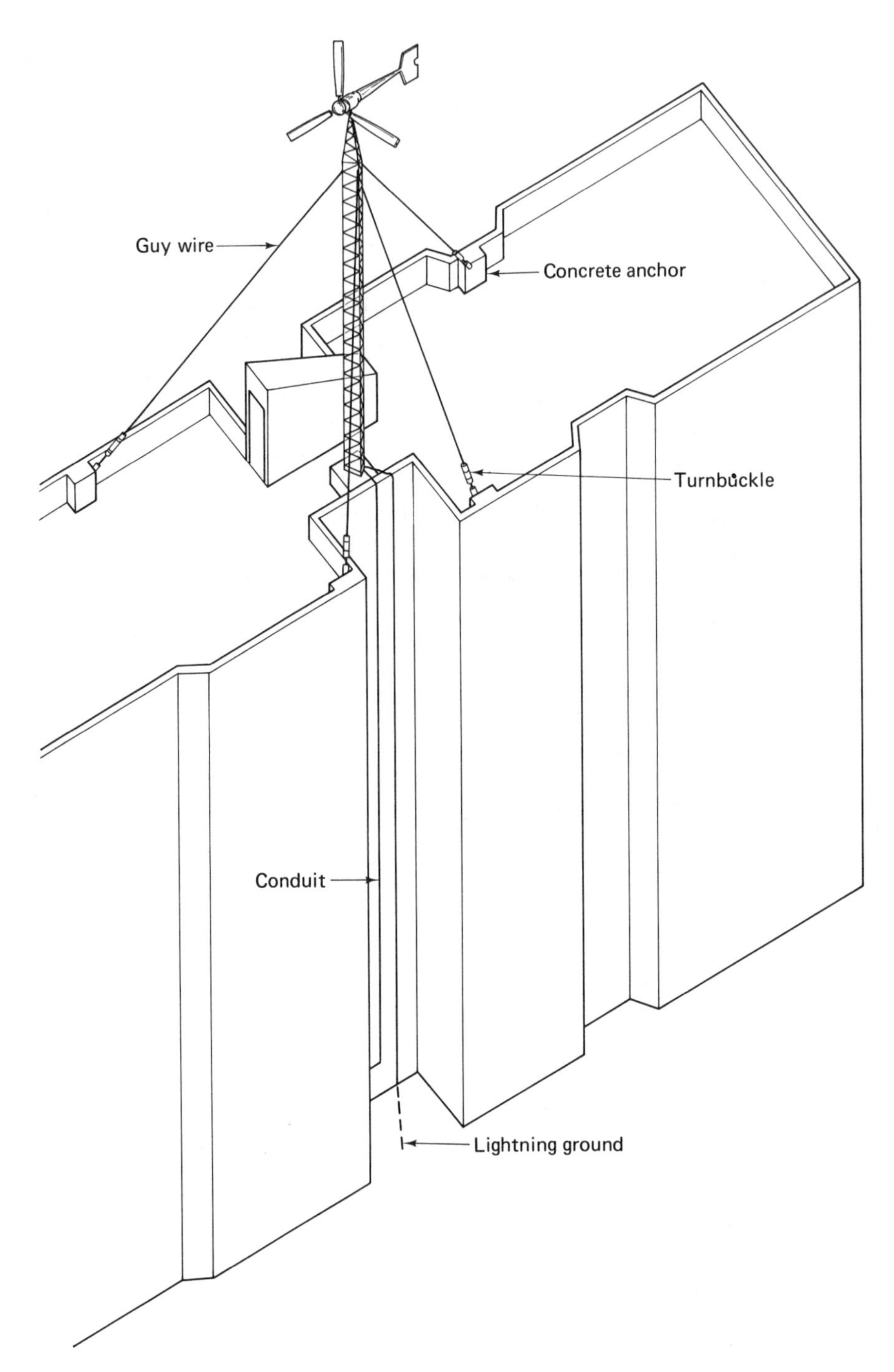

Figure 9.21 Guyed tower installed on top of a building. Whereas guyed towers can be used anywhere a self-supporting tower can be installed, the reverse is not always possible. Guyed towers are somewhat less expensive than their self-supporting counterparts. The guy wires must be inspected periodically and the tension adjusted properly to maintain the integrity of the installation.

(a)

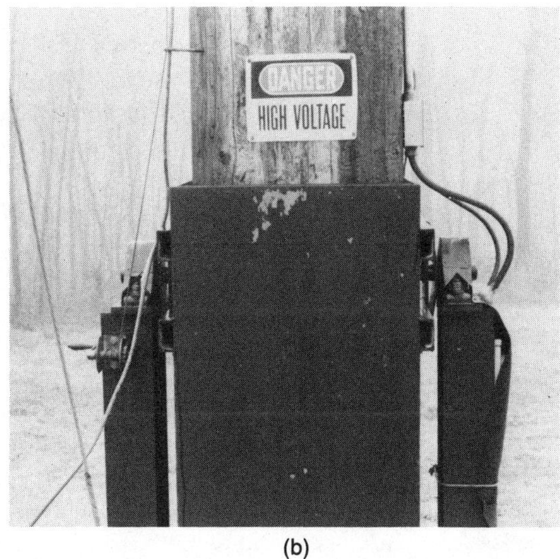

(b)

Figure 9.22 Hinged pole-mounted wind generator. (a) Note guy wire and gin pole assembly used to raise and lower machine. (b) Hinged assembly at base of pole. Base is constructed from heavy-gauge steel and pillow blocks.

should be designed and built only by people with engineering and construction abilities sophisticated enough for this type of installation. Installating the generator is easily accomplished by fabricating a conversion mounting on the top of the pole, which acts as a transition from the pole to the generator castings.

All pole mounts require the use of guy wires for rigidity and support. Also, the pole should be checked periodically to ensure that no cracks or splits develop which would undermine its integrity. Only those poles that have been pressure treated for all-weather installation and classified for utility service should be used. Most pole installations also employ guy wires located approximately every 20 ft of pole height to maintain rigidity of the pole and generator during operation. A ground wire should be installed from the generator to a ground rod adjacent to the pole for lightning protection.

ENERGY STORAGE AND CONVERSION

The power produced from a wind generator is usually stored or converted in one of three ways: stored in storage batteries, changed to synchronous line voltage by using a synchronous inverter, or generated by synchronous alternators in phase with utility power. The type of storage or conversion device used depends on the answer to several questions concerning energy supply and demand within the residence. For example, does the home have utility power, or is the generator the only source of electricity? How much money is available for a storage system? Will the wind generator be used in conjunction with some other type of energy storage or conversion system? We will look at the major components of various storage systems, along with some of the factors that might influence a decision as to which method is best in a particular situation.

Wiring and Utility Company Considerations

While all residential wiring falls under the auspices of the *National Electrical Code®* and interpretations by the local electrical inspector, wind generator wiring sometimes falls into a nondistinct category. In some situations, all the wiring between the generator and the generator control device falls outside of Code regulations; in other instances, depending on the local electrical and building inspector, it may not. Therefore, both the local utility and the electrical inspector must be consulted prior to undertaking the installation of any system, in order to plan for any wiring modifications and procedures that may be required.

Where possible, all wiring should be enclosed in rigid conduit, protected from the elements and from damage, and should be done in accordance with procedures outlined in the *National Electrical Code®* (see Chapter 1). If the installation is to include monitoring equipment such as anemometers and generator output sensing devices, the wiring for these devices should be run separately from that of the generator to prevent induced magnetic fields in the generator wire from interfering with the accuracy of the monitoring equipment.

If the wind system uses a generator rather than an alternator, higher system amperages are likely to be encountered; therefore, the wire sizing must reflect maximum current and voltage expectations together with a built-in safety margin included when sizing the wire.

Even if the homeowner rather than a licensed electrician is to do the wiring for the system, an electrician should be called on to inspect the completed wiring prior to a site visitation by the electrical inspector or utility company.

If the wind generator is to operate in conjunction with the local utility grid, the *utility must be notified* and consulted on every aspect of the system configuration. In battery storage systems where there is no direct or indirect interface with either the grid or house wiring (often a separate dc wiring circuit is run from the battery storage to supply dc and universally wired appliances), the utility will probably not play a role in either the installation or wiring of the system. However, when synchronous alternators or synchronous inversion devices are to be used, utility approval must be granted in order to hook up to the grid. Also, the utility will usually require a licensed electrician to work on any wind installation that involves an interface with the grid.

Federal law stipulates that the local utility must allow an interconnection of a wind generator (or small-scale cogeneration unit) with the grid; however, it is left to state and local authorities to determine and pass judgment on what is acceptable wiring, metering, and installation practices and approved equipment.

It is good practice to begin negotiations with utility officials well in advance of the expected date of generator installation. This will ensure that all equipment, wiring procedures and configurations, and metering practices will be decided on without last-minute surprises.

Since many wind generators operate on a feedback principle where excess power supplied by the unit is fed back into the grid, local laws and statutes will govern the payback rate by the utility to the homeowner. If the utility has been involved with wind generator installations before, they will usually have an information packet available to facilitate installation and metering procedures. This information consists of wiring guidelines, approved equipment lists, and a contract between the homeowner and utility company that spells out all procedures, payback rates, and the legal obligations of both parties.

Battery Storage Systems

Batteries have been used most widely for storing electrical power produced by wind generators. All early wind systems were designed around the battery storage systems that were used in conjunction with gas-driven electrical generators.

Wind generators require special batteries if the power is to be stored efficiently for extended periods. Batteries are classified in many different ways: cold cranking power, delivered power in ampere hours, and cycling characteristics. Each of these factors must be carefully weighed when designing a battery storage system for the wind system.

Battery storage basics. The conventional lead-acid battery is the type that is best suited for wind generator applications. The conventional battery consists of a series of lead-alloy plates that are suspended in an electrolytic solution (an electrolytic solution, or electrolyte, is a solution capable of conducting an electrical current). The plates are alternately charged positive and negative. All the positive plates are connected together, as are the negatively charged plates. Both the positive and negative plate connections end at two battery terminal posts. The capacity of the battery both to store and to deliver power is determined by the type and number of plates, their thickness, and overall size.

Batteries are also classified according to their cycling characteristics. A battery cycle describes the process of going from a fully charged state to a fully discharged state. Battery cycling characteristics are classified as light, medium, or deep cycle. Light-cycle batteries are those used in automotive applications. Medium-cycle batteries are used in electric vehicles. Deep-cycle batteries are those found in telephone systems and wind generator applications. In automobile batteries, for example, this cycling process can take place only a few times before the battery is no longer capable of retaining a charge. In medium-cycle batteries such as those used in electric vehicles, this cycling can occur several hundred times before the battery can no longer be charged. Deep-cycle batteries, sometimes referred to as house lighting batteries, are capable of sustaining many thousands of cycles with no adverse affect on battery performance. The thickness of the plates and their overall size, the space between the bottom of the plates and the bottom of the battery case, and the design and integrity of the plate insulators all contribute to the cycling characteristics of the battery. House lighting, deep-cycle batteries are far more expensive than their automotive counterparts, and in the larger ampere/hour capacities are commonly sold in a 2-V size.

During the cycling of any lead-acid battery, considerable amounts of hydrogen gas are given off. If a typical wind generator system contains sixty 2-V batteries (to obtain 120 V of power), provision must be made not only to accommodate the space required by the batteries but also to vent the hydrogen gas. Batteries are usually installed on specially constructed heavy-duty wooden racks. A vent line must be run from the battery room to the outdoors to prevent the buildup of hydrogen gas. Batteries should not be placed near appliances such as oil burners or gas-fired heating equipment. When batteries are stored in areas adjacent to this type of heating equipment, adequate ventilation must be provided in addition to the venting system for the battery storage area. Since storage batteries operate more efficiently at higher temperatures than lower ones, they should be located in a room that is temperature controlled. Many batteries have built-in charge indicators, such as floating pilot balls or color code solutions, that show the state of charge of the battery. If they do not, the voltage per cell will indicate its state of charge (see Table 9.4). When installing the battery system, space should be left between the batteries to arrange the wiring neatly and safely.

Sizing the storage system. Battery storage systems are sized according to the amount of power they must deliver to the home on a daily basis, governed by the charging rate of the wind generator and the number of days that the batteries can supply power to the residence without being charged. When this is determined, the ampere-hour capacity of the battery bank can be selected.

TABLE 9.4 PERCENTAGE OF BATTERY CHARGE, BASED ON CELL VOLTAGE MEASUREMENTS

Volts per cell	Approximate percent charge[a]
2.00	100
1.99	90
1.98	80
1.97	70
1.96	60
1.94	50
1.93	40
1.91	30
1.87	20
1.83	10
1.75	0

[a] Figures will vary based on temperature of battery.

The maximum charging rate of the generator should be approximately 10% of the ampere-hour capacity of the battery. For example, a generator with a maximum charging rate of 20 A would be well matched to a 200-A-hr battery. If the ampere-hour capacity of the battery is too large for the generator, the battery bank will not charge fully in a reasonable amount of time.

Once a decision has been made as to the ampere-hour capacity of the battery system, it can be determined how long the batteries can deliver power to the home without being recharged. A 150-A-hr battery storage system can deliver 1 A for 150 hr, 150 A for 1 hr, or anything in between. By averaging the electrical consumption in the house on an hour-by-hour basis, the length of time the storage system can furnish power to the home before becoming completely discharged can be calculated. In practice, 2 to 4 days of standby battery power is designed into the system. These figures will depend on the number of appliances in the home and their electrical current draw. Wind generators should not be expected to provide electric power for resistance space heating. The required draw on the system with this type of heating is far too excessive for any small-scale residential machine to furnish economically.

Battery-Driven Power Conversion

Storage batteries offer the following options for utilizing electric power: straight-wired dc power, rotary inverters, and electronic inverters.

Many appliances in the home are rated as **universal**; that is, they can operate on either ac or dc power. Universal electric motors will be identified on their nameplate. Virtually all resistive appliances such as irons, toasters, and lighting can use either ac or dc power. In these situations it makes sense to install a second wiring system in the home for dc only, connected directly to the battery storage bank. The outlets in this circuit should be modified and carefully labeled, so there is no confusion as to their voltage. All applicable codes apply to this type of wiring and should be thoroughly researched prior to installation.

A rotary inverter consists of a dc motor that is connected to an inverter. An inverter is a device that changes direct current (dc) to alternating current (ac). Since many devices such as televisions, radios, stereos, and many types of electric motors require ac electricity to operate, the direct current in the batteries must be changed to ac. The dc motor in the rotary inverter mechanically drives an alternator, which converts the direct current in the batteries to alternating current. Rotary inverters produce what is known as square-wave alternating current, as opposed to the sine-wave alternating current that is usually necessary for optimum appliance performance (Figure 9.23).

Although rotary inverters may be satisfactory for many ac appliances, their overall efficiency is limited to approximately 60 to 65%, limiting their effectiveness. A solid-state inverter that produces sine-wave ac solves this problem by providing power conversion electronically, greatly reducing standby power and conversion losses.

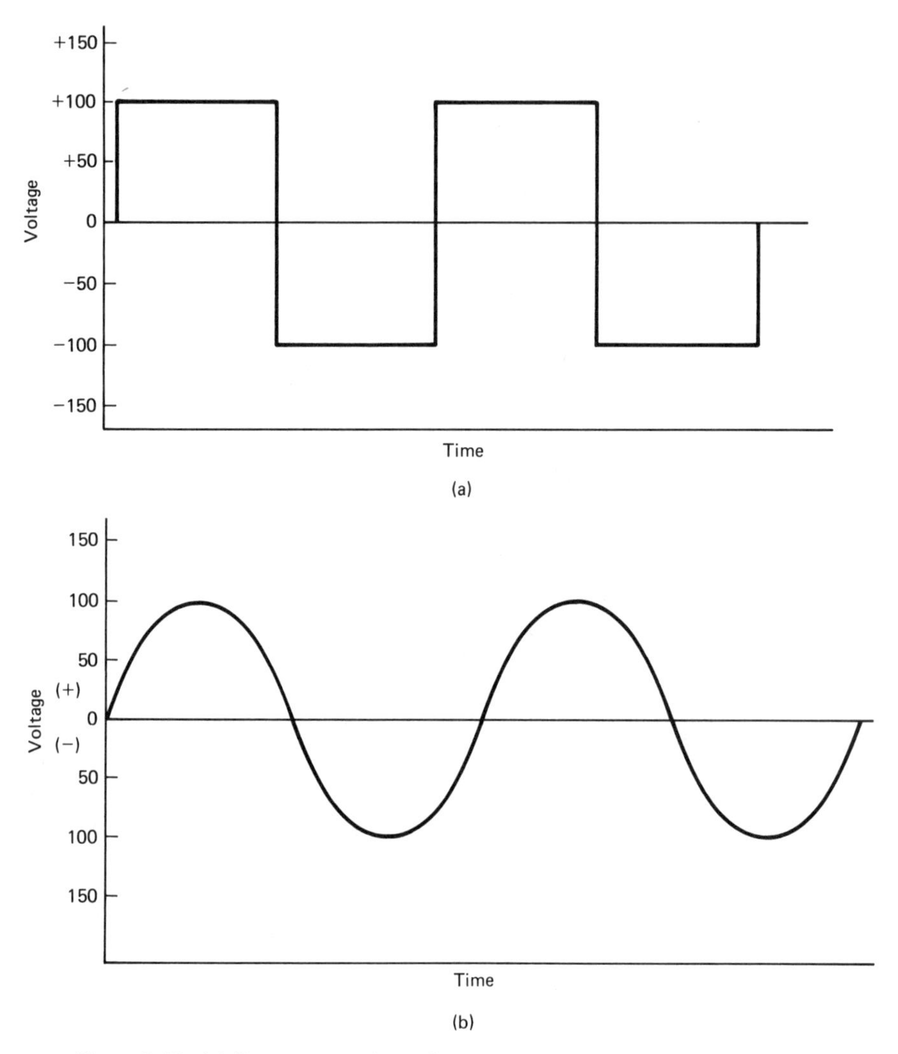

Figure 9.23 (a) Square-wave alternating current (two cycles shown); (b) sine-wave alternating current (two cycles shown). While square-wave alternating current is acceptable for various appliances, many, such as televisions and stereo receivers, require sine-wave power in order to operate properly.

A final consideration of battery storage systems relates to their cost: They are expensive. A 2-V 200-A-hr house lighting battery can cost anywhere between $350 and $600. Keeping in mind that 60 of these batteries are required for the average wind generator leads to a total battery cost above $18,000. This figure does not include the price of storage room accessories or their installation. It is sometimes possible to obtain these batteries from surplus businesses that deal with this type of equipment and can offer these batteries at significantly reduced prices.

Synchronous Generation and Inverters

The device that opened up the wind generator market to widespread residential use was the synchronous inverter. This device takes the variable dc input of the wind generator and inverts in into 220-V 60-Hz current in phase with the incoming utility electrical power (Figure 9.24). When the wind generator is operating, the

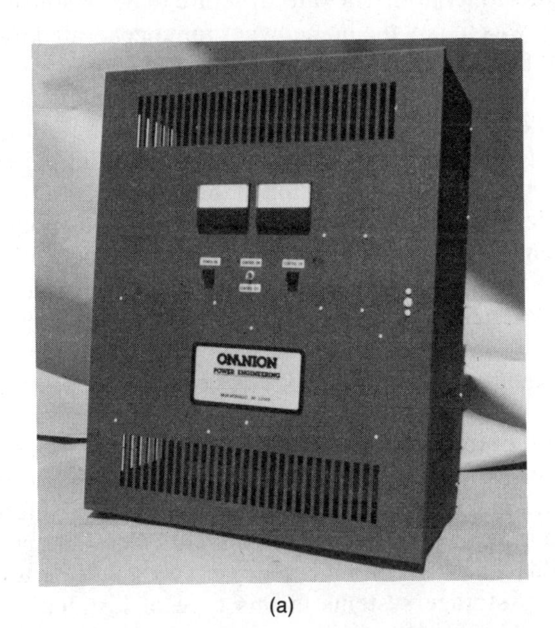

(a)

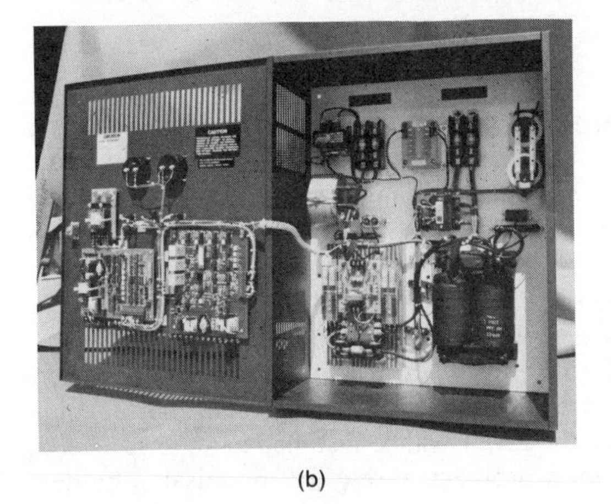

(b)

Figure 9.24 Solid-state synchronous inverter. These devices accept varying direct-current inputs and invert this to 60-Hz current, using the utility power as the phasing device. The use of synchronous inversion has greatly reduced the cost of installing residential wind energy systems. (Courtesy of Omnion Power Systems, Inc.)

synchronous inverter treats the grid as an infinite storage battery, allowing wind-generated electricity to be used either in the home or, in the case of excess generator capacity, to be fed back into the grid. In so doing, the household utility meter is turned backward, since net electrical flow is toward the utility from the residence rather than from the utility to the residence. As one would expect, this type of operation caused quite a flurry when it was first introduced. Several lawsuits regarding its nature of operation soon followed. It was quickly determined, however, that the individual homeowner had the right to cogenerate electricity while interconnected to the utility grid. The rate structure to be established, which stipulated what the utility was to pay the homeowner for cogenerated power, was left to the state public utility commissions to decide on.

The synchronous inverter works in harmony with the incoming electrical power and superimposes electricity from the wind system on the grid. One side of the inverter is connected to the wind generator; the other side is connected into the home circuit breaker panel. The operation of the inverter is fully automatic and operationally invisible to the homeowner. The inverter acts as both a control panel to regulate wind generator output and as a central switching station to feed cogenerated electricity to the household appliances or to the grid. In the absence of wind, all household electricity requirements are automatically supplied from the utility system.

Prices on typical synchronous inversion devices range from $3000 to $7500. The use of a synchronous inverter does not preclude the use of storage batteries as an auxiliary power source. In some instances the use of a small storage battery system designed to provide power for perhaps one day is advantageous since the synchronous inverter does not function during a power failure (a safety feature is designed into the synchronous inverter, which prevents electrical power from being fed into "dead utility lines" when emergency crews might be working on them). Also, small battery storage systems in this type of installation can use medium-cycle rather than deep-cycle batteries and can often be assembled for as little as $200 to $400.

MISCELLANEOUS CONSIDERATIONS OF WIND GENERATOR PLANTS

We have examined the necessity of obtaining required permission and clearances from the local utility and electrical inspections prior to undertaking a wind system installation. There are other considerations that must be taken into account prior to installing the wind generator, most of which deal with local authorities regarding building permits and the granting of zoning variances.

If the generator is to be installed in a rural residential area, the only requirements that might be applicable are restrictions on the maximum height of the windmill tower. Since the tower will be installed in an area where neighboring property or housing is not a problem, rarely will permission be denied to erect the unit. However, a detailed site plan as well as the procedures to be used in anchoring the tower might be required by the building inspector.

If, on the other hand, the unit is to be installed in a residential suburban area, the adjoining property owners must be notified and approval given by them before the installation can proceed. The proposed installation of many wind generators has been stopped due to objections of neighboring property owners that focus on what is viewed as either a potential hazard or the aesthetic appearance of the system. Since most zoning ordinances do not cover the installation of wind generators, their installation and use constitutes a variation of property use that requires the granting of a variance.

Careful consideration of all these factors will lead to a well-designed, cost-effective installation that can be counted on to supply a cost-effective alternative to utility-generated power.

APPENDIXES

APPENDIX A CONDUCTIVITIES, CONDUCTANCE, AND RESISTANCE OF BUILDING AND INSULATING MATERIALS[a]

Description	Customary Unit						SI Unit	
	Density (lb/ft^3)	Conductivity (k)	Conductance (C)	Resistance[b] (R) Per inch thickness ($1/k$)	Resistance[b] (R) For thickness listed ($1/C$)	Specific Heat, Btu/(lb) (deg F)	Resistance[b] (R) $\frac{(m \cdot K)}{W}$	Resistance[b] (R) $\frac{(m^2 \cdot K)}{W}$
BUILDING BOARD								
Boards, Panels, Subflooring, Sheathing								
Woodboard Panel Products								
Asbestos-cement board	120	4.0	—	0.25	—	0.24	1.73	
Asbestos-cement board 0.125 in.	120	—	33.00	—	0.03			0.005
Asbestos-cement board 0.25 in.	120	—	16.50	—	0.06			0.01
Gypsum or plaster board 0.375 in.	50	—	3.10	—	0.32	0.26		0.06
Gypsum or plaster board 0.5 in.	50	—	2.22	—	0.45			0.08
Gypsum or plaster board 0.625 in.	50	—	1.78	—	0.56			0.10
Plywood (Douglas Fir)[c]	34	0.80	—	1.25	—	0.29	8.66	
Plywood (Douglas Fir) 0.25 in.	34	—	3.20	—	0.31			0.05
Plywood (Douglas Fir) 0.375 in.	34	—	2.13	—	0.47			0.08
Plywood (Douglas Fir) 0.5 in.	34	—	1.60	—	0.62			0.11
Plywood (Douglas Fir) 0.625 in.	34	—	1.29	—	0.77			0.19
Plywood or wood panels 0.75 in.	34	—	1.07	—	0.93	0.29		0.16
Vegetable Fiber Board								
Sheathing, regular density 0.5 in.	18	—	0.76	—	1.32	0.31		0.23
0.78125 in.	18	—	0.49	—	2.06			0.36
Sheathing intermediate density 0.5 in.	22	—	0.82	—	1.22	0.31		0.21
Nail-base sheathing 0.5 in.	25	—	0.88	—	1.14	0.31		0.20
Shingle backer 0.375 in.	18	—	1.06	—	0.94	0.31		0.17
Shingle backer 0.3125 in.	18	—	1.28	—	0.78			0.14
Sound deadening board 0.5 in.	15	—	0.74	—	1.35	0.30		0.24
Tile and lay-in panels, plain or acoustic	18	0.40	—	2.50	—	0.14	17.33	
0.5 in.	18	—	0.80	—	1.25			0.22
0.75 in.	18	—	0.53	—	1.89			0.33
Laminated paperboard	30	0.50	—	2.00	—	0.33	13.86	
Homogeneous board from repulped paper	30	0.50	—	2.00	—	0.28	13.86	
Hardboard								
Medium density	50	0.73	—	1.37	—	0.31	9.49	
High density, service temp. service underlay	55	0.82	—	1.22	—	0.32	8.46	
High density, std. tempered	63	1.00	—	1.00	—	0.32	6.93	
Particleboard								
Low density	37	0.54	—	1.85	—	0.31	12.82	
Medium density	50	0.94	—	1.06	—	0.31	7.35	
High density	62.5	1.18	—	0.85	—	0.31	5.89	
Underlayment 0.625 in.	40	—	1.22	—	0.82	0.29		0.14
Wood subfloor 0.75 in.		—	1.06	—	0.94	0.33		0.17
BUILDING MEMBRANE								
Vapor—permeable felt	—	—	16.70	—	0.06			0.01
Vapor—seal, 2 layers of mopped 15-lb felt	—	—	8.35	—	0.12			0.02
Vapor—seal, plastic film	—	—	—	—	Negl.			
FINISH FLOORING MATERIALS								
Carpet and fibrous pad	—	—	0.48	—	2.08	0.34		0.37
Carpet and rubber pad	—	—	0.81	—	1.23	0.33		0.22
Cork tile 0.125 in.	—	—	3.60	—	0.28	0.48		0.05
Terrazzo 1 in.	—	—	12.50	—	0.08	0.19		0.01
Tile—asphalt, linoleum, vinyl, rubber	—	—	20.00	—	0.05	0.30		0.01
vinyl asbestos						0.24		
ceramic						0.19		
Wood, hardwood finish 0.75 in.			1.47		0.68			0.12
INSULATING MATERIALS								
BLANKET AND BATT[d]								
Mineral Fiber, fibrous form processed from rock, slag, or glass								
approx.[e] 3–3.5 in.	0.3-2.0	—	0.091	—	11[d]			1.94
approx.[e] 5.50–6.5	0.3-2.0	—	0.053	—	19[d]			3.35
approx.[e] 6–7 in.	0.3-2.0		0.045		22[d]			3.87
approx.[e] 8.5–9 in.	0.3-2.0		0.033		30[d]			5.28
approx. 12 in.	0.3-2.0	—	0.026		38[d]			6.69

Source: ASHRAE Handbook of Fundamentals.

APPENDIX A CONDUCTIVITIES, CONDUCTANCE, AND RESISTANCE OF BUILDING AND INSULATING MATERIALS[a] (*continued*)

Description	Customary Unit						SI Unit	
	Density (lb/ft³)	Conductivity (k)	Conductance (C)	Resistance[b] (R) Per inch thickness (1/k)	Resistance[b] (R) For thickness listed (1/C)	Specific Heat, Btu/(lb) (deg F)	Resistance[b] (R) (m·K)/W	Resistance[b] (R) (m²·K)/W
BOARD AND SLABS								
Cellular glass	8.5	0.35	—	2.86	—	0.18	19.81	
Glass fiber, organic bonded	4–9	0.25	—	4.00	—	0.23	27.72	
Expanded perlite, organic bonded	1.0	0.36	—	2.78	—	0.30	19.26	
Expanded rubber (rigid)	4.5	0.22	—	4.55	—	0.40	31.53	
Expanded polystyrene extruded								
Cut cell surface	1.8	0.25	—	4.00	—	0.29	27.72	
Smooth skin surface	1.8–3.5	0.20	—	5.00	—	0.29	34.65	
Expanded polystyrene, molded beads	1.0	0.26	—	—	—	—	26.3	3.8
	1.25	0.25	—	—	—	—	27.8	4.0
	1.5	0.24	—	—	—	—	29.1	4.2
	1.75	0.24	—	—	—	—	29.1	4.2
	2.0	0.23	—	—	—	—	29.8	4.3
Cellular polyurethane[f] (R-11 exp.)(unfaced)	1.5	0.16	—	6.25	—	0.38	43.82	
(Thickness 1 in. or greater)	2.5							
(Thickness 1 in. or greater—high resistance to gas permeation facing)	1.5	0.14						
Foil-faced, glass fiber-reinforced cellular								
Polyisocyanurate (R-11 exp.)[n]	2	0.14	—	7.04	—	0.22	48.79	
Nominal 0.5 in.		—	0.278	—	3.6			0.63
Nominal 1.0 in.		—	0.139	—	7.2			1.27
Nominal 2.0 in.		—	0.069	—	14.4			2.53
Mineral fiber with resin binder	15	0.29	—	3.45	—	0.17	23.91	
Mineral fiberboard, wet felted								
Core or roof insulation	16–17	0.34	—	2.94	—		20.38	
Acoustical tile	18	0.35	—	2.86	—	0.19	19.82	
Acoustical tile	21	0.37	—	2.70	—		18.71	
Mineral fiberboard, wet molded								
Acoustical tile[g]	23	0.42	—	2.38	—	0.14	16.49	
Wood or cane fiberboard								
Acoustical tile[g] 0.5 in.	—	—	0.80	—	1.25	0.31		0.22
Acoustical tile[g] 0.75 in.	—	—	0.53	—	1.89			0.33
Interior finish (plank, tile)	15	0.35	—	2.86	—	0.32	19.82	
Cement fiber slabs (shredded wood with Portland cement binder)	25–27	0.50–0.53	—	2.0–1.89	—	—	13.87	
Cement fiber slabs (shredded wood with magnesia oxysulfide binder)	22	0.57	—	1.75	—	0.31	12.16	
LOOSE FILL								
Cellulosic insulation (milled paper or wood pulp)	2.3–3.2	0.27–0.32	—	3.13–3.70	—	0.33	21.69–25.64	
Sawdust or shavings	8.0–15.0	0.45	—	2.22	—	0.33	15.39	
Wood fiber, softwoods	2.0–3.5	0.30	—	3.33	—	0.33	23.08	
Perlite, expanded			—	2.70	—	0.26	18.71	
	2.0–4.1	0.27–0.31	3.7–3.3					
	4.1–7.4	0.31–0.36	3.3–2.8					
	7.4–11.0	0.36–0.42	2.8–2.4					
Mineral fiber (rock, slag or glass)								
approx.[e] 3.75–5 in.	0.6–2.0	—	—		11	0.17		1.94
approx.[e] 6.5–8.75 in.	0.6–2.0	—	—		19			3.35
approx.[e] 7.5–10 in.	0.6–2.0	—	—		22			3.87
approx.[e] 10.25–13.75 in.	0.6–2.0	—	—		30			5.28
Vermiculite, exfoliated	7.0–8.2	0.47	—	2.13	—	3.20	14.76	
	4.0–6.0	0.44	—	2.27	—		15.73	
ROOF INSULATION[h]								
Preformed, for use above deck. Different roof insulations are available in different thicknesses to provide the design C values listed.[h] Consult individual manufacturers for actual thickness of their material			0.36 to 0.05		2.7 to 20		—	0.49 to 3.52
MASONRY MATERIALS								
CONCRETES								
Cement mortar	116	5.0	—	0.20	—		1.39	
Gypsum-fiber concrete 87.5% gypsum, 12.5% wood chips	51	1.66	—	0.60	—	0.21	4.16	

APPENDIX A CONDUCTIVITIES, CONDUCTANCE, AND RESISTANCE OF BUILDING AND INSULATING MATERIALS[a] *(continued)*

Description	Customary Unit						SI Unit	
	Density (lb/ft³)	Conductivity (k)	Conductance (C)	Resistance[b] (R)		Specific Heat, Btu/(lb) (deg F)	Resistance[b] (R)	
				Per inch thickness (1/k)	For thickness listed (1/C)		(m·K)/W	(m²·K)/W
Lightweight aggregates including expanded shale, clay or slate; expanded slags; cinders; pumice; vermiculite; also cellular concretes	120	5.2	—	0.19	—		1.32	
	100	3.6	—	0.28	—		1.94	
	80	2.5	—	0.40	—		2.77	
	60	1.7	—	0.59	—		4.09	
	40	1.15	—	0.86	—		5.96	
	30	0.90	—	1.11	—		7.69	
	20	0.70		1.43			9.91	
Perlite, expanded	40	0.93		1.08			7.48	
	30	0.71		1.41			9.77	
	20	0.50		2.00		0.32	13.86	
Sand and gravel or stone aggregate (oven dried)	140	9.0	—	0.11		0.22	0.76	
Sand and gravel or stone aggregate (not dried)	140	12.0	—	0.08			0.55	
Stucco	116	5.0	—	0.20			1.39	
MASONRY UNITS								
Brick, common[i]	120	5.0	—	0.20	—	0.19	1.39	
Brick, face[i]	130	9.0	—	0.11	—		0.76	
Clay tile, hollow:								
1 cell deep ... 3 in.	—	—	1.25	—	0.80	0.21		0.14
1 cell deep ... 4 in.	—	—	0.90	—	1.11			0.20
2 cells deep ... 6 in.	—	—	0.66	—	1.52			0.27
2 cells deep ... 8 in.	—	—	0.54	—	1.85			0.33
2 cells deep ... 10 in.	—	—	0.45	—	2.22			0.39
3 cells deep ... 12 in.	—	—	0.40	—	2.50			0.44
Concrete blocks, three oval core:								
Sand and gravel aggregate ... 4 in.	—	—	1.40	—	0.71	0.22		0.13
... 8 in.	—	—	0.90	—	1.11			0.20
... 12 in.	—	—	0.78	—	1.28			0.23
Cinder aggregate ... 3 in.	—	—	1.16	—	0.86	0.21		0.15
... 4 in.	—	—	0.90	—	1.11			0.20
... 8 in.	—	—	0.58	—	1.72			0.30
... 12 in.	—	—	0.53	—	1.89			0.33
Lightweight aggregate ... 3 in.	—	—	0.79	—	1.27	0.21		0.22
(expanded shale, clay, slate ... 4 in.	—	—	0.67	—	1.50			0.26
or slag; pumice) ... 8 in.	—	—	0.50	—	2.00			0.35
... 12 in.	—	—	0.44	—	2.27			0.40
Concrete blocks, rectangular core.*[j]								
Sand and gravel aggregate								
2 core, 8 in. 36 lb.[k]*	—	—	0.96	—	1.04	0.22		0.18
Same with filled cores[l]*	—	—	0.52	—	1.93	0.22		0.34
Lightweight aggregate (expanded shale, clay, slate or slag, pumice):								
3 core, 6 in. 19 lb.[k]*	—	—	0.61	—	1.65	0.21		0.29
Same with filled cores[l]*	—	—	0.33	—	2.99			0.53
2 core, 8 in. 24 lb.[k]*	—	—	0.46	—	2.18			0.38
Same with filled cores[l]*	—	—	0.20	—	5.03			0.89
3 core, 12 in. 38 lb.[k]*	—	—	0.40	—	2.48			0.44
Same with filled cores[l]*	—	—	0.17	—	5.82			1.02
Stone, lime or sand	—	12.50	—	0.08	—	0.19	0.55	
Gypsum partition tile:								
3 × 12 × 30 in. solid	—	—	0.79	—	1.26	0.19		0.22
3 × 12 × 30 in. 4-cell	—	—	0.74	—	1.35			0.24
4 × 12 × 30 in. 3-cell	—	—	0.60	—	1.67			0.29
METALS (See Chapter 39, Table 3)								
PLASTERING MATERIALS								
Cement plaster, sand aggregate	116	5.0	—	0.20	—	0.20	1.39	
Sand aggregate ... 0.375 in.	—	—	13.3	—	0.08	0.20		0.01
Sand aggregate ... 0.75 in.	—	—	6.66	—	0.15	0.20		0.03
Gypsum plaster:								
Lightweight aggregate ... 0.5 in.	45	—	3.12	—	0.32			0.06
Lightweight aggregate ... 0.625 in.	45	—	2.67	—	0.39			0.07
Lightweight agg. on metal lath ... 0.75 in.	—	—	2.13	—	0.47			0.08
Perlite aggregate	45	1.5	—	0.67	—	0.32	4.64	
Sand aggregate	105	5.6	—	0.18	—	0.20	1.25	
Sand aggregate ... 0.5 in.	105	—	11.10	—	0.09			0.02
Sand aggregate ... 0.625 in.	105	—	9.10	—	0.11			0.02
Sand aggregate on metal lath ... 0.75 in.	—	—	7.70	—	0.13			0.02
Vermiculite aggregate	45	1.7	—	0.59	—		4.09	

APPENDIX A CONDUCTIVITIES, CONDUCTANCE, AND RESISTANCE OF BUILDING AND INSULATING MATERIALS[a] (*continued*)

Description	Density (lb/ft³)	Conductivity (k)	Conductance (C)	Customary Unit: Resistance[b] (R): Per inch thickness (1/k)	Customary Unit: Resistance[b] (R): For thickness listed (1/C)	Specific Heat, Btu/(lb) (deg F)	SI Unit: Resistance[b] (R): (m·K)/W	SI Unit: Resistance[b] (R): (m²·K)/W
ROOFING								
Asbestos-cement shingles	120	—	4.76	—	*0.21*	0.24		*0.04*
Asphalt roll roofing	70	—	6.50	—	*0.15*	0.36		*0.03*
Asphalt shingles	70	—	2.27	—	*0.44*	0.30		0.08
Built-up roofing 0.375 in.	70	—	3.00	—	*0.33*	0.35		*0.06*
Slate 0.5 in.	—	—	20.00	—	*0.05*	0.30		*0.01*
Wood shingles, plain and plastic film faced	—	—	1.06	—	*0.94*	0.31		*0.17*
SIDING MATERIALS (ON FLAT SURFACE)								
Shingles								
Asbestos-cement	120	—	4.75	—	*0.21*			*0.04*
Wood, 16 in., 7.5 exposure	—	—	1.15	—	*0.87*	0.31		*0.15*
Wood, double, 16-in., 12-in. exposure	—	—	0.84	—	*1.19*	0.28		*0.21*
Wood, plus insul. backer board, 0.3125 in.	—	—	0.71	—	*1.40*	0.31		*0.25*
Siding								
Asbestos-cement, 0.25 in., lapped	—	—	4.76	—	*0.21*	0.24		*0.04*
Asphalt roll siding	—	—	6.50	—	*0.15*	0.35		*0.03*
Asphalt insulating siding (0.5 in. bed.)	—	—	0.69	—	*1.46*	0.35		*0.26*
Hardboard siding, 0.4375 in.	40	1.49	—	*0.67*		0.28	*4.65*	
Wood, drop, 1 × 8 in.	—	—	1.27	—	*0.79*	0.28		*0.14*
Wood, bevel, 0.5 × 8 in., lapped	—	—	1.23	—	*0.81*	0.28		*0.14*
Wood, bevel, 0.75 × 10 in., lapped	—	—	0.95	—	*1.05*	0.28		*0.18*
Wood, plywood, 0.375 in., lapped	—	—	1.59	—	*0.59*	0.29		*0.10*
Aluminum or Steel[m], over sheathing								
Hollow-backed	—	—	1.61	—	*0.61*	0.29		*0.11*
Insulating-board backed nominal 0.375 in.	—	—	0.55	—	*1.82*	0.32		*0.32*
Insulating-board backed nominal 0.375 in., foil backed			0.34		*2.96*			*0.52*
Architectural glass	—	—	10.00	—	*0.10*	0.20		*0.02*
WOODS[o,p]								
Maple, oak, and similar hardwoods	45	1.10	—	*0.91*	—	0.30	*6.31*	—
Fir, pine, etc.	32	0.80	—	*1.25*	—	0.33	*8.66*	—
0.75 in.	32	—	1.06	—	*0.94*		—	*0.17*
1.5 in.		—	0.53	—	*1.88*		—	*0.33*
2.5 in.		—	0.32	—	*3.12*		—	*0.55*
3.5 in.		—	0.23	—	*4.38*		—	*0.77*
5.5 in.		—	0.14	—	*7.14*		—	*1.26*
7.25 in.		—	-.11	—	*9.09*		—	*1.60*
9.25 in.		—	0.09	—	*11.11*		—	*1.96*
11.25 in.		—	0.07	—	*14.28*		—	*2.15*

Notes for Table

[a]Representative values for dry materials were selected by ASHRAE TC 4.4, Thermal Insulation and Moisture Retarders (Total Thermal Performance Design Criteria). They are intended as design (not specification) values for materials in normal use. Insulation materials in actual service may have thermal values which vary from design values depending on their in-situ properties such as density and moisture content. For properties of a particular product, use the value supplied by the manufacturer or by unbiased tests.

[b]Resistance values are the reciprocals of *C* before rounding off *C* to two decimal places.

[c]Also see Insulating Materials, Board.

[d]Does not include paper backing and facing, if any. Where insulation forms a boundary (reflective or otherwise) of an air space, see Tables 1 and 2 for the insulating value of air space for the appropriate effective emittance and temperature conditions of the space.

[e]Conductivity varies with fiber diameter. (See Chapter 21, Thermal Conductivity section, and Fig. 1) Insulation is produced by different densities; therefore, there is a wide variation in thickness for the same *R*-value among manufacturers. No effort should be made to relate any specific *R*-value to any specific thickness. Commercial thicknesses generally available range from 2 to 8.5.

[f]Values are for aged, unfaced, board stock. For change in conductivity with age of expanded urethane, see Chapter 20, Factors Affecting Thermal Conductivity.

[g]Insulating values of acoustical tile vary, depending on density of the board and on type, size, and depth of perforations.

[h]ASTM C-855-77 recognizes the specification of roof insulation on the basis of the *C*-values shown. Roof insulation is made in thicknesses to meet these values.

[i]Face brick and common brick do not always have these specific densities. When density is different from that shown, there will be a change in thermal conductivity.

[j]Data on rectangular core concrete blocks differ from the above data on oval core blocks, due to core configuration, different mean temperatures, and possibly differences in unit weights. Weight data on the oval core blocks tested are not available.

[k]Weights of units approximately 7.625 in. high and 15.75 in. long. These weights are given as a means of describing the blocks tested, but conductance values are all for 1 ft^2 of area.

[l]Vermiculite, perlite, or mineral wool insulation. Where insulation is used, vapor barriers or other precautions must be considered to keep insulation dry.

[m]Values for metal siding applied over flat surfaces vary widely, depending on amount of ventilation of air space beneath the siding; whether air space is reflective or nonreflective; and on thickness, type, and application of insulating backing-board used. Values given are averages for use as design guides, and were obtained from several guarded hotbox tests (ASTM C236) or calibrated hotbox (BSS 77) on hollow-backed types and types made using backing-boards of wood fiber, foamed plastic, and glass fiber. Departures of±50% or more from the values given may occur.

[n]Time-aged values for board stock with gas-barrier quality (0.001 in. thickness or greater) aluminum foil facers on two major surfaces.

[o]Forest Products Laboratory Wood Handbook, U.S. Dept. of Agriculture #72, 1974, Tables 3 and 4.

[p]L. Adams: Supporting cryogenic equipment with wood (*Chemical Engineering*, May 17, 1971).

APPENDIX B THERMAL CONDUCTIVITY OF INDUSTRIAL INSULATION

Expressed in Btu per (hour)(square foot)(degree Fahrenheit temperature difference per in.)

Form / Material Composition	Accepted Max Temp for Use, F*	Typical Density (lb/ft³)	Typical Conductivity k at Mean Temp F: −100	−75	−50	−25	0	25	50	75	100	200	300	500	700	900
BLANKETS & FELTS																
MINERAL FIBER (Rock, slag or glass)																
Blanket, metal reinforced	1200	6–12									0.26	0.32	0.39	0.54		
	1000	2.5-6									0.24	0.31	0.40	0.61		
Mineral fiber, glass Blanket, flexible, fine-fiber organic bonded	350	less than 0.75				0.25	0.26	0.28	0.30	0.33	0.36	0.53				
		0.75				0.24	0.25	0.27	0.29	0.32	0.34	0.48				
		1.0				0.23	0.24	0.25	0.27	0.29	0.32	0.43				
		1.5				0.21	0.22	0.23	0.25	0.27	0.28	0.37				
		2.0				0.20	0.21	0.22	0.23	0.25	0.26	0.33				
		3.0				0.19	0.20	0.21	0.22	0.23	0.24	0.31				
Blanket, flexible, textile-fiber organic bonded	350	0.65				0.27	0.28	0.29	0.30	0.31	0.32	0.50	0.68			
		0.75				0.26	0.27	0.28	0.29	0.31	0.32	0.48	0.66			
		1.0				0.24	0.25	0.26	0.27	0.29	0.31	0.45	0.60			
		1.5				0.22	0.23	0.24	0.25	0.27	0.29	0.39	0.51			
		3.0				0.20	0.21	0.22	0.23	0.24	0.25	0.32	0.41			
Felt, semirigid organic bonded	400	3–8						0.24	0.25	0.26	0.27	0.35	0.44			
	850	3	0.16	0.17	0.18	0.19	0.20	0.21	0.22	0.23	0.24	0.35	0.55			
Laminated & felted Without binder	1200	7.5											0.35	0.45	0.60	
VEGETABLE & ANIMAL FIBER																
Hair Felt or Hair Felt plus Jute	180	10						0.26	0.28	0.29	0.30					
BLOCKS, BOARDS & PIPE INSULATION																
ASBESTOS																
Laminated asbestos paper	700	30									0.40	0.45	0.50	0.60		
Corrugated & laminated asbestos Paper																
4-ply	300	11–13								0.54	0.57	0.68				
6-ply	300	15–17								0.49	0.51	0.59				
8-ply	300	18–20								0.47	0.49	0.57				
MOLDED AMOSITE & BINDER	1500	15–18									0.32	0.37	0.42	0.52	0.62	0.72
85% MAGNESIA	600	11–12									0.35	0.38	0.42			
CALCIUM SILICATE	1200	11–15									0.38	0.41	0.44	0.52	0.62	0.72
	1800	12–15												0.63	0.74	0.95
CELLULAR GLASS	900	8.5	0.27	0.28	0.29	0.30	0.31	0.32	0.33	0.35	0.36	0.42	0.49	0.70	1.03	
DIATOMACEOUS SILICA	1600	21–22												0.64	0.68	0.72
	1900	23–25												0.70	0.75	0.80
MINERAL FIBER																
Glass, Organic bonded, block and boards	400	3–10	0.16	0.17	0.18	0.19	0.20	0.22	0.24	0.25	0.26	0.33	0.40			
Nonpunking binder	1000	3–10									0.26	0.31	0.38	0.52		
Pipe insulation, slag or glass	350	3–4					0.20	0.21	0.22	0.23	0.24	0.29				
	500	3–10					0.20	0.22	0.24	0.25	0.26	0.33	0.40			
Inorganic bonded-block	1000	10–15									0.33	0.38	0.45	0.55		
	1800	15–24									0.32	0.37	0.42	0.52	0.62	0.74
Pipe insulation slag or glass	1000	10–15									0.33	0.38	0.45	0.55		
MINERAL FIBER Resin binder		15			0.23	0.24	0.25	0.26	0.28	0.29						
RIGID POLYSTYRENE																
Extruded, Refrigerant 12 exp, smooth skin surface	170	2.2	0.16	0.16	0.17	0.16	0.17	0.18	0.19	0.20						
Extruded cut cell surface	170	1.8	0.17	0.18	0.19	0.20	0.21	0.23	0.24	0.25	0.27					
Molded beads	170	1	0.17	0.19	0.20	0.21	0.22	0.24	0.25	0.26	0.28					
		1.25	0.17	0.18	0.19	0.20	0.22	0.23	0.24	0.25	0.27					
		1.5	0.16	0.17	0.19	0.20	0.21	0.22	0.23	0.24	0.26					
		1.75	0.16	0.17	0.18	0.19	0.20	0.22	0.23	0.24	0.25					
		2.0	0.15	0.16	0.18	0.19	0.20	0.21	0.22	0.23	0.24					
RIGID POLYISOCYANDRATE** Cellular, foil-faced glass fiber reinforced, Refrigerant 11 exp	250	2						0.12	0.13	0.14	0.15					
POLYURETHANE*** Refrigerant 11 exp (unfaced)	210	1.5–2.5	0.16	0.17	0.18	0.18	0.18	0.17	0.16	0.16	0.17					
RUBBER, Rigid Foamed	150	4.5						0.20	0.21	0.22	0.23					
VEGETABLE & ANIMAL FIBER Wool felt (pipe insulation)	180	20						0.28	0.30	0.31	0.33					
INSULATING CEMENTS																
MINERAL FIBER (Rock, slag, or glass)																
With colloidal clay binder	1800	24–30									0.49	0.55	0.61	0.73	0.85	
With hydraulic setting binder	1200	30–40									0.75	0.80	0.85	0.95		

Source: ASHRAE Handbook of Fundamentals.

APPENDIX C ALLOWABLE CURRENT-CARRYING CAPACITY (AMPERES) OF COPPER CONDUCTORS

	In raceway or cable[a]		In free air	
Size of conductor AWG	Rubber types R, RW, RU, RUW, and thermoplastic types T, TW	Rubber type RH	Rubber types R, RW, RU, RUW, and thermoplastic types T, TW	Rubber type RH
14	15	15	20	20
12	20	20	25	25
10	30	30	40	40
8	40	45	55	65
6	55	65	80	95
4	70	85	105	125
2	95	115	140	170
1	110	130	165	195

Source: From *Home Wiring* by R. Graf and G. Whalen. Published by Craftsman Book Company, Box 6500, Carlsbad, CA 92008.

Note: Based on room temperature of 30°C (86°F).

[a] Not more than three conductors in raceway or cable; if the number of conductors is four, the allowable carrying capacity is 80% of the values given.

APPENDIX D FRICTION LOSSES FOR DUCT PIPE

Duct Design

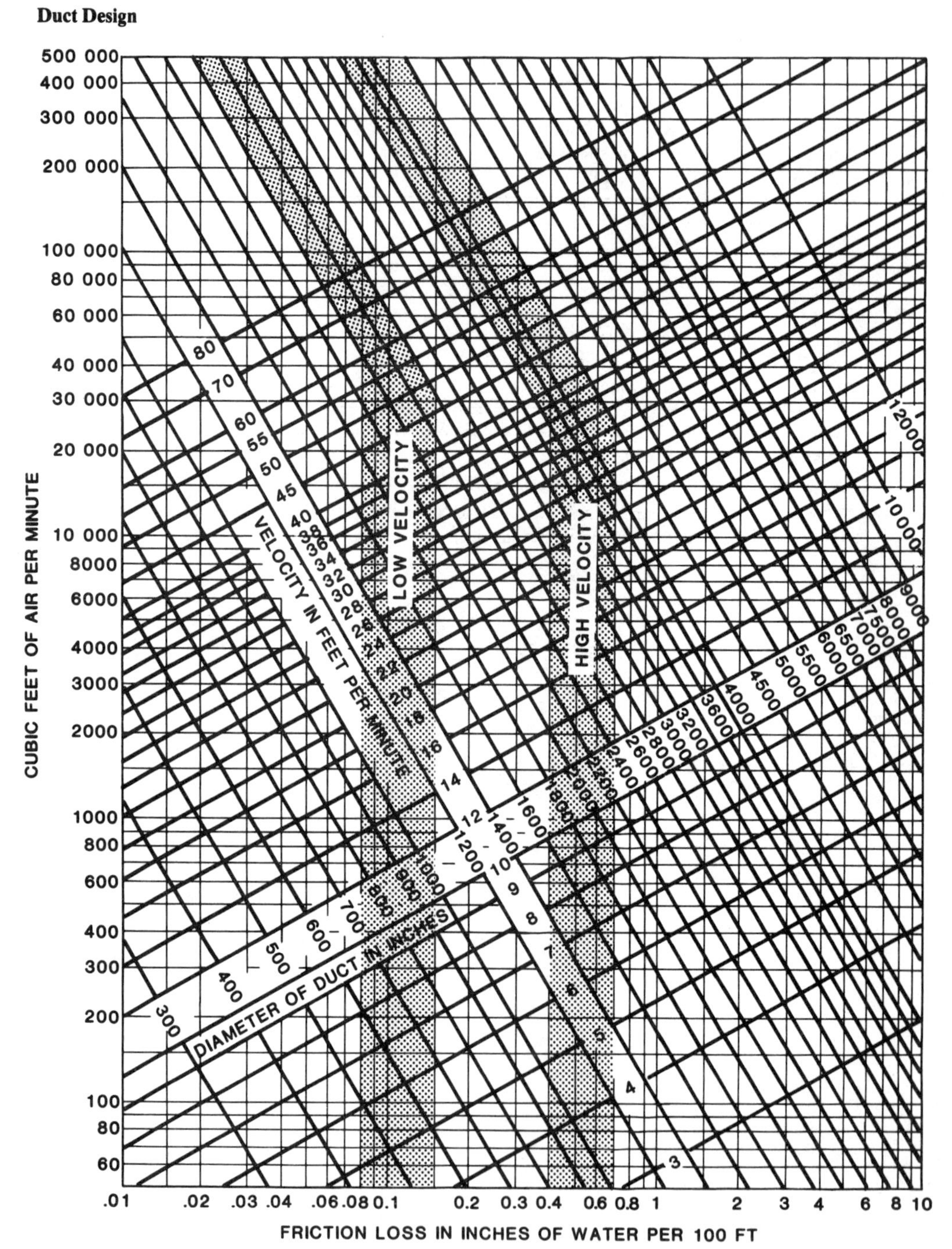

Source: ASHRAE Handbook of Fundamentals.

APPENDIX E DESIGN TEMPERATURE VALUES FOR THE UNITED STATES AND CANADA[a]

Col. 1 State and Station[b]	Col. 2 Latitude[c] °	′	Col. 3 Elev,[d] Ft	Winter: Col. 4 Median of Annual Extremes	Col. 4 99%	Col. 4 97½%	Col. 5 Coincident Wind Velocity[e]	Summer: Col. 6 Design Dry-Bulb 1%	Col. 6 2½%	Col. 6 5%	Col. 7 Outdoor Daily Range[f]	Col. 8 Design Wet-Bulb 1%	Col. 8 2½%	Col. 8 5%
ALABAMA														
Alexander City	33	0	660	12	16	20	L	96	94	93	21	79	78	77
Anniston AP	33	4	599	12	17	19	L	96	94	93	21	79	78	77
Auburn	32	4	730	17	21	25	L	98	96	95	21	80	79	78
Birmingham AP	33	3	610	14	19	22	L	97	94	93	21	79	78	77
Decatur	34	4	580	10	15	19	L	97	95	94	22	79	78	77
Dothan AP	31	2	321	19	23	27	L	97	95	94	20	81	80	79
Florence AP	34	5	528	8	13	17	L	97	95	94	22	79	78	77
Gadsden	34	0	570	11	16	20	L	96	94	93	22	78	77	76
Huntsville AP	34	4	619	−6	13	17	L	97	95	94	23	78	77	76
Mobile AP	30	4	211	21	26	29	M	95	93	91	18	80	79	79
Mobile CO	30	4	119	24	28	32	M	96	94	93	16	80	79	79
Montgomery AP	32	2	195	18	22	26	L	98	95	93	21	80	79	78
Selma-Craig AFB	32	2	207	18	23	27	L	98	96	94	21	81	80	79
Talladega	33	3	565	11	15	19	L	97	95	94	21	79	78	77
Tuscaloosa AP	33	1	170r	14	19	23	L	98	96	95	22	81	80	79
ALASKA														
Anchorage AP	61	1	90	−29	−25	−20	VL	73	70	67	15	63	61	59
Barrow	71	2	22	−49	−45	−42	M	58	54	50	12	54	51	48
Fairbanks AP	64	5	436	−59	−53	−50	VL	82	78	75	24	64	63	61
Juneau AP	58	2	17	−11	− 7	− 4	L	75	71	68	15	66	64	62
Kodiak	57	3	21	4	8	12	M	71	66	63	10	62	60	58
Nome AP	64	3	13	−37	−32	−28	L	66	62	59	10	58	56	54
ARIZONA†														
Douglas AP	31	3	4098	13	18	22	VL	100	98	96	31	70	69	68
Flagstaff AP	35	1	6973	−10	0	5	VL	84	82	80	31	61	60	59
Fort Huachuca AP	31	3	4664	18	25	28	VL	95	93	91	27	69	68	67
Kingman AP	35	2	3446	18	25	29	VL	103	100	97	30	70	69	69
Nogales	31	2	3800	15	20	24	VL	100	98	96	31	72	71	70
Phoenix AP	33	3	1117	25	31	34	VL	108	106	104	27	77	76	75
Prescott AP	34	4	5014	7	15	19	VL	96	94	91	30	67	66	65
Tucson AP	33	1	2584	23	29	32	VL	105	102	100	26	74	73	72
Winslow AP	35	0	4880	2	9	13	VL	97	95	92	32	66	65	64
Yuma AP	32	4	199	32	37	40	VL	111	109	107	27	79	78	77
ARKANSAS														
Blytheville AFB	36	0	264	6	12	17	L	98	96	93	21	80	79	78
Camden	33	4	116	13	19	23	L	99	97	96	21	81	80	79
El Dorado AP	33	1	252	13	19	23	L	98	96	95	21	81	80	79
Fayetteville AP	36	0	1253	3	9	13	M	97	95	93	23	77	76	75
Fort Smith AP	35	2	449	9	15	19	M	101	99	96	24	79	78	77
Hot Springs Nat. Pk.	34	3	710	12	18	22	M	99	97	96	22	79	78	77
Jonesboro	35	5	345	8	14	18	M	98	96	95	21	80	79	78
Little Rock AP	34	4	257	13	19	23	M	99	96	94	22	80	79	78
Pine Bluff AP	34	1	204	14	20	24	L	99	96	95	22	81	80	79
Texarkana AP	33	3	361	16	22	26	M	99	97	96	21	80	79	78
CALIFORNIA†														
Bakersfield AP	35	2	495	26	31	33	VL	103	101	99	32	72	71	70
Barstow AP	34	5	2142	18	24	28	VL	104	102	99	37	73	72	71
Blythe AP	33	4	390	26	31	35	VL	111	109	106	28	78	77	76
Burbank AP	34	1	699	30	36	38	VL	97	94	91	25	72	70	69
Chico	39	5	205	23	29	33	VL	102	100	97	36	71	70	69
Concord	38	0	195	27	32	36	VL	96	92	88	32	69	67	66

* Data for U. S. stations extracted from *Evaluated Weather Data for Cooling Equipment Design, Addendum No. 1, Winter and Summer Data*, with the permission of the publisher, Fluor Products Company, Inc., Box 1267, Santa Rosa, California.

[a] Data compiled from official weather stations, where hourly weather observations are made by trained observers, and from other sources. Table 1 prepared by ASHRAE Technical Committee 2.2, Weather Data and Design Conditions. Percentage of *winter* design data show the percent of 3-month period, December through February. Canadian data are based on January only. Percentage of *summer* design data show the percent of 4-month period, June through September. Canadian data are based on July only. Also see References 1 to 7.

[b] When airport temperature observations were used to develop design data, "AP" follows station name, and "AFB" follows Air Force Bases. Data for stations followed by "CO" came from office locations within an urban area and generally reflect an influence of the surrounding area. Stations without designation can be considered semirural and may be directly compared with most airport data.

[c] Latitude is given to the nearest 10 minutes, for use in calculating solar loads. For example, the latitude for Anniston, Alabama is given as 33 4, or 33°40′.

[d] Elevations are ground elevations for each station as of 1964. Temperature readings are generally made at an elevation of 5 ft above ground, except for locations marked r, indicating roof exposure of thermometer.

[e] Coincident wind velocities derived from approximately coldest 600 hours out of 20,000 hours of December through February data per station. Also see References 5 and 6. The four classifications are:
VL = Very Light, 70 percent or more of cold extreme hours ≤7 mph. M = Moderate, 50 to 74 percent cold extreme hours >7 mph.
L = Light, 50 to 69 percent cold extreme hours ≤7 mph. H = High, 75 percent or more cold extreme hours >7 mph, and 50 percent are >12 mph.

[f] The difference between the average maximum and average minimum temperatures during the warmest month.

† More detailed data on Arizona, California, and Nevada may be found in *Recommended Design Temperatures, Northern California*, published by the Golden Gate Chapter; and *Recommended Design Temperatures, Southern California, Arizona, Nevada*, published by the Southern California Chapter.

Source: ASHRAE Handbook of Fundamentals.

APPENDIX E DESIGN TEMPERATURE VALUES FOR THE UNITED STATES AND CANADA[a] *(continued)*

Col. 1 State and Station[b]	Col. 2 Latitude[c] °	′	Col. 3 Elev,[d] Ft	Winter: Col. 4 Median of Annual Extremes	Winter: Col. 4 99%	Winter: Col. 4 97½%	Winter: Col. 5 Coincident Wind Velocity[e]	Summer: Col. 6 Design Dry-Bulb 1%	Col. 6 2½%	Col. 6 5%	Summer: Col. 7 Outdoor Daily Range[f]	Summer: Col. 8 Design Wet-Bulb 1%	Col. 8 2½%	Col. 8 5%
CALIFORNIA† (continued)														
Covina	34	0	575	32	38	41	VL	100	97	94	31	73	72	71
Crescent City AP	41	5	50	28	33	36	L	72	69	65	18	61	60	59
Downey	34	0	116	30	35	38	VL	93	90	87	22	72	71	70
El Cajon	32	4	525	26	31	34	VL	98	95	92	30	74	73	72
El Centro AP	32	5	−30	26	31	35	VL	111	109	106	34	81	80	79
Escondido	33	0	660	28	33	36	VL	95	92	89	30	73	72	71
Eureka/Arcata AP	41	0	217	27	32	35	L	67	65	63	11	60	59	58
Fairfield-Travis AFB	38	2	72	26	32	34	VL	98	94	90	34	71	69	67
Fresno AP	36	5	326	25	28	31	VL	101	99	97	34	73	72	71
Hamilton AFB	38	0	3	28	33	35	VL	89	85	81	28	71	68	66
Laguna Beach	33	3	35	32	37	39	VL	83	80	77	18	69	68	67
Livermore	37	4	545	23	28	30	VL	99	97	94	24	70	69	68
Lompoc, Vandenburg AFB	34	4	552	32	36	38	VL	82	79	76	20	65	63	61
Long Beach AP	33	5	34	31	36	38	VL	87	84	81	22	72	70	69
Los Angeles AP	34	0	99	36	41	43	VL	86	83	80	15	69	68	67
Los Angeles CO	34	0	312	38	42	44	VL	94	90	87	20	72	70	69
Merced-Castle AFB	37	2	178	24	30	32	VL	102	99	96	36	73	72	70
Modesto	37	4	91	26	32	36	VL	101	98	96	36	72	71	70
Monterey	36	4	38	29	34	37	VL	82	79	76	20	64	63	61
Napa	38	2	16	26	31	34	VL	94	92	89	30	69	68	67
Needles AP	34	5	913	27	33	37	VL	112	110	107	27	76	75	74
Oakland AP	37	4	3	30	35	37	VL	85	81	77	19	65	63	62
Oceanside	33	1	30	33	38	40	VL	84	81	78	13	69	68	67
Ontario	34	0	995	26	32	34	VL	100	97	94	36	72	71	70
Oxnard AFB	34	1	43	32	35	37	VL	84	80	78	19	70	69	67
Palmdale AP	34	4	2517	18	24	27	VL	103	101	98	35	70	68	67
Palm Springs	33	5	411	27	32	36	VL	110	108	105	35	79	78	77
Pasadena	34	1	864	31	36	39	VL	96	93	90	29	72	70	69
Petaluma	38	1	27	24	29	32	VL	94	90	87	31	70	68	67
Pomona CO	34	0	871	26	31	34	VL	99	96	93	36	73	72	71
Redding AP	40	3	495	25	31	35	VL	103	101	98	32	70	69	67
Redlands	34	0	1318	28	34	37	VL	99	96	93	33	72	71	70
Richmond	38	0	55	28	35	38	VL	85	81	77	17	66	64	63
Riverside-March AFB	33	5	1511	26	32	34	VL	99	96	94	37	72	71	69
Sacramento AP	38	3	17	24	30	32	VL	100	97	94	36	72	70	69
Salinas AP	36	4	74	27	32	35	VL	87	85	82	24	67	65	64
San Bernardino, Norton AFB	34	1	1125	26	31	33	VL	101	98	96	38	75	73	71
San Diego AP	32	4	19	38	42	44	VL	86	83	80	12	71	70	68
San Fernando	34	1	977	29	34	37	VL	100	97	94	38	73	72	71
San Francisco AP	37	4	8	32	35	37	L	83	79	75	20	65	63	62
San Francisco CO	37	5	52	38	42	44	VL	80	77	73	14	64	62	61
San Jose AP	37	2	70r	30	34	36	VL	90	88	85	26	69	67	65
San Luis Obispo	35	2	315	30	35	37	VL	89	85	82	26	65	64	63
Santa Ana AP	33	4	115r	28	33	36	VL	92	89	86	28	72	71	70
Santa Barbara CO	34	3	100	30	34	36	VL	87	84	81	24	67	66	65
Santa Cruz	37	0	125	28	32	34	VL	87	84	80	28	66	65	63
Santa Maria AP	34	5	238	28	32	34	VL	85	82	79	23	65	64	63
Santa Monica CO	34	0	57	38	43	45	VL	80	77	74	16	69	68	67
Santa Paula	34	2	263	28	33	36	VL	91	89	86	36	72	71	70
Santa Rosa	38	3	167	24	29	32	VL	95	93	90	34	70	68	67
Stockton AP	37	5	28	25	30	34	VL	101	98	96	37	72	70	69
Ukiah	39	1	620	22	27	30	VL	98	96	93	40	70	69	67
Visalia	36	2	354	26	32	36	VL	102	100	97	38	73	72	70
Yreka	41	4	2625	7	13	17	VL	96	94	91	38	68	66	65
Yuba City	39	1	70	24	30	34	VL	102	100	97	36	71	70	69
COLORADO														
Alamosa AP	37	3	7536	−26	−17	−13	VL	84	82	79	35	62	61	60
Boulder	40	0	5385	− 5	4	8	L	92	90	87	27	64	63	62
Colorado Springs AP	38	5	6173	− 9	− 1	4	L	90	88	86	30	63	62	61
Denver AP	39	5	5283	− 9	− 2	3	L	92	90	89	28	65	64	63
Durango	37	1	6550	−10	0	4	VL	88	86	83	30	64	63	62

APPENDIX E DESIGN TEMPERATURE VALUES FOR THE UNITED STATES AND CANADA[a] (*continued*)

Col. 1 State and Station[b]	Col. 2 Latitude[c] °	′	Col. 3 Elev,[d] Ft	Winter: Col. 4 Median of Annual Extremes	Col. 4 99%	Col. 4 97½%	Col. 5 Coincident Wind Velocity[e]	Summer: Col. 6 Design Dry-Bulb 1%	Col. 6 2½%	Col. 6 5%	Col. 7 Outdoor Daily Range[f]	Col. 8 Design Wet-Bulb 1%	Col. 8 2½%	Col. 8 5%
COLORADO (continued)														
Fort Collins	40	4	5001	−18	−9	−5	L	91	89	86	28	63	62	61
Grand Junction AP	39	1	4849	−2	8	11	VL	96	94	92	29	64	63	62
Greeley	40	3	4648	−18	−9	−5	L	94	92	89	29	65	64	63
La Junta AP	38	0	4188	−14	−6	−2	M	97	95	93	31	72	71	69
Leadville	39	2	10177	−18	−9	−4	VL	76	73	70	30	56	55	54
Pueblo AP	38	2	4639	−14	−5	−1	L	96	94	92	31	68	67	66
Sterling	40	4	3939	−15	−6	−2	M	95	93	90	30	67	66	65
Trinidad AP	37	2	5746	−9	1	5	L	93	91	89	32	66	65	64
CONNECTICUT														
Bridgeport AP	41	1	7	−1	4	8	M	90	88	85	18	77	76	75
Hartford, Brainard Field	41	5	15	−4	1	5	M	90	88	85	22	77	76	74
New Haven AP	41	2	6	0	5	9	H	88	86	83	17	77	76	75
New London	41	2	60	0	4	8	H	89	86	83	16	77	75	74
Norwalk	41	1	37	−5	0	4	M	91	89	86	19	77	76	75
Norwich	41	3	20	−7	−2	2	M	88	86	83	18	77	76	75
Waterbury	41	3	605	−5	0	4	M	90	88	85	21	77	76	75
Windsor Locks, Bradley Field	42	0	169	−7	−2	2	M	90	88	85	22	76	75	73
DELAWARE														
Dover AFB	39	0	38	8	13	15	M	93	90	88	18	79	78	77
Wilmington AP	39	4	78	6	12	15	M	93	90	87	20	79	77	76
DISTRICT OF COLUMBIA														
Andrews AFB	38	5	279	9	13	16	M	94	91	88	18	79	77	76
Washington National AP	38	5	14	12	16	19	M	94	92	90	18	78	77	76
FLORIDA														
Belle Glade	26	4	16	31	35	39	M	93	91	90	16	80	79	79
Cape Kennedy AP	28	1	16	33	37	40	L	90	89	88	15	81	80	79
Daytona Beach AP	29	1	31	28	32	36	L	94	92	91	15	81	80	79
Fort Lauderdale	26	0	13	37	41	45	M	91	90	89	15	81	80	79
Fort Myers AP	26	4	13	34	38	42	M	94	92	91	18	80	80	79
Fort Pierce	27	3	10	33	37	41	M	93	91	90	15	81	80	79
Gainesville AP	29	4	155	24	28	32	L	96	94	93	18	80	79	79
Jacksonville AP	30	3	24	26	29	32	L	96	94	92	19	80	79	79
Key West AP	24	3	6	50	55	58	M	90	89	88	9	80	79	79
Lakeland CO	28	0	214	31	35	39	M	95	93	91	17	80	79	78
Miami AP	25	5	7	39	44	47	M	92	90	89	15	80	79	79
Miami Beach CO	25	5	9	40	45	48	M	91	89	88	10	80	79	79
Ocala	29	1	86	25	29	33	L	96	94	93	18	80	79	79
Orlando AP	28	3	106r	29	33	37	L	96	94	93	17	80	79	78
Panama City, Tyndall AFB	30	0	22	28	32	35	M	92	91	90	14	81	80	80
Pensacola CO	30	3	13	25	29	32	M	92	90	89	14	82	81	80
St. Augustine	29	5	15	27	31	35	L	94	92	90	16	81	80	79
St. Petersburg	28	0	35	35	39	42	M	93	91	90	16	81	80	79
Sanford	28	5	14	29	33	37	L	95	93	92	17	80	79	79
Sarasota	27	2	30	31	35	39	M	93	91	90	17	80	80	79
Tallahassee AP	30	2	58	21	25	29	L	96	94	93	19	80	79	79
Tampa AP	28	0	19	32	36	39	M	92	91	90	17	81	80	79
West Palm Beach AP	26	4	15	36	40	44	M	92	91	89	16	81	80	80
GEORGIA														
Albany, Turner AFB	31	3	224	21	26	30	L	98	96	94	20	80	79	78
Americus	32	0	476	18	22	25	L	98	96	93	20	80	79	78
Athens	34	0	700	12	17	21	L	96	94	91	21	78	77	76
Atlanta AP	33	4	1005	14	18	23	H	95	92	90	19	78	77	76
Augusta AP	33	2	143	17	20	23	L	98	95	93	19	80	79	78
Brunswick	31	1	14	24	27	31	L	97	95	92	18	81	80	79
Columbus, Lawson AFB	32	3	242	19	23	26	L	98	96	94	21	80	79	78
Dalton	34	5	720	10	15	19	L	97	95	92	22	78	77	76
Dublin	32	0	215	17	21	25	L	98	96	93	20	80	79	78
Gainesville	34	2	1254	11	16	20	L	94	92	90	21	78	77	76
Griffin	33	1	980	13	17	22	L	95	93	90	21	79	78	77
La Grange	33	0	715	12	16	20	L	96	94	92	21	79	78	77
Macon AP	32	4	356	18	23	27	L	98	96	94	22	80	79	78
Marietta, Dobbins AFB	34	0	1016	12	17	21	L	95	93	91	21	78	77	76

APPENDIX E DESIGN TEMPERATURE VALUES FOR THE UNITED STATES AND CANADA[a] *(continued)*

Col. 1 State and Station[b]	Col. 2 Latitude[c] °	′	Col. 3 Elev,[d] Ft	Winter: Col. 4 Median of Annual Extremes	Winter: Col. 4 99%	Winter: Col. 4 97½%	Winter: Col. 5 Coincident Wind Velocity[e]	Summer: Col. 6 Design Dry-Bulb 1%	Col. 6 2½%	Col. 6 5%	Summer: Col. 7 Outdoor Daily Range[f]	Summer: Col. 8 Design Wet-Bulb 1%	Col. 8 2½%	Col. 8 5%
GEORGIA (continued)														
Moultrie	31	1	340	22	26	30	L	97	95	93	20	80	79	78
Rome AP	34	2	637	11	16	20	L	97	95	93	23	78	77	76
Savannah-Travis AP	32	1	52	21	24	27	L	96	94	92	20	81	80	79
Valdosta-Moody AFB	31	0	239	24	28	31	L	96	94	92	20	80	79	78
Waycross	31	2	140	20	24	28	L	97	95	93	20	80	79	78
HAWAII														
Hilo AP	19	4	31	56	59	61	L	85	83	82	15	74	73	72
Honolulu AP	21	2	7	58	60	62	L	87	85	84	12	75	74	73
Kaneohe	21	2	198	58	60	61	L	85	83	82	12	74	73	73
Wahiawa	21	3	215	57	59	61	L	86	84	83	14	75	74	73
IDAHO														
Boise AP	43	3	2842	0	4	10	L	96	93	91	31	68	66	65
Burley	42	3	4180	−5	4	8	VL	95	93	89	35	68	66	64
Coeur d'Alene AP	47	5	2973	−4	2	7	VL	94	91	88	31	66	65	63
Idaho Falls AP	43	3	4730r	−17	−12	−6	VL	91	88	85	38	65	64	62
Lewiston AP	46	2	1413	1	6	12	VL	98	96	93	32	67	66	65
Moscow	46	4	2660	−11	−3	1	VL	91	89	86	32	64	63	61
Mountain Home AFB	43	0	2992	−3	2	9	L	99	96	93	36	68	66	64
Pocatello AP	43	0	4444	−12	−8	−2	VL	94	91	88	35	65	63	62
Twin Falls AP	42	3	4148	−5	4	8	L	96	94	91	34	66	64	63
ILLINOIS														
Aurora	41	5	744	−13	−7	−3	M	93	91	88	20	78	77	75
Belleville, Scott AFB	38	3	447	0	6	10	M	97	95	92	21	79	78	77
Bloomington	40	3	775	−7	−1	3	M	94	92	89	21	79	78	77
Carbondale	37	5	380	1	7	11	M	98	96	94	21	80	79	78
Champaign/Urbana	40	0	743	−6	0	4	M	96	94	91	21	79	78	77
Chicago, Midway AP	41	5	610	−7	−4	1	M	95	92	89	20	78	76	75
Chicago, O'Hare AP	42	0	658	−9	−4	0	M	93	90	87	20	77	75	74
Chicago, CO	41	5	594	−5	−3	1	M	94	91	88	15	78	76	75
Danville	40	1	558	−6	−1	4	M	96	94	91	21	79	78	76
Decatur	39	5	670	−6	0	4	M	96	93	91	21	79	78	77
Dixon	41	5	696	−13	−7	−3	M	93	91	89	23	78	77	76
Elgin	42	0	820	−14	−8	−4	M	92	90	87	21	78	76	75
Freeport	42	2	780	−16	−10	−6	M	92	90	87	24	78	77	75
Galesburg	41	0	771	−10	−4	0	M	95	92	89	22	79	78	76
Greenville	39	0	563	−3	3	7	M	96	94	92	21	79	78	77
Joliet AP	41	3	588	−11	−5	−1	M	94	92	89	20	78	77	75
Kankakee	41	1	625	−10	−4	1	M	94	92	89	21	78	77	76
La Salle/Peru	41	2	520	−9	−3	1	M	94	93	90	22	78	77	76
Macomb	40	3	702	−5	−3	1	M	95	93	90	22	79	78	77
Moline AP	41	3	582	−12	−7	−3	M	94	91	88	23	79	77	76
Mt. Vernon	38	2	500	0	6	10	M	97	95	92	21	79	78	77
Peoria AP	40	4	652	−8	−2	2	M	94	92	89	22	78	77	76
Quincy AP	40	0	762	−8	−2	2	M	97	95	92	22	80	79	77
Rantoul, Chanute AFB	40	2	740	−7	−1	3	M	94	92	89	21	78	77	76
Rockford	42	1	724	−13	−7	−3	M	92	90	87	24	77	76	75
Springfield AP	39	5	587	−7	−1	4	M	95	92	90	21	79	78	77
Waukegan	42	2	680	−11	−5	−1	M	92	90	87	21	77	76	75
INDIANA														
Anderson	40	0	847	−5	0	5	M	93	91	88	22	78	77	76
Bedford	38	5	670	−3	3	7	M	95	93	90	22	79	78	77
Bloomington	39	1	820	−3	3	7	M	95	92	90	22	79	78	76
Columbus, Bakalar AFB	39	2	661	−3	3	7	M	95	92	90	22	79	78	76
Crawfordsville	40	0	752	−8	−2	2	M	95	93	90	22	79	77	76
Evansville AP	38	0	381	1	6	10	M	96	94	91	22	79	78	77
Fort Wayne AP	41	0	791	−5	0	5	M	93	91	88	24	77	76	75
Goshen AP	41	3	823	−10	−4	0	M	92	90	87	23	77	76	74
Hobart	41	3	600	−10	−4	0	M	93	91	88	21	78	76	75
Huntington	40	4	802	−8	−2	2	M	94	92	89	23	78	76	75
Indianapolis AP	39	4	793	−5	0	4	M	93	91	88	22	78	77	76
Jeffersonville	38	2	455	3	9	13	M	96	94	91	23	79	78	77
Kokomo	40	3	790	−6	0	4	M	94	92	89	22	78	76	75
Lafayette	40	2	600	−7	−1	3	M	94	92	89	22	78	77	76

APPENDIX E DESIGN TEMPERATURE VALUES FOR THE UNITED STATES AND CANADA[a] (*continued*)

Col. 1 State and Station[b]	Col. 2 Latitude[c] ° ′	Col. 3 Elev,[d] Ft	Winter: Col. 4 Median of Annual Extremes	Winter: Col. 4 99%	Winter: Col. 4 97½%	Winter: Col. 5 Coincident Wind Velocity[e]	Summer: Col. 6 Design Dry-Bulb 1%	Summer: Col. 6 Design Dry-Bulb 2½%	Summer: Col. 6 Design Dry-Bulb 5%	Summer: Col. 7 Outdoor Daily Range[f]	Summer: Col. 8 Design Wet-Bulb 1%	Summer: Col. 8 Design Wet-Bulb 2½%	Summer: Col. 8 Design Wet-Bulb 5%
INDIANA (continued)													
La Porte	41 3	810	−10	−4	0	M	93	91	88	22	77	76	74
Marion	40 3	791	−8	−2	2	M	93	91	88	23	78	76	75
Muncie	40 1	955	−8	−2	2	M	93	91	88	22	78	77	75
Peru, Bunker Hill AFB	40 4	804	−9	−3	1	M	91	89	86	22	77	76	74
Richmond AP	39 5	1138	−7	−1	3	M	93	91	88	22	78	77	75
Shelbyville	39 3	765	−4	2	6	M	94	92	89	22	78	77	76
South Bend AP	41 4	773	−6	−2	3	M	92	89	87	22	77	76	74
Terre Haute AP	39 3	601	−3	3	7	M	95	93	91	22	79	78	77
Valparaiso	41 2	801	−12	−6	−2	M	92	90	87	22	78	76	75
Vincennes	38 4	420	−1	5	9	M	96	94	91	22	79	78	77
IOWA													
Ames	42 0	1004	−17	−11	−7	M	94	92	89	23	79	78	76
Burlington AP	40 5	694	−10	−4	0	M	95	92	89	22	80	78	77
Cedar Rapids AP	41 5	863	−14	−8	−4	M	92	90	87	23	78	76	75
Clinton	41 5	595	−13	−7	−3	M	92	90	87	23	78	77	76
Council Bluffs	41 2	1210	−14	−7	−3	M	97	94	91	22	79	78	76
Des Moines AP	41 3	948r	−13	−7	−3	M	95	92	89	23	79	77	76
Dubuque	42 2	1065	−17	−11	−7	M	92	90	87	22	78	76	75
Fort Dodge	42 3	1111	−18	−12	−8	M	94	92	89	23	78	77	75
Iowa City	41 4	645	−14	−8	−4	M	94	91	88	22	79	77	76
Keokuk	40 2	526	−9	−3	1	M	95	93	90	22	79	78	77
Marshalltown	42 0	898	−16	−10	−6	M	93	91	88	23	79	77	76
Mason City AP	43 1	1194	−20	−13	−9	M	91	88	85	24	77	75	74
Newton	41 4	946	−15	−9	−5	M	95	93	90	23	79	77	76
Ottumwa AP	41 1	842	−12	−6	−2	M	95	93	90	22	79	78	76
Sioux City AP	42 2	1095	−17	−10	−6	M	96	93	90	24	79	77	76
Waterloo	42 3	868	−18	−12	−8	M	91	89	86	23	78	76	75
KANSAS													
Atchison	39 3	945	−9	−2	2	M	97	95	92	23	79	78	77
Chanute AP	37 4	977	−3	3	7	H	99	97	95	23	79	78	77
Dodge City AP	37 5	2594	−5	3	7	M	99	97	95	25	74	73	72
El Dorado	37 5	1282	−3	4	8	H	101	99	96	24	78	77	76
Emporia	38 2	1209	−4	3	7	H	99	97	94	25	78	77	76
Garden City AP	38 0	2882	−10	−1	3	M	100	98	96	28	74	73	72
Goodland AP	39 2	3645	−10	−2	4	M	99	96	93	31	71	70	69
Great Bend	38 2	1940	−5	2	6	M	101	99	96	28	77	76	75
Hutchinson AP	38 0	1524	−5	2	6	H	101	99	96	28	77	76	75
Liberal	37 0	2838	−4	4	8	M	102	100	99	28	74	73	71
Manhattan, Fort Riley	39 0	1076	−7	−1	4	H	101	98	95	24	79	78	77
Parsons	37 2	908	−2	5	9	H	99	97	94	23	79	78	77
Russell AP	38 5	1864	−7	0	4	M	102	100	97	29	78	76	75
Salina	38 5	1271	−4	3	7	H	101	99	96	26	78	76	75
Topeka AP	39 0	877	−4	3	6	M	99	96	94	24	79	78	77
Wichita AP	37 4	1321	−1	5	9	H	102	99	96	23	77	76	75
KENTUCKY													
Ashland	38 3	551	1	6	10	L	94	92	89	22	77	76	75
Bowling Green AP	37 0	535	1	7	11	L	97	95	93	21	79	78	77
Corbin AP	37 0	1175	0	5	9	L	93	91	89	23	79	77	76
Covington AP	39 0	869	−3	3	8	L	93	90	88	22	77	76	75
Hopkinsville, Campbell AFB	36 4	540	4	10	14	L	97	95	92	21	79	78	77
Lexington AP	38 0	979	0	6	10	M	94	92	90	22	78	77	76
Louisville AP	38 1	474	1	8	12	L	96	93	91	23	79	78	77
Madisonville	37 2	439	1	7	11	L	96	94	92	22	79	78	77
Owensboro	37 5	420	0	6	10	L	96	94	92	23	79	78	77
Paducah AP	37 0	398	4	10	14	L	97	95	94	20	80	79	78
LOUISIANA													
Alexandria AP	31 2	92	20	25	29	L	97	95	94	20	80	80	79
Baton Rouge AP	30 3	64	22	25	30	L	96	94	92	19	81	80	79
Bogalusa	30 5	103	20	24	28	L	96	94	93	19	80	79	78
Houma	29 3	13	25	29	33	L	94	92	91	15	81	80	79
Lafayette AP	30 1	38	23	28	32	L	95	93	92	18	81	81	80
Lake Charles AP	30 1	14	25	29	33	M	95	93	91	17	80	79	79
Minden	32 4	250	17	22	26	L	98	96	95	20	81	80	79

APPENDIX E DESIGN TEMPERATURE VALUES FOR THE UNITED STATES AND CANADA[a] *(continued)*

Col. 1 State and Station[b]	Col. 2 Latitude[c] ° ′	Col. 3 Elev,[d] Ft	Winter: Col. 4 Median of Annual Extremes	Winter: Col. 4 99%	Winter: Col. 4 97½%	Winter: Col. 5 Coincident Wind Velocity[e]	Summer: Col. 6 Design Dry-Bulb 1%	Summer: Col. 6 Design Dry-Bulb 2½%	Summer: Col. 6 Design Dry-Bulb 5%	Summer: Col. 7 Outdoor Daily Range[f]	Summer: Col. 8 Design Wet-Bulb 1%	Summer: Col. 8 Design Wet-Bulb 2½%	Summer: Col. 8 Design Wet-Bulb 5%
LOUISIANA (continued)													
Monroe AP	32 3	78	18	23	27	L	98	96	95	20	81	81	80
Natchitoches	31 5	120	17	22	26	L	99	97	96	20	81	80	79
New Orleans AP	30 0	3	29	32	35	M	93	91	90	16	81	80	79
Shreveport AP	32 3	252	18	22	26	M	99	96	94	20	81	80	79
MAINE													
Augusta AP	44 2	350	−13	− 7	− 3	M	88	86	83	22	74	73	71
Bangor, Dow AFB	44 5	162	−14	− 8	− 4	M	88	85	81	22	75	73	71
Caribou AP	46 5	624	−24	−18	−14	L	85	81	78	21	72	70	68
Lewiston	44 0	182	−14	− 8	− 4	M	88	86	83	22	74	73	71
Millinocket AP	45 4	405	−22	−16	−12	L	87	85	82	22	74	72	70
Portland AP	43 4	61	−14	− 5	0	L	88	85	81	22	75	73	71
Waterville	44 3	89	−15	− 9	− 5	M	88	86	82	22	74	73	71
MARYLAND													
Baltimore AP	39 1	146	8	12	15	M	94	91	89	21	79	78	77
Baltimore CO	39 2	14	12	16	20	M	94	92	89	17	79	78	77
Cumberland	39 4	945	0	5	9	L	94	92	89	22	76	75	74
Frederick AP	39 2	294	2	7	11	M	94	92	89	22	78	77	76
Hagerstown	39 4	660	1	6	10	L	94	92	89	22	77	76	75
Salisbury	38 2	52	10	14	18	M	92	90	87	18	79	78	77
MASSACHUSETTS													
Boston AP	42 2	15	− 1	6	10	H	91	88	85	16	76	74	73
Clinton	42 2	398	− 8	− 2	2	M	87	85	82	17	75	74	72
Fall River	41 4	190	− 1	5	9	H	88	86	83	18	75	74	73
Framingham	42 2	170	− 7	− 1	3	M	91	89	86	17	76	74	73
Gloucester	42 3	10	− 4	2	6	H	86	84	81	15	74	73	72
Greenfield	42 3	205	−12	− 6	− 2	M	89	87	84	23	75	74	73
Lawrence	42 4	57	− 9	− 3	1	M	90	88	85	22	76	74	72
Lowell	42 3	90	− 7	− 1	3	M	91	89	86	21	76	74	72
New Bedford	41 4	70	3	9	13	H	86	84	81	19	75	73	72
Pittsfield AP	42 3	1170	−11	− 5	− 1	M	86	84	81	23	74	72	71
Springfield, Westover AFB	42 1	247	− 8	− 3	2	M	91	88	85	19	76	74	73
Taunton	41 5	20	− 9	− 4	0	H	88	86	83	18	76	75	74
Worcester AP	42 2	986	− 8	− 3	1	M	89	87	84	18	75	73	71
MICHIGAN													
Adrian	41 5	754	− 6	0	4	M	93	91	88	23	76	75	74
Alpena AP	45 0	689	−11	− 5	− 1	M	87	85	82	27	74	73	71
Battle Creek AP	42 2	939	− 6	1	5	M	92	89	86	23	76	74	73
Benton Harbor AP	42 1	649	− 7	− 1	3	M	90	88	85	20	76	74	73
Detroit Met. CAP	42 2	633	0	4	8	M	92	88	85	20	76	75	74
Escanaba	45 4	594	−13	− 7	− 3	M	82	80	77	17	73	71	69
Flint AP	43 0	766	− 7	− 1	3	M	89	87	84	25	76	75	74
Grand Rapids AP	42 5	681	− 3	2	6	M	91	89	86	24	76	74	73
Holland	42 5	612	− 4	2	6	M	90	88	85	22	76	74	73
Jackson AP	42 2	1003	− 6	0	4	M	92	89	86	23	76	75	74
Kalamazoo	42 1	930	− 5	1	5	M	92	89	86	23	76	75	74
Lansing AP	42 5	852	− 4	2	6	M	89	87	84	24	76	75	73
Marquette CO	46 3	677	−14	− 8	− 4	L	88	86	83	18	73	71	69
Mt. Pleasant	43 4	796	− 9	− 3	1	M	89	87	84	24	75	74	73
Muskegon AP	43 1	627	− 2	4	8	M	87	85	82	21	75	74	73
Pontiac	42 4	974	− 6	0	4	M	90	88	85	21	76	75	73
Port Huron	43 0	586	− 6	− 1	3	M	90	88	85	21	76	74	73
Saginaw AP	43 3	662	− 7	− 1	3	M	88	86	83	23	76	75	73
Sault Ste. Marie AP	46 3	721	−18	−12	− 8	L	83	81	78	23	73	71	69
Traverse City AP	44 4	618	− 6	0	4	M	89	86	83	22	75	73	72
Ypsilanti	42 1	777	− 3	− 1	5	M	92	89	86	22	76	74	73
MINNESOTA													
Albert Lea	43 4	1235	−20	−14	−10	M	91	89	86	24	77	76	74
Alexandria AP	45 5	1421	−26	−19	−15	L	90	88	85	24	76	74	72
Bemidji AP	47 3	1392	−38	−32	−28	L	87	84	81	24	73	72	71
Brainerd	46 2	1214	−31	−24	−20	L	88	85	82	24	74	73	72
Duluth AP	46 5	1426	−25	−19	−15	M	85	82	79	22	73	71	69
Faribault	44 2	1190	−23	−16	−12	L	90	88	85	24	77	75	74
Fergus Falls	46 1	1210	−28	−21	−17	L	92	89	86	24	75	74	72
International Falls AP	48 3	1179	−35	−29	−24	L	86	82	79	26	72	69	68

APPENDIX E DESIGN TEMPERATURE VALUES FOR THE UNITED STATES AND CANADA[a] *(continued)*

Col. 1 State and Station[b]	Col. 2 Latitude[c] °	′	Col. 3 Elev,[d] Ft	Winter: Col. 4 Median of Annual Extremes	99%	97½%	Col. 5 Coincident Wind Velocity[e]	Summer: Col. 6 Design Dry-Bulb 1%	2½%	5%	Col. 7 Outdoor Daily Range[f]	Col. 8 Design Wet-Bulb 1%	2½%	5%
MINNESOTA (continued)														
Mankato	44	1	785	−23	−16	−12	L	91	89	86	24	77	75	74
Minneapolis/St. Paul AP	44	5	822	−19	−14	−10	L	92	89	86	22	77	75	74
Rochester AP	44	0	1297	−23	−17	−13	M	90	88	85	24	77	75	74
St. Cloud AP	45	4	1034	−26	−20	−16	L	90	88	85	24	77	75	73
Virginia	47	3	1435	−32	−25	−21	L	86	83	80	23	73	71	69
Willmar	45	1	1133	−25	−18	−14	L	91	88	85	24	77	75	73
Winona	44	1	652	−19	−12	− 8	M	91	89	86	24	77	76	74
MISSISSIPPI														
Biloxi, Keesler AFB	30	2	25	26	30	32	M	93	92	90	16	82	81	80
Clarksdale	34	1	178	14	20	24	L	98	96	95	21	81	80	79
Columbus AFB	33	4	224	13	18	22	L	97	95	93	22	79	79	78
Greenville AFB	33	3	139	16	21	24	L	98	96	94	21	81	80	79
Greenwood	33	3	128	14	19	23	L	98	96	94	21	81	80	79
Hattiesburg	31	2	200	18	22	26	L	97	95	94	21	80	79	78
Jackson AP	32	2	330	17	21	24	L	98	96	94	21	79	78	78
Laurel	31	4	264	18	22	26	L	97	95	94	21	80	79	78
McComb AP	31	2	458	18	22	26	L	96	94	93	18	80	79	79
Meridian AP	32	2	294	15	20	24	L	97	95	94	22	80	79	78
Natchez	31	4	168	18	22	26	L	96	94	93	21	80	80	79
Tupelo	34	2	289	13	18	22	L	98	96	95	22	80	79	78
Vicksburg CO	32	2	234	18	23	26	L	97	95	94	21	80	80	79
MISSOURI														
Cape Girardeau	37	1	330	2	8	12	M	98	96	94	21	80	79	78
Columbia AP	39	0	778	− 4	2	6	M	97	95	92	22	79	78	77
Farmington AP	37	5	928	− 2	4	8	M	97	95	93	22	79	78	77
Hannibal	39	4	489	− 7	− 1	4	M	96	94	91	22	79	78	77
Jefferson City	38	4	640	− 4	2	6	M	97	95	93	23	79	78	77
Joplin AP	37	1	982	1	7	11	M	97	95	93	24	79	78	77
Kansas City AP	39	1	742	− 2	4	8	M	100	97	94	20	79	77	76
Kirksville AP	40	1	966	−13	− 7	− 3	M	96	94	91	24	79	78	77
Mexico	39	1	775	− 7	− 1	3	M	96	94	91	22	79	78	77
Moberly	39	3	850	− 8	− 2	2	M	96	94	91	23	79	78	77
Poplar Bluff	36	5	322	3	9	13	M	98	96	94	22	80	79	78
Rolla	38	0	1202	− 3	3	7	M	97	95	93	22	79	78	77
St. Joseph AP	39	5	809	− 8	− 1	3	M	97	95	92	23	79	78	77
St. Louis AP	38	5	535	− 2	4	8	M	98	95	92	21	79	78	77
St. Louis CO	38	4	465	1	7	11	M	96	94	92	18	79	78	77
Sedalia, Whiteman AFB	38	4	838	− 2	4	9	M	97	94	92	22	79	77	76
Sikeston	36	5	318	4	10	14	L	98	96	94	21	80	79	78
Springfield AP	37	1	1265	0	5	10	M	97	94	91	23	78	77	76
MONTANA														
Billings AP	45	5	3567	−19	−10	− 6	L	94	91	88	31	68	66	65
Bozeman	45	5	4856	−25	−15	−11	L	88	85	82	32	61	60	59
Butte AP	46	0	5526r	−34	−24	−16	VL	86	83	80	35	60	59	57
Cut Bank AP	48	4	3838r	−32	−23	−17	L	89	86	82	35	65	63	61
Glasgow AP	48	1	2277	−33	−25	−20	L	96	93	89	29	69	67	65
Glendive	47	1	2076	−28	−20	−16	L	96	93	90	29	71	69	68
Great Falls AP	47	3	3664r	−29	−20	−16	L	91	88	85	28	64	63	61
Havre	48	3	2488	−32	−22	−15	M	91	87	84	33	66	64	63
Helena AP	46	4	3893	−27	−17	−13	L	90	87	84	32	65	63	61
Kalispell AP	48	2	2965	−17	− 7	− 3	VL	88	84	81	34	65	63	62
Lewiston AP	47	0	4132	−27	−18	−14	L	89	86	83	30	65	63	62
Livingston AP	45	4	4653	−26	−17	−13	L	91	88	85	32	63	62	61
Miles City AP	46	3	2629	−27	−19	−15	L	97	94	91	30	71	69	68
Missoula AP	46	5	3200	−16	− 7	− 3	VL	92	89	86	36	65	63	61
NEBRASKA														
Beatrice	40	2	1235	−10	− 3	1	M	99	97	94	24	78	77	76
Chadron AP	42	5	3300	−21	−13	− 9	M	97	95	92	30	72	70	69
Columbus	41	3	1442	−14	− 7	− 3	M	98	96	93	25	78	76	75
Fremont	41	3	1203	−14	− 7	− 3	M	99	97	94	22	78	77	76
Grand Island AP	41	0	1841	−14	− 6	− 2	M	98	95	92	28	76	75	74
Hastings	40	4	1932	−11	− 3	1	M	98	96	94	27	77	75	74
Kearney	40	4	2146	−14	− 6	− 2	M	97	95	92	28	76	75	74
Lincoln CO	40	5	1150	−10	− 4	0	M	100	96	93	24	78	77	76

APPENDIX E DESIGN TEMPERATURE VALUES FOR THE UNITED STATES AND CANADA[a] (*continued*)

Col. 1 State and Station[b]	Col. 2 Latitude[c] °	′	Col. 3 Elev,[d] Ft	Winter: Col. 4 Median of Annual Extremes	99%	97½%	Col. 5 Coincident Wind Velocity[e]	Summer: Col. 6 Design Dry-Bulb 1%	2½%	5%	Col. 7 Outdoor Daily Range[f]	Col. 8 Design Wet-Bulb 1%	2½%	5%
NEBRASKA (continued)														
McCook	40	1	2565	−12	−4	0	M	99	97	94	28	74	72	71
Norfolk	42	0	1532	−18	−11	−7	M	97	95	92	30	78	76	75
North Platte AP	41	1	2779	−13	−6	−2	M	97	94	90	28	74	73	72
Omaha AP	41	2	978	−12	−5	−1	M	97	94	91	22	79	78	76
Scottsbluff AP	41	5	3950	−16	−8	−4	M	96	94	91	31	70	69	67
Sidney AP	41	1	4292	−15	−7	−2	M	95	92	89	31	70	69	67
NEVADA†														
Carson City	39	1	4675	−4	3	7	VL	93	91	88	42	62	61	60
Elko AP	40	5	5075	−21	−13	−7	VL	94	92	90	42	64	62	61
Ely AP	39	1	6257	−15	−6	−2	VL	90	88	86	39	60	59	58
Las Vegas AP	36	1	2162	18	23	26	VL	108	106	104	30	72	71	70
Lovelock AP	40	0	3900	0	7	11	VL	98	96	93	42	65	64	62
Reno AP	39	3	4404	−2	2	7	VL	95	92	90	45	64	62	61
Reno CO	39	3	4490	8	12	17	VL	94	92	89	45	64	62	61
Tonopah AP	38	0	5426	2	9	13	VL	95	92	90	40	64	63	62
Winnemucca AP	40	5	4299	−8	1	5	VL	97	95	93	42	64	62	61
NEW HAMPSHIRE														
Berlin	44	3	1110	−25	−19	−15	L	87	85	82	22	73	71	70
Claremont	43	2	420	−19	−13	−9	L	89	87	84	24	74	73	72
Concord AP	43	1	339	−17	−11	−7	M	91	88	85	26	75	73	72
Keene	43	0	490	−17	−12	−8	M	90	88	85	24	75	73	72
Laconia	43	3	505	−22	−16	−12	M	89	87	84	25	74	73	72
Manchester, Grenier AFB	43	0	253	−11	−5	1	M	92	89	86	24	76	74	73
Portsmouth, Pease AFB	43	1	127	−8	−2	3	M	88	86	83	22	75	73	72
NEW JERSEY														
Atlantic City CO	39	3	11	10	14	18	H	91	88	85	18	78	77	76
Long Branch	40	2	20	4	9	13	H	93	91	88	18	77	76	75
Newark AP	40	4	11	6	11	15	M	94	91	88	20	77	76	75
New Brunswick	40	3	86	3	8	12	M	91	89	86	19	77	76	75
Paterson	40	5	100	3	8	12	M	93	91	88	21	77	76	75
Phillipsburg	40	4	180	1	6	10	L	93	91	88	21	77	76	75
Trenton CO	40	1	144	7	12	16	M	92	90	87	19	78	77	76
Vineland	39	3	95	7	12	16	M	93	90	87	19	78	77	76
NEW MEXICO														
Alamagordo, Holloman AFB	32	5	4070	12	18	22	L	100	98	96	30	70	69	68
Albuquerque AP	35	0	5310	6	14	17	L	96	94	92	27	66	65	64
Artesia	32	5	3375	9	16	19	L	101	99	97	30	71	70	69
Carlsbad AP	32	2	3234	11	17	21	L	101	99	97	28	72	71	70
Clovis AP	34	3	4279	2	14	17	L	99	97	95	28	70	69	68
Farmington AP	36	5	5495	−3	6	9	VL	95	93	91	30	66	65	64
Gallup	35	3	6465	−13	−5	−1	VL	92	90	87	32	64	63	62
Grants	35	1	6520	−15	−7	−3	VL	91	89	86	32	64	63	62
Hobbs AP	32	4	3664	9	15	19	L	101	99	96	29	72	71	70
Las Cruces	32	2	3900	13	19	23	L	102	100	97	30	70	69	68
Los Alamos	35	5	7410	−4	5	9	L	88	86	83	32	64	63	62
Raton AP	36	5	6379	−11	−2	2	L	92	90	88	34	66	65	64
Roswell, Walker AFB	33	2	3643	5	16	19	L	101	99	97	33	71	70	69
Santa Fe CO	35	4	7045	−2	7	11	L	90	88	85	28	65	63	62
Silver City AP	32	4	5373	8	14	18	VL	95	93	91	30	68	67	66
Socorro AP	34	0	4617	6	13	17	L	99	97	94	30	67	66	65
Tucumcari AP	35	1	4053	1	9	13	L	99	97	95	28	71	70	69
NEW YORK														
Albany AP	42	5	277	−14	−5	0	L	91	88	85	23	76	74	73
Albany CO	42	5	19	−5	1	5	L	91	89	86	20	76	74	73
Auburn	43	0	715	−10	−2	2	M	89	87	84	22	75	73	72
Batavia	43	0	900	−7	−1	3	M	89	87	84	22	75	74	72
Binghamton CO	42	1	858	−8	−2	2	L	91	89	86	20	74	72	71
Buffalo AP	43	0	705r	−3	3	6	M	88	86	83	21	75	73	72
Cortland	42	4	1129	−11	−5	−1	L	90	88	85	23	75	73	72
Dunkirk	42	3	590	−2	4	8	M	88	86	83	18	75	74	72
Elmira AP	42	1	860	−5	1	5	L	92	90	87	24	75	73	72
Geneva	42	5	590	−8	−2	2	M	91	89	86	22	75	73	72
Glens Falls	43	2	321	−17	−11	−7	L	88	86	83	23	74	72	71
Gloversville	43	1	770	−12	−6	−2	L	89	87	84	23	75	73	71
Hornell	42	2	1325	−15	−9	−5	L	87	85	82	24	74	72	71

APPENDIX E DESIGN TEMPERATURE VALUES FOR THE UNITED STATES AND CANADA[a] (*continued*)

Col. 1 State and Station[b]	Col. 2 Latitude[c] °	′	Col. 3 Elev,[d] Ft	Winter: Col. 4 Median of Annual Extremes	99%	97½%	Col. 5 Coincident Wind Velocity[e]	Summer: Col. 6 Design Dry-Bulb 1%	2½%	5%	Col. 7 Outdoor Daily Range[f]	Col. 8 Design Wet-Bulb 1%	2½%	5%
NEW YORK (continued)														
Ithaca	42	3	950	−10	−4	0	L	91	88	85	24	75	73	72
Jamestown	42	1	1390	−5	1	5	M	88	86	83	20	75	73	72
Kingston	42	0	279	−8	−2	2	L	92	90	87	22	76	74	73
Lockport	43	1	520	−4	2	6	M	87	85	82	21	75	74	72
Massena AP	45	0	202r	−22	−16	−12	M	86	84	81	20	75	74	72
Newburgh-Stewart AFB	41	3	460	−4	2	6	M	92	89	86	21	78	76	74
NYC-Central Park	40	5	132	6	11	15	H	94	91	88	17	77	76	75
NYC-Kennedy AP	40	4	16	12	17	21	H	91	87	84	16	77	76	75
NYC-LaGuardia AP	40	5	19	7	12	16	H	93	90	87	16	77	76	75
Niagara Falls AP	43	1	596	−2	4	7	M	88	86	83	20	75	74	73
Olean	42	1	1420	−13	−8	−3	L	87	85	82	23	74	72	71
Oneonta	42	3	1150	−13	−7	−3	L	89	87	84	24	74	72	71
Oswego CO	43	3	300	−4	2	6	M	86	84	81	20	75	74	72
Plattsburg AFB	44	4	165	−16	−10	−6	L	86	84	81	22	74	73	71
Poughkeepsie	41	4	103	−6	−1	3	L	93	90	87	21	77	75	74
Rochester AP	43	1	543	−5	2	5	M	91	88	85	22	75	74	72
Rome-Griffiss AFB	43	1	515	−13	−7	−3	L	90	87	84	22	76	74	73
Schenectady	42	5	217	−11	−5	−1	L	90	88	85	22	75	73	72
Suffolk County AFB	40	5	57	4	9	13	H	87	84	81	16	76	75	74
Syracuse AP	43	1	424	−10	−2	2	M	90	87	85	20	76	74	73
Utica	43	1	714	−12	−6	−2	L	89	87	84	22	75	73	72
Watertown	44	0	497	−20	−14	−10	M	86	84	81	20	75	74	72
NORTH CAROLINA														
Asheville AP	35	3	2170r	8	13	17	L	91	88	86	21	75	74	73
Charlotte AP	35	1	735	13	18	22	L	96	94	92	20	78	77	76
Durham	36	0	406	11	15	19	L	94	92	89	20	78	77	76
Elizabeth City AP	36	2	10	14	18	22	M	93	91	89	18	80	79	78
Fayetteville, Pope AFB	35	1	95	13	17	20	L	97	94	92	20	80	79	78
Goldsboro, Seymour-Johnson AFB	35	2	88	14	18	21	M	95	92	90	18	80	79	78
Greensboro AP	36	1	897	9	14	17	L	94	91	89	21	77	76	75
Greenville	35	4	25	14	18	22	M	95	93	90	19	81	80	79
Henderson	36	2	510	8	12	16	L	94	92	89	20	79	78	77
Hickory	35	4	1165	9	14	18	L	93	91	88	21	77	76	75
Jacksonville	34	5	24	17	21	25	M	94	92	89	18	81	80	79
Lumberton	34	4	132	14	18	22	L	95	93	90	20	81	80	79
New Bern AP	35	1	17	14	18	22	L	94	92	89	18	81	80	79
Raleigh/Durham AP	35	5	433	13	16	20	L	95	92	90	20	79	78	77
Rocky Mount	36	0	81	12	16	20	L	95	93	90	19	80	79	78
Wilmington AP	34	2	30	19	23	27	L	93	91	89	18	82	81	80
Winston-Salem AP	36	1	967	9	14	17	L	94	91	89	20	77	76	75
NORTH DAKOTA														
Bismarck AP	46	5	1647	−31	−24	−19	VL	95	91	88	27	74	72	70
Devil's Lake	48	1	1471	−30	−23	−19	M	93	89	86	25	73	71	69
Dickinson AP	46	5	2595	−31	−23	−19	L	96	93	90	25	72	70	68
Fargo AP	46	5	900	−28	−22	−17	L	92	88	85	25	76	74	72
Grand Forks AP	48	0	832	−30	−26	−23	L	91	87	84	25	74	72	70
Jamestown AP	47	0	1492	−29	−22	−18	L	95	91	88	26	75	73	71
Minot AP	48	2	1713	−31	−24	−20	M	91	88	84	25	72	70	68
Williston	48	1	1877	−28	−21	−17	M	94	90	87	25	71	69	67
OHIO														
Akron/Canton AP	41	0	1210	−5	1	6	M	89	87	84	21	75	73	72
Ashtabula	42	0	690	−3	3	7	M	89	87	84	18	76	75	74
Athens	39	2	700	−3	3	7	M	93	91	88	22	77	76	75
Bowling Green	41	3	675	−7	−1	3	M	93	91	88	23	77	75	74
Cambridge	40	0	800	−6	0	4	M	91	89	86	23	77	76	75
Chillicothe	39	2	638	−1	5	9	M	93	91	88	22	77	76	75
Cincinnati CO	39	1	761	2	8	12	L	94	92	90	21	78	77	76
Cleveland AP	41	2	777r	−2	2	7	M	91	89	86	22	76	75	74
Columbus AP	40	0	812	−1	2	7	M	92	88	86	24	77	76	75
Dayton AP	39	5	997	−2	0	6	M	92	90	87	20	77	75	74
Defiance	41	2	700	−7	−1	1	M	93	91	88	24	77	76	74
Findlay AP	41	0	797	−6	0	4	M	92	90	88	24	77	76	75

APPENDIX E DESIGN TEMPERATURE VALUES FOR THE UNITED STATES AND CANADA[a] (*continued*)

Col. 1 State and Station[b]	Col. 2 Latitude[c] °	′	Col. 3 Elev,[d] Ft	Winter: Col. 4 Median of Annual Extremes	99%	97½%	Col. 5 Coincident Wind Velocity[e]	Summer: Col. 6 Design Dry-Bulb 1%	2½%	5%	Col. 7 Outdoor Daily Range[f]	Col. 8 Design Wet-Bulb 1%	2½%	5%
OHIO (continued)														
Fremont	41	2	600	−7	−1	3	M	92	90	87	24	76	75	74
Hamilton	39	2	650	−2	4	8	M	94	92	90	22	78	77	76
Lancaster	39	4	920	−5	1	5	M	93	91	88	23	77	76	75
Lima	40	4	860	−6	0	4	M	93	91	88	24	77	76	75
Mansfield AP	40	5	1297	−7	1	3	M	91	89	86	22	76	75	74
Marion	40	4	920	−5	1	6	M	93	91	88	23	77	76	75
Middletown	39	3	635	−3	3	7	M	93	91	88	22	77	76	75
Newark	40	1	825	−7	−1	3	M	92	90	87	23	77	76	75
Norwalk	41	1	720	−7	−1	3	M	92	90	87	22	76	75	74
Portsmouth	38	5	530	0	5	9	L	94	92	89	22	77	76	75
Sandusky CO	41	3	606	−2	4	8	M	92	90	87	21	76	75	74
Springfield	40	0	1020	−3	3	7	M	93	90	88	21	77	76	75
Steubenville	40	2	992	−2	4	9	M	91	89	86	22	76	75	74
Toledo AP	41	4	676r	−5	1	5	M	92	90	87	25	77	75	74
Warren	41	2	900	−6	0	4	M	90	88	85	23	75	74	73
Wooster	40	5	1030	−7	−1	3	M	90	88	85	22	76	75	74
Youngstown AP	41	2	1178	−5	1	6	M	89	86	84	23	75	74	73
Zanesville AP	40	0	881	−7	−1	3	M	92	89	87	23	77	76	75
OKLAHOMA														
Ada	34	5	1015	6	12	16	H	102	100	98	23	79	78	77
Altus AFB	34	4	1390	7	14	18	H	103	101	99	25	77	76	75
Ardmore	34	2	880	9	15	19	H	103	101	99	23	79	78	77
Bartlesville	36	5	715	−1	5	9	H	101	99	97	23	79	78	77
Chickasha	35	0	1085	5	12	16	H	103	101	99	24	77	76	75
Enid-Vance AFB	36	2	1287	3	10	14	H	103	100	98	24	78	77	76
Lawton AP	34	3	1108	6	13	16	H	103	101	98	24	78	77	76
McAlester	34	5	760	7	13	17	H	102	100	98	23	79	78	77
Muskogee AP	35	4	610	6	12	16	M	102	99	96	23	79	78	77
Norman	35	1	1109	5	11	15	H	101	99	97	24	78	77	76
Oklahoma City AP	35	2	1280	4	11	15	H	100	97	95	23	78	77	76
Ponca City	36	4	996	1	8	12	H	102	100	97	24	78	77	76
Seminole	35	2	865	6	12	16	H	102	100	98	23	78	77	76
Stillwater	36	1	884	2	9	13	H	101	99	97	24	78	77	76
Tulsa AP	36	1	650	4	12	16	H	102	99	96	22	79	78	77
Woodward	36	3	1900	−3	4	8	H	103	101	98	26	76	74	73
OREGON														
Albany	44	4	224	17	23	27	VL	91	88	84	31	69	67	65
Astoria AP	46	1	8	22	27	30	M	79	76	72	16	61	60	59
Baker AP	44	5	3368	−10	−3	1	VL	94	92	89	30	66	65	63
Bend	44	0	3599	−7	0	4	VL	89	87	84	33	64	62	61
Corvallis	44	3	221	17	23	27	VL	91	88	84	31	69	67	65
Eugene AP	44	1	364	16	22	26	VL	91	88	84	31	69	67	65
Grants Pass	42	3	925	16	22	26	VL	94	92	89	33	68	66	65
Klamath Falls AP	42	1	4091	−5	1	5	VL	89	87	84	36	63	62	61
Medford AP	42	2	1298	15	21	23	VL	98	94	91	35	70	68	66
Pendleton AP	45	4	1492	−2	3	10	VL	97	94	91	29	66	65	63
Portland AP	45	4	21	17	21	24	L	89	85	81	23	69	67	66
Portland CO	45	3	57	21	26	29	L	91	88	84	21	69	68	67
Roseburg AP	43	1	505	19	25	29	VL	93	91	88	30	69	67	65
Salem AP	45	0	195	15	21	25	VL	92	88	84	31	69	67	66
The Dalles	45	4	102	7	13	17	VL	93	91	88	28	70	68	67
PENNSYLVANIA														
Allentown AP	40	4	376	−2	3	5	M	92	90	87	22	77	75	74
Altoona CO	40	2	1468	−4	1	5	L	89	87	84	23	74	73	72
Butler	40	4	1100	−8	−2	2	L	91	89	86	22	75	74	73
Chambersburg	40	0	640	0	5	9	L	94	92	89	23	76	75	74
Erie AP	42	1	732	1	7	11	M	88	85	82	18	76	74	73
Harrisburg AP	40	1	335	4	9	13	L	92	89	86	21	76	75	74
Johnstown	40	2	1214	−4	1	5	L	91	87	85	23	74	73	72
Lancaster	40	1	255	−3	2	6	L	92	90	87	22	77	76	75
Meadville	41	4	1065	−6	0	4	M	88	86	83	21	75	73	72
New Castle	41	0	825	−7	−1	4	M	91	89	86	23	75	74	73

APPENDIX E DESIGN TEMPERATURE VALUES FOR THE UNITED STATES AND CANADA[a] *(continued)*

Col. 1 State and Station[b]	Col. 2 Latitude[c] ° ′	Col. 3 Elev,[d] Ft	Winter: Col. 4 Median of Annual Extremes	Col. 4 99%	Col. 4 97½%	Col. 5 Coincident Wind Velocity[e]	Summer: Col. 6 Design Dry-Bulb 1%	Col. 6 2½%	Col. 6 5%	Col. 7 Outdoor Daily Range[f]	Col. 8 Design Wet-Bulb 1%	Col. 8 2½%	Col. 8 5%
PENNSYLVANIA (continued)													
Philadelphia AP	39 5	7	7	11	15	M	93	90	87	21	78	77	76
Pittsburgh AP	40 3	1137	− 1	5	9	M	90	87	85	22	75	74	73
Pittsburgh CO	40 3	749r	1	7	11	M	90	88	85	19	75	74	73
Reading CO	40 2	226	1	6	9	M	92	90	87	19	77	76	75
Scranton/Wilkes-Barre	41 2	940	− 3	2	6	L	89	87	84	19	75	74	73
State College	40 5	1175	− 3	2	6	L	89	87	84	23	74	73	72
Sunbury	40 5	480	− 2	3	7	L	91	89	86	22	76	75	74
Uniontown	39 5	1040	− 1	4	8	L	90	88	85	22	75	74	73
Warren	41 5	1280	− 8	− 3	1	L	89	87	84	24	75	73	72
West Chester	40 0	440	4	9	13	M	92	90	87	20	77	76	75
Williamsport AP	41 1	527	− 5	1	5	L	91	89	86	23	76	75	74
York	40 0	390	− 1	4	8	L	93	91	88	22	77	76	75
RHODE ISLAND													
Newport	41 3	20	1	5	11	H	86	84	81	16	75	74	73
Providence AP	41 4	55	0	6	10	M	89	86	83	19	76	75	74
SOUTH CAROLINA													
Anderson	34 3	764	13	18	22	L	96	94	91	21	77	76	75
Charleston AFB	32 5	41	19	23	27	L	94	92	90	18	81	80	79
Charleston CO	32 5	9	23	26	30	L	95	93	90	13	81	80	79
Columbia AP	34 0	217	16	20	23	L	98	96	94	22	79	79	78
Florence AP	34 1	146	16	21	25	L	96	94	92	21	80	79	78
Georgetown	33 2	14	19	23	26	L	93	91	88	18	81	80	79
Greenville AP	34 5	957	14	19	23	L	95	93	91	21	77	76	75
Greenwood	34 1	671	15	19	23	L	97	95	92	21	78	77	76
Orangeburg	33 3	244	17	21	25	L	97	95	92	20	80	79	78
Rock Hill	35 0	470	13	17	21	L	97	95	92	20	78	77	76
Spartanburg AP	35 0	816	13	18	22	L	95	93	90	20	77	76	75
Sumter-Shaw AFB	34 0	291	18	23	26	L	96	94	92	21	80	79	78
SOUTH DAKOTA													
Aberdeen AP	45 3	1296	−29	−22	−18	L	95	92	89	27	77	75	74
Brookings	44 2	1642	−26	−19	−15	M	93	90	87	25	77	75	74
Huron AP	44 3	1282	−24	−16	−12	L	97	93	90	28	77	75	74
Mitchell	43 5	1346	−22	−15	−11	M	96	94	91	28	77	76	74
Pierre AP	44 2	1718r	−21	−13	− 9	M	98	96	93	29	76	74	73
Rapid City AP	44 0	3165	−17	− 9	− 6	M	96	94	91	28	72	71	69
Sioux Falls AP	43 4	1420	−21	−14	−10	M	95	92	89	24	77	75	74
Watertown AP	45 0	1746	−27	−20	−16	L	93	90	87	26	76	74	73
Yankton	43 0	1280	−18	−11	− 7	M	96	94	91	25	78	76	75
TENNESSEE													
Athens	33 3	940	10	14	18	L	96	94	91	22	77	76	75
Bristol-Tri City AP	36 3	1519	−1	11	16	L	92	90	88	22	76	75	74
Chattanooga AP	35 0	670	11	15	19	L	97	94	92	22	78	78	77
Clarksville	36 4	470	6	12	16	L	98	96	94	21	79	78	77
Columbia	35 4	690	8	13	17	L	97	95	93	21	79	78	77
Dyersburg	36 0	334	7	13	17	L	98	96	94	21	80	79	78
Greenville	35 5	1320	5	10	14	L	93	91	88	22	76	75	74
Jackson AP	35 4	413	8	14	17	L	97	95	94	21	80	79	78
Knoxville AP	35 5	980	9	13	17	L	95	92	90	21	77	76	75
Memphis AP	35 0	263	11	17	21	L	98	96	94	21	80	79	78
Murfreesboro	35 5	608	7	13	17	L	97	94	92	22	79	78	77
Nashville AP	36 1	577	6	12	16	L	97	95	92	21	79	78	77
Tullahoma	35 2	1075	7	13	17	L	96	94	92	22	79	78	77
TEXAS													
Abilene AP	32 3	1759	12	17	21	M	101	99	97	22	76	75	74
Alice AP	27 4	180	26	30	34	M	101	99	97	20	81	80	79
Amarillo AP	35 1	3607	2	8	12	M	98	96	93	26	72	71	70
Austin AP	30 2	597	19	25	29	M	101	98	96	22	79	78	77
Bay City	29 0	52	25	29	33	M	95	93	91	16	81	80	79
Beaumont	30 0	18	25	29	33	M	96	94	93	19	81	80	79
Beeville	28 2	225	24	28	32	M	99	97	96	18	81	80	79
Big Spring AP	32 2	2537	12	18	22	M	100	98	96	26	75	73	72
Brownsville AP	25 5	16	32	36	40	M	94	92	91	18	80	80	79
Brownwood	31 5	1435	15	20	25	M	102	100	98	22	76	75	74
Bryan AP	30 4	275	22	27	31	M	100	98	96	20	79	78	78

APPENDIX E DESIGN TEMPERATURE VALUES FOR THE UNITED STATES AND CANADA[a] (*continued*)

Col. 1 State and Station[b]	Col. 2 Latitude[c] °	Col. 2 Latitude[c] ′	Col. 3 Elev,[d] Ft	Winter: Col. 4 Median of Annual Extremes	Winter: Col. 4 99%	Winter: Col. 4 97½%	Winter: Col. 5 Coincident Wind Velocity[e]	Summer: Col. 6 Design Dry-Bulb 1%	Summer: Col. 6 Design Dry-Bulb 2½%	Summer: Col. 6 Design Dry-Bulb 5%	Summer: Col. 7 Outdoor Daily Range[f]	Summer: Col. 8 Design Wet-Bulb 1%	Summer: Col. 8 Design Wet-Bulb 2½%	Summer: Col. 8 Design Wet-Bulb 5%
TEXAS (continued)														
Corpus Christi AP	27	5	43	28	32	36	M	95	93	91	19	81	80	80
Corsicana	32	0	425	16	21	25	M	102	100	98	21	79	78	77
Dallas AP	32	5	481	14	19	24	H	101	99	97	20	79	78	78
Del Rio, Laughlin AFB	29	2	1072	24	28	31	M	101	99	98	24	79	77	76
Denton	33	1	655	12	18	22	H	102	100	98	22	79	78	77
Eagle Pass	28	5	743	23	27	31	L	106	104	102	24	80	79	78
El Paso AP	31	5	3918	16	21	25	L	100	98	96	27	70	69	68
Fort Worth AP	32	5	544r	14	20	24	H	102	100	98	22	79	78	77
Galveston AP	29	2	5	28	32	36	M	91	89	88	10	82	81	81
Greenville	33	0	575	13	19	24	H	101	99	97	21	79	78	78
Harlingen	26	1	37	30	34	38	M	96	95	94	19	80	80	79
Houston AP	29	4	50	23	28	32	M	96	94	92	18	80	80	79
Houston CO	29	5	158r	24	29	33	M	96	94	92	18	80	80	79
Huntsville	30	4	494	22	27	31	M	99	97	96	20	80	79	78
Killeen-Gray AFB	31	0	1021	17	22	26	M	100	99	97	22	78	77	76
Lamesa	32	5	2965	7	14	18	M	100	98	96	26	74	73	72
Laredo AFB	27	3	503	29	32	36	L	103	101	100	23	79	78	78
Longview	32	2	345	16	21	25	M	100	98	96	20	81	80	79
Lubbock AP	33	4	3243	4	11	15	M	99	97	94	26	73	72	71
Lufkin AP	31	1	286	19	24	28	M	98	96	95	20	81	80	79
McAllen	26	1	122	30	34	38	M	102	100	98	21	80	79	78
Midland AP	32	0	2815r	13	19	23	M	100	98	96	26	74	73	72
Mineral Wells AP	32	5	934	12	18	22	H	102	100	98	22	78	77	76
Palestine CO	31	5	580	16	21	25	M	99	97	96	20	80	79	78
Pampa	35	3	3230	0	7	11	M	100	98	95	26	73	72	71
Pecos	31	2	2580	10	15	19	L	102	100	97	27	72	71	70
Plainview	34	1	3400	3	10	14	M	100	98	95	26	73	72	71
Port Arthur AP	30	0	16	25	29	33	M	94	92	91	19	81	80	80
San Angelo, Goodfellow AFB	31	2	1878	15	20	25	M	101	99	97	24	76	75	74
San Antonio AP	29	3	792	22	25	30	L	99	97	96	19	77	77	76
Sherman-Perrin AFB	33	4	763	12	18	23	H	101	99	97	22	79	78	77
Snyder	32	4	2325	9	15	19	M	102	100	97	26	75	74	73
Temple	31	1	675	18	23	27	M	101	99	97	22	79	78	77
Tyler AP	32	2	527	15	20	24	M	99	97	96	21	80	79	78
Vernon	34	1	1225	7	14	18	H	103	101	99	24	77	76	75
Victoria AP	28	5	104	24	28	32	M	98	96	95	18	80	79	79
Waco AP	31	4	500	16	21	26	M	101	99	98	22	79	78	78
Wichita Falls AP	34	0	994	9	15	19	H	103	100	98	24	77	76	75
UTAH														
Cedar City AP	37	4	5613	−10	−1	6	VL	94	91	89	32	65	64	62
Logan	41	4	4775	−7	3	7	VL	93	91	89	33	66	65	63
Moab	38	5	3965	2	12	16	VL	100	98	95	30	66	65	64
Ogden CO	41	1	4400	−3	7	11	VL	94	92	89	33	66	65	64
Price	39	4	5580	−7	3	7	L	93	91	88	33	65	64	63
Provo	40	1	4470	−6	2	6	L	96	93	91	32	67	66	65
Richfield	38	5	5300	−10	−1	3	L	94	92	89	34	66	65	64
St. George CO	37	1	2899	13	22	26	VL	104	102	99	33	71	70	69
Salt Lake City AP	40	5	4220	−2	5	9	L	97	94	92	32	67	66	65
Vernal AP	40	3	5280	−20	−10	−6	VL	90	88	84	32	64	63	62
VERMONT														
Barre	44	1	1120	−23	−17	−13	L	86	84	81	23	73	72	70
Burlington AP	44	3	331	−18	−12	−7	M	88	85	83	23	74	73	71
Rutland	43	3	620	−18	−12	−8	L	87	85	82	23	74	73	71
VIRGINIA														
Charlottsville	38	1	870	7	11	15	L	93	90	88	23	79	77	76
Danville AP	36	3	590	9	13	17	L	95	92	90	21	78	77	76
Fredericksburg	38	2	50	6	10	14	M	94	92	89	21	79	78	76
Harrisonburg	38	3	1340	0	5	9	L	92	90	87	23	78	77	76
Lynchburg AP	37	2	947	10	15	19	L	94	92	89	21	77	76	75
Norfolk AP	36	5	26	18	20	23	M	94	91	89	18	79	78	78
Petersburg	37	1	194	10	15	18	L	96	94	91	20	80	79	78

APPENDIX E DESIGN TEMPERATURE VALUES FOR THE UNITED STATES AND CANADA[a] (*continued*)

Col. 1 State and Station[b]	Col. 2 Latitude[c] °	′	Col. 2 Elev.[d] Ft	Winter: Col. 4 Median of Annual Extremes	Col. 4 99%	Col. 4 97½%	Col. 5 Coincident Wind Velocity[e]	Summer: Col. 6 Design Dry-Bulb 1%	2½%	5%	Col. 6 Outdoor Daily Range[f]	Col. 8 Design Wet-Bulb 1%	2½%	5%
VIRGINIA (continued)														
Richmond AP	37	3	162	10	14	18	L	96	93	91	21	79	78	77
Roanoke AP	37	2	1174r	9	15	18	L	94	91	89	23	76	75	74
Staunton	38	2	1480	3	8	12	L	92	90	87	23	78	77	75
Winchester	39	1	750	1	6	10	L	94	92	89	21	78	76	75
WASHINGTON														
Aberdeen	47	0	12	19	24	27	M	83	80	77	16	62	61	60
Bellingham AP	48	5	150	8	14	18	L	76	74	71	19	67	65	63
Bremerton	47	3	162	17	24	29	L	85	81	77	20	68	66	65
Ellensburg AP	47	0	1729	−5	2	6	VL	91	89	86	34	67	65	63
Everett-Paine AFB	47	5	598	13	19	24	L	82	78	74	20	67	65	63
Kennewick	46	0	392	4	11	15	VL	98	96	93	30	69	68	66
Longview	46	1	12	14	20	24	L	88	86	83	30	68	66	65
Moses Lake, Larson AFB	47	1	1183	−14	−7	−1	VL	96	93	90	32	68	66	65
Olympia AP	47	0	190	15	21	25	L	85	83	80	32	67	65	63
Port Angeles	48	1	99	20	26	29	M	75	73	70	18	60	58	57
Seattle-Boeing Fld	47	3	14	17	23	27	L	82	80	77	24	67	65	64
Seattle CO	47	4	14	22	28	32	L	81	79	76	19	67	65	64
Seattle-Tacoma AP	47	3	386	14	20	24	L	85	81	77	22	66	64	63
Spokane AP	47	4	2357	−5	−2	4	VL	93	90	87	28	66	64	63
Tacoma-McChord AFB	47	1	350	14	20	24	L	85	81	78	22	68	66	64
Walla Walla AP	46	1	1185	5	12	16	VL	98	96	93	27	69	68	66
Wenatchee	47	2	634	−2	5	9	VL	95	92	89	32	68	66	64
Yakima AP	46	3	1061	−1	6	10	VL	94	92	89	36	69	67	65
WEST VIRGINIA														
Beckley	37	5	2330	−4	0	6	L	91	88	86	22	74	73	72
Bluefield AP	37	2	2850	1	6	10	L	88	86	83	22	74	73	72
Charleston AP	38	2	939	1	9	14	L	92	90	88	20	76	75	74
Clarksburg	39	2	977	−2	3	7	L	92	90	87	21	76	75	74
Elkins AP	38	5	1970	−4	1	5	L	87	84	82	22	74	73	72
Huntington CO	38	2	565r	4	10	14	L	95	93	91	22	77	76	75
Martinsburg AP	39	2	537	1	6	10	L	96	94	91	21	78	77	76
Morgantown AP	39	4	1245	−2	3	7	L	90	88	85	22	76	74	73
Parkersburg CO	39	2	615r	2	8	12	L	93	91	88	21	77	76	75
Wheeling	40	1	659	0	5	9	L	91	89	86	21	76	75	74
WISCONSIN														
Appleton	44	2	742	−16	−10	−6	M	89	87	84	23	75	74	72
Ashland	46	3	650	−27	−21	−17	L	85	83	80	23	73	71	69
Beloit	42	3	780	−13	−7	−3	M	92	90	87	24	77	76	75
Eau Claire AP	44	5	888	−21	−15	−11	L	90	88	85	23	76	74	72
Fond du Lac	43	5	760	−17	−11	−7	M	89	87	84	23	76	74	73
Green Bay AP	44	3	683	−16	−12	−7	M	88	85	82	23	75	73	72
La Crosse AP	43	5	652	−18	−12	−8	M	90	88	85	22	78	76	75
Madison AP	43	1	858	−13	−9	−5	M	92	88	85	22	77	75	73
Manitowoc	44	1	660	−11	−5	−1	M	88	86	83	21	75	74	72
Marinette	45	0	605	−14	−8	−4	M	88	86	83	20	74	72	70
Milwaukee AP	43	0	672	−11	−6	−2	M	90	87	84	21	77	75	73
Racine	42	4	640	−10	−4	0	M	90	88	85	21	77	75	73
Sheboygan	43	4	648	−10	−4	0	M	89	87	84	20	76	74	72
Stevens Point	44	3	1079	−22	−16	−12	M	89	87	84	23	75	73	71
Waukesha	43	0	860	−12	−6	−2	M	91	89	86	22	77	75	74
Wausau AP	44	6	1196	−24	−18	−14	M	89	86	83	23	74	72	70
WYOMING														
Casper AP	42	5	5319	−20	−11	−5	L	92	90	87	31	63	62	60
Cheyenne AP	41	1	6126	−15	−6	−2	M	89	86	83	30	63	62	61
Cody AP	44	3	5090	−23	−13	−9	L	90	87	84	32	61	60	59
Evanston	41	2	6860	−22	−12	−8	VL	84	82	79	32	58	57	56
Lander AP	42	5	5563	−26	−16	−12	VL	92	90	87	32	63	62	60
Laramie AP	41	2	7266	−17	−6	−2	M	82	80	77	28	61	59	58
Newcastle	43	5	4480	−18	−9	−5	M	92	89	86	30	68	67	66
Rawlins	41	5	6736	−24	−15	−11	L	86	84	81	40	62	61	60
Rock Springs AP	41	4	6741	−16	−6	−1	VL	86	84	82	32	58	57	56
Sheridan AP	44	5	3942	−21	−12	−7	L	95	92	89	32	67	65	64
Torrington	42	0	4098	−20	−11	−7	M	94	92	89	30	68	67	66

APPENDIX E DESIGN TEMPERATURE VALUES FOR THE UNITED STATES AND CANADA[a] (*continued*)

CANADA

Col. 1 Province and Station[b]	Col. 2 Latitude[c] °	′	Col. 3 Elev,[d] Ft	Winter Col. 4 Average Annual Minimum	Winter Col. 4 99%	Winter Col. 4 97½%	Winter Col. 5 Coincident Wind Velocity[e]	Summer Col. 6 Design Dry-Bulb 1%	Col. 6 2½%	Col. 6 5%	Summer Col. 7 Outdoor Daily Range[f]	Summer Col. 8 Design Wet-Bulb 1%	Col. 8 2½%	Col. 8 5%
ALBERTA														
Calgary AP	51	1	3540	−30	−29	−25	M	87	85	82	26	66	64	63
Edmonton AP	53	3	2219	−30	−29	−26	VL	86	83	80	23	69	67	65
Grande Prairie AP	55	1	2190	−44	−43	−37	VL	84	81	78	23	66	64	63
Jasper CO	52	5	3480	−38	−32	−28	VL	87	84	81	28	66	64	63
Lethbridge AP	49	4	3018	−31	−31	−24	M	91	88	85	28	68	66	64
McMurray AP	56	4	1216	−44	−42	−39	VL	87	84	81	28	69	67	65
Medicine Hat AP	50	0	2365	−33	−30	−26	M	96	93	90	28	72	69	67
Red Deer AP	52	1	2965	−38	−33	−28	VL	88	86	83	25	67	65	64
BRITISH COLUMBIA														
Dawson Creek	55	5	2200	−47	−40	−35	L	84	81	78	25	66	64	63
Fort Nelson AP	58	5	1230	−43	−44	−41	VL	87	84	81	23	66	64	63
Kamloops CO	50	4	1150	−15	−16	−10	VL	97	94	91	31	71	69	68
Nanaimo CO	49	1	100	16	17	20	VL	81	78	75	20	66	64	62
New Westminster CO	49	1	50	12	15	19	VL	86	84	82	20	68	66	65
Penticton AP	49	3	1121	0	− 1	3	L	94	91	88	31	71	69	68
Prince George AP	53	5	2218	−38	−37	−31	VL	85	82	79	26	68	65	63
Prince Rupert CO	54	2	170	9	11	15	L	73	71	69	13	62	60	59
Trail	49	1	1400	− 3	− 2	3	VL	94	91	88	30	70	68	67
Vancouver AP	49	1	16	13	15	19	L	80	78	76	17	68	66	65
Victoria CO	48	3	228	20	20	23	M	80	76	72	16	64	62	60
MANITOBA														
Brandon CO	49	5	1200	−36	−29	−26	M	90	87	84	26	75	73	71
Churchill AP	58	5	115	−43	−40	−38	H	79	75	72	18	68	66	63
Dauphin AP	51	1	999	−35	−29	−26	M	89	86	83	24	74	72	70
Flin Flon CO	54	5	1098	−38	−40	−36	L	85	81	78	19	71	69	67
Portage la Prairie AP	49	5	867	−28	−25	−22	M	90	87	84	22	75	74	72
The Pas AP	54	0	894	−41	−35	−32	M	85	81	78	20	73	71	69
Winnipeg AP	49	5	786	−31	−28	−25	M	90	87	84	23	75	74	72
NEW BRUNSWICK														
Campbellton CO	48	0	25	−20	−18	−14	L	87	84	81	20	74	71	69
Chatham AP	47	0	112	−17	−15	−10	M	90	87	84	22	74	71	69
Edmundston CO	47	2	500	−29	−20	−16	M	84	81	78	21	75	72	70
Fredericton AP	45	5	74	−19	−16	−10	L	89	86	83	23	73	70	68
Moncton AP	46	1	248	−16	−12	− 7	H	88	85	82	21	74	71	69
Saint John AP	45	2	352	−15	−12	− 7	M	81	79	77	18	71	68	66
NEWFOUNDLAND														
Corner Brook CO	49	0	40	− 9	−10	− 5	H	84	81	79	18	69	68	66
Gander AP	49	0	482	− 5	− 5	− 1	H	85	82	79	20	69	68	66
Goose Bay AP	53	2	144	−28	−27	−25	M	86	81	77	18	69	67	65
St. John's AP	47	4	463	1	2	6	H	79	77	75	17	69	68	66
Stephenville	48	3	44	− 4	− 6	− 1	H	79	76	74	13	69	68	66
NORTHWEST TERRITORIES														
Fort Smith AP	60	0	665	−51	−49	−46	VL	85	83	80	25	67	65	64
Frobisher Bay AP	63	5	68	−45	−45	−42	H	63	59	56	14			
Inuvik	68	2	75	−54	−50	−48	VL	80	77	75	23	63	61	60
Resolute AP	74	4	209	−52	−49	−47	M	54	51	49	10			
Yellowknife AP	62	3	682	−51	−49	−47	VL	78	76	74	17	65	63	62
NOVA SCOTIA														
Amherst	45	5	63	−15	−10	− 5	H	85	82	79	21	72	70	68
Halifax AP	44	4	136	− 4	0	4	H	83	80	77	16	69	68	67
Kentville CO	45	0	50	− 8	− 4	0	M	86	83	80	23	72	70	69
New Glasgow	45	4	317	−16	−10	− 5	H	84	81	79	21	72	70	68
Sydney AP	46	1	197	− 3	0	5	H	84	82	80	20	72	70	68
Truro CO	45	2	77	−17	−12	− 7	M	84	81	79	22	72	70	69
Yarmouth AP	43	5	136	2	5	9	H	76	73	71	15	69	68	67
ONTARIO														
Belleville CO	44	1	250	−15	−11	− 7	M	89	86	84	21	77	75	73
Chatham CO	42	2	600	− 1	3	6	M	92	90	88	20	77	75	74
Cornwall	45	0	210	−22	−14	− 9	M	89	86	84	23	77	75	74
Fort William AP	48	2	644	−31	−27	−23	L	86	83	80	23	72	70	68
Hamilton	43	2	303	− 2	0	3	M	91	88	86	21	77	75	73
Kapuskasing AP	49	3	752	−37	−31	−28	M	87	84	81	23	73	71	69
Kenora AP	49	5	1345	−33	−31	−28	M	86	83	80	20	75	73	71
Kingston CO	44	2	300	−16	−10	− 7	M	85	82	80	20	77	75	73

APPENDIX F HEAT GAIN THROUGH FLAT ROOFS

Description of Roof Construction[a,b]	Wt, lb per sq ft	U value Btu/(hr)(ft²)(F°)	Sun Time A.M. 8 D	8 L	10 D	10 L	12 D	12 L	P.M. 2 D	2 L	4 D	4 L	6 D	6 L	8 D	8 L	10 D	10 L	12 D	12 L	λ	δ
Light Construction Roofs—Exposed to Sun																						
1″ insulation + steel siding	7.4	0.213	28	11	65	31	90	48	95	53	78	45	43	27	8	6	1	1	−3	−3	1.0	0
2″ insulation + steel siding	7.8	0.125	24	8	61	29	88	46	96	53	81	46	48	30	10	8	2	2	−3	−3	0.99	1
1″ insulation + 1″ wood[c]	8.4	0.206	12	2	47	21	77	39	92	50	86	48	61	36	25	16	7	5	0	−1	0.93	2
2″ insulation + 1″ wood[c]	8.5	0.122	8	0	41	18	72	36	90	48	88	49	65	38	30	19	9	7	1	0	0.93	2
1″ insulation + 2.5″ wood[c]	12.7	0.193	2	−2	23	8	48	23	70	36	79	42	71	40	50	29	29	17	15	9	0.73	3
2″ insulation + 2.5″ wood[c]	13.1	0.117	1	−2	19	6	43	20	65	33	76	41	72	40	53	31	33	20	18	11	0.68	4
Medium Construction Roofs—Exposed to Sun																						
1″ insulation + 4″ wood[c]	17.3	0.183	5	0	14	5	31	14	49	24	62	32	65	35	56	31	41	24	29	17	0.51	5
2″ insulation + 4″ wood[c]	17.8	0.113	6	1	13	4	28	12	45	22	58	30	63	34	56	31	43	25	32	18	0.48	5
1″ insulation + 2″ h.w. concrete	28.3	0.206	4	−1	27	11	54	26	74	39	81	44	70	40	45	27	24	15	12	7	0.75	3
2″ insulation + 2″ h.w. concrete	28.8	0.122	2	−2	23	9	49	23	70	36	79	43	71	40	49	29	28	17	15	9	0.73	3
4″ l.w. concrete	17.8	0.213	1	−3	28	11	59	28	82	43	88	48	74	42	44	27	19	12	6	4	0.87	3
6″ l.w. concrete	24.5	0.157	−2	−4	9	2	31	13	55	27	72	38	76	41	64	36	42	25	25	15	0.67	5
8″ l.w. concrete	31.2	0.125	6	2	6	1	16	6	32	14	49	24	61	32	63	34	55	31	41	24	0.50	6
Heavy Construction Roofs—Exposed to Sun																						
1″ insulation + 4″ h.w. concrete	51.6	0.199	7	1	17	6	33	15	50	25	61	32	63	34	53	30	40	23	28	16	0.48	5
2″ insulation + 4″ h.w. concrete	52.1	0.120	7	2	15	6	30	13	46	23	58	30	61	33	54	30	41	23	31	17	0.45	5
1″ insulation + 6″ h.w. concrete	75.0	0.193	13	6	17	7	26	12	38	18	48	25	53	28	51	27	43	24	35	19	0.33	6
2″ insulation + 6″ h.w. concrete	75.4	0.117	15	7	17	7	25	11	36	17	46	23	51	27	50	27	43	24	36	20	0.30	6

Roofs Covered with Water—Exposed to Sun

	Outside Air Dew Point (F)	Water Layer Thickness (in.)	A.M. 8	10	12	P.M. 2	4	6	8	10	12
Light Construction	60	6	−6	−6	−1	6	13	17	17	13	7
		1	−12	−6	4	15	21	22	17	8	0
		0	−12	−4	7	17	23	22	16	5	−3
	70	6	−1	0	4	11	18	21	21	17	12
		1	−5	0	10	19	25	26	21	12	5
		0	−5	2	12	21	26	26	19	9	2
Heavy Construction	60	6	−3	−4	−1	4	9	13	15	13	10
		1	−8	−6	1	8	15	18	17	11	6
		0	−9	−5	2	10	16	19	16	10	4
	70	6	2	2	4	9	14	18	20	18	15
		1	−2	0	6	14	20	23	21	16	11
		0	−2	1	8	16	21	23	21	15	9

[a] Includes outside surface resistance, ½″ slag, membrane and ⅜″ felt on the top (code number A4 of Table 41) and inside surface resistance on the bottom (code number E0 of Table 41). The property data for components are listed in Table 41.

[b] Dark roof, $\alpha/h_o = 0.30$; light roof, $\alpha/h_o = 0.15$.

[c] Nominal thickness of wood.

Explanation: {Total heat transmission from solar radiation and temperature difference between outdoor and room air. Btu per (hr) (sq ft) of roof area} = {Equivalent Temperature Differential from above table} × {Heat transmission coefficient for summer, Btu per (hr) (sq ft) (F deg)}

1. *Application.* These values may be used for all normal air conditioning estimated; usually without correction (except as noted below) in latitude 0 deg to 50 deg north or south when the load is calculated for the hottest weather.

2. *Corrections.* The values in the table were calculated for an inside temperature of 75 F and an outdoor maximum temperature of 95 F with an outdoor daily range of 21 F deg. The table remains approximately correct for other outdoor maximums (93–102 F) and other outdoor daily ranges (16–34 F deg) provided the outdoor daily average temperature remains approximately 85 F. If the room air temperature is different from 75 F and/or the outdoor daily average temperature is different from 85 F, Equation 43 can be used for computing new values or the following rules can be applied:

a. For room air temperature less than 75 F, add the difference between 75 F and room air temperature; if greater than 75 F, subtract the difference.

b. For outdoor daily average temperature less than 85 F, subtract the difference between 85 F and the daily average temperature; if greater than 85 F, add the difference.

3. *Attics or other spaces between the roof and ceiling.* If the ceiling is insulated and a fan is used for positive ventilation in the space between the ceiling and roof, the total temperature differential for calculating the room load may be decreased by 25 percent.

If the attic space contains a return duct or other air plenum, care should be taken in determining the portion of the heat gain that reaches the ceiling.

4. *Light Colors.* Credit should not be taken for light colored roofs except where the permanence of light color is established by experience, as in rural areas or where there is little smoke.

5. *For solar transmission in other months.* The table values of temperature differentials that were calculated for July 21 will be approximately correct for a roof in the following months:

North Latitude: Latitude (deg)	North Latitude: Months	South Latitude: Latitude (deg)	South Latitude: Months
0	All Months	0	All Months
10	All Months	10	All Months
20	All Months except Nov., Dec, Jan.	20	All Months except May, June, July
30	Mar., Apr., May, June, July, Aug., Sept.	30	Sept., Oct., Nov., Dec., Jan., Feb., March
40	April, May, June, July, Aug.	40	Oct., Nov., Dec., Jan., Feb.
50	May, June, July	50	Nov., Dec., Jan.

Source: ASHRAE Handbook of Fundamentals.

APPENDIX G SOLAR INSOLATION IN BTU/FT^2

SOLAR POSITION AND INSOLATION VALUES FOR 24 DEGREES NORTH LATITUDE

DATE	SOLAR TIME AM	SOLAR TIME PM	SOLAR POSITION ALT	SOLAR POSITION AZM	BTUH/SQ. FT. TOTAL INSOLATION ON SURFACES NORMAL	HORIZ.	SOUTH FACING SURFACE ANGLE WITH HORIZ. 14	24	34	54	90
JAN 21	7	5	4.8	65.6	71	10	17	21	25	28	31
	8	4	16.9	58.3	239	83	110	126	137	145	127
	9	3	27.9	48.8	288	151	188	207	221	228	176
	10	2	37.2	36.1	308	204	246	268	282	287	207
	11	1	43.6	19.6	317	237	283	306	319	324	226
	12		46.0	0.0	320	249	296	319	332	336	232
	SURFACE DAILY TOTALS				2766	1622	1984	2174	2300	2360	1766
FEB 21	7	5	9.3	74.6	158	35	44	49	53	56	46
	8	4	22.3	67.2	263	116	135	145	150	151	102
	9	3	34.4	57.6	298	187	213	225	230	228	141
	10	2	45.1	44.2	314	241	273	286	291	287	168
	11	1	53.0	25.0	321	276	310	324	328	323	185
	12		56.0	0.0	324	288	323	337	341	335	191
	SURFACE DAILY TOTALS				3036	1998	2276	2396	2446	2424	1476
MAR 21	7	5	13.7	83.8	194	60	63	64	62	59	27
	8	4	27.2	76.8	267	141	150	152	149	142	64
	9	3	40.2	67.9	295	212	226	229	225	214	95
	10	2	52.3	54.8	309	266	285	288	283	270	120
	11	1	61.9	33.4	315	300	322	326	320	305	135
	12		66.0	0.0	317	312	334	339	333	317	140
	SURFACE DAILY TOTALS				3078	2270	2428	2456	2412	2298	1022
APR 21	6	6	4.7	100.6	40	7	5	4	4	3	2
	7	5	18.3	94.9	203	83	77	70	62	51	10
	8	4	32.0	89.0	256	160	157	149	137	122	16
	9	3	45.6	81.9	280	227	227	220	206	186	41
	10	2	59.0	71.8	292	278	282	275	259	237	61
	11	1	71.1	51.6	298	310	316	309	293	269	74
	12		77.6	0.0	299	321	328	321	305	280	79
	SURFACE DAILY TOTALS				3036	2454	2458	2374	2228	2016	488
MAY 21	6	6	8.0	108.4	86	22	15	10	9	9	5
	7	5	21.2	103.2	203	98	85	73	59	44	12
	8	4	34.6	98.5	248	171	159	145	127	106	15
	9	3	48.3	93.6	269	233	224	210	190	165	16
	10	2	62.0	87.7	280	281	275	261	239	211	22
	11	1	75.5	76.9	286	311	307	293	270	240	34
	12		86.0	0.0	288	322	317	304	281	250	37
	SURFACE DAILY TOTALS				3032	2556	2447	2286	2072	1800	246
JUN 21	6	6	9.3	111.6	97	29	20	12	12	11	7
	7	5	22.3	106.8	201	103	87	73	58	41	13
	8	4	35.5	102.6	242	173	158	142	122	99	16
	9	3	49.0	98.7	263	234	221	204	182	155	18
	10	2	62.6	95.0	274	280	269	253	229	199	18
	11	1	76.3	90.8	279	309	300	283	259	227	19
	12		89.4	0.0	281	319	310	294	269	236	22
	SURFACE DAILY TOTALS				2994	2574	2422	2230	1992	1700	204
JUL 21	6	6	8.2	109.0	81	23	16	11	10	9	6
	7	5	21.4	103.8	195	98	85	73	59	44	13
	8	4	34.8	99.2	239	169	157	143	125	104	16
	9	3	48.4	94.5	261	231	221	207	187	161	18
	10	2	62.1	89.0	272	278	270	256	235	206	21
	11	1	75.7	79.2	278	307	302	287	265	235	32
	12		86.6	0.0	280	317	312	298	275	245	36
	SURFACE DAILY TOTALS				2932	2526	2412	2250	2036	1766	246
AUG 21	6	6	5.0	101.3	35	7	5	4	4	4	2
	7	5	18.5	95.6	186	82	76	69	60	50	11
	8	4	32.2	89.7	241	158	154	146	134	118	16
	9	3	45.9	82.9	265	223	222	214	200	181	39
	10	2	59.3	73.0	278	273	275	268	252	230	58
	11	1	71.6	53.2	284	304	309	301	285	261	71
	12		78.3	0.0	286	315	320	313	296	272	75
	SURFACE DAILY TOTALS				2864	2408	2402	2316	2168	1958	470
SEP 21	7	5	13.7	83.8	173	57	60	60	59	56	26
	8	4	27.2	76.8	248	136	144	146	143	136	62
	9	3	40.2	67.9	278	205	218	221	217	206	93
	10	2	52.3	54.8	292	258	275	278	273	261	116
	11	1	61.9	33.4	299	291	311	315	309	295	131
	12		66.0	0.0	301	302	323	327	321	306	136
	SURFACE DAILY TOTALS				2878	2194	2342	2366	2322	2212	992
OCT 21	7	5	9.1	74.1	138	32	40	45	48	50	42
	8	4	22.0	66.7	247	111	129	139	144	145	99
	9	3	34.1	57.1	284	180	206	217	223	221	138
	10	2	44.7	43.8	301	234	265	277	282	279	165
	11	1	52.5	24.7	309	268	301	315	319	314	182
	12		55.5	0.0	311	279	314	328	332	327	188
	SURFACE DAILY TOTALS				2868	1928	2198	2314	2364	2346	1442
NOV 21	7	5	4.9	65.8	67	10	16	20	24	27	29
	8	4	17.0	58.4	232	82	108	123	135	142	124
	9	3	28.0	48.9	282	150	186	205	217	224	172
	10	2	37.3	36.3	303	203	244	265	278	283	204
	11	1	43.8	19.7	312	236	280	302	316	320	222
	12		46.2	0.0	315	247	293	315	328	332	228
	SURFACE DAILY TOTALS				2706	1610	1962	2146	2268	2324	1730
DEC 21	7	5	3.2	62.6	30	3	7	9	11	12	14
	8	4	14.9	55.3	225	71	99	116	129	139	130
	9	3	25.5	46.0	281	137	176	198	214	223	184
	10	2	34.3	33.7	304	189	234	258	275	283	217
	11	1	40.4	18.2	314	221	270	295	312	320	236
	12		42.6	0.0	317	232	282	308	325	332	243
	SURFACE DAILY TOTALS				2624	1474	1852	2058	2204	2286	1808

NOTE: 1) BASED ON DATA IN TABLE 1, pp 387 in 1972 ASHRAE HANDBOOK OF FUNDAMENTALS; 0% GROUND REFLECTANCE; 1.0 CLEARNESS FACTOR.
2) SEE FIG. 4, pp 394 in 1972 ASHRAE HANDBOOK OF FUNDAMENTALS FOR TYPICAL REGIONAL CLEARNESS FACTORS.
3) GROUND REFLECTION NOT INCLUDED ON NORMAL OR HORIZONTAL SURFACES.

"Reprinted from ASHRAE TRANSACTIONS 1974, Volume 80, Part II, by permission of the American Society of Heating, Refrigerating and Air-Conditioning Engineers, Inc."

Source: ASHRAE Handbook of Fundamentals.

APPENDIX G SOLAR INSOLATION IN BTU/FT^2 (*continued*)

SOLAR POSITION AND INSOLATION VALUES FOR 32 DEGREES NORTH LATITUDE

DATE	SOLAR TIME		SOLAR POSITION		BTUH/SQ. FT. TOTAL INSOLATION ON SURFACES						
							SOUTH FACING SURFACE ANGLE WITH HORIZ.				
	AM	PM	ALT	AZM	NORMAL	HORIZ.	22	32	42	52	90
JAN 21	7	5	1.4	65.2	1	0	0	0	0	1	1
	8	4	12.5	56.5	203	56	93	106	116	123	115
	9	3	22.5	46.0	269	118	175	193	206	212	181
	10	2	30.6	33.1	295	167	235	256	269	274	221
	11	1	36.1	17.5	306	198	273	295	308	312	245
	12		38.0	0.0	310	209	285	308	321	324	253
	SURFACE DAILY TOTALS				2458	1288	1839	2008	2118	2166	1779
FEB 21	7	5	7.1	73.5	121	22	34	37	40	42	38
	8	4	19.0	64.4	247	95	127	136	140	141	108
	9	3	29.9	53.4	288	161	206	217	222	220	158
	10	2	39.1	39.4	306	212	266	278	283	279	193
	11	1	45.6	21.4	315	244	304	317	321	315	214
	12		48.0	0.0	317	255	316	330	334	328	222
	SURFACE DAILY TOTALS				2872	1724	2188	2300	2345	2322	1644
MAR 21	7	5	12.7	81.9	185	54	60	60	59	56	32
	8	4	25.1	73.0	260	129	146	147	144	137	78
	9	3	36.8	62.1	290	194	222	224	220	209	119
	10	2	47.3	47.5	304	245	280	283	278	265	150
	11	1	55.0	26.8	311	277	317	321	315	300	170
	12		58.0	0.0	313	287	329	333	327	312	177
	SURFACE DAILY TOTALS				3012	2084	2378	2403	2358	2246	1276
APR 21	6	6	6.1	99.9	66	14	9	6	6	5	3
	7	5	18.8	92.2	206	86	78	71	62	51	10
	8	4	31.5	84.0	255	158	156	148	136	120	35
	9	3	43.9	74.2	278	220	225	217	203	183	68
	10	2	55.7	60.3	290	267	279	272	256	234	95
	11	1	65.4	37.5	295	297	313	306	290	265	112
	12		69.6	0.0	297	307	325	318	301	276	118
	SURFACE DAILY TOTALS				3076	2390	2444	2356	2206	1994	764
MAY 21	6	6	10.4	107.2	119	36	21	13	13	12	7
	7	5	22.8	100.1	211	107	88	75	60	44	13
	8	4	35.4	92.9	250	175	159	145	127	105	15
	9	3	48.1	84.7	269	233	223	209	188	163	33
	10	2	60.6	73.3	280	277	273	259	237	208	56
	11	1	72.0	51.9	285	305	305	290	268	237	72
	12		78.0	0.0	286	315	315	301	278	247	77
	SURFACE DAILY TOTALS				3112	2582	2454	2284	2064	1788	469
JUN 21	6	6	12.2	110.2	131	45	26	16	15	14	9
	7	5	24.3	103.4	210	115	91	76	59	41	14
	8	4	36.9	96.8	245	180	159	143	122	99	16
	9	3	49.6	89.4	264	236	221	204	181	153	19
	10	2	62.2	79.7	274	279	268	251	227	197	41
	11	1	74.2	60.9	279	306	299	282	257	224	56
	12		81.5	0.0	280	315	309	292	267	234	60
	SURFACE DAILY TOTALS				3084	2634	2436	2234	1990	1690	370
JUL 21	6	6	10.7	107.7	113	37	22	14	13	12	8
	7	5	23.1	100.6	203	107	87	75	60	44	14
	8	4	35.7	93.6	241	174	158	143	125	104	16
	9	3	48.4	85.5	261	231	220	205	185	159	31
	10	2	60.9	74.3	271	274	269	254	232	204	54
	11	1	72.4	53.3	277	302	300	285	262	232	69
	12		78.6	0.0	279	311	310	296	273	242	74
	SURFACE DAILY TOTALS.				3012	2558	2422	2250	2030	1754	458
AUG 21	6	6	6.5	100.5	59	14	9	7	6	6	4
	7	5	19.1	92.8	190	85	77	69	60	50	12
	8	4	31.8	84.7	240	156	152	144	132	116	33
	9	3	44.3	75.0	263	216	220	212	197	178	65
	10	2	56.1	61.3	276	262	272	264	249	226	91
	11	1	66.0	38.4	282	292	305	298	281	257	107
	12		70.3	0.0	284	302	317	309	292	268	113
	SURFACE DAILY TOTALS				2902	2352	2388	2296	2144	1934	736
SEP 21	7	5	12.7	81.9	163	51	56	56	55	52	30
	8	4	25.1	73.0	240	124	140	141	138	131	75
	9	3	36.8	62.1	272	188	213	215	211	201	114
	10	2	47.3	47.5	287	237	270	273	268	255	145
	11	1	55.0	26.8	294	268	306	309	303	289	164
	12		58.0	0.0	296	278	318	321	315	300	171
	SURFACE DAILY TOTALS				2808	2014	2288	2308	2264	2154	1226
OCT 21	7	5	6.8	73.1	99	19	29	32	34	36	32
	8	4	18.7	64.0	229	90	120	128	133	134	104
	9	3	29.5	53.0	273	155	198	208	213	212	153
	10	2	38.7	39.1	293	204	257	269	273	270	188
	11	1	45.1	21.1	302	236	294	307	311	306	209
	12		47.5	0.0	304	247	306	320	324	318	217
	SURFACE DAILY TOTALS				2696	1654	2100	2208	2252	2232	1588
NOV 21	7	5	1.5	65.4	2	0	0	0	1	1	1
	8	4	[illegible]	56.6	196	55	91	104	113	119	111
	9	3	22.6	46.1	263	118	173	190	202	208	176
	10	2	30.8	33.2	289	166	233	252	265	270	217
	11	1	36.2	17.6	301	197	270	291	303	307	241
	12		38.2	0.0	304	207	282	304	316	320	249
	SURFACE DAILY TOTALS				2406	1280	1816	1980	2084	2130	1742
DEC 21	8	4	10.3	53.8	176	41	77	90	101	108	107
	9	3	19.8	43.6	257	102	161	180	195	204	183
	10	2	27.6	31.2	288	150	221	244	259	267	226
	11	1	32.7	16.4	301	180	258	282	298	305	251
	12		34.6	0.0	304	190	271	295	311	318	259
	SURFACE DAILY TOTALS				2348	1136	1704	1888	2016	2086	1794

NOTE: 1) BASED ON DATA IN TABLE 1, pp 387 in 1972 ASHRAE HANDBOOK OF FUNDAMENTALS; 0% GROUND REFLECTANCE; 1.0 CLEARNESS FACTOR.
2) SEE FIG. 4, pp 394 in 1972 ASHRAE HANDBOOK OF FUNDAMENTALS FOR TYPICAL REGIONAL CLEARNESS FACTORS.
3) GROUND REFLECTION NOT INCLUDED ON NORMAL OR HORIZONTAL SURFACES

APPENDIX G SOLAR INSOLATION IN BTU/FT^2 (*continued*)

SOLAR POSITION AND INSOLATION VALUES FOR 40 DEGREES NORTH LATITUDE

DATE	SOLAR TIME		SOLAR POSITION		BTUH/SQ. FT. TOTAL INSOLATION ON SURFACES						
							SOUTH FACING SURFACE ANGLE WITH HORIZ.				
	AM	PM	ALT	AZM	NORMAL	HORIZ.	30	40	50	60	90
JAN 21	8	4	8.1	55.3	142	28	65	74	81	85	84
	9	3	16.8	44.0	239	83	155	171	182	187	171
	10	2	23.8	30.9	274	127	218	237	249	254	223
	11	1	28.4	16.0	289	154	257	277	290	293	253
	12		30.0	0.0	294	164	270	291	303	306	263
	SURFACE DAILY TOTALS				2182	948	1660	1810	1906	1944	1726
FEB 21	7	5	4.8	72.7	69	10	19	21	23	24	22
	8	4	15.4	62.2	224	73	114	122	126	127	107
	9	3	25.0	50.2	274	132	195	205	209	208	167
	10	2	32.8	35.9	295	178	256	267	271	267	210
	11	1	38.1	18.9	305	206	293	306	310	304	236
	12		40.0	0.0	308	216	306	319	323	317	245
	SURFACE DAILY TOTALS				2640	1414	2060	2162	2202	2176	1730
MAR 21	7	5	11.4	80.2	171	46	55	55	54	51	35
	8	4	22.5	69.6	250	114	140	141	138	131	89
	9	3	32.8	57.3	282	173	215	217	213	202	138
	10	2	41.6	41.9	297	218	273	276	271	258	176
	11	1	47.7	22.6	305	247	310	313	307	293	200
	12		50.0	0.0	307	257	322	326	320	305	208
	SURFACE DAILY TOTALS				2916	1852	2308	2330	2284	2174	1484
APR 21	6	6	7.4	98.9	89	20	11	8	7	7	4
	7	5	18.9	89.5	206	87	77	70	61	50	12
	8	4	30.3	79.3	252	152	153	145	133	117	53
	9	3	41.3	67.2	274	207	221	213	199	179	93
	10	2	51.2	51.4	286	250	275	267	252	229	126
	11	1	58.7	29.2	292	277	308	301	285	260	147
	12		61.6	0.0	293	287	320	313	296	271	154
	SURFACE DAILY TOTALS				3092	2274	2412	2320	2168	1956	1022
MAY 21	5	7	1.9	114.7	1	0	0	0	0	0	0
	6	6	12.7	105.6	144	49	25	15	14	13	9
	7	5	24.0	96.6	216	214	89	76	60	44	13
	8	4	35.4	87.2	250	175	158	144	125	104	25
	9	3	46.8	76.0	267	227	221	206	186	160	60
	10	2	57.5	60.9	277	267	270	255	233	205	89
	11	1	66.2	37.1	283	293	301	287	264	234	108
	12		70.0	0.0	284	301	312	297	274	243	114
	SURFACE DAILY TOTQLS				3160	2552	2442	2264	2040	1760	724
JUN 21	5	7	4.2	117.3	22	4	3	3	2	2	1
	6	6	14.8	108.4	155	60	30	18	17	16	10
	7	5	26.0	99.7	216	123	92	77	59	41	14
	8	4	37.4	90.7	246	182	159	142	121	97	16
	9	3	48.8	80.2	263	233	219	202	179	151	47
	10	2	59.8	65.8	272	272	266	248	224	194	74
	11	1	69.2	41.9	277	296	296	278	253	221	92
	12		73.5	0.0	279	304	306	289	263	230	98
	SURFACE DAILY TOTALS				3180	2648	2434	2224	1974	1670	610
JUL 21	5	7	2.3	115.2	2	0	0	0	0	0	0
	6	6	13.1	106.1	138	50	26	17	15	14	9
	7	5	24.3	97.2	208	114	89	75	60	44	14
	8	4	35.8	87.8	241	174	157	142	124	102	24
	9	3	47.2	76.7	259	225	218	203	182	157	58
	10	2	57.9	61.7	269	265	266	251	229	200	86
	11	1	66.7	37.9	275	290	296	281	258	228	104
	12		70.6	0.0	276	298	307	292	269	238	111
	SURFACE DAILY TOTALS				3062	2534	2409	2230	2006	1728	702
AUG 21	6	6	7.9	99.5	81	21	12	9	8	7	5
	7	5	19.3	90.0	191	87	76	69	60	49	12
	8	4	30.7	79.9	237	150	150	141	129	113	50
	9	3	41.8	67.9	260	205	216	207	193	173	89
	10	2	51.7	52.1	272	246	267	259	244	221	120
	11	1	59.3	29.7	278	273	300	292	276	252	140
	12		62.3	0.0	280	282	311	303	287	262	147
	SURFACE DAILY TOTALS				2916	2244	2354	2258	2104	1894	978
SEP 21	7	5	11.4	80.2	149	43	51	51	49	47	32
	8	4	22.5	69.6	230	109	133	134	131	124	84
	9	3	32.8	57.3	263	167	206	208	203	193	132
	10	2	41.6	41.9	280	211	262	265	260	247	168
	11	1	47.7	22.6	287	239	298	301	295	281	192
	12		50.0	0.0	290	249	310	313	307	292	200
	SURFACE DAILY TOTALS				2708	1788	2210	2228	2182	2074	1416
OCT 21	7	5	4.5	72.3	48	7	14	15	17	17	16
	8	4	15.0	61.9	204	68	106	113	117	118	100
	9	3	24.5	49.8	257	126	185	195	200	198	160
	10	2	32.4	35.6	280	170	245	257	261	257	203
	11	1	37.6	18.7	291	199	283	295	299	294	229
	12		39.5	0.0	294	208	295	308	312	306	238
	SURFACE DAILY TOTALS				2454	1348	1962	2060	2098	2074	1654
NOV 21	8	4	8.2	55.4	136	28	63	72	78	82	81
	9	3	17.0	44.1	232	82	152	167	178	183	167
	10	2	24.0	31.0	268	126	215	233	245	249	219
	11	1	28.6	16.1	283	153	254	273	285	288	248
	12		30.2	0.0	288	163	267	287	298	301	258
	SURFACE DAILY TOTALS				2128	942	1636	1778	1870	1908	1686
DEC	8	4	5.5	53.0	89	14	39	45	50	54	56
	9	3	14.0	41.9	217	65	135	152	164	171	163
	10	2	20.7	29.4	261	107	200	221	235	242	221
	11	1	25.0	15.2	280	134	239	262	276	283	252
	12		26.6	0.0	285	143	253	275	290	296	263
	SURFACE DAILY TOTALS				1978	782	1480	1634	1740	1796	1646

NOTE: 1) BASED ON DATA IN TABLE 1, pp 387 in 1972 ASHRAE HANDBOOK OF FUNDAMENTALS; 0% GROUND REFLECTANCE; 1.0 CLEARNESS FACTOR.
2) SEE FIG. 4, pp 394 in 1972 ASHRAE HANDBOOK OF FUNDAMENTALS FOR TYPICAL REGIONAL CLEARNESS FACTORS.
3) GROUND REFLECTION NOT INCLUDED ON NORMAL OR HORIZONTAL SURFACES.

APPENDIX G SOLAR INSOLATION IN BTU/FT^2 (*continued*)

SOLAR POSITION AND INSOLATION VALUES FOR 48 DEGREES NORTH LATITUDE

DATE	SOLAR TIME		SOLAR POSITION		BTUH/SQ. FT. TOTAL INSOLATION ON SURFACES						
							SOUTH FACING SURFACE ANGLE WITH HORIZ.				
	AM	PM	ALT	AZM	NORMAL	HORIZ.	38	48	58	68	90
JAN 21	8	4	3.5	54.6	37	4	17	19	21	22	22
	9	3	11.0	42.6	185	46	120	132	140	145	139
	10	2	16.9	29.4	239	83	190	206	216	220	206
	11	1	20.7	15.1	261	107	231	249	260	263	243
	12		22.0	0.0	267	115	245	264	275	278	255
	SURFACE DAILY TOTALS				1710	596	1360	1478	1550	1578	1478
FEB 21	7	5	2.4	72.2	12	1	3	4	4	4	4
	8	4	11.6	60.5	188	49	95	102	105	106	96
	9	3	19.7	47.7	251	100	178	187	191	190	167
	10	2	26.2	33.3	278	139	240	251	255	251	217
	11	1	30.5	17.2	290	165	278	290	294	288	247
	12		32.0	0.0	293	173	291	304	307	301	258
	SURFACE DAILY TOTALS				2330	1080	1880	1972	2024	1978	1720
MAR 21	7	5	10.0	78.7	153	37	49	49	47	45	35
	8	4	19.5	66.8	236	96	131	132	129	122	96
	9	3	28.2	53.4	270	147	205	207	203	193	152
	10	2	35.4	37.8	287	187	263	266	261	248	195
	11	1	40.3	19.8	295	212	300	303	297	283	223
	12		42.0	0.0	298	220	312	315	309	294	232
	SURFACE DAILY TOTALS				2780	1578	2208	2228	2182	2074	1632
APR 21	6	6	8.6	97.8	108	27	13	9	8	7	5
	7	5	18.6	86.7	205	85	76	69	59	48	21
	8	4	28.5	74.9	247	142	149	141	129	113	69
	9	3	37.8	61.2	268	191	216	208	194	174	115
	10	2	45.8	44.6	280	228	268	260	245	223	152
	11	1	51.5	24.0	286	252	301	294	278	254	177
	12		53.6	0.0	288	260	313	305	289	264	185
	SURFACE DAILY TOTALS				3076	2106	2358	2266	2114	1902	1262
MAY 21	5	7	5.2	114.3	41	9	4	4	4	3	2
	6	6	14.7	103.7	162	61	27	16	15	13	10
	7	5	24.6	93.0	219	118	89	75	60	43	13
	8	4	34.7	81.6	248	171	156	142	123	101	45
	9	3	44.3	68.3	264	217	217	202	182	156	86
	10	2	53.0	51.3	274	252	265	251	229	200	120
	11	1	59.5	28.6	279	274	296	281	258	228	141
	12		62.0	0.0	280	281	306	292	269	238	149
	SURFACE DAILY TOTALS				3254	2482	2418	2234	2010	1728	982
JUN 21	5	7	7.9	116.5	77	21	9	9	8	7	5
	6	6	17.2	106.2	172	74	33	19	18	16	12
	7	5	27.0	95.8	220	129	93	77	59	39	15
	8	4	37.1	84.6	246	181	157	140	119	95	35
	9	3	46.9	71.6	261	225	216	198	175	147	74
	10	2	55.8	54.8	269	259	262	244	220	189	105
	11	1	62.7	31.2	274	280	291	273	248	216	126
	12		65.5	0.0	275	287	301	283	258	225	133
	SURFACE DAILY TOTALS				3312	2626	2420	2204	1950	1644	874
JUL 21	5	7	5.7	114.7	43	10	5	5	4	4	3
	6	6	15.2	104.1	156	62	28	18	16	15	11
	7	5	25.1	93.5	211	118	89	75	59	42	14
	8	4	35.1	82.1	240	171	154	140	121	99	43
	9	3	44.8	68.8	256	215	214	199	178	153	83
	10	2	53.5	51.9	266	250	261	246	224	195	116
	11	1	60.1	29.0	271	272	291	276	253	223	137
	12		62.6	0.0	272	279	301	286	263	232	144
	SURFACE DAILY TOTALS				3158	2474	2386	2200	1974	1694	956
AUG 21	6	6	9.1	98.3	99	28	14	10	9	8	6
	7	5	19.1	87.2	190	85	75	67	58	47	20
	8	4	29.0	75.4	232	141	145	137	125	109	65
	9	3	38.4	61.8	254	189	210	201	187	168	110
	10	2	46.4	45.1	266	225	260	252	237	214	146
	11	1	52.2	24.3	272	248	293	285	268	244	169
	12		54.3	0.0	274	256	304	296	279	255	177
	SURFACE DAILY TOTALS				2898	2086	2300	2200	2046	1836	1208
SEP 21	7	5	10.0	78.7	131	35	44	44	43	40	31
	8	4	19.5	66.8	215	92	124	124	121	115	90
	9	3	28.2	53.4	251	142	196	197	193	183	143
	10	2	35.4	37.8	269	181	251	254	248	236	185
	11	1	40.3	19.8	278	205	287	289	284	269	212
	12		42.0	0.0	280	213	299	302	296	281	221
	SURFACE DAILY TOTALS				2568	1522	2102	2118	2070	1966	1546
OCT 21	7	5	2.0	71.9	4	0	1	1	1	1	1
	8	4	11.2	60.2	165	44	86	91	95	95	87
	9	3	19.3	47.4	233	94	167	176	180	178	157
	10	2	25.7	33.1	262	133	228	239	242	239	207
	11	1	30.0	17.1	274	157	266	277	281	276	237
	12		31.5	0.0	278	166	279	291	294	288	247
	SURFACE DAILY TOTALS				2154	1022	1774	1860	1890	1866	1626
NOV 21	8	4	3.6	54.7	36	5	17	19	21	22	22
	9	3	11.2	42.7	179	46	117	129	137	141	135
	10	2	17.1	29.5	233	83	186	202	212	215	201
	11	1	20.9	15.1	255	107	227	245	255	258	238
	12		22.2	0.0	261	115	241	259	270	272	250
	SURFACE DAILY TOTALS				1668	596	1336	1448	1518	1544	1442
DEC 21	9	3	8.0	40.9	140	27	87	98	105	110	109
	10	2	13.6	28.2	214	63	164	180	192	197	190
	11	1	17.3	14.4	242	86	207	226	239	244	231
	12		18.6	0.0	250	94	222	241	254	260	244
	SURFACE DAILY TOTALS				1444	446	1136	1250	1326	1364	1304

NOTE: 1) BASED ON DATA IN TABLE 1, pp 387 in 1972 ASHRAE HANDBOOK OF FUNDAMENTALS; 0% GROUND REFLECTANCE; 1.0 CLEARNESS FACTOR.
2) SEE FIG. 4, pp 394 in 1972 ASHRAE HANDBOOK OF FUNDAMENTALS FOR TYPICAL REGIONAL CLEARNESS FACTORS.
3) GROUND REFLECTION NOT INCLUDED ON NORMAL OR HORIZONTAL SURFACES.

APPENDIX G SOLAR INSOLATION IN BTU/FT² (*continued*)

SOLAR POSITION AND INSOLATION VALUES FOR 56 DEGREES NORTH LATITUDE

DATE	SOLAR TIME AM	SOLAR TIME PM	SOLAR POSITION ALT	SOLAR POSITION AZM	BTUH/SQ. FT. TOTAL INSOLATION ON SURFACES NORMAL	HORIZ.	SOUTH FACING SURFACE ANGLE WITH HORIZ. 46	56	66	76	90
JAN 21	9	3	5.0	41.8	78	11	50	55	59	60	60
	10	2	9.9	28.5	170	39	135	146	154	156	153
	11	1	12.9	14.5	207	58	183	197	206	208	201
	12		14.0	0.0	217	65	198	214	222	225	217
	SURFACE DAILY TOTALS				1126	282	934	1010	1058	1074	1044
FEB 21	8	4	7.6	59.4	129	25	65	69	72	72	69
	9	3	14.2	45.9	214	65	151	159	162	161	151
	10	2	19.4	31.5	250	98	215	225	228	224	208
	11	1	22.8	16.1	266	119	254	265	268	263	243
	12		24.0	0.0	270	126	268	279	282	276	255
	SURFACE DAILY TOTALS				1986	740	1640	1716	1742	1716	1598
MAR 21	7	5	8.3	77.5	128	28	40	40	39	37	32
	8	4	16.2	64.4	215	75	119	120	117	111	97
	9	3	23.3	50.3	253	118	192	193	189	180	154
	10	2	29.0	34.9	272	151	249	251	246	234	205
	11	1	32.7	17.9	282	172	285	288	282	268	236
	12		34.0	0.0	284	179	297	300	294	280	246
	SURFACE DAILY TOTALS				2586	1268	2066	2084	2040	1938	1700
APR 21	5	7	1.4	108.8	0	0	0	0	0	0	0
	6	6	9.6	96.5	122	32	14	9	8	7	6
	7	5	18.0	84.1	201	81	74	66	57	46	29
	8	4	26.1	70.9	239	129	143	135	123	108	82
	9	3	33.6	56.3	260	169	208	200	186	167	133
	10	2	39.9	39.7	272	201	259	251	236	214	174
	11	1	44.1	20.7	278	220	292	284	268	245	200
	12		45.6	0.0	280	227	303	295	279	255	209
	SURFACE DAILY TOTALS				3024	1892	2282	2186	2038	1830	1458
MAY 21	4	8	1.2	125.5	0	0	0	0	0	0	0
	5	7	8.5	113.4	93	25	10	9	8	7	6
	6	6	16.5	101.5	175	71	28	17	15	13	11
	7	5	24.8	89.3	219	119	88	74	58	41	16
	8	4	33.1	76.3	244	163	153	138	119	98	63
	9	3	40.9	61.6	259	201	212	197	176	151	109
	10	2	47.6	44.2	268	231	259	244	222	194	146
	11	1	52.3	23.4	273	249	288	274	251	222	170
	12		54.0	0.0	275	255	299	284	261	231	178
	SURFACE DAILY TOTALS				3340	2374	2374	2188	1962	1682	1218
JUN 21	4	8	4.2	127.2	21	4	2	2	2	2	1
	5	7	11.4	115.3	122	40	14	13	11	10	8
	6	6	19.3	103.6	185	86	34	19	17	15	12
	7	5	27.6	91.7	222	132	92	76	57	38	15
	8	4	35.9	78.8	243	175	154	137	116	92	55
	9	3	43.8	64.1	257	212	211	193	170	143	98
	10	2	50.7	46.4	265	240	255	238	214	184	133
	11	1	55.6	24.9	269	258	284	267	242	210	156
	12		57.5	0.0	271	264	294	276	251	219	164
	SURFACE DAILY TOTALS				3438	2562	2388	2166	1910	1606	1120
JUL 21	4	8	1.7	125.8	0	0	0	0	0	0	0
	5	7	9.0	113.7	91	27	11	10	9	8	6
	6	6	17.0	101.9	169	72	30	18	16	14	12
	7	5	25.3	89.7	212	119	88	74	58	41	15
	8	4	33.6	76.7	237	163	151	136	117	96	61
	9	3	41.4	62.0	252	201	208	193	173	147	106
	10	2	48.2	44.6	261	230	254	239	217	189	142
	11	1	52.9	23.7	265	248	283	268	245	216	165
	12		54.6	0.0	267	254	293	278	255	225	173
	SURFACE DAILY TOTALS				3240	2372	2342	2152	1926	1646	1186
AUG 21	5	7	2.0	109.2	1	0	0	0	0	0	0
	6	6	10.2	97.0	112	34	16	11	10	9	7
	7	5	18.5	84.5	187	82	73	65	56	45	28
	8	4	26.7	71.3	225	128	140	131	119	104	78
	9	3	34.3	56.7	246	168	202	193	179	160	126
	10	2	40.5	40.0	258	199	251	242	227	206	166
	11	1	44.8	20.9	264	218	282	274	258	235	191
	12		46.3	0.0	266	225	293	285	269	245	200
	SURFACE DAILY TOTALS				2850	1884	2218	2118	1966	1760	1392
SEP 21	7	5	8.3	77.5	107	25	36	36	34	32	28
	8	4	16.2	64.4	194	72	111	111	108	102	89
	9	3	23.3	50.3	233	114	181	182	178	168	147
	10	2	29.0	34.9	253	146	236	237	232	221	193
	11	1	32.7	17.9	263	166	271	273	267	254	223
	12		34.0	0.0	266	173	283	285	279	265	233
	SURFACE DAILY TOTALS				2368	1220	1950	1962	1918	1820	1594
OCT 21	8	4	7.1	59.1	104	20	53	57	59	59	57
	9	3	13.8	45.7	193	60	138	145	148	147	138
	10	2	19.0	31.3	231	92	201	210	213	210	195
	11	1	22.3	16.0	248	112	240	250	253	248	230
	12		23.5	0.0	253	119	253	263	266	261	241
	SURFACE DAILY TOTALS				1804	688	1516	1586	1612	1588	1480
NOV 21	9	3	5.2	41.9	76	12	49	54	57	59	58
	10	2	10.0	28.5	165	39	132	143	149	152	148
	11	1	13.1	14.5	201	58	179	193	201	203	196
	12		14.2	0.0	211	65	194	209	217	219	211
	SURFACE DAILY TOTALS				1094	284	914	986	1032	1046	1016
DEC 21	9	3	1.9	40.5	5	0	3	4	4	4	4
	10	2	6.6	27.5	113	19	86	95	101	104	103
	11	1	9.5	13.9	166	37	141	154	163	167	164
	12		10.6	0.0	180	43	159	173	182	186	182
	SURFACE DAILY TOTALS				748	156	620	678	716	734	722

NOTE: 1) BASED ON DATA IN TABLE 1, pp 387 in 1972 ASHRAE HANDBOOK OF FUNDAMENTALS; 0% GROUND REFLECTANCE; 1.0 CLEARNESS FACTOR.
2) SEE FIG. 4, pp 394 in 1972 ASHRAE HANDBOOK OF FUNDAMENTALS FOR TYPICAL REGIONAL CLEARNESS FACTORS.
3) GROUND REFLECTION NOT INCLUDED ON NORMAL OR HORIZONTAL SURFACES.

APPENDIX G SOLAR INSOLATION IN BTU/FT² (*continued*)

SOLAR POSITION AND INSOLATION VALUES FOR 64 DEGREES NORTH LATITUDE

DATE	SOLAR TIME		SOLAR POSITION		BTUH/SQ. FT. TOTAL INSOLATION ON SURFACES						
							SOUTH FACING SURFACE ANGLE WITH HORIZ.				
	AM	PM	ALT.	AZM	NORMAL	HORIZ.	54	64	74	84	90
JAN 21	10	2	2.8	28.1	22	2	17	19	20	20	20
	11	1	5.2	14.1	81	12	72	77	80	81	81
	12		6.0	0.0	100	16	91	98	102	103	103
	SURFACE DAILY TOTALS				306	45	268	290	302	306	304
FEB 21	8	4	3.4	58.7	35	4	17	19	19	19	19
	9	3	8.6	44.8	147	31	103	108	111	110	107
	10	2	12.6	30.3	199	55	170	178	181	178	173
	11	1	15.1	15.3	222	71	212	220	223	219	213
	12		16.0	0.0	228	77	225	235	237	232	226
	SURFACE DAILY TOTALS				1432	400	1230	1286	1302	1282	1252
MAR 21	7	5	6.5	76.5	95	18	30	29	29	27	25
	8	4	20.7	62.6	185	54	101	102	99	94	89
	9	3	18.1	48.1	227	87	171	172	169	160	153
	10	2	22.3	32.7	249	112	227	229	224	213	203
	11	1	25.1	16.6	260	129	262	265	259	246	235
	12		26.0	0.0	263	134	274	277	271	258	246
	SURFACE DAILY TOTALS				2296	932	1856	1870	1830	1736	1656
APR 21	5	7	4.0	108.5	27	5	2	2	2	1	1
	6	6	10.4	95.1	133	37	15	9	8	7	6
	7	5	17.0	81.6	194	76	70	63	54	43	37
	8	4	23.3	67.5	228	112	136	128	116	102	91
	9	3	29.0	52.3	248	144	197	189	176	158	145
	10	2	33.5	36.0	260	169	246	239	224	203	188
	11	1	36.5	18.4	266	184	278	270	255	233	216
	12		97.6	0.0	268	190	289	281	266	243	225
	SURFACE DAILY TOTALS				2982	1644	2176	2082	1936	1736	1594
MAY 21	4	8	5.8	125.1	51	11	5	4	4	3	3
	5	7	11.6	112.1	132	42	13	11	10	9	8
	6	6	17.9	99.1	185	79	29	16	14	12	11
	7	5	24.5	85.7	218	117	86	72	56	39	28
	8	4	30.9	71.5	239	152	148	133	115	94	80
	9	3	36.8	56.1	252	182	204	190	170	145	128
	10	2	41.6	38.9	261	205	249	235	213	186	167
	11	1	44.9	20.1	265	219	278	264	242	213	193
	12		46.0	0.0	267	224	288	274	251	222	201
	SURFACE DAILY TOTALS				3470	2236	2312	2124	1898	1624	1436
JUN 21	3	9	4.2	139.4	21	4	2	2	2	2	1
	4	8	9.0	126.4	93	27	10	9	8	7	6
	5	7	14.7	113.6	154	60	16	15	13	11	10
	6	6	21.0	100.8	194	96	34	19	17	14	13
	7	5	27.5	87.5	221	132	91	74	55	36	23
	8	4	34.0	73.3	239	166	150	133	112	88	73
	9	3	39.9	57.8	251	195	204	187	164	137	119
	10	2	44.9	40.4	258	217	247	230	206	177	157
	11	1	48.3	20.9	262	231	275	258	233	202	181
	12		49.5	0.0	263	235	284	267	242	211	189
	SURFACE DAILY TOTALS				3650	2488	2342	2118	1862	1558	1356
JUL 21	4	8	6.4	125.3	53	13	6	5	5	4	4
	5	7	12.1	112.4	128	44	14	13	11	10	9
	6	6	18.4	99.4	179	81	30	17	16	13	12
	7	5	25.0	86.0	211	118	86	72	56	38	28
	8	4	31.4	71.8	231	152	146	131	113	91	77
	9	3	37.3	56.3	245	182	201	186	166	141	124
	10	2	42.2	39.2	253	204	245	230	208	181	162
	11	1	45.4	20.2	257	218	273	258	236	207	187
	12		46.6	0.0	259	223	282	267	245	216	195
	SURFACE DAILY TOTALS				3372	2248	2280	2090	1864	1588	1400
AUG 21	5	7	4.6	108.8	29	6	3	3	2	2	2
	6	6	11.0	95.5	123	39	16	11	10	8	7
	7	5	17.6	81.9	181	77	69	61	52	42	35
	8	4	23.9	67.8	214	113	132	123	112	97	87
	9	3	29.6	52.6	234	144	190	182	169	150	138
	10	2	34.2	36.2	246	168	237	229	215	194	179
	11	1	37.2	18.5	252	183	268	260	244	222	205
	12		38.3	0.0	254	188	278	270	255	232	215
	SURFACE DAILY TOTALS				2808	1646	2108	1008	1860	1662	1522
SEP 21	7	5	6.5	76.5	77	16	25	25	24	23	21
	8	4	12.7	72.6	163	51	92	92	90	85	81
	9	3	18.1	48.1	206	83	159	159	156	147	141
	10	2	22.3	32.7	229	108	212	213	209	198	189
	11	1	25.1	16.6	240	124	246	248	243	230	220
	12		26.0	0.0	244	129	258	260	254	241	230
	SURFACE DAILY TOTALS				2074	892	1726	1736	1696	1608	1532
OCT 21	8	4	3.0	58.5	17	2	9	9	10	10	10
	9	3	8.1	44.6	122	26	86	91	93	92	90
	10	2	12.1	30.2	176	50	152	159	161	159	155
	11	1	14.6	15.2	201	65	193	201	203	200	295
	12		15.5	0.0	208	71	207	215	217	213	208
	SURFACE DAILY TOTALS				1238	358	1088	1136	1152	1134	1106
NOV 21	10	2	3.0	28.1	23	3	18	20	21	21	21
	11	1	5.4	14.2	79	12	70	76	79	80	79
	12		6.2	0.0	97	17	89	96	100	101	100
	SURFACE DAILY TOTALS				302	46	266	286	298	302	300
DEC 21	11	1	1.8	13.7	4	0	3	4	4	4	4
	12		2.6	0.0	16	2	14	15	16	17	17
	SURFACE DAILY TOTALS				24	2	20	22	24	24	24

NOTE: 1) BASED ON DATE IN TABLE 1, pp 387 in 1972 ASHRAE HANDBOOK OF FUNDAMENTALS; 0% GROUND REFLECTANCE; 1.0 CLEARNESS FACTOR.
2) SEE FIG. 4, pp 394 in 1972 ASHRAE HANDBOOK OF FUNDAMENTALS FOR TYPICAL REGIONAL CLEARNESS FACTORS.
3) GROUND REFLECTION NOT INCLUDED ON NORMAL OR HORIZONTAL SURFACES.

APPENDIX H DEGREE-DAYS FOR VARIOUS LOCATIONS IN THE UNITED STATES[a,b,c]

State	Station	Avg. Winter Temp[d]	July	Aug.	Sept.	Oct.	Nov.	Dec.	Jan.	Feb.	Mar.	Apr.	May	June	Yearly Total
Ala.	Birmingham .... A	54.2	0	0	6	93	363	555	592	462	363	108	9	0	2551
	Huntsville .... A	51.3	0	0	12	127	426	663	694	557	434	138	19	0	3070
	Mobile .... A	59.9	0	0	0	22	213	357	415	300	211	42	0	0	1560
	Montgomery .... A	55.4	0	0	0	68	330	527	543	417	316	90	0	0	2291
Alaska	Anchorage .... A	23.0	245	291	516	930	1284	1572	1631	1316	1293	879	592	315	10864
	Fairbanks .... A	6.7	171	332	642	1203	1833	2254	2359	1901	1739	1068	555	222	14279
	Juneau .... A	32.1	301	338	483	725	921	1135	1237	1070	1073	810	601	381	9075
	Nome .... A	13.1	481	496	693	1094	1455	1820	1879	1666	1770	1314	930	573	14171
Ariz.	Flagstaff .... A	35.6	46	68	201	558	867	1073	1169	991	911	651	437	180	7152
	Phoenix .... A	58.5	0	0	0	22	234	415	474	328	217	75	0	0	1765
	Tucson .... A	58.1	0	0	0	25	231	406	471	344	242	75	6	0	1800
	Winslow .... A	43.0	0	0	6	245	711	1008	1054	770	601	291	96	0	4782
	Yuma .... A	64.2	0	0	0	0	108	264	307	190	90	15	0	0	974
Ark.	Fort Smith .... A	50.3	0	0	12	127	450	704	781	596	456	144	22	0	3292
	Little Rock .... A	50.5	0	0	9	127	465	716	756	577	434	126	9	0	3219
	Texarkana .... A	54.2	0	0	0	78	345	561	626	468	350	105	0	0	2533
Calif.	Bakersfield .... A	55.4	0	0	0	37	282	502	546	364	267	105	19	0	2122
	Bishop .... A	46.0	0	0	48	260	576	797	874	680	555	306	143	36	4275
	Blue Canyon .... A	42.2	28	37	108	347	594	781	896	795	806	597	412	195	5596
	Burbank .... A	58.6	0	0	6	43	177	301	366	277	239	138	81	18	1646
	Eureka C	49.9	270	257	258	329	414	499	546	470	505	438	372	285	4643
	Fresno .... A	53.3	0	0	0	84	354	577	605	426	335	162	62	6	2611
	Long Beach .... A	57.8	0	0	9	47	171	316	397	311	264	171	93	24	1803
	Los Angeles .... A	57.4	28	28	42	78	180	291	372	302	288	219	158	81	2061
	Los Angeles C	60.3	0	0	6	31	132	229	310	230	202	123	68	18	1349
	Mt. Shasta C	41.2	25	34	123	406	696	902	983	784	738	525	347	159	5722
	Oakland .... A	53.5	53	50	45	127	309	481	527	400	353	255	180	90	2870
	Red Bluff .... A	53.8	0	0	0	53	318	555	605	428	341	168	47	0	2515
	Sacramento .... A	53.9	0	0	0	56	321	546	583	414	332	178	72	0	2502
	Sacramento C	54.4	0	0	0	62	312	533	561	392	310	173	76	0	2419
	Sandberg C	46.8	0	0	30	202	480	691	778	661	620	426	264	57	4209
	San Diego .... A	59.5	9	0	21	43	135	236	298	235	214	135	90	42	1458
	San Francisco .... A	53.4	81	78	60	143	306	462	508	395	363	279	214	126	3015
	San Francisco C	55.1	192	174	102	118	231	388	443	336	319	279	239	180	3001
	Santa Maria .... A	54.3	99	93	96	146	270	391	459	370	363	282	233	165	2967
Colo.	Alamosa .... A	29.7	65	99	279	639	1065	1420	1476	1162	1020	696	440	168	8529
	Colorado Springs .... A	37.3	9	25	132	456	825	1032	1128	938	893	582	319	84	6423
	Denver .... A	37.6	6	9	117	428	819	1035	1132	938	887	558	288	66	6283
	Denver C	40.8	0	0	90	366	714	905	1004	851	800	492	254	48	5524
	Grand Junction .... A	39.3	0	0	30	313	786	1113	1209	907	729	387	146	21	5641
	Pueblo .... A	40.4	0	0	54	326	750	986	1085	871	772	429	174	15	5462
Conn.	Bridgeport .... A	39.9	0	0	66	307	615	986	1079	966	853	510	208	27	5617
	Hartford .... A	37.3	0	12	117	394	714	1101	1190	1042	908	519	205	33	6235
	New Haven .... A	39.0	0	12	87	347	648	1011	1097	991	871	543	245	45	5897
Del.	Wilmington .... A	42.5	0	0	51	270	588	927	980	874	735	387	112	6	4930
D.C.	Washington .... A	45.7	0	0	33	217	519	834	871	762	626	288	74	0	4224
Fla.	Apalachicola C	61.2	0	0	0	16	153	319	347	260	180	33	0	0	1308
	Daytona Beach .... A	64.5	0	0	0	0	75	211	248	190	140	15	0	0	879
	Fort Myers .... A	68.6	0	0	0	0	24	109	146	101	62	0	0	0	442
	Jacksonville .... A	61.9	0	0	0	12	144	310	332	246	174	21	0	0	1239
	Key West .... A	73.1	0	0	0	0	0	28	40	31	9	0	0	0	108
	Lakeland C	66.7	0	0	0	0	57	164	195	146	99	0	0	0	661
	Miami .... A	71.1	0	0	0	0	0	65	74	56	19	0	0	0	214

[a] Data for United States cities from a publication of the United States Weather Bureau, *Monthly Normals of Temperature, Precipitation and Heating Degree Days*, 1962, are for the period 1931 to 1960 inclusive. These data also include information from the 1963 revisions to this publication, where available.
[b] Data for airport stations, A, and city stations, C, are both given where available.
[c] Data for Canadian cities were computed by the Climatology Division, Department of Transport from normal monthly mean temperatures, and the monthly values of heating degree days data were obtained using the National Research Council computer and a method devised by H. C. S. Thom of the United States Weather Bureau. The heating degree days are based on the period from 1931 to 1960.
[d] For period October to April, inclusive.

Source: ASHRAE Handbook of Fundamentals.

APPENDIX H DEGREE-DAYS FOR VARIOUS LOCATIONS IN THE UNITED STATES[a,b,c] *(continued)*

State	Station	Avg. Winter Temp[d]	July	Aug.	Sept.	Oct.	Nov.	Dec.	Jan.	Feb.	Mar.	Apr.	May	June	Yearly Total
Fla. (Cont'd)	Miami Beach C	72.5	0	0	0	0	0	40	56	36	9	0	0	0	141
	Orlando A	65.7	0	0	0	0	72	198	220	165	105	6	0	0	766
	Pensacola A	60.4	0	0	0	19	195	353	400	277	183	36	0	0	1463
	Tallahassee A	60.1	0	0	0	28	198	360	375	286	202	36	0	0	1485
	Tampa A	66.4	0	0	0	0	60	171	202	148	102	0	0	0	683
	West Palm Beach A	68.4	0	0	0	0	6	65	87	64	31	0	0	0	253
Ga.	Athens A	51.8	0	0	12	115	405	632	642	529	431	141	22	0	2929
	Atlanta A	51.7	0	0	18	124	417	648	636	518	428	147	25	0	2961
	Augusta A	54.5	0	0	0	78	333	552	549	445	350	90	0	0	2397
	Columbus A	54.8	0	0	0	87	333	543	552	434	338	96	0	0	2383
	Macon A	56.2	0	0	0	71	297	502	505	403	295	63	0	0	2136
	Rome A	49.9	0	0	24	161	474	701	710	577	468	177	34	0	3326
	Savannah A	57.8	0	0	0	47	246	437	437	353	254	45	0	0	1819
	Thomasville C	60.0	0	0	0	25	198	366	394	305	208	33	0	0	1529
Hawaii	Lihue A	72.7	0	0	0	0	0	0	0	0	0	0	0	0	0
	Honolulu A	74.2	0	0	0	0	0	0	0	0	0	0	0	0	0
	Hilo A	71.9	0	0	0	0	0	0	0	0	0	0	0	0	0
Idaho	Boise A	39.7	0	0	132	415	792	1017	1113	854	722	438	245	81	5809
	Lewiston A	41.0	0	0	123	403	756	933	1063	815	694	426	239	90	5542
	Pocatello A	34.8	0	0	172	493	900	1166	1324	1058	905	555	319	141	7033
Ill.	Cairo C	47.9	0	0	36	164	513	791	856	680	539	195	47	0	3821
	Chicago (O'Hare) A	35.8	0	12	117	381	807	1166	1265	1086	939	534	260	72	6639
	Chicago (Midway) A	37.5	0	0	81	326	753	1113	1209	1044	890	480	211	48	6155
	Chicago C	38.9	0	0	66	279	705	1051	1150	1000	868	489	226	48	5882
	Moline A	36.4	0	9	99	335	774	1181	1314	1100	918	450	189	39	6408
	Peoria A	38.1	0	6	87	326	759	1113	1218	1025	849	426	183	33	6025
	Rockford A	34.8	6	9	114	400	837	1221	1333	1137	961	516	236	60	6830
	Springfield A	40.6	0	0	72	291	696	1023	1135	935	769	354	136	18	5429
Ind.	Evansville A	45.0	0	0	66	220	606	896	955	767	620	237	68	0	4435
	Fort Wayne A	37.3	0	9	105	378	783	1135	1178	1028	890	471	189	39	6205
	Indianapolis A	39.6	0	0	90	316	723	1051	1113	949	809	432	177	39	5699
	South Bend A	36.6	0	6	111	372	777	1125	1221	1070	933	525	239	60	6439
Iowa	Burlington A	37.6	0	0	93	322	768	1135	1259	1042	859	426	177	33	6114
	Des Moines A	35.5	0	6	96	363	828	1225	1370	1137	915	438	180	30	6588
	Dubuque A	32.7	12	31	156	450	906	1287	1420	1204	1026	546	260	78	7376
	Sioux City A	34.0	0	9	108	369	867	1240	1435	1198	989	483	214	39	6951
	Waterloo A	32.6	12	19	138	428	909	1296	1460	1221	1023	531	229	54	7320
Kans.	Concordia A	40.4	0	0	57	276	705	1023	1163	935	781	372	149	18	5479
	Dodge City A	42.5	0	0	33	251	666	939	1051	840	719	354	124	9	4986
	Goodland A	37.8	0	6	81	381	810	1073	1166	955	884	507	236	42	6141
	Topeka A	41.7	0	0	57	270	672	980	1122	893	722	330	124	12	5182
	Wichita A	44.2	0	0	33	229	618	905	1023	804	645	270	87	6	4620
Ky.	Covington A	41.4	0	0	75	291	669	983	1035	893	756	390	149	24	5265
	Lexington A	43.8	0	0	54	239	609	902	946	818	685	325	105	0	4683
	Louisville A	44.0	0	0	54	248	609	890	930	818	682	315	105	9	4660
La.	Alexandria A	57.5	0	0	0	56	273	431	471	361	260	69	0	0	1921
	Baton Rouge A	59.8	0	0	0	31	216	369	409	294	208	33	0	0	1560
	Lake Charles A	60.5	0	0	0	19	210	341	381	274	195	39	0	0	1459
	New Orleans A	61.0	0	0	0	19	192	322	363	258	192	39	0	0	1385
	New Orleans C	61.8	0	0	0	12	165	291	344	241	177	24	0	0	1254
	Shreveport A	56.2	0	0	0	47	297	477	552	426	304	81	0	0	2184
Me.	Caribou A	24.4	78	115	336	682	1044	1535	1690	1470	1308	858	468	183	9767
	Portland A	33.0	12	53	195	508	807	1215	1339	1182	1042	675	372	111	7511
Md.	Baltimore A	43.7	0	0	48	264	585	905	936	820	679	327	90	0	4654
	Baltimore C	46.2	0	0	27	189	486	806	859	762	629	288	65	0	4111
	Frederich A	42.0	0	0	66	307	624	955	995	876	741	384	127	12	5087
Mass.	Boston A	40.0	0	9	60	316	603	983	1088	972	846	513	208	36	5634
	Nantucket A	40.2	12	22	93	332	573	896	992	941	896	621	384	129	5891
	Pittsfield A	32.6	25	59	219	524	831	1231	1339	1196	1063	660	326	105	7578
	Worcester A	34.7	6	34	147	450	774	1172	1271	1123	998	612	304	78	6969

(continued)

APPENDIX H DEGREE-DAYS FOR VARIOUS LOCATIONS IN THE UNITED STATES[a,b,c] (continued)

State	Station	Avg. Winter Temp[d]	July	Aug.	Sept.	Oct.	Nov.	Dec.	Jan.	Feb.	Mar.	Apr.	May	June	Yearly Total
Mich.	Alpena .... A	29.7	68	105	273	580	912	1268	1404	1299	1218	777	446	156	8506
	Detroit (City) .... A	37.2	0	0	87	360	738	1088	1181	1058	936	522	220	42	6232
	Detroit (Wayne) .... A	37.1	0	0	96	353	738	1088	1194	1061	933	534	239	57	6293
	Detroit (Willow Run) .... A	37.2	0	0	90	357	750	1104	1190	1053	921	519	229	45	6258
	Escanaba C	29.6	59	87	243	539	924	1293	1445	1296	1203	777	456	159	8481
	Flint .... A	33.1	16	40	159	465	843	1212	1330	1198	1066	639	319	90	7377
	Grand Rapids .... A	34.9	9	28	135	434	804	1147	1259	1134	1011	579	279	75	6894
	Lansing .... A	34.8	6	22	138	431	813	1163	1262	1142	1011	579	273	69	6909
	Marquette C	30.2	59	81	240	527	936	1268	1411	1268	1187	771	468	177	8393
	Muskegon .... A	36.0	12	28	120	400	762	1088	1209	1100	995	594	310	78	6696
	Sault Ste. Marie .... A	27.7	96	105	279	580	951	1367	1525	1380	1277	810	477	201	9048
Minn.	Duluth .... A	23.4	71	109	330	632	1131	1581	1745	1518	1355	840	490	198	10000
	Minneapolis .... A	28.3	22	31	189	505	1014	1454	1631	1380	1166	621	288	81	8382
	Rochester .... A	28.8	25	34	186	474	1005	1438	1593	1366	1150	630	301	93	8295
Miss.	Jackson .... A	55.7	0	0	0	65	315	502	546	414	310	87	0	0	2239
	Meridian .... A	55.4	0	0	0	81	339	518	543	417	310	81	0	0	2289
	Vicksburg C	56.9	0	0	0	53	279	462	512	384	282	69	0	0	2041
Mo.	Columbia .... A	42.3	0	0	54	251	651	967	1076	874	716	324	121	12	5046
	Kansas City .... A	43.9	0	0	39	220	612	905	1032	818	682	294	109	0	4711
	St. Joseph .... A	40.3	0	6	60	285	708	1039	1172	949	769	348	133	15	5484
	St. Louis .... A	43.1	0	0	60	251	627	936	1026	848	704	312	121	15	4900
	St. Louis C	44.8	0	0	36	202	576	884	977	801	651	270	87	0	4484
	Springfield .... A	44.5	0	0	45	223	600	877	973	781	660	291	105	6	4900
Mont.	Billings .... A	34.5	6	15	186	487	897	1135	1296	1100	970	570	285	102	7049
	Glasgow .... A	26.4	31	47	270	608	1104	1466	1711	1439	1187	648	335	150	8996
	Great Falls .... A	32.8	28	53	258	543	921	1169	1349	1154	1063	642	384	186	7750
	Havre .... A	28.1	28	53	306	595	1065	1367	1584	1364	1181	657	338	162	8700
	Havre C	29.8	19	37	252	539	1014	1321	1528	1305	1116	612	304	135	8182
	Helena .... A	31.1	31	59	294	601	1002	1265	1438	1170	1042	651	381	195	8129
	Kalispell .... A	31.4	50	99	321	654	1020	1240	1401	1134	1029	639	397	207	8191
	Miles City .... A	31.2	6	6	174	502	972	1296	1504	1252	1057	579	276	99	7723
	Missoula .... A	31.5	34	74	303	651	1035	1287	1420	1120	970	621	391	219	8125
Neb.	Grand Island .... A	36.0	0	6	108	381	834	1172	1314	1089	908	462	211	45	6530
	Lincoln C	38.8	0	6	75	301	726	1066	1237	1016	834	402	171	30	5864
	Norfolk .... A	34.0	9	0	111	397	873	1234	1414	1179	983	498	233	48	6979
	North Platte .... A	35.5	0	6	123	440	885	1166	1271	1039	930	519	248	57	6684
	Omaha .... A	35.6	0	12	105	357	828	1175	1355	1126	939	465	208	42	6612
	Scottsbluff .... A	35.9	0	0	138	459	876	1128	1231	1008	921	552	285	75	6673
	Valentine .... A	32.6	9	12	165	493	942	1237	1395	1176	1045	579	288	84	7425
Nev.	Elko .... A	34.0	9	34	225	561	924	1197	1314	1036	911	621	409	192	7433
	Ely .... A	33.1	28	43	234	592	939	1184	1308	1075	977	672	456	225	7733
	Las Vegas .... A	53.5	0	0	0	78	387	617	688	487	335	111	6	0	2709
	Reno .... A	39.3	43	87	204	490	801	1026	1073	823	729	510	357	189	6332
	Winnemucca .... A	36.7	0	34	210	536	876	1091	1172	916	837	573	363	153	6761
N.H.	Concord .... A	33.0	6	50	177	505	822	1240	1358	1184	1032	636	298	75	7383
	Mt. Washington Obsv.	15.2	493	536	720	1057	1341	1742	1820	1663	1652	1260	930	603	13817
N.J.	Atlantic City .... A	43.2	0	0	39	251	549	880	936	848	741	420	133	15	4812
	Newark .... A	42.8	0	0	30	248	573	921	983	876	729	381	118	0	4589
	Trenton C	42.4	0	0	57	264	576	924	989	885	753	399	121	12	4980
N. M.	Albuquerque .... A	45.0	0	0	12	229	642	868	930	703	595	288	81	0	4348
	Clayton .... A	42.0	0	6	66	310	699	899	986	812	747	429	183	21	5158
	Raton .... A	38.1	9	28	126	431	825	1048	1116	904	834	543	301	63	6228
	Roswell .... A	47.5	0	0	18	202	573	806	840	641	481	201	31	0	3793
	Silver City .... A	48.0	0	0	6	183	525	729	791	605	518	261	87	0	3705
N.Y.	Albany .... A	34.6	0	19	138	440	777	1194	1311	1156	992	564	239	45	6875
	Albany C	37.2	0	9	102	375	699	1104	1218	1072	908	498	186	30	6201
	Binghamton .... A	33.9	22	65	201	471	810	1184	1277	1154	1045	645	313	99	7286
	Binghamton C	36.6	0	28	141	406	732	1107	1190	1081	949	543	229	45	6451
	Buffalo .... A	34.5	19	37	141	440	777	1156	1256	1145	1039	645	329	78	7062
	New York (Cent. Park) C	42.8	0	0	30	233	540	902	986	885	760	408	118	9	4871
	New York (La Guardia) .... A	43.1	0	0	27	223	528	887	973	879	750	414	124	6	4811

APPENDIX H DEGREE-DAYS FOR VARIOUS LOCATIONS IN THE UNITED STATES[a,b,c] *(continued)*

State	Station	Avg. Winter Temp[d]	July	Aug.	Sept.	Oct.	Nov.	Dec.	Jan.	Feb.	Mar.	Apr.	May	June	Yearly Total
	New York (Kennedy) .... A	41.4	0	0	36	248	564	933	1029	935	815	480	167	12	5219
	Rochester .... A	35.4	9	31	126	415	747	1125	1234	1123	1014	597	279	48	6748
	Schenectady C	35.4	0	22	123	422	756	1159	1283	1131	970	543	211	30	6650
	Syracuse .... A	35.2	6	28	132	415	744	1153	1271	1140	1004	570	248	45	6756
N. C.	Asheville C	46.7	0	0	48	245	555	775	784	683	592	273	87	0	4042
	Cape Hatteras	53.3	0	0	0	78	273	521	580	518	440	177	25	0	2612
	Charlotte .... A	50.4	0	0	6	124	438	691	691	582	481	156	22	0	3191
	Greensboro .... A	47.5	0	0	33	192	513	778	784	672	552	234	47	0	3805
	Raleigh .... A	49.4	0	0	21	164	450	716	725	616	487	180	34	0	3393
	Wilmington .... A	54.6	0	0	0	74	291	521	546	462	357	96	0	0	2347
	Winston-Salem .... A	48.4	0	0	21	171	483	747	753	652	524	207	37	0	3595
N. D.	Bismarck .... A	26.6	34	28	222	577	1083	1463	1708	1442	1203	645	329	117	8851
	Devils Lake C	22.4	40	53	273	642	1191	1634	1872	1579	1345	753	381	138	9901
	Fargo .... A	24.8	28	37	219	574	1107	1569	1789	1520	1262	690	332	99	9226
	Williston .... A	25.2	31	43	261	601	1122	1513	1758	1473	1262	681	357	141	9243
Ohio	Akron-Canton .... A	38.1	0	9	96	381	726	1070	1138	1016	871	489	202	39	6037
	Cincinnati C	45.1	0	0	39	208	558	862	915	790	642	294	96	6	4410
	Cleveland .... A	37.2	9	25	105	384	738	1088	1159	1047	918	552	260	66	6351
	Columbus .... A	39.7	0	6	84	347	714	1039	1088	949	809	426	171	27	5660
	Columbus C	41.5	0	0	57	285	651	977	1032	902	760	396	136	15	5211
	Dayton .... A	39.8	0	6	78	310	696	1045	1097	955	809	429	167	30	5622
	Mansfield .... A	36.9	9	22	114	397	768	1110	1169	1042	924	543	245	60	6403
	Sandusky C	39.1	0	6	66	313	684	1032	1107	991	868	495	198	36	5796
	Toledo .... A	36.4	0	16	117	406	792	1138	1200	1056	924	543	242	60	6494
	Youngstown .... A	36.8	6	19	120	412	771	1104	1169	1047	921	540	248	60	6417
Okla.	Oklahoma City .... A	48.3	0	0	15	164	498	766	868	664	527	189	34	0	3725
	Tulsa .... A	47.7	0	0	18	158	522	787	893	683	539	213	47	0	3860
Ore.	Astoria .... A	45.6	146	130	210	375	561	679	753	622	636	480	363	231	5186
	Burns C	35.9	12	37	210	515	867	1113	1246	988	856	570	366	177	6957
	Eugene .... A	45.6	34	34	129	366	585	719	803	627	589	426	279	135	4726
	Meacham .... A	34.2	84	124	288	580	918	1091	1209	1005	983	726	527	339	7874
	Medford .... A	43.2	0	0	78	372	678	871	918	697	642	432	242	78	5008
	Pendleton .... A	42.6	0	0	111	350	711	884	1017	773	617	396	205	63	5127
	Portland .... A	45.6	25	28	114	335	597	735	825	644	586	396	245	105	4635
	Portland C	47.4	12	16	75	267	534	679	769	594	536	351	198	78	4109
	Roseburg .... A	46.3	22	16	105	329	567	713	766	608	570	405	267	123	4491
	Salem .... A	45.4	37	31	111	338	594	729	822	647	611	417	273	144	4754
Pa.	Allentown .... A	38.9	0	0	90	353	693	1045	1116	1002	849	471	167	24	5810
	Erie .... A	36.8	0	25	102	391	714	1063	1169	1081	973	585	288	60	6451
	Harrisburg .... A	41.2	0	0	63	298	648	992	1045	907	766	396	124	12	5251
	Philadelphia .... A	41.8	0	0	60	297	620	965	1016	889	747	392	118	40	5144
	Philadelphia C	44.5	0	0	30	205	513	856	924	823	691	351	93	0	4486
	Pittsburgh .... A	38.4	0	9	105	375	726	1063	1119	1002	874	480	195	39	5987
	Pittsburgh C	42.2	0	0	60	291	615	930	983	885	763	390	124	12	5053
	Reading C	42.4	0	0	54	257	597	939	1001	885	735	372	105	0	4945
	Scranton .... A	37.2	0	19	132	434	762	1104	1156	1028	893	498	195	33	6254
	Williamsport .... A	38.5	0	9	111	375	717	1073	1122	1002	856	468	177	24	5934
R. I.	Block Island .... A	40.1	0	16	78	307	594	902	1020	955	877	612	344	99	5804
	Providence .... A	38.8	0	16	96	372	660	1023	1110	988	868	534	236	51	5954
S. C.	Charleston .... A	56.4	0	0	0	59	282	471	487	389	291	54	0	0	2033
	Charleston C	57.9	0	0	0	34	210	425	443	367	273	42	0	0	1794
	Columbia .... A	54.0	0	0	0	84	345	577	570	470	357	81	0	0	2484
	Florence .... A	54.5	0	0	0	78	315	552	552	459	347	84	0	0	2387
	Greenville-Spartenburg .... A	51.6	0	0	6	121	399	651	660	546	446	132	19	0	2980
S. D.	Huron .... A	28.8	9	12	165	508	1014	1432	1628	1355	1125	600	288	87	8223
	Rapid City .... A	33.4	22	12	165	481	897	1172	1333	1145	1051	615	326	126	7345
	Sioux Falls .... A	30.6	19	25	168	462	972	1361	1544	1285	1082	573	270	78	7839
Tenn.	Bristol .... A	46.2	0	0	51	236	573	828	828	700	598	261	68	0	4143
	Chattanooga .... A	50.3	0	0	18	143	468	698	722	577	453	150	25	0	3254
	Knoxville .... A	49.2	0	0	30	171	489	725	732	613	493	198	43	0	3494
	Memphis .... A	50.5	0	0	18	130	447	698	729	585	456	147	22	0	3232

(continued)

APPENDIX H DEGREE-DAYS FOR VARIOUS LOCATIONS IN THE UNITED STATES[a,b,c] (*continued*)

State or Prov.	Station	Avg. Winter Temp[d]	July	Aug.	Sept.	Oct.	Nov.	Dec.	Jan.	Feb.	Mar.	Apr.	May	June	Yearly Total
	Memphis ... C	51.6	0	0	12	102	396	648	710	568	434	129	16	0	3015
	Nashville ... A	48.9	0	0	30	158	495	732	778	644	512	189	40	0	3578
	Oak Ridge ... C	47.7	0	0	39	192	531	772	778	669	552	228	56	0	3817
Tex.	Abilene ... A	53.9	0	0	0	99	366	586	642	470	347	114	0	0	2624
	Amarillo ... A	47.0	0	0	18	205	570	797	877	664	546	252	56	0	3985
	Austin ... A	59.1	0	0	0	31	225	388	468	325	223	51	0	0	1711
	Brownsville ... A	67.7	0	0	0	0	66	149	205	106	74	0	0	0	600
	Corpus Christi ... A	64.6	0	0	0	0	120	220	291	174	109	0	0	0	914
	Dallas ... A	55.3	0	0	0	62	321	524	601	440	319	90	6	0	2363
	El Paso ... A	52.9	0	0	0	84	414	648	685	445	319	105	0	0	2700
	Fort Worth ... A	55.1	0	0	0	65	324	536	614	448	319	99	0	0	2405
	Galveston ... A	62.2	0	0	0	6	147	276	360	263	189	33	0	0	1274
	Galveston ... C	62.0	0	0	0	0	138	270	350	258	189	30	0	0	1235
	Houston ... A	61.0	0	0	0	6	183	307	384	288	192	36	0	0	1396
	Houston ... C	62.0	0	0	0	0	165	288	363	258	174	30	0	0	1278
	Laredo ... A	66.0	0	0	0	0	105	217	267	134	74	0	0	0	797
	Lubbock ... A	48.8	0	0	18	174	513	744	800	613	484	201	31	0	3578
	Midland ... A	53.8	0	0	0	87	381	592	651	468	322	90	0	0	2591
	Port Arthur ... A	60.5	0	0	0	22	207	329	384	274	192	39	0	0	1447
	San Angelo ... A	56.0	0	0	0	68	318	536	567	412	288	66	0	0	2255
	San Antonio ... A	60.1	0	0	0	31	204	363	428	286	195	39	0	0	1546
	Victoria ... A	62.7	0	0	0	6	150	270	344	230	152	21	0	0	1173
	Waco ... A	57.2	0	0	0	43	270	456	536	389	270	66	0	0	2030
	Wichita Falls ... A	53.0	0	0	0	99	381	632	698	518	378	120	6	0	2832
Utah	Milford ... A	36.5	0	0	99	443	867	1141	1252	988	822	519	279	87	6497
	Salt Lake City ... A	38.4	0	0	81	419	849	1082	1172	910	763	459	233	84	6052
	Wendover ... A	39.1	0	0	48	372	822	1091	1178	902	729	408	177	51	5778
Vt.	Burlington ... A	29.4	28	65	207	539	891	1349	1513	1333	1187	714	353	90	8269
Va.	Cape Henry ... C	50.0	0	0	0	112	360	645	694	633	536	246	53	0	3279
	Lynchburg ... A	46.0	0	0	51	223	540	822	849	731	605	267	78	0	4166
	Norfolk ... A	49.2	0	0	0	136	408	698	738	655	533	216	37	0	3421
	Richmond ... A	47.3	0	0	36	214	495	784	815	703	546	219	53	0	3865
	Roanoke ... A	46.1	0	0	51	229	549	825	834	722	614	261	65	0	4150
Wash.	Olympia ... A	44.2	68	71	198	422	636	753	834	675	645	450	307	177	5236
	Seattle-Tacoma ... A	44.2	56	62	162	391	633	750	828	678	657	474	295	159	5145
	Seattle ... C	46.9	50	47	129	329	543	657	738	599	577	396	242	117	4424
	Spokane ... A	36.5	9	25	168	493	879	1082	1231	980	834	531	288	135	6655
	Walla Walla ... C	43.8	0	0	87	310	681	843	986	745	589	342	177	45	4805
	Yakima ... A	39.1	0	12	144	450	828	1039	1163	868	713	435	220	69	5941
W. Va.	Charleston ... A	44.8	0	0	63	254	591	865	880	770	648	300	96	9	4476
	Elkins ... A	40.1	9	25	135	400	729	992	1008	896	791	444	198	48	5675
	Huntington ... A	45.0	0	0	63	257	585	856	880	764	636	294	99	12	4446
	Parkersburg ... C	43.5	0	0	60	264	606	905	942	826	691	339	115	6	4754
Wisc.	Green Bay ... A	30.3	28	50	174	484	924	1333	1494	1313	1141	654	335	99	8029
	La Crosse ... A	31.5	12	19	153	437	924	1339	1504	1277	1070	540	245	69	7589
	Madison ... A	30.9	25	40	174	474	930	1330	1473	1274	1113	618	310	102	7863
	Milwaukee ... A	32.6	43	47	174	471	876	1252	1376	1193	1054	642	372	135	7635
Wyo.	Casper ... A	33.4	6	16	192	524	942	1169	1290	1084	1020	657	381	129	7410
	Cheyenne ... A	34.2	28	37	219	543	909	1085	1212	1042	1026	702	428	150	7381
	Lander ... A	31.4	6	19	204	555	1020	1299	1417	1145	1017	654	381	153	7870
	Sheridan ... A	32.5	25	31	219	539	948	1200	1355	1154	1051	642	366	150	7680
Alta.	Banff ... C	—	220	295	498	797	1185	1485	1624	1364	1237	855	589	402	10551
	Calgary ... A	—	109	186	402	719	1110	1389	1575	1379	1268	798	477	291	9703
	Edmonton ... A	—	74	180	411	738	1215	1603	1810	1520	1330	765	400	222	10268
	Lethbridge ... A	—	56	112	318	611	1011	1277	1497	1291	1159	696	403	213	8644
B. C.	Kamloops ... A	—	22	40	189	546	894	1138	1314	1057	818	462	217	102	6799
	Prince George* ... A	—	236	251	444	747	1110	1420	1612	1319	1122	747	468	279	9755
	Prince Rupert ... C	—	273	248	339	539	708	868	936	808	812	648	493	357	7029
	Vancouver* ... A	—	81	87	219	456	657	787	862	723	676	501	310	156	5515
	Victoria* ... A	—	136	140	225	462	663	775	840	718	691	504	341	204	5699
	Victoria ... C	—	172	184	243	426	607	723	805	668	660	487	354	250	5579

APPENDIX H DEGREE-DAYS FOR VARIOUS LOCATIONS IN THE UNITED STATES[a,b,c] *(continued)*

State or Prov.	Station		Avg. Winter Temp[d]	July	Aug.	Sept.	Oct.	Nov.	Dec.	Jan.	Feb.	Mar.	Apr.	May	June	Yearly Total
Man.	Brandon*	A	—	47	90	357	747	1290	1792	2034	1737	1476	837	431	198	11036
	Churchill	A	—	360	375	681	1082	1620	2248	2558	2277	2130	1569	1153	675	16728
	The Pas	C	—	59	127	429	831	1440	1981	2232	1853	1624	969	508	228	12281
	Winnipeg	A	—	38	71	322	683	1251	1757	2008	1719	1465	813	405	147	10679
N. B.	Fredericton*	A	—	78	68	234	592	915	1392	1541	1379	1172	753	406	141	8671
	Moncton	C	—	62	105	276	611	891	1342	1482	1336	1194	789	468	171	8727
	St. John	C	—	109	102	246	527	807	1194	1370	1229	1097	756	490	249	8219
Nfld.	Argentia	A	—	260	167	294	564	750	1001	1159	1085	1091	879	707	483	8440
	Corner Brook	C	—	102	133	324	642	873	1194	1358	1283	1212	885	639	333	8978
	Gander	A	—	121	152	330	670	909	1231	1370	1266	1243	939	657	366	9254
	Goose*	A	—	130	205	444	843	1227	1745	1947	1689	1494	1074	741	348	11887
	St. John's*	A	—	186	180	342	651	831	1113	1262	1170	1187	927	710	432	8991
N. W. T.	Aklavik	C	—	273	459	807	1414	2064	2530	2632	2336	2282	1674	1063	483	18017
	Fort Norman	C	—	164	341	666	1234	1959	2474	2592	2209	2058	1386	732	294	16109
	Resolution Island	C	—	843	831	900	1113	1311	1724	2021	1850	1817	1488	1181	942	16021
N. S.	Halifax	C	—	58	51	180	457	710	1074	1213	1122	1030	742	487	237	7361
	Sydney	A	—	62	71	219	518	765	1113	1262	1206	1150	840	567	276	8049
	Yarmouth	A	—	102	115	225	471	696	1029	1156	1065	1004	726	493	258	7340
Ont.	Cochrane	C	—	96	180	405	760	1233	1776	1978	1701	1528	963	570	222	11412
	Fort William	A	—	90	133	366	694	1140	1597	1792	1557	1380	876	543	237	10405
	Kapuskasing	C	—	74	171	405	756	1245	1807	2037	1735	1562	978	580	222	11572
	Kitchener	C	—	16	59	177	505	855	1234	1342	1226	1101	663	322	66	7566
	London	A	—	12	43	159	477	837	1206	1305	1198	1066	648	332	66	7349
	North Bay	C	—	37	90	267	608	990	1507	1680	1463	1277	780	400	120	9219
	Ottawa	C	—	25	81	222	567	936	1469	1624	1441	1231	708	341	90	8735
	Toronto	C	—	7	18	151	439	760	1111	1233	1119	1013	616	298	62	6827
P.E.I.	Charlottetown	C	—	40	53	198	518	804	1215	1380	1274	1169	813	496	204	8164
	Summerside	C	—	47	84	216	546	840	1246	1438	1291	1206	841	518	216	8488
Que.	Arvida	C	—	102	136	327	682	1074	1659	1879	1619	1407	891	521	231	10528
	Montreal*	A	—	9	43	165	521	882	1392	1566	1381	1175	684	316	69	8203
	Montreal	C	—	16	28	165	496	864	1355	1510	1328	1138	657	288	54	7899
	Quebec*	A	—	56	84	273	636	996	1516	1665	1477	1296	819	428	126	9372
	Quebec	C	—	40	68	243	592	972	1473	1612	1418	1228	780	400	111	8937
Sasks	Prince Albert	A	—	81	136	414	797	1368	1872	2108	1763	1559	867	446	219	11630
	Regina	A	—	78	93	360	741	1284	1711	1965	1687	1473	804	409	201	10806
	Saskatoon	C	—	56	87	372	750	1302	1758	2006	1689	1463	798	403	186	10870
Y. T.	Dawson	C	—	164	326	645	1197	1875	2415	2561	2150	1838	1068	570	258	15067
	Mayo Landing	C	—	208	366	648	1135	1794	2325	2427	1992	1665	1020	580	294	14454

*The data for these normals were from the full ten-year period 1951–1960, adjusted to the standard normal period 1931–1960.

INDEX